Lecture Notes in Mathematics 2008

Editors:
J.-M. Morel, Cachan
F. Takens, Groningen
B. Teissier, Paris

Paul Frank Baum · Guillermo Cortiñas
Ralf Meyer · Rubén Sánchez-García
Marco Schlichting · Bertrand Toën

Topics in Algebraic and Topological K-Theory

Editor: Guillermo Cortiñas

 Springer

Paul Frank Baum
The Pennsylvania State University
Mathematics Department
312, McAllister Building
University Park, PA 16802
USA
baum@math.psu.edu

Guillermo Cortiñas
Universidad de Buenos Aires
Facultad de Ciencias Exactas y
Naturales
Mathematics Department
C1428EGA Ciudad Autónoma de
Buenos Aires
Argentina
gcorti@dm.uba.ar

Ralf Meyer
Universität Göttingen
Mathematisches Institut
Bunsenstr. 3-5
37073 Göttingen
Germany
rameyer@uni-math.gwdg.de

Rubén Sánchez-García
Heinrich-Heine-Universität
Düsseldorf
Mathematisches Institut
Universitätsstrasse 1
40225 Düsseldorf
Germany
sanchez@math.uni-duesseldorf.de

Marco Schlichting
University of Warwick
Mathematics Institute
Zeeman Building
Coventry CV4 7AL
United Kingdom
m.schlichting@warwick.ac.uk

Bertrand Toën
Université Montpellier II
Institut de Mathématiques et de
Modélisation de Montpellier
Place Eugéne Bataillon
34095 Montpellier
France
btoen@math.univ-montp2.fr

ISBN: 978-3-642-15707-3   e-ISBN: 978-3-642-15708-0
DOI: 10.1007/978-3-642-15708-0
Springer Heidelberg Dordrecht London New York

Lecture Notes in Mathematics ISSN print edition: 0075-8434
                             ISSN electronic edition: 1617-9692

Library of Congress Control Number: 2010938612

Mathematics Subject Classification (2010): 19-XX

Cover photo with kind permission by Eugenia Ellis.
It shows a snowman built by the winter school participants.

Cover design: SPi Publisher Services

Printed on acid-free paper

Springer is part of Springer Science+Business Media (www.springer.com)

Foreword

The five articles of this volume evolved from the lecture notes of Swisk[1], the Sedano Winter School on K-theory held in Sedano, Spain, during the week January 22–27 of 2007. Lectures were delivered by Paul F. Baum, Carlo Mazza, Ralf Meyer, Marco Schlichting, Betrand Toën and myself, for a public of 45 participants. The school was supported by the Ministerio de Educación y Ciencia and the Proyecto Consolider Mathematica of Spain. Funding to cover expenses of US based participants was provided by NSF, through a grant to C.A. Weibel, who was responsible, first of all, for preparing a successful funding application, and then for managing the funds. The local committee, composed of N. Abad and E. Ellis, were in charge of conference logistics. Marco Schlichting collaborated in the Scientific Committee. As organizer of the school and editor of this volume, I am indebted to all these people and institutions for their support, and to my fellow coauthors for their contributions.

Guillermo Cortiñas

---

[1] The webpage of the school can be found at http://cms.dm.uba.ar/Members/gcorti/workgroup.swisk/index_html.html

Preface

This book evolved from the lecture notes of Swisk, the Sedano Winter School on K-theory held in Sedano, Spain, during the week January 22–27 of 2007. It intends to be an introduction to K-theory, both algebraic and topological, with emphasis on their interconnections. While a wide range of topics is covered, an effort has been made to keep the exposition as elementary and self-contained as possible.

Since its beginning in the celebrated work of Grothendieck on the Riemmann-Roch theorem, applications of K-theory have been found in a variety of subjects, including algebraic geometry, number theory, algebraic and geometric topology, representation theory and geometric and functional analysis. Because of this, mathematicians from each of these areas have become interested in the subject, and they all look at it from their own perspective. On the one hand, this is the richness and appeal of K-theory. On the other hand, it makes it hard to see a global perspective. For example it is not often that an algebraic K-theorist, coming, say, from the algebraic geometry side of the subject, and a topological K-theorist, coming from the functional analysis side, meet together in the same K-theory conference. Thus it is not uncommon to find that algebraic and topological K-theory are regarded as distinct subjects altogether. These notes modestly attempt to illustrate current developments in both branches of the subject, and to emphasize their contacts.

The book is divided into five articles.

The first two are concerned with Kasparov's bivariant K-theory of C^*-algebras and its role in the Baum–Connes conjecture. If G is a locally compact group and A and B are two separable C^*-algebras equipped with a G-action, the Kasparov bivariant K-theory group $KK^G(A,B)$ is defined as the homotopy classes of G-equivariant Hilbert (A,B) bimodules equipped with a suitable Fredholm operator. Kasparov defines an associative product

$$K^G(A,B) \otimes K^G(B,C) \to K^G(A,C)$$

There is an additive category KK^G whose objects are the separable G–C^*-algebras, so that $KK^G(A,B) = \hom_{KK^G}(A,B)$ and composition is given by the Kasparov product. This category is related to usual category G–C^*–Alg of G–C^*-algebras and equivariant $*$-homomorphism by means of a functor $\iota : G\text{–}C^*\text{–Alg} \to KK^G$. The functor ι has the following properties:

- (Stability) If $A \in G\text{–}C^*\text{–Alg}$, and H_1, H_2 are nonzero G-Hilbert spaces, then $\iota(A \otimes K(H_1) \to A \otimes K(H_1 \oplus H_2))$ is an isomorphism.
- (Split Exactness) If $A \xrightarrow{j} B \xrightarrow{p} C$ is a short exact sequence of $G\text{–}C^*$-algebras, split by a G-equivariant homomorphism $s : C \to B$, then $(\iota(j), \iota(s)) : \iota(A) \oplus \iota(C) \to \iota(B)$ is an isomorphism.

Moreover ι is universal (initial) among stable, split exact functors to additive categories. Kasparov theory has many other important properties. To mention one, consider the case when $G = \{1\}$ is the trivial group, and we take $\mathbb{C}$ as the first variable; then

$$K_0(B) = KK(\mathbb{C}, B)$$

is the usual Grothendieck group. Because topological K_1 of a C^*-algebra B is just K_0 of the suspension of B, we also have

$$K_1^{\text{top}}(B) = KK(\mathbb{C}, SB)$$

Thus the whole topological K-theory is recovered from KK, since K^{top} is 2-periodic.

Another application of equivariant KK is in the definition of equivariant K-homology, which plays a fundamental role in the Baum–Connes conjecture. A Hausdorff, locally compact, second countable space X equipped with an action of G by homeomorphisms is called *proper* if the map

$$G \times X \to X \times X, \qquad (g, x) \mapsto (gx, x)$$

is proper, that is, if the inverse image of any compact subspace is compact. A G-subspace $\Delta \subset X$ G is called G-compact if it is proper and the quotient $G \backslash \Delta$ is compact. The equivariant K-homology of a proper G-space X is

$$K_*^G(X) = \operatorname*{colim}_{\Delta \subset X} KK_*^G(C_0(\Delta), \mathbb{C})$$

Here the colimit is taken over all G-compact subspaces $\Delta \subset X$; C_0 is the C^*-algebra of continous functions vanishing at infinity, and $KK_*^G(A, \mathbb{C}) = KK^G(SA, \mathbb{C})$.

The Baum–Connes conjecture proposes a description of the topological K-theory of the reduced C^*-algebra $C_r(G)$ of a locally compact, Hausdorff, second countable group G in terms of G-equivariant K-homology of the universal (final) proper G-space. There is a map

$$K_*^G(\underline{E}G) \to K_*^{\text{top}}(C_r(G))$$

called the assembly map, and the conjecture says it is an isomorphism. The proper G-space $\underline{E}G$ is characterized up to homotopy by the property that any proper G-space X maps to $\underline{E}G$ and that any two such maps are homotopic. The reduced C^*-algebra is defined as follows. If G is a locally compact, second countable group, and μ is a left invariant Haar measure on G, then one can form the separable Hilbert space $H = L^2(G, \mu)$ of square-integrable functions on G. The algebra $C_c(G)$

of compactly supported continuous functions $G \to \mathbb{C}$ with convolution product is faithfully represented inside the algebra $\mathscr{B}(H)$ of bounded operators on H, and $C_r(G)$ is the norm completion of $C_c(G)$.

The conjecture is known to be true for wide classes of groups; no counterexamples are known. There is also a more general version of the conjecture relating the equivarient K-homology of $\underline{E}G$ with coefficients in a separable G–C^*-algebra A with the topological K-theory of the reduced C^*-algebra of G with coefficients in A, $C_r(G,A)$. The latter conjecture is also known for large classes of groups, and is expected to be true in many cases.

The Baum–Connes conjecture is related to a great number of conjectures in functional analysis, algebra, geometry and topology. Most of these conjectures follow from either the injectivity or the surjectivity of the assembly map. A significant example is the Novikov conjecture on the homotopy invariance of higher signatures of closed, connected, oriented, smooth manifolds. This conjecture follows from the injectivity of the rationalized assembly map.

The first article of this volume, K-theory for group algebras, written by P. Baum and R. Sánchez-García, introduces the subject step by step, beginning with the definition of a C^*-algebra, passing through K-theory of C^*-algebras and its connection with Atiyah–Hirzebruch theory, to the general formulation of the Baum–Connes conjecture with coefficients, and of Kasparov's equivariant KK-theory. The latter is introduced in terms of homotopy classes of Hilbert bimodules.

Universal Coefficient Theorems and assembly maps in KK-theory, by R. Meyer, looks at KK-theory and the Baum–Connes conjecture from the point of view of triangulated categories. Equivariant Kasparov theory is introduced using its universal property, and it is explained how this category can be triangulated. The Baum–Connes assembly map is constructed by localising the Kasparov category at a suitable subcategory. Then a general machinery to construct derived functors and spectral sequences in triangulated categories is explained. This produces various generalizations of the Rosenberg–Schochet Universal Coefficient Theorem.

The next article, Algebraic versus topological K-theory: a friendly match, by G. Cortiñas, attempts to be a bridge between the algebraic and topological branches. It presents various variants of algebraic K-theory of rings, including Quillen's, Karoubi–Villamayor's, and Weibel's homotopy algebraic K-theory, denoted respectively K, KV and KH. These variants of algebraic K-theory differ in their behavior with respect to homotopy and excision. Both KV and KH are invariant under polynomial homotopy; if A is any ring, we have $KV_*(A[t]) = KV_*(A)$ and similarly for KH. On the other hand the identity $K_*(A) = K_*(A[t])$ holds in particular cases (e.g. when A is noetherian regular) but not in general. As to excision, if

$$0 \to A \to B \to C \to 0$$

is an exact sequence of (nonunital) rings, then there is a long exact sequence ($n \in \mathbb{Z}$)

$$KH_{n+1}(C) \longrightarrow KH_n(A) \longrightarrow KH_n(B) \longrightarrow KH_n(C)$$

A similar sequence holds for KV under the additional assumption that the sequence be split by a ring homomorphism $C \to B$. The sequence

$$K_{n+1}(C) \to K_n(A) \to K_n(B) \to K_n(C)$$

is exact for $n \leq 0$, but not for $n \geq 1$, in general. Topological K-theory of topological algebras also has several variants, essentially depending on the type of algebras considered. The topological K-theory of Banach algebras is invariant under continuous homotopies; that for locally convex algebras is invariant under C^{∞}-homotopies. Both satisfy excision and (when suitably stabilized) Bott periodicity: $K_n^{\text{top}} = K_{n+2}^{\text{top}}$.

If A is a topological algebra, there is a comparison map

$$K_*(A) \to K_*^{\text{top}}(A)$$

which is not an isomorphism in general.

Cortiñas' article emphasizes the connections –both formal and concrete– between the algebraic and topological counterparts. For example, Bott periodicity for topological K-theory and the fundamental theorem in algebraic K-theory (which computes the K-groups of the Laurent polynomials) are introduced in a way that makes it clear that each of them is the counterpart of the other. As a concrete connection between algebraic and topological K-theory, the question of whether the comparison map $K_*(A) \to K_*^{\text{top}}(A)$ between the algebraic and topological K-theory of a given topological algebra A is an isomorphism is discussed; Karoubi's conjecture (Suslin–Wodzicki's theorem) establishes that the answer is affirmative for stable C^*-algebras. Proofs of this theorem and of some of its variants are given.

The last two articles approach algebraic K-theory from a categorical point of view.

Higher algebraic K-theory (after Quillen, Thomason and others), by M. Schlichting, introduces higher algebraic K-theory of schemes; emphasis is on the modern point of view where structure theorems on derived categories of sheaves are used to compute higher algebraic K-groups. There are many results in the literature about the structure of triangulated categories, and virtually all of them translate into results about higher algebraic K-groups. The link is provided by an abstract localization theorem due to Thomason and Waldhausen, which –omitting hypothesis– says that a short exact sequence of triangulated categories gives rise to a long exact sequence of algebraic K-groups. This theorem, and its applications, are the heart of the article. Among the main applications presented in the article is Thomason's Mayer–Vietoris theorem, which says that if X is a quasi-compact, quasi-separated scheme and U and V are open quasi-compact subschemes, then there is a long exact sequence

$$K_{n+1}(U \cap V) \to K_n(X) \to K_n(U) \oplus K_n(V) \to K_n(U \cap V) \to K_{n-1}(X)$$

Although the particular case of this result for regular noetherian separated schemes follows from Quillen's early work in the 1970s, the full generality was obtained only twenty years later, by Thomason. The use of derived categories is essential in its proof. Another application is Thomason's blow-up formula. If $Y \subset X$ is a regular embedding of pure codimension d with X quasi-compact and separated, and X' is the blow-up of X along Y, then

$$K_*(X') = K_*(X) \oplus \bigoplus_{i=1}^{d-1} K_*(Y)$$

The methods explained in Schlichting's paper can also be applied to any of the other (co-) homology theories which satisfy an analog of Thomason–Waldhausen's localization theorem; these include Hochschild homology, (negative, periodic, ordinary) cyclic homology, topological Hochschild (and cyclic) homology, triangular Witt groups and higher Grothendieck–Witt groups (the last two when 2 is invertible).

Lectures on dg-categories, by B. Toën, provides an introduction to this theory, which is deeply intertwined with K-theory. The connection comes from the fact that the categories of complexes of sheaves on a scheme are dg-categories. The approach to the subject emphasizes the localization problem, in the sense of category theory. In the same way that the notion of complexes is introduced for the need of derived functors, dg-categories are introduced here for the need of a "derived version" of the localization construction. The existence and properties of this localization are then studied. The notion of triangulated dg-categories, which is a refined version of the usual notion of triangulated categories, is presented, and it is shown that many invariants (such as K-theory, Hochschild homology, . . .) are invariants of dg-categories, though it is known that they are not invariants of triangulated categories. Finally the notion of saturated dg-categories is given and it is explained how they can be used in order to define a secondary K-theory.

June, 2010

Paul F. Baum (University Park)
Guillermo Cortiñas (Buenos Aires)
Ralf Meyer (Göttingen)
Rubén J. Sánchez-García (Düsseldorf)
Marco Schlichting (Coventry)
Betrand Toën (Montpellier)

Contents

Lectures on DG-Categories

K-Theory for Group C^*-algebras

Paul F. Baum[1] and Rubén J. Sánchez-García[2]

[1] Mathematics Department,
206 McAllister Building,
The Pennsylvania State University,
University Park, PA 16802,
USA,
baum@math.psu.edu
[2] Mathematisches Institut,
Heinrich-Heine-Universität Düsseldorf,
Universitätsstr. 1,
40225 Düsseldorf,
Germany,
sanchez@math.uni-duesseldorf.de

1 Introduction

These notes are based on a lecture course given by the first author in the Sedano Winter School on K-theory held in Sedano, Spain, on January 22–27th of 2007. They aim at introducing K-theory of C^*-algebras, equivariant K-homology and KK-theory in the context of the Baum–Connes conjecture.

We start by giving the main definitions, examples and properties of C^*-algebras in Sect. 2. A central construction is the reduced C^*-algebra of a locally compact, Hausdorff, second countable group G. In Sect. 3 we define K-theory for C^*-algebras, state the Bott periodicity theorem and establish the connection with Atiyah–Hirzebruch topological K-theory.

Our main motivation will be to study the K-theory of the reduced C^*-algebra of a group G as above. The Baum–Connes conjecture asserts that these K-theory groups are isomorphic to the equivariant K-homology groups of a certain G-space, by means of the index map. The G-space is the universal example for proper actions of G, written $\underline{E}G$. Hence we proceed by discussing proper actions in Sect. 4 and the universal space $\underline{E}G$ in Sect. 5.

Equivariant K-homology is explained in Sect. 6. This is an equivariant version of the dual of Atiyah–Hirzebruch K-theory. Explicitly, we define the groups $K_j^G(X)$ for $j = 0, 1$ and X a proper G-space with compact, second countable quotient $G\backslash X$. These are quotients of certain equivariant K-cycles by homotopy, although the precise definition of homotopy is postponed. We then address the problem of extending the definition to $\underline{E}G$, whose quotient by the G-action may not be compact.

P.F. Baum et al., *Topics in Algebraic and Topological K-Theory*,
Lecture Notes in Mathematics 2008, DOI 10.1007/978-3-642-15708-0_1,
© Springer-Verlag Berlin Heidelberg 2011

In Sect. 7 we concentrate on the case when G is a discrete group, and in Sect. 8 on the case G compact. In Sect. 9 we introduce KK-theory for the first time. This theory, due to Kasparov, is a generalization of both K-theory of C^*-algebras and K-homology. Here we define $KK_G^j(A, \mathbb{C})$ for a separable C^*-algebra A and $j = 0, 1$, although we again postpone the exact definition of homotopy. The already defined $K_j^G(X)$ coincides with this group when $A = C_0(X)$.

At this point we introduce a generalization of the conjecture called the Baum–Connes conjecture with coefficients, which consists in adding coefficients in a G-C^*-algebra (Sect. 10). To fully describe the generalized conjecture we need to introduce Hilbert modules and the reduced crossed-product (Sect. 11), and to define KK-theory for pairs of C^*-algebras. This is done in the non-equivariant situation in Sect. 12 and in the equivariant setting in Sect. 13. In addition we give at this point the missing definition of homotopy. Finally, using equivariant KK-theory, we can insert coefficients in equivariant K-homology, and then extend it again to $\underline{E}G$.

The only ingredient of the conjecture not yet accounted for is the index map. It is defined in Sect. 14 via the Kasparov product and descent maps in KK-theory. We finish with a brief exposition of the history of K-theory and a discussion of Karoubi's conjecture, which symbolizes the unity of K-theory, in Sect. 15.

We thank the editor G. Cortiñas for his colossal patience while we were preparing this manuscript, and the referee for her or his detailed scrutiny.

2 C^*-algebras

We start with some definitions and basic properties of C^*-algebras. Good references for C^*-algebra theory are [1, 15, 39] or [41].

2.1 Definitions

Definition 1. *A* Banach algebra *is an (associative, not necessarily unital) algebra A over $\mathbb{C}$ with a given norm $\| \ \|$*

$$\| \ \| : A \longrightarrow [0, \infty)$$

such that A is a complete normed algebra, that is, for all $a, b \in A$, $\lambda \in \mathbb{C}$,

(a) $\|\lambda a\| = |\lambda| \|a\|$
(b) $\|a + b\| \le \|a\| + \|b\|$
(c) $\|a\| = 0 \Leftrightarrow a = 0$
(d) $\|ab\| \le \|a\| \|b\|$
(e) Every Cauchy sequence is convergent in A (with respect to the metric $d(a, b) = \|a - b\|$)

A C^*-algebra is a Banach algebra with an involution satisfying the C^*-*algebra identity.*

Definition 2. *A C**-algebra $A = (A, \| \ \|, *)$ is a Banach algebra $(A, \| \ \|)$ with a map* $* : A \to A, a \mapsto a^*$ *such that for all $a, b \in A$, $\lambda \in \mathbb{C}$*

(a) $(a+b)^* = a^* + b^*$
(b) $(\lambda a)^* = \overline{\lambda} a^*$
(c) $(ab)^* = b^* a^*$
(d) $(a^*)^* = a$
(e) $\|aa^*\| = \|a\|^2$ (C*-algebra identity)*

Note that in particular $\|a\| = \|a^*\|$ for all $a \in A$: for $a = 0$ this is clear; if $a \neq 0$ then $\|a\| \neq 0$ and $\|a\|^2 = \|aa^*\| \leq \|a\| \|a^*\|$ implies $\|a\| \leq \|a^*\|$, and similarly $\|a^*\| \leq \|a\|$.

A *C**-algebra is *unital* if it has a multiplicative unit $1 \in A$. A *sub-C**-algebra is a non-empty subset of A which is a *C**-algebra with the operations and norm given on A.

Definition 3. *A $*$-homomorphism is an algebra homomorphism $\varphi : A \to B$ such that $\varphi(a^*) = (\varphi(a))^*$, for all $a \in A$.*

Proposition 1. *If $\varphi : A \to B$ is a $*$-homomorphism then $\|\varphi(a)\| \leq \|a\|$ for all $a \in A$. In particular, φ is a (uniformly) continuous map.*

For a proof see, for instance, [41, Thm. 1.5.7].

2.2 Examples

We give three examples of *C**-algebras.

Example 1. Let X be a Hausdorff, locally compact topological space. Let $X^+ = X \cup \{p_\infty\}$ be its one-point compactification. (Recall that X^+ is Hausdorff if and only if X is Hausdorff and locally compact.)

Define the *C**-algebra

$$C_0(X) = \left\{ \alpha : X^+ \to \mathbb{C} \,\middle|\, \alpha \text{ continuous}, \alpha(p_\infty) = 0 \right\},$$

with operations: for all $\alpha, \beta \in C_0(X), p \in X^+, \lambda \in \mathbb{C}$

$$(\alpha + \beta)(p) = \alpha(p) + \beta(p),$$
$$(\lambda \alpha)(p) = \lambda \alpha(p),$$
$$(\alpha\beta)(p) = \alpha(p)\beta(p),$$
$$\alpha^*(p) = \overline{\alpha(p)},$$
$$\|\alpha\| = \sup_{p \in X} |\alpha(p)|.$$

Note that if X is compact Hausdorff, then

$$C_0(X) = C(X) = \{\alpha : X \to \mathbb{C} \mid \alpha \text{ continuous}\}.$$

Example 2. Let H be a Hilbert space. A Hilbert space is *separable* if it admits a countable (or finite) orthonormal basis. (We shall deal with separable Hilbert spaces unless explicit mention is made to the contrary.)

Let $\mathscr{L}(H)$ be the set of bounded linear operators on H, that is, linear maps $T :$ $H \to H$ such that

$$\|T\| = \sup_{\|u\|=1} \|Tu\| < \infty,$$

where $\|u\| = \langle u,u \rangle^{1/2}$. It is a complex algebra with

$$(T+S)u = Tu + Su,$$
$$(\lambda T)u = \lambda(Tu),$$
$$(TS)u = T(Su),$$

for all $T, S \in \mathscr{L}(H)$, $u \in H$, $\lambda \in \mathbb{C}$. The norm is the operator norm $\|T\|$ defined above, and T^* is the adjoint operator of T, that is, the unique bounded operator such that

$$\langle Tu,v \rangle = \langle u, T^*v \rangle$$

for all $u,v \in H$.

Example 3. Let $\mathscr{L}(H)$ be as above. A bounded operator is *compact* if it is a norm limit of operators with finite-dimensional image, that is,

$$\mathscr{K}(H) = \{T \in \mathscr{L}(H) \,|\, T \text{ compact operator}\} = \overline{\{T \in \mathscr{L}(H) \,|\, \dim_{\mathbb{C}} T(H) < \infty\}},$$

where the overline denotes closure with respect to the operator norm. $\mathscr{K}(H)$ is a sub-C^*-algebra of $\mathscr{L}(H)$. Moreover, it is an ideal of $\mathscr{L}(H)$ and, in fact, the only norm-closed ideal except 0 and $\mathscr{L}(H)$.

2.3 The Reduced C^*-algebra of a Group

Let G be a topological group which is locally compact, Hausdorff and second countable (i.e. as a topological space it has a countable basis). There is a C^*-algebra associated to G, called the *reduced C^*-algebra of G*, defined as follows.

Remark 1. We need G to be locally compact and Hausdorff to guarantee the existence of a Haar measure. The countability assumption makes the Hilbert space $L^2(G)$ separable and also avoids some technical difficulties when later defining Kasparov's KK-theory.

Fix a left-invariant Haar measure dg on G. By left-invariant we mean that if $f: G \to \mathbb{C}$ is continuous with compact support then

$$\int_G f(\gamma g)dg = \int_G f(g)dg \qquad \text{for all } \gamma \in G.$$

Define the Hilbert space L^2G as

$$L^2G = \left\{ u : G \to \mathbb{C} \mid \int_G |u(g)|^2 dg < \infty \right\},$$

with scalar product

$$\langle u, v \rangle = \int_G \overline{u(g)}\, v(g)\, dg$$

for all $u, v \in L^2G$.

Let $\mathscr{L}(L^2G)$ be the C^*-algebra of all bounded linear operators $T : L^2G \to L^2G$. On the other hand, define

$$C_c G = \{f : G \to \mathbb{C} \mid f \text{ continuous with compact support}\}.$$

It is an algebra with

$$
\begin{aligned}
(f + h)(g) &= f(g) + h(g), \\
(\lambda f)(g) &= \lambda f(g),
\end{aligned}
$$

for all $f, h \in C_c G$, $\lambda \in \mathbb{C}$, $g \in G$, and multiplication given by *convolution*

$$(f * h)(g_0) = \int_G f(g) h(g^{-1} g_0)\, dg \quad \text{for all } g_0 \in G.$$

Remark 2. When G is discrete, $\int_G f(g) dg = \sum_G f(g)$ is a Haar measure, $C_c G$ is the complex group algebra $\mathbb{C}[G]$ and $f * h$ is the usual product in $\mathbb{C}[G]$.

There is an injection of algebras

$$
\begin{array}{ccc}
0 \longrightarrow C_c G & \longrightarrow & \mathscr{L}(L^2G) \\
f & \longmapsto & T_f
\end{array}
$$

where

$$
\begin{aligned}
T_f(u) &= f * u & u &\in L^2G, \\
(f * u)(g_0) &= \int_G f(g) u(g^{-1} g_0) dg & g_0 &\in G.
\end{aligned}
$$

Note that $C_c G$ is not necessarily a sub-C^*-algebra of $\mathscr{L}(L^2G)$ since it may not be complete. We define $C_r^*(G)$, the *reduced C^*-algebra of G*, as the norm closure of $C_c G$ in $\mathscr{L}(L^2G)$:

$$C_r^*(G) = \overline{C_c G} \subset \mathscr{L}(L^2G).$$

Remark 3. There are other possible completions of $C_c G$. This particular one, i.e. $C_r^*(G)$, uses only the left regular representation of G (cf. [41, Chap. 7]).

2.4 Two Classical Theorems

We recall two classical theorems about C^*-algebras. The first one says that any C^*-algebra is (non-canonically) isomorphic to a C^*-algebra of operators, in the sense of the following definition.

Definition 4. *A subalgebra A of $\mathscr{L}(H)$ is a C^*-algebra of operators if*

(a) A is closed with respect to the operator norm
(b) if $T \in A$ then the adjoint operator $T^* \in A$

That is, A is a sub-C^*-algebra of $\mathscr{L}(H)$, for some Hilbert space H.

Theorem 1 (I. Gelfand and V. Naimark). *Any C^*-algebra is isomorphic, as a C^*-algebra, to a C^*-algebra of operators.*

The second result states that any commutative C^*-algebra is (canonically) isomorphic to $C_0(X)$, for some topological space X.

Theorem 2 (I. Gelfand). *Let A be a commutative C^*-algebra. Then A is (canonically) isomorphic to $C_0(X)$ for X the space of maximal ideals of A.*

Remark 4. The topology on X is the *Jacobson topology* or *hull-kernel topology* [39, p. 159].

Thus a non-commutative C^*-algebra can be viewed as a 'non-commutative, locally compact, Hausdorff topological space.'

2.5 The Categorical Viewpoint

Example 1 gives a functor between the category of locally compact, Hausdorff, topological spaces and the category of C^*-algebras, given by $X \mapsto C_0(X)$. Theorem 2 tells us that its restriction to commutative C^*-algebras is an equivalence of categories,

$$\begin{pmatrix} \text{commutative} \\ C^*\text{-algebras} \end{pmatrix} \simeq \begin{pmatrix} \text{locally compact, Hausdorff,} \\ \text{topological spaces} \end{pmatrix}^{op}$$
$$C_0(X) \longleftarrow X$$

On one side we have C^*-algebras and $*$-homomorphisms, and on the other locally compact, Hausdorff topological spaces with morphisms from Y to X being continuous maps $f \colon X^+ \to Y^+$ such that $f(p_\infty) = q_\infty$. (The symbol op means the opposite or dual category, in other words, the functor is contravariant.)

Remark 5. This is not the same as continuous proper maps $f \colon X \to Y$ since we do not require that the map $f \colon X^+ \to Y^+$ maps X to Y.

3 *K*-Theory of *C**-algebras

In this section we define the *K*-theory groups of an arbitrary *C**-algebra. We first give the definition for a *C**-algebra with unit and then extend it to the non-unital case. We also discuss Bott periodicity and the connection with topological *K*-theory of spaces. More details on *K*-theory of *C**-algebras is given in Sect. 3 of Cortiñas' notes [12], including a proof of Bott periodicity. Other references are [39, 42, 49].

Our main motivation is to study the *K*-theory of $C_r^*(G)$, the reduced *C**-algebra of *G*. From Bott periodicity, it suffices to compute $K_j(C_r^*(G))$ for $j = 0, 1$. In 1980, Paul Baum and Alain Connes conjectured that these *K*-theory groups are isomorphic to the *equivariant K-homology* (Sect. 6) of a certain *G*-space. This *G*-space is the *universal example for proper actions of G* (Sects. 4 and 5), written $\underline{E}G$. Moreover, the conjecture states that the isomorphism is given by a particular map called the *index map* (Sect. 14).

Conjecture 1 (P. Baum and A. Connes, 1980). Let *G* be a locally compact, Hausdorff, second countable, topological group. Then the index map

$$\mu : K_j^G(\underline{E}G) \longrightarrow K_j(C_r^*(G)) \quad j = 0, 1$$

is an isomorphism.

3.1 Definition for Unital *C**-algebras

Let *A* be a *C**-algebra with unit 1_A. Consider $GL(n, A)$, the group of invertible *n* by *n* matrices with coefficients in *A*. It is a topological group, with topology inherited from *A*. We have a standard inclusion

$$GL(n, A) \hookrightarrow GL(n + 1, A)$$

$$\begin{pmatrix} a_{11} & \cdots & a_{1n} \\ \vdots & \cdots & \vdots \\ a_{n1} & \cdots & a_{nn} \end{pmatrix} \longmapsto \begin{pmatrix} a_{11} & \cdots & a_{1n} & 0 \\ \vdots & \cdots & \vdots & \vdots \\ a_{n1} & \cdots & a_{nn} & 0 \\ 0 & \cdots & 0 & 1_A \end{pmatrix}.$$

Define $GL(A)$ as the direct limit with respect to these inclusions

$$GL(A) = \bigcup_{n=1}^{\infty} GL(n, A).$$

It is a topological group with the *direct limit topology*: a subset θ is open if and only if $\theta \cap GL(n, A)$ is open for every $n \geq 1$. In particular, $GL(A)$ is a topological space, and hence we can consider its homotopy groups.

Definition 5 (*K*-theory of a unital *C**-algebra).

$$K_j(A) = \pi_{j-1}(GL(A)) \quad j = 1, 2, 3, \ldots$$

Finally, we define $K_0(A)$ as the *algebraic K*-theory group of the ring A, that is, the Grothendieck group of finitely generated (left) projective A-modules (cf. [12, Remark 2.1.9]),

$$K_0(A) = K_0^{\text{alg}}(A).$$

Remark 6. Note that $K_0(A)$ only depends on the ring structure of A and so we can 'forget' the norm and the involution. The definition of $K_1(A)$ does require the norm but not the involution, so in fact we are defining *K*-theory of Banach algebras with unit. Everything we say in Sect. 3.2 below, including Bott periodicity, is true for Banach algebras.

3.2 Bott Periodicity

The fundamental result is *Bott periodicity*. It says that the homotopy groups of $GL(A)$ are periodic modulo 2 or, more precisely, that the double loop space of $GL(A)$ is homotopy equivalent to itself,

$$\Omega^2 GL(A) \simeq GL(A).$$

As a consequence, the *K*-theory of the C^*-algebra A is periodic modulo 2

$$K_j(A) = K_{j+2}(A) \quad j \geq 0.$$

Hence from now on we will only consider $K_0(A)$ and $K_1(A)$.

3.3 Definition for Non-Unital C^*-algebras

If A is a C^*-algebra without a unit, we formally adjoin one. Define $\widetilde{A} = A \oplus \mathbb{C}$ as a complex algebra with multiplication, involution and norm given by

$$(a, \lambda) \cdot (b, \mu) = (ab + \mu a + \lambda b, \lambda \mu),$$

$$(a, \lambda)^* = (a^*, \overline{\lambda}),$$

$$\|(a, \lambda)\| = \sup_{\|b\|=1} \|ab + \lambda b\|.$$

This makes $\widetilde{A}$ a unital C^*-algebra with unit $(0, 1)$. We have an exact sequence

$$0 \longrightarrow A \longrightarrow \widetilde{A} \longrightarrow \mathbb{C} \longrightarrow 0.$$

Definition 6. *Let A be a non-unital C^*-algebra. Define $K_0(A)$ and $K_1(A)$ as*

$$K_0(A) = \ker\left(K_0(\widetilde{A}) \to K_0(\mathbb{C})\right)$$

$$K_1(A) = K_1(\widetilde{A}).$$

This definition agrees with the previous one when A has a unit. It also satisfies Bott periodicity (see Cortiñas' notes [12, Sect. 3.2]).

Remark 7. Note that the C^*-algebra $C_r^*(G)$ is unital if and only if G is discrete, with unit the Dirac function on 1_G.

Remark 8. There is algebraic K-theory of rings (see [12]). Althought a C^*-algebra is in particular a ring, the two K-theories are different; algebraic K-theory does not satisfy Bott periodicity and K_1 is in general a quotient of K_1^{alg}. We shall compare both definitions in Sect. 15.3 (see also [12, Sect. 7]).

3.4 Functoriality

Let A, B be C^*-algebras (with or without units), and $\varphi : A \to B$ a $*$-homomorphism. Then φ induces a homomorphism of abelian groups

$$\varphi_* : K_j(A) \longrightarrow K_j(B) \quad j = 0, 1.$$

This makes $A \mapsto K_j(A)$, $j = 0, 1$, covariant functors from C^*-algebras to abelian groups [42, Sects. 4.1 and 8.2].

Remark 9. When A and B are unital and $\varphi(1_A) = 1_B$, the map φ_* is the one induced by $GL(A) \to GL(B)$, $(a_{ij}) \mapsto (\varphi(a_{ij}))$ on homotopy groups.

3.5 More on Bott Periodicity

In the original article [9], Bott computed the stable homotopy of the classical groups and, in particular, the homotopy groups $\pi_j(GL(n, \mathbb{C}))$ when $n \gg j$.

Fig. 1. Raoul Bott

Theorem 3 (R. Bott [9]). *The homotopy groups of GL(n, ℂ) are*

$$\pi_j(GL(n,\mathbb{C})) = \begin{cases} 0 & j \text{ even} \\ \mathbb{Z} & j \text{ odd} \end{cases}$$

for all $j = 0, 1, 2, \ldots, 2n - 1$.

As a corollary of the previous theorem, we obtain the K-theory of $\mathbb{C}$, considered as a C^*-algebra.

Theorem 4 (R. Bott).

$$K_j(\mathbb{C}) = \begin{cases} \mathbb{Z} & j \text{ even,} \\ 0 & j \text{ odd.} \end{cases}$$

Sketch of proof. Since $\mathbb{C}$ is a field, $K_0(\mathbb{C}) = K_0^{alg}(\mathbb{C}) = \mathbb{Z}$. By the polar decomposition, $GL(n, \mathbb{C})$ is homotopy equivalent to $U(n)$. The homotopy long exact sequence of the fibration $U(n) \to U(n+1) \to S^{2n+1}$ gives $\pi_j(U(n)) = \pi_j(U(n+1))$ for all $j \leq 2n+1$. Hence $K_j(\mathbb{C}) = \pi_{j-1}(GL(\mathbb{C})) = \pi_{j-1}(GL(2j-1, \mathbb{C}))$ and apply the previous theorem.   □

Remark 10. Compare this result with $K_1^{alg}(\mathbb{C}) = \mathbb{C}^*$ (since $\mathbb{C}$ is a field, see [12, Ex. 3.1.6]). Higher algebraic K-theory groups for $\mathbb{C}$ are only partially understood.

3.6 Topological K-Theory

There is a close connection between K-theory of C^*-algebras and topological K-theory of spaces.

Let X be a locally compact, Hausdorff, topological space. Atiyah and Hirzebruch [3] defined abelian groups $K^0(X)$ and $K^1(X)$ called *topological K-theory with compact supports*. For instance, if X is compact, $K^0(X)$ is the Grothendieck group of complex vector bundles on X.

Theorem 5. *Let X be a locally compact, Hausdorff, topological space. Then*

$$K^j(X) = K_j(C_0(X)), \quad j = 0, 1.$$

Remark 11. This is known as Swan's theorem when $j = 0$ and X compact.

In turn, topological K-theory can be computed up to torsion via a Chern character. Let X be as above. There is a *Chern character* from topological K-theory to rational cohomology with compact supports

$$ch : K^j(X) \longrightarrow \bigoplus_{l \geq 0} H_c^{j+2l}(X; \mathbb{Q}), \quad j = 0, 1.$$

Here the target cohomology theory $H_c^*(-; \mathbb{Q})$ can be Čech cohomology with compact supports, Alexander–Spanier cohomology with compact supports or representable Eilenberg–MacLane cohomology with compact supports.

This map becomes an isomorphism when tensored with the rationals.

Theorem 6. *Let X be a locally compact, Hausdorff, topological space. The Chern character is a rational isomorphism, that is,*

$$K^j(X) \otimes_{\mathbb{Z}} \mathbb{Q} \longrightarrow \bigoplus_{l \geq 0} H_c^{j+2l}(X; \mathbb{Q}), \quad j = 0, 1$$

is an isomorphism.

Remark 12. This theorem is still true for singular cohomology when *X* is a locally finite CW-complex.

4 Proper *G*-Spaces

In the following three sections, we will describe the left-hand side of the Baum–Connes conjecture (Conjecture 1). The space $\underline{E}G$ appearing on the topological side of the conjecture is the *universal example for proper actions for G*. Hence we will start by studying proper *G*-spaces.

Recall the definition of *G*-space, *G*-map and *G*-homotopy.

Definition 7. *A G-space is a topological space X with a given continuous action of G*

$$G \times X \longrightarrow X.$$

A G-map is a continuous map f : X → Y between G-spaces such that

$$f(gp) = gf(p) \text{ for all } (g, p) \in G \times X.$$

Two G-maps $f_0, f_1 : X \to Y$ are G-homotopic if they are homotopic through G-maps, that is, there exists a homotopy $\{f_t\}_{0 \leq t \leq 1}$ with each f_t a G-map.

We will require proper *G*-spaces to be Hausdorff and paracompact. Recall that a space *X* is *paracompact* if every open cover of *X* has a locally finite open refinement or, alternatively, a locally finite partition of unity subordinate to any given open cover.

Remark 13. Any metrizable space (i.e. there is a metric with the same underlying topology) or any CW-complex (in its usual CW-topology) is Hausdorff and paracompact.   ·

Definition 8. *A G-space X is* proper *if*

- *X is Hausdorff and paracompact*
- *The quotient space G\X (with the quotient topology) is Hausdorff and paracompact*
- *For each $p \in X$ there exists a triple (U, H, ρ) such that*
  (a) U is an open neighborhood of p in X with $gu \in U$ for all $(g, u) \in G \times U$.
  (b) H is a compact subgroup of G.
  (c) $\rho : U \to G/H$ is a G-map.

Note that, in particular, the stabilizer stab(*p*) is a closed subgroup of a conjugate of *H* and hence compact.

Remark 14. The converse is not true in general; the action of $\mathbb{Z}$ on S^1 by an irrational rotation is free but it is not a proper $\mathbb{Z}$-space.

Remark 15. If X is a *G-CW-complex* then it is a proper G-space (even in the weaker definition below) if and only if all the cell stabilizers are compact, see Thm. 1.23 in [30].

Our definition is stronger than the usual definition of proper G-space, which requires the map $G \times X \to X \times X$, $(g,x) \mapsto (gx,x)$ to be proper, in the sense that the pre-image of a compact set is compact. Nevertheless, both definitions agree for locally compact, Hausdorff, second countable G-spaces.

Proposition 2 (J. Chabert, S. Echterhoff, R. Meyer [11]). *If X is a locally compact, Hausdorff, second countable G-space, then X is proper if and only if the map*

$$G \times X \longrightarrow X \times X$$
$$(g,x) \longmapsto (gx,x)$$

is proper.

Remark 16. For a more general comparison among these and other definitions of proper actions see [7].

5 Classifying Space for Proper Actions

Now we are ready for the definition of the space $\underline{E}G$ appearing in the statement of the Baum–Connes Conjecture. Most of the material in this section is based on Sects. 1 and 2 of [5].

Definition 9. *A* universal example for proper actions *of G, denoted $\underline{E}G$, is a proper G-space such that:*

- *If X is any proper G-space, then there exists a G-map $f : X \to \underline{E}G$ and any two G-maps from X to $\underline{E}G$ are G-homotopic.*

$\underline{E}G$ exists for every topological group G [5, Appendix 1] and it is unique up to G-homotopy, as follows. Suppose that $\underline{E}G$ and $(\underline{E}G)'$ are both universal examples for proper actions of G. Then there exist G-maps

$$f : \underline{E}G \longrightarrow (\underline{E}G)'$$
$$f' : (\underline{E}G)' \longrightarrow \underline{E}G$$

and $f' \circ f$ and $f \circ f'$ must be G-homotopic to the identity maps of $\underline{E}G$ and $(\underline{E}G)'$ respectively.

The following are equivalent axioms for a space Y to be $\underline{E}G$ [5, Appendix 2].

(a) Y is a proper G-space.
(b) If H is any compact subgroup of G, then there exists $p \in Y$ with $hp = p$ for all $h \in H$.
(c) Consider $Y \times Y$ as a G-space via $g(y_0, y_1) = (gy_0, gy_1)$, and the maps

$$\rho_0, \rho_1 : Y \times Y \longrightarrow Y$$
$$\rho_0(y_0, y_1) = y_0, \quad \rho_1(y_0, y_1) = y_1.$$

Then ρ_0 and ρ_1 are G-homotopic.

Remark 17. It is possible to define a universal space for any family of (closed) subgroups of G closed under conjugation and finite intersections [32]. Then $\underline{E}G$ is the universal space for the family of compact subgroups of G.

Remark 18. The space $\underline{E}G$ can always be assumed to be a G-CW-complex. Then there is a homotopy characterization: a proper G-CW-complex X is an $\underline{E}G$ if and only if for each compact subgroup H of G the fixed point subcomplex X^H is contractible (see [32]).

5.1 Examples

(a) If G is compact, $\underline{E}G$ is just a one-point space.
(b) If G is a Lie group with finitely many connected components then $\underline{E}G = G/H$, where H is a maximal compact subgroup (i.e. maximal among compact subgroups).
(c) If G is a p-adic group then $\underline{E}G = \beta G$ the affine Bruhat-Tits building for G. For example, $\beta SL(2, \mathbb{Q}_p)$ is the $(p+1)$-regular tree, that is, the unique tree with exactly $p+1$ edges at each vertex (see Fig. 2) (cf. [46]).

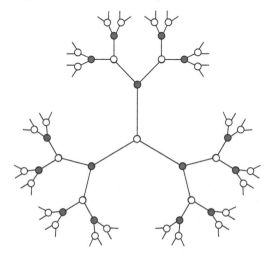

Fig. 2. The $(p+1)$-regular tree is $\beta SL(2, \mathbb{Q}_p)$

(d) If Γ is an arbitrary (countable) discrete group, there is an explicit construction,

$$\underline{E}\Gamma = \left\{ f : \Gamma \to [0,1] \,\middle|\, f \text{ finite support}, \sum_{\gamma \in \Gamma} f(\gamma) = 1 \right\},$$

that is, the space of all finite probability measures on Γ, topologized by the metric $d(f,h) = \sqrt{\sum_{\gamma \in \Gamma} |f(\gamma) - h(\gamma)|^2}$.

6 Equivariant K-Homology

K-homology is the dual theory to Atiyah–Hirzebruch K-theory (Sect. 3.6). Here we define an equivariant generalization due to Kasparov [24, 25]. If X is a proper G-space with compact, second countable quotient then $K_i^G(X)$, $i = 0, 1$, are abelian groups defined as homotopy classes of K-cycles for X. These K-cycles can be viewed as G-equivariant abstract elliptic operators on X.

Remark 19. For a discrete group G, there is a topological definition of equivariant K-homology and the index map via equivariant spectra [14]. This and other constructions of the index map are shown to be equivalent in [18].

6.1 Definitions

Let G be a locally compact, Hausdorff, second countable, topological group.

Let H be a separable Hilbert space. Write $\mathscr{U}(H)$ for the set of unitary operators

$$\mathscr{U}(H) = \{ U \in \mathscr{L}(H) \,|\, UU^* = U^*U = I \}.$$

Definition 10. *A unitary representation of G on H is a group homomorphism $\pi : G \to \mathscr{U}(H)$ such that for each $v \in H$ the map $\pi_v : G \to H$, $g \mapsto \pi(g)v$ is a continuous map from G to H.*

Definition 11. *A G-C^*-algebra is a C^*-algebra A with a given continuous action of G*

$$G \times A \longrightarrow A$$

such that G acts by C^*-algebra automorphisms.

The continuity condition is that, for each $a \in A$, the map $G \to A$, $g \mapsto ga$ is a continuous map. We also have that, for each $g \in G$, the map $A \to A$, $a \mapsto ga$ is a C^*-algebra automorphism.

Example 4. Let X be a locally compact, Hausdorff G-space. The action of G on X gives an action of G on $C_0(X)$,

$$(g\alpha)(x) = \alpha(g^{-1}x),$$

where $g \in G$, $\alpha \in C_0(X)$ and $x \in X$. This action makes $C_0(X)$ into a G-C^*-algebra.

Recall that a C^*-algebra is *separable* if it has a countable dense subset.

Definition 12. *Let A be a separable G-C*-algebra. A* representation *of A is a triple* (H, ψ, π) *with:*

- *H is a separable Hilbert space*
- $\psi: A \to \mathscr{L}(H)$ *is a *-homomorphism*
- $\pi: G \to \mathscr{U}(H)$ *is a unitary representation of G on H*
- $\psi(ga) = \pi(g)\psi(a)\pi(g^{-1})$ *for all* $(g, a) \in G \times A$.

Remark 20. We are using a slightly non-standard notation; in the literature this is usually called a *covariant representation*.

Definition 13. *Let X be a proper G-space with compact, second countable quotient space G\X. An* equivariant odd K-cycle *for X is a 4-tuple* (H, ψ, π, T) *such that:*

- (H, ψ, π) *is a representation of the G-C*-algebra* $C_0(X)$
- $T \in \mathscr{L}(H)$
- $T = T^*$
- $\pi(g)T - T\pi(g) = 0$ *for all* $g \in G$
- $\psi(\alpha)T - T\psi(\alpha) \in \mathscr{K}(H)$ *for all* $\alpha \in C_0(X)$
- $\psi(\alpha)(I - T^2) \in \mathscr{K}(H)$ *for all* $\alpha \in C_0(X)$.

Remark 21. If G is a locally compact, Hausdorff, second countable topological group and X a proper G-space with locally compact quotient then X is also locally compact and hence $C_0(X)$ is well-defined.

Write $\mathscr{E}_1^G(X)$ for the set of equivariant odd K-cycles for X. This concept was introduced by Kasparov as an abstraction of an equivariant self-adjoint elliptic operator and goes back to Atiyah's theory of elliptic operators [2].

Example 5. Let $G = \mathbb{Z}$, $X = \mathbb{R}$ with the action $\mathbb{Z} \times \mathbb{R} \to \mathbb{R}$, $(n, t) \mapsto n + t$. The quotient space is S^1, which is compact. Consider $H = L^2(\mathbb{R})$ the Hilbert space of complex-valued square integrable functions with the usual Lebesgue measure. Let $\psi: C_0(\mathbb{R}) \to \mathscr{L}(L^2(\mathbb{R}))$ be defined as $\psi(\alpha)u = \alpha u$, where $\alpha u(t) = \alpha(t)u(t)$, for all $\alpha \in C_0(\mathbb{R})$, $u \in L^2(\mathbb{R})$ and $t \in \mathbb{R}$. Finally, let $\pi: \mathbb{Z} \to \mathscr{U}(L^2(\mathbb{R}))$ be the map $(\pi(n)u)(t) = u(t - n)$ and consider the operator $\left(-i\frac{d}{dt}\right)$. This operator is self-adjoint but *not* bounded on $L^2(\mathbb{R})$. We "normalize" it to obtain a bounded operator

$$T = \left(\frac{x}{\sqrt{1 + x^2}}\right)\left(-i\frac{d}{dt}\right).$$

This notation means that the function $\frac{x}{\sqrt{1+x^2}}$ is applied using functional calculus to the operator $\left(-i\frac{d}{dt}\right)$. Note that the operator $\left(-i\frac{d}{dt}\right)$ is essentially self adjoint. Thus the function $\frac{x}{\sqrt{1+x^2}}$ can be applied to the unique self-adjoint extension of $\left(-i\frac{d}{dt}\right)$.

Equivalently, T can be constructed using Fourier transform. Let $\mathscr{M}_x$ be the operator "multiplication by x"

$$\mathscr{M}_x(f(x)) = xf(x).$$

The Fourier transform $\mathscr{F}$ converts $-i\frac{d}{dt}$ to $\mathscr{M}_x$, i.e. there is a commutative diagram

$$
\begin{array}{ccc}
L^2(\mathbb{R}) & \xrightarrow{\ \mathscr{F}\ } & L^2(\mathbb{R}) \\
{\scriptstyle -i\frac{d}{dt}}\big\downarrow & & \big\downarrow{\scriptstyle \mathscr{M}_x} \\
L^2(\mathbb{R}) & \xrightarrow[\ \mathscr{F}\ ]{} & L^2(\mathbb{R}).
\end{array}
$$

Let $\mathscr{M}_{\frac{x}{\sqrt{1+x^2}}}$ be the operator "multiplication by $\frac{x}{\sqrt{1+x^2}}$"

$$
\mathscr{M}_{\frac{x}{\sqrt{1+x^2}}}(f(x)) = \frac{x}{\sqrt{1+x^2}}f(x).
$$

T is the unique bounded operator on $L^2(\mathbb{R})$ such that the following diagram is commutative

$$
\begin{array}{ccc}
L^2(\mathbb{R}) & \xrightarrow{\ \mathscr{F}\ } & L^2(\mathbb{R}) \\
{\scriptstyle T}\big\downarrow & & \big\downarrow{\scriptstyle \mathscr{M}_{\frac{x}{\sqrt{1+x^2}}}} \\
L^2(\mathbb{R}) & \xrightarrow[\ \mathscr{F}\ ]{} & L^2(\mathbb{R}).
\end{array}
$$

Then we have an equivariant odd K-cycle $(L^2(\mathbb{R}), \psi, \pi, T) \in \mathscr{E}_1^{\mathbb{Z}}(\mathbb{R})$.

Let X be a proper G-space with compact, second countable quotient $G\backslash X$ and $\mathscr{E}_1^G(X)$ defined as above. The *equivariant K-homology group* $K_1^G(X)$ is defined as the quotient

$$
K_1^G(X) = \mathscr{E}_1^G(X)/\sim,
$$

where $\sim$ represents *homotopy*, in a sense that will be made precise later (Sect. 12). It is an abelian group with addition and inverse given by

$$
(H, \psi, \pi, T) + (H', \psi', \pi', T') = (H \oplus H', \psi \oplus \psi', \pi \oplus \pi', T \oplus T'),
$$
$$
-(H, \psi, \pi, T) = (H, \psi, \pi, -T).
$$

Remark 22. The K-cycles defined above differ slightly from the K-cycles used by Kasparov [25]. However, the abelian group $K_1^G(X)$ is isomorphic to the Kasparov group $KK_G^1(C_0(X), \mathbb{C})$, where the isomorphism is given by the evident map which views one of our K-cycles as one of Kasparov's K-cycles. In other words, the K-cycles we are using are more special than the K-cycles used by Kasparov, however the obvious map of abelian groups is an isomorphism.

We define *even K-cycles* in a similar way, just dropping the condition of T being self-adjoint.

Definition 14. *Let X be a proper G-space with compact, second countable quotient space $G\backslash X$. An equivariant even K-cycle for X is a 4-tuple (H, ψ, π, T) such that:*

- (H, ψ, π) is a representation of the G-C^*-algebra $C_0(X)$
- $T \in \mathcal{L}(H)$
- $\pi(g)T - T\pi(g) = 0$ for all $g \in G$
- $\psi(\alpha)T - T\psi(\alpha) \in \mathcal{K}(H)$ for all $\alpha \in C_0(X)$
- $\psi(\alpha)(I - T^*T) \in \mathcal{K}(H)$ for all $\alpha \in C_0(X)$
- $\psi(\alpha)(I - TT^*) \in \mathcal{K}(H)$ for all $\alpha \in C_0(X)$

Write $\mathcal{E}_0^G(X)$ for the set of such equivariant even K-cycles.

Remark 23. In the literature the definition is somewhat more complicated. In particular, the Hilbert space H is required to be $\mathbb{Z}/2$-graded. However, at the level of abelian groups, the abelian group $K_0^G(X)$ obtained from the equivariant even K-cycles defined here will be isomorphic to the Kasparov group $KK_G^0(C_0(X), \mathbb{C})$ [25].

    More precisely, let $(H, \psi, \pi, T, \omega)$ be a K-cycle in Kasparov's sense, where ω is a $\mathbb{Z}/2$-grading of the Hilbert space $H = H_0 \oplus H_1$, $\psi = \psi_0 \oplus \psi_1$, $\pi = \pi_0 \oplus \pi_1$ and T is self-adjoint but off-diagonal

$$T = \begin{pmatrix} 0 & T_- \\ T_+ & 0 \end{pmatrix}.$$

To define the isomorphism from $KK_G^0(C_0(X), \mathbb{C})$ to $K_0^G(X)$, we map a Kasparov cycle $(H, \psi, \pi, T, \omega)$ to (H', ψ', π', T') where

$$H' = \ldots H_0 \oplus H_0 \oplus H_0 \oplus H_1 \oplus H_1 \oplus H_1 \ldots$$
$$\psi' = \ldots \psi_0 \oplus \psi_0 \oplus \psi_0 \oplus \psi_1 \oplus \psi_1 \oplus \psi_1 \ldots$$
$$\pi' = \ldots \pi_0 \oplus \pi_0 \oplus \pi_0 \oplus \pi_1 \oplus \pi_1 \oplus \pi_1 \ldots$$

and T' is the obvious right-shift operator, where we use T_+ to map the last copy of H_0 to the first copy of H_1. The isomorphism from $\mathcal{E}_0^G(X)$ to $KK_G^0(C_0(X), \mathbb{C})$ is given by

$$(H, \psi, \pi, T) \mapsto (H \oplus H, \psi \oplus \psi, \pi \oplus \pi, \begin{pmatrix} 0 & T^* \\ T & 0 \end{pmatrix}).$$

    Let X be a proper G-space with compact, second countable quotient $G \backslash X$ and $\mathcal{E}_0^G(X)$ as above. The *equivariant K-homology group* $K_0^G(X)$ is defined as the quotient

$$K_0^G(X) = \mathcal{E}_0^G(X) / \sim,$$

where $\sim$ is *homotopy*, in a sense that will be made precise later. It is an abelian group with addition and inverse given by

$$(H, \psi, \pi, T) + (H', \psi', \pi', T') = (H \oplus H', \psi \oplus \psi', \pi \oplus \pi', T \oplus T'),$$
$$-(H, \psi, \pi, T) = (H, \psi, \pi, T^*).$$

Remark 24. Since the even K-cycles are more general, we have $\mathcal{E}_1^G(X) \subset \mathcal{E}_0^G(X)$. However, this inclusion induces the zero map from $K_1^G(X)$ to $K_0^G(X)$.

6.2 Functoriality

Equivariant K-homology gives a (covariant) functor between the category proper G-spaces with compact quotient and the category of abelian groups. Indeed, given a continuous G-map $f \colon X \to Y$ between proper G-spaces with compact quotient, it induces a map $\widetilde{f} \colon C_0(Y) \to C_0(X)$ by $\widetilde{f}(\alpha) = \alpha \circ f$ for all $\alpha \in C_0(Y)$. Then, we obtain homomorphisms of abelian groups

$$K_j^G(X) \longrightarrow K_j^G(Y) \quad j = 0, 1$$

by defining, for each $(H, \psi, \pi, T) \in \mathscr{E}_j^G(X)$,

$$(H, \psi, \pi, T) \mapsto (H, \psi \circ \widetilde{f}, \pi, T).$$

6.3 The Index Map

Let X be a proper second countable G-space with compact quotient $G \backslash X$. There is a map of abelian groups

$$K_j^G(X) \longrightarrow K_j(C_r^*(G))$$
$$(H, \psi, \pi, T) \mapsto \operatorname{Index}(T)$$

for $j = 0, 1$. It is called the *index map* and will be defined in Sect. 14.

This map is natural, that is, if X and Y are proper second countable G-spaces with compact quotient and if $f \colon X \to Y$ is a continuous G-equivariant map, then the following diagram commutes:

$$
\begin{array}{ccc}
K_j^G(X) & \xrightarrow{\;f_*\;} & K_j^G(Y) \\
{\scriptstyle \text{Index}}\big\downarrow & & \big\downarrow{\scriptstyle \text{Index}} \\
K_j^G(C_r^*(G)) & \xrightarrow{\;=\;} & K_j^G(C_r^*(G)).
\end{array}
$$

We would like to define equivariant K-homology and the index map for $\underline{E}G$. However, the quotient of $\underline{E}G$ by the G-action might not be compact. The solution will be to consider all proper second countable G-subspaces with compact quotient.

Definition 15. *Let Z be a proper G-space. We call $\Delta \subseteq Z$ G-compact if*

(a) $gx \in \Delta$ for all $g \in G$, $x \in \Delta$
(b) Δ is a proper G-space
(c) The quotient space $G \backslash \Delta$ is compact

That is, Δ is a G-subspace which is proper as a G-space and has compact quotient $G \backslash \Delta$.

Remark 25. Since we are always assuming that G is locally compact, Hausdorff and second countable, we may also assume without loss of generality that any G-compact subset of $\underline{E}G$ is second countable. From now on we shall assume that $\underline{E}G$ has this property.

We define the *equivariant K-homology of $\underline{E}G$ with G-compact supports* as the direct limit

$$K_j^G(\underline{E}G) = \varinjlim_{\substack{\Delta \subseteq \underline{E}G \\ G\text{-compact}}} K_j^G(\Delta).$$

There is then a well-defined index map on the direct limit

$$\mu \colon K_j^G(\underline{E}G) \longrightarrow K_j(C_r^*G)$$
$$(H, \psi, \pi, T) \longmapsto \text{Index}(T), \tag{1}$$

as follows. Suppose that $\Delta \subset \Omega$ are G-compact. By the naturality of the functor $K_j^G(-)$, there is a commutative diagram

$$
\begin{array}{ccc}
K_j^G(\Delta) & \longrightarrow & K_j^G(\Omega) \\
{\scriptstyle \text{Index}}\downarrow & & \downarrow{\scriptstyle \text{Index}} \\
K_j(C_r^*G) & \xlongequal{\quad} & K_j(C_r^*G),
\end{array}
$$

and thus the index map is defined on the direct limit.

7 The Discrete Case

We discuss several aspects of the Baum–Connes conjecture when the group is discrete.

7.1 Equivariant *K*-Homology

For a discrete group Γ, there is a simple description of $K_j^\Gamma(\underline{E}\Gamma)$ up to torsion, in purely algebraic terms, given by a Chern character. Here we follow Sect. 7 in [5].

Let Γ be a (countable) discrete group. Define $F\Gamma$ as the set of finite formal sums

$$F\Gamma = \left\{ \sum_{\text{finite}} \lambda_\gamma[\gamma] \text{ where } \gamma \in \Gamma, \text{order}(\gamma) < \infty, \lambda_\gamma \in \mathbb{C} \right\}.$$

$F\Gamma$ is a complex vector space and also a Γ-module with Γ-action:

$$g \cdot \left(\sum_{\lambda \in \Gamma} \lambda_\gamma[\gamma] \right) = \sum_{\lambda \in \Gamma} \lambda_\gamma[g\gamma g^{-1}].$$

Note that the identity element of the group has order 1 and therefore $F\Gamma \neq 0$.

Consider $H_j(\Gamma; F\Gamma)$, $j \geq 0$, the homology groups of Γ with coefficients in the Γ-module $F\Gamma$.

Remark 26. This is standard homological algebra, with no topology involved (Γ is a discrete group and $F\Gamma$ is a non-topologized module over Γ). They are classical homology groups and have a purely algebraic description (cf. [10]). In general, if M is a Γ-module then $H_*(\Gamma;M)$ is isomorphic to $H_*(B\Gamma;\underline{M})$, where $\underline{M}$ means the local system on $B\Gamma$ obtained from the Γ-module M.

Let us write $K_j^{\mathrm{top}}(\Gamma)$ for $K_j^{\Gamma}(\underline{E}\Gamma)$, $j = 0,1$. There is a Chern character ch: $K_*^{\mathrm{top}}(\Gamma) \to H_*(\Gamma;F\Gamma)$ which maps into odd, respectively even, homology

$$\mathrm{ch}\colon K_j^{\mathrm{top}}(\Gamma) \to \bigoplus_{l \geq 0} H_{j+2l}(\Gamma;F\Gamma) \quad j = 0,1.$$

This map becomes an isomorphism when tensored with $\mathbb{C}$ (cf. [4] or [31]).

Proposition 3. *The map*

$$\mathrm{ch} \otimes_{\mathbb{Z}} \mathbb{C}\colon \; K_j^{\mathrm{top}}(\Gamma) \otimes_{\mathbb{Z}} \mathbb{C} \longrightarrow \bigoplus_{l \geq 0} H_{j+2l}(\Gamma;F\Gamma) \quad j = 0,1$$

is an isomorphism of vector spaces over $\mathbb{C}$.

Remark 27. If G is finite, the rationalized Chern character becomes the character map from $R(G)$, the complex representation ring of G, to class functions, given by $\rho \mapsto \chi(\rho)$ in the even case, and the zero map in the odd case.

If the Baum–Connes conjecture is true for Γ, then Proposition 3 computes the tensored topological K-theory of the reduced C^*-algebra of Γ.

Corollary 1. *If the Baum–Connes conjecture is true for Γ then*

$$K_j(C_r^*\Gamma) \otimes_{\mathbb{Z}} \mathbb{C} \cong \bigoplus_{l \geq 0} H_{j+2l}(\Gamma;F\Gamma) \quad j = 0,1.$$

7.2 Some Results on Discrete Groups

We recollect some results on discrete groups which satisfy the Baum–Connes conjecture.

Theorem 7 (N. Higson, G. Kasparov [21]). *If Γ is a discrete group which is amenable (or, more generally, a-T-menable) then the Baum–Connes conjecture is true for Γ.*

Theorem 8 (I. Mineyev, G. Yu [37]; independently V. Lafforgue [28]). *If Γ is a discrete group which is hyperbolic (in Gromov's sense) then the Baum–Connes conjecture is true for Γ.*

Theorem 9 (Schick [45]).
   Let B_n be the braid group on n strands, for any positive integer n. Then the Baum–Connes conjecture is true for B_n.

Theorem 10 (Matthey, Oyono–Oyono, Pitsch [35]). *Let M be a connected orientable 3-dimensional manifold (possibly with boundary). Let Γ be the fundamental group of M. Then the Baum–Connes conjecture is true for Γ.*

The Baum–Connes index map has been shown to be injective or rationally injective for some classes of groups. For example, it is injective for countable subgroups of $GL(n, K)$, K any field [17], and injective for

- Closed subgroups of connected Lie groups [26]
- Closed subgroups of reductive *p*-adic groups [27]

More results on groups satisfying the Baum–Connes conjecture can be found in [34].

The Baum–Connes conjecture remains a widely open problem. For example, it is not known for $SL(n, \mathbb{Z})$, $n \geq 3$. These infinite discrete groups have Kazhdan's property (T) and hence they are not a-T-menable. On the other hand, it is known that the index map is injective for $SL(n, \mathbb{Z})$ (see above) and the groups $K_j^G(\underline{E}G)$ for $G = SL(3, \mathbb{Z})$ have been calculated [44].

Remark 28. The conjecture might be too general to be true for all groups. Nevertheless, we expect it to be true for a large family of groups, in particular for all exact groups (a group *G* is *exact* if the functor $C_r^*(G, -)$, as defined in Sect. 11.2, is exact).

7.3 Corollaries of the Baum–Connes Conjecture

The Baum–Connes conjecture is related to a great number of conjectures in functional analysis, algebra, geometry and topology. Most of these conjectures follow from either the injectivity or the surjectivity of the index map. A significant example is the Novikov conjecture on the homotopy invariance of higher signatures of closed, connected, oriented, smooth manifolds. This conjecture follows from the injectivity of the rationalized index map [5]. For more information on conjectures related to Baum–Connes, see the appendix in [38].

Remark 29. By a "corollary" of the Baum–Connes conjecture we mean: if the Baum–Connes conjecture is true for a group *G* then the corollary is true for that group *G*. (For instance, in the Novikov conjecture *G* is the fundamental group of the manifold.)

8 The Compact Case

If *G* is compact, we can take $\underline{E}G$ to be a one-point space. On the other hand, $K_0(C_r^*G) = R(G)$ the (complex) representation ring of *G*, and $K_1(C_r^*G) = 0$ (see Remark below). Recall that $R(G)$ is the Grothendieck group of the category of finite dimensional (complex) representations of *G*. It is a free abelian group with one generator for each distinct (i.e. non-equivalent) irreducible representation of *G*.

Remark 30. When G is compact, the reduced C^*-algebra of G is a direct sum (in the C^*-algebra sense) over the irreducible representations of G, of matrix algebras of dimension equal to the dimension of the representation. The K-theory functor commutes with direct sums and $K_j(M_n(\mathbb{C})) \cong K_j(\mathbb{C})$, which is $\mathbb{Z}$ for j even and 0 otherwise (Theorem 4).

Hence the index map takes the form

$$\mu \colon K_G^0(point) \longrightarrow R(G),$$

for $j = 0$ and is the zero map for $j = 1$.

Given $(H, \psi, T, \pi) \in \mathscr{E}_G^0(point)$, we may assume within the equivalence relation on $\mathscr{E}_G^0(point)$ that

$$\psi(\lambda) = \lambda I \quad \text{for all } \lambda \in C_0(point) = \mathbb{C},$$

where I is the identity operator of the Hilbert space H. Hence the non-triviality of (H, ψ, T, π) is coming from

$$I - TT^* \in \mathscr{K}(H), \text{ and } I - T^*T \in \mathscr{K}(H),$$

that is, T is a Fredholm operator. Therefore

$$\dim_{\mathbb{C}}(\ker(T)) < \infty,$$
$$\dim_{\mathbb{C}}(\operatorname{coker}(T)) < \infty,$$

hence $\ker(T)$ and $\operatorname{coker}(T)$ are finite dimensional representations of G (recall that G is acting via $\pi \colon G \to \mathscr{L}(H)$). Then

$$\mu(H, \psi, T, \pi) = \operatorname{Index}(T) = \ker(T) - \operatorname{coker}(T) \in R(G).$$

Remark 31. The assembly map for G compact just described is an isomorphism (exercise).

Remark 32. In general, for G non-compact, the elements of $K_0^G(X)$ can be viewed as generalized elliptic operators on $\underline{E}G$, and the index map μ assigns to such an operator its 'index', $\ker(T) - \operatorname{coker}(T)$, in some suitable sense [5]. This should be made precise later using Kasparov's descent map and an appropriate Kasparov product (Sect. 14).

9 Equivariant K-Homology for G-C^*-algebras

We have defined equivariant K-homology for G-spaces in Sect. 6. Now we define equivariant K-homology for a separable G-C^*-algebra A as the KK-theory groups $K_G^j(A, \mathbb{C})$, $j = 0, 1$. This generalises the previous construction since $K_j^G(X) = KK_G^j(C_0(X), \mathbb{C})$. Later on we shall define KK-theory groups in full generality (Sects. 12 and 13).

Definition 16. *Let A be a separable G-C*-algebra. Define $\mathcal{E}_G^1(A)$ to be the set of 4-tuples*

$$\{(H,\psi,\pi,T)\}$$

such that (H,ψ,π) is a representation of the G-C*-algebra A, $T \in \mathcal{L}(H)$, and the following conditions are satisfied:

- $T = T^*$
- $\pi(g)T - T\pi(g) \in \mathcal{K}(H)$
- $\psi(a)T - T\psi(a) \in \mathcal{K}(H)$
- $\psi(a)(I - T^2) \in \mathcal{K}(H)$

for all $g \in G$, $a \in A$.

Remark 33. Note that this is not quite $\mathcal{E}_1^G(X)$ when $A = C_0(X)$ and X is a proper G-space with compact quotient, since the third condition is more general than before. However, the inclusion $\mathcal{E}_1^G(X) \subset \mathcal{E}_G^1(C_0(X))$ gives an isomorphism of abelian groups so that $K_1^G(X) = KK_G^1(C_0(X),\mathbb{C})$ (as defined below). The point is that, for a proper G-space with compact quotient, an averaging argument using a cut-off function and the Haar measure of the group G allows us to assume that the operator T is G-equivariant.

Given a separable G-C^*-algebra A, we define the KK-group $KK_G^1(A,\mathbb{C})$ as $\mathcal{E}_G^1(A)$ modulo an equivalence relation called *homotopy*, which will be made precise later. Addition in $KK_G^1(A,\mathbb{C})$ is given by direct sum

$$(H,\psi,\pi,T) + (H',\psi',\pi',T') = (H \oplus H', \psi \oplus \psi', \pi \oplus \pi', T \oplus T')$$

and the negative of an element by

$$-(H,\psi,\pi,T) = (H,\psi,\pi,-T).$$

Remark 34. We shall later define $KK_G^1(A,B)$ for a separable G-C^*-algebras A and an arbitrary G-C^*-algebra B (Sect. 12).

Let A, B be separable G-C^*-algebras. A G-equivariant $*$-homomorphism $\phi: A \to B$ gives a map $\mathcal{E}_G^1(B) \to \mathcal{E}_G^1(A)$ by

$$(H,\psi,\pi,T) \mapsto (H,\psi \circ \phi,\pi,T),$$

and this induces a map $KK_G^1(B,\mathbb{C}) \to KK_G^1(A,\mathbb{C})$. That is, $KK_G^1(A,\mathbb{C})$ is a contravariant functor in A.

For the even case, the operator T is not required to be self-adjoint.

Definition 17. *Let A be a separable G-C*-algebra. Define $\mathcal{E}_G^0(A)$ as the set of 4-tuples*

$$\{(H,\psi,\pi,T)\}$$

such that (H,ψ,π) is a representation of the G-C*-algebra A, $T \in \mathcal{L}(H)$ and the following conditions are satisfied:

- $\pi(g)T - T\pi(g) \in \mathcal{K}(H)$
- $\psi(a)T - T\psi(a) \in \mathcal{K}(H)$
- $\psi(a)(I - T^*T) \in \mathcal{K}(H)$
- $\psi(a)(I - TT^*) \in \mathcal{K}(H)$

for all $g \in G$, $a \in A$.

Remark 35. Again, if X is a proper G-space with compact quotient, the inclusion $\mathcal{E}_0^G(X) \subset \mathcal{E}_G^0(C_0(X))$ gives an isomorphism in K-homology, so we can write $K_0^G(X) = KK^0(C_0(X), \mathbb{C})$ (as defined below). The issue of the $\mathbb{Z}/2$-grading (which is present in the Kasparov definition but not in our definition) is dealt with as in Remark 23.

We define the KK-groups $KK_G^0(A, \mathbb{C})$ as $\mathcal{E}_G^0(A)$ modulo an equivalence relation called *homotopy*, which will be made precise later. Addition in $KK_G^1(A, \mathbb{C})$ is given by direct sum

$$(H, \psi, \pi, T) + (H', \psi', \pi', T') = (H \oplus H', \psi \oplus \psi', \pi \oplus \pi', T \oplus T')$$

and the negative of an element by

$$-(H, \psi, \pi, T) = (H, \psi, \pi, T^*).$$

Remark 36. We shall later define in general $KK_G^0(A, B)$ for a separable G-C^*-algebras A and an arbitrary G-C^*-algebra B (Sect. 13).

Let A, B be separable G-C^*-algebras. A G-equivariant $*$-homomorphism $\phi: A \to B$ gives a map $\mathcal{E}_G^0(B) \to \mathcal{E}_G^0(A)$ by

$$(H, \psi, \pi, T) \mapsto (H, \psi \circ \phi, \pi, T),$$

and this induces a map $KK_G^0(B, \mathbb{C}) \to KK_G^0(A, \mathbb{C})$. That is, $KK_G^0(A, \mathbb{C})$ is a contravariant functor in A.

10 The Conjecture with Coefficients

There is a generalized version of the Baum–Connes conjecture, known as the *Baum–Connes conjecture with coefficients*, which adds coefficients in a G-C^*-algebra. We recall the definition of G-C^*-algebra.

Definition 18. *A G-C^*-algebra is a C^*-algebra A with a given continuous action of G*

$$G \times A \longrightarrow A$$

such that G acts by C^*-algebra automorphisms. Continuity means that, for each $a \in A$, the map $G \to A$, $g \mapsto ga$ is a continuous map.

Remark 37. Observe that the only ∗-homomorphism of ℂ as a *C**-algebra is the identity. Hence the only *G-C**-algebra structure on ℂ is the one with trivial *G*-action.

Let *A* be a *G-C**-algebra. Later we shall define the reduced crossed-product *C**-algebra $C_r^*(G,A)$, and the equivariant *K*-homology group with coefficients $K_j^G(\underline{E}G,A)$. These constructions reduce to $C_r^*(G)$, respectively $K_j^G(\underline{E}G)$, when $A = \mathbb{C}$. Moreover, the index map extends to this general setting and is also conjectured to be an isomorphism.

Conjecture 2 (P. Baum, A. Connes, 1980). Let *G* be a locally compact, Hausdorff, second countable, topological group, and let *A* be any *G-C**-algebra. Then

$$\mu \colon K_j^G(\underline{E}G,A) \longrightarrow K_j(C_r^*(G,A)) \quad j = 0,1$$

is an isomorphism.

Conjecture 1 follows as a particular case when $A = \mathbb{C}$. A fundamental difference is that the conjecture with coefficients is subgroup closed, that is, if it is true for a group *G* for *any* coefficients then it is true, for *any* coefficients, for any closed subgroup of *G*.

The conjecture with coefficients has been proved for:

- Compact groups
- Abelian groups
- Groups acting simplicially on a tree with all vertex stabilizers satisfying the conjecture with coefficients [40]
- Amenable groups and, more generally, a-T-menable groups (groups with the Haagerup property) [22]
- The Lie group $Sp(n,1)$ [23]
- 3-manifold groups [35]

For more examples of groups satisfying the conjecture with coefficients see [34].

Expander Graphs

Suppose that Γ is a finitely generated, discrete group which contains an expander family [13] in its Cayley graph as a subgraph. Such a Γ is a counter-example to the conjecture with coefficients [19]. M. Gromov outlined a proof that such Γ exists. A number of mathematicians are now filling in the details. It seems quite likely that this group exists.

11 Hilbert Modules

In this section we introduce the concept of Hilbert module over a *C**-algebra. It generalises the definition of Hilbert space by allowing the inner product to take values in a *C**-algebra. Our main application will be the definition of the reduced crossed-product *C**-algebra in Sect. 11.2. For a concise reference on Hilbert modules see [29].

11.1 Definitions and Examples

Let A be a C^*-algebra.

Definition 19. *An element $a \in A$ is* positive *(notation: $a \geq 0$) if there exists $b \in A$ with $bb^* = a$.*

The subset of positive elements, A^+, is a convex cone (closed under positive linear combinations) and $A^+ \cap (-A^+) = \{0\}$ [15, 1.6.1]. Hence we have a well-defined partial ordering in A given by $x \geq y \iff x - y \geq 0$.

Definition 20. *A* pre-Hilbert A-module *is a right A-module $\mathcal{H}$ with a given A-valued inner product $\langle \, , \, \rangle$ such that:*

- $\langle u, v_1 + v_2 \rangle = \langle u, v_1 \rangle + \langle u, v_2 \rangle$
- $\langle u, va \rangle = \langle u, v \rangle a$
- $\langle u, v \rangle = \langle v, u \rangle^*$
- $\langle u, u \rangle \geq 0$
- $\langle u, u \rangle = 0 \iff u = 0$

for all $u, v, v_1, v_2 \in \mathcal{H}$, $a \in A$.

Definition 21. *A* Hilbert A-module *is a pre-Hilbert A-module which is complete with respect to the norm*
$$\|u\| = \|\langle u, u \rangle\|^{1/2}.$$

Remark 38. If $\mathcal{H}$ is a Hilbert A-module and A has a unit 1_A then $\mathcal{H}$ is a complex vector space with
$$u\lambda = u(\lambda 1_A) \qquad u \in \mathcal{H}, \lambda \in \mathbb{C}.$$

If A does not have a unit, then by using an *approximate identity* [41] in A, it is also a complex vector space.

Example 6. Let A be a C^*-algebra and $n \geq 1$. Then $A^n = A \oplus \ldots \oplus A$ is a Hilbert A-module with operations
$$(a_1, \ldots, a_n) + (b_1, \ldots, b_n) = (a_1 + b_1, \ldots, a_n + b_n),$$
$$(a_1, \ldots, a_n)a = (a_1 a, \ldots, a_n a),$$
$$\langle (a_1, \ldots, a_n), (b_1, \ldots, b_n) \rangle = a_1^* b_1 + \ldots + a_n^* b_n,$$

for all $a_j, b_j, a \in A$.

Example 7. Let A be a C^*-algebra. Define
$$\mathcal{H} = \left\{ (a_1, a_2, \ldots) \,\middle|\, a_j \in A, \ \sum_{j=1}^{\infty} a_j^* a_j \text{ is norm-convergent in } A \right\},$$

with operations

$$(a_1, a_2, \ldots) + (b_1, b_2 \ldots) = (a_1 + b_1, a_2 + b_2, \ldots),$$
$$(a_1, a_2, \ldots)a = (a_1 a, a_2 a, \ldots),$$
$$\langle (a_1, a_2, \ldots), (b_1, b_2 \ldots) \rangle = \sum_{j=1}^{\infty} a_j^* b_j.$$

The previous examples can be generalized. Note that a C^*-algebra A is a Hilbert module over itself with inner product $\langle a, b \rangle = a^* b$.

Example 8. If $\mathcal{H}_1, \ldots, \mathcal{H}_n$ are Hilbert A-modules then the direct sum $\mathcal{H}_1 \oplus \ldots \oplus \mathcal{H}_n$ is a Hilbert A-module with

$$\langle (x_1, \ldots, x_n), (y_1, \ldots, y_n) \rangle = \sum_i x_i^* y_i.$$

We write $\mathcal{H}^n$ for the direct sum of n copies of a Hilbert A-module $\mathcal{H}$.

Example 9. If $\{\mathcal{H}_i\}_{i \in \mathbb{N}}$ is a countable family of Hilbert A-modules then

$$\mathcal{H} = \left\{ (x_1, x_2, \ldots) \,\middle|\, x_i \in \mathcal{H}_i, \ \sum_{j=1}^{\infty} \langle x_j, x_j \rangle \text{ is norm-convergent in } A \right\}$$

is a Hilbert A-module with inner product $\langle x, y \rangle = \sum_{j=1}^{\infty} \langle x_j, y_j \rangle$.

The following is our key example.

Example 10. Let G be a locally compact, Hausdorff, second countable, topological group. Fix a left-invariant Haar measure dg for G. Let A be a G-C^*-algebra. Then $L^2(G, A)$ is a Hilbert A-module defined as follows. Denote by $C_c(G, A)$ the set of all continuous compactly supported functions from G to A. On $C_c(G, A)$ consider the norm

$$\|f\| = \left\| \int_G g^{-1} (f(g)^* f(g)) \, dg \right\|.$$

$L^2(G, A)$ is the completion of $C_c(G, A)$ in this norm. It is a Hilbert A-module with operations

$$(f + h)g = f(g) + h(g),$$
$$(fa)g = f(g)(ga),$$
$$\langle f, h \rangle = \int_G g^{-1} (f(g)^* h(g)) \, dg.$$

Note that when $A = \mathbb{C}$ the group action is trivial and we get $L^2(G)$ (cf. Remark 37).

Definition 22. *Let $\mathcal{H}$ be a Hilbert A-module. An A-module map $T \colon \mathcal{H} \to \mathcal{H}$ is adjointable if there exists an A-module map $T^* \colon \mathcal{H} \to \mathcal{H}$ with*

$$\langle Tu, v \rangle = \langle u, T^* v \rangle \quad \text{for all } u, v \in \mathcal{H}.$$

If T^* exists, it is unique, and $\sup_{\|u\|=1} \|Tu\| < \infty$. Set

$$\mathscr{L}(\mathscr{H}) = \{T : \mathscr{H} \to \mathscr{H} \mid T \text{ is adjointable}\}.$$

Then $\mathscr{L}(\mathscr{H})$ is a C^*-algebra with operations

$$(T+S)u = Tu + Su,$$
$$(ST)u = S(Tu),$$
$$(T\lambda)u = (Tu)\lambda$$
$$T^* \text{ as above,}$$
$$\|T\| = \sup_{\|u\|=1} \|Tu\|.$$

11.2 The Reduced Crossed-Product $C^*_r(G,A)$

Let A be a G-C^*-algebra. Define

$$C_c(G,A) = \{f : G \to A \mid f \text{ continuous with compact support}\}.$$

Then $C_c(G,A)$ is a complex algebra with

$$(f+h)g = f(g) + h(g),$$
$$(f\lambda)g = f(g)\lambda,$$
$$(f * h)(g_0) = \int_G f(g)\left(gh(g^{-1}g_0)\right) dg,$$

for all $g, g_0 \in G$, $\lambda \in \mathbb{C}$, $f, h \in C_c(G,A)$. The product $*$ is called *twisted convolution*. Consider the Hilbert A-module $L^2(G,A)$. There is an injection of algebras

$$C_c(G,A) \hookrightarrow \mathscr{L}(L^2(G,A))$$
$$f \mapsto T_f$$

where $T_f(u) = f * u$ is twisted convolution as above. We define $C^*_r(G,A)$ as the C^*-algebra obtained by completing $C_c(G,A)$ with respect to the norm $\|f\| = \|T_f\|$. When $A = \mathbb{C}$, the G-action must be trivial and $C^*_r(G,A) = C^*_r(G)$.

Example 11. Let G be a finite group, and A a G-C^*-algebra. Let dg be the Haar measure such that each $g \in G$ has mass 1. Then

$$C^*_r(G,A) = \left\{ \sum_{\gamma \in G} a_\gamma[\gamma] \,\middle|\, a_\gamma \in A \right\}$$

with operations

$$\left(\sum_{\gamma\in G} a_\gamma[\gamma]\right) + \left(\sum_{\gamma\in G} b_\gamma[\gamma]\right) = \sum_{\gamma\in G}(a_\gamma + b_\gamma)[\gamma],$$

$$\left(\sum_{\gamma\in G} a_\gamma[\gamma]\right)\lambda = \sum_{\gamma\in G}(a_\gamma\lambda)[\gamma],$$

$$(a_\alpha[\alpha])(b_\beta[\beta]) = a_\alpha(\alpha b_\beta)[\alpha\beta] \quad \text{(twisted convolution)},$$

$$\left(\sum_{\gamma\in G} a_\gamma[\gamma]\right)^* = \sum_{\gamma\in G}(\gamma^{-1}a_\gamma^*)[\gamma^{-1}].$$

Here $a_\gamma[\gamma]$ denotes the function from G to A which has the value a_γ at γ and 0 at $g \neq \gamma$.

Let X be a Hausdorff, locally compact G-space. We know that $C_0(X)$ becomes a G-C^*-algebra with G-action

$$(gf)(x) = f(g^{-1}x),$$

for $g \in G$, $f \in C_0(X)$ and $x \in X$. The reduced crossed-product $C_r^*(G,C_0(X))$ will be denoted $C_r^*(G,X)$.

A natural question is to calculate the K-theory of this C^*-algebra. If G is compact, this is the Atiyah–Segal group $K_G^j(X)$, $j = 0,1$. Hence for G non-compact, $K_j(C_r^*(G,X))$ is the natural extension of the Atiyah–Segal theory to the case when G is non-compact.

Definition 23. *We call a G-space G-compact if the quotient space $G\backslash X$ (with the quotient topology) is compact.*

Let X be a proper, G-compact G-space. Then a G-equivariant $\mathbb{C}$-vector bundle E on X determines an element

$$[E] \in K_0(C_r^*(G,X)).$$

Remark 39. From E, a Hilbert module over $C_r^*(G,X)$ is constructed. This Hilbert $C_r^*(G,X)$-module determines an element in $KK_0(\mathbb{C},C_r^*(G,X)) \cong K_0(C_r^*(G,X))$. Note that, quite generally, a Hilbert A-module determines an element in $K_0(A)$ if and only if it is finitely generated.

Recall that a G-equivariant vector bundle E over X is a (complex) vector bundle $\pi: E \to X$ together with a G-action on E such that π is G-equivariant and, for each $p \in X$, the map on the fibers $E_p \to E_{gp}$ induced by multiplication by g is linear.

Theorem 11 (W. Lück and B. Oliver [33]). *If Γ is a (countable) discrete group and X is a proper Γ-compact Γ-space, then*

$$K_0(C_r^*(\Gamma,X)) = \text{Grothendieck group of } \Gamma\text{-equivariant } \mathbb{C}\text{-vector bundles on } X.$$

Remark 40. In [33] this theorem is not explicitly stated. However, it follows from their results. For clarification see [6].

Remark 41. Let X be a proper G-compact G-space. Let $\mathbb{I}$ be the trivial G-equivariant complex vector bundle on X,

$$\mathbb{I} = X \times \mathbb{C}, \quad g(x, \lambda) = (gx, \lambda),$$

for all $g \in G$, $x \in X$ and $\lambda \in \mathbb{C}$. Then $[\mathbb{I}] \in K_0(C_r^*(G, X))$.

11.3 Push-Forward of Hilbert Modules

Let A, B be C^*-algebras, $\varphi \colon A \to B$ a $*$-homomorphism and $\mathcal{H}$ a Hilbert A-module. We shall define a Hilbert B-module $\mathcal{H} \otimes_A B$, called the *push-forward of $\mathcal{H}$ with respect to φ* or *interior tensor product* ([29, Chap. 4]). First, form the algebraic tensor product $\mathcal{H} \odot_A B = \mathcal{H} \otimes_A^{alg} B$ (B is an A-module via φ). This is an abelian group and also a (right) B-module

$$(h \otimes b)b' = h \otimes bb' \quad \text{for all } h \in \mathcal{H}, b, b' \in B.$$

Define a B-valued inner product on $\mathcal{H} \odot_A B$ by

$$\langle h \otimes b, h' \otimes b' \rangle = b^* \varphi(\langle h, h' \rangle)b'.$$

Set

$$\mathcal{N} = \{\xi \in \mathcal{H} \odot_A B \mid \langle \xi, \xi \rangle = 0\}.$$

$\mathcal{N}$ is a B-sub-module of $\mathcal{H} \odot_A B$ and $(\mathcal{H} \odot_A B)/\mathcal{N}$ is a pre-Hilbert B-module.

Definition 24. *Define $\mathcal{H} \otimes_A B$ to be the Hilbert B-module obtained by completing $(\mathcal{H} \odot_A B)/\mathcal{N}$.*

Example 12. Let X be a locally compact, Hausdorff space. Let $A = C_0(X)$, $B = \mathbb{C}$ and $ev_p \colon C_0(X) \to \mathbb{C}$ the evaluation map at a point $p \in X$. Then we can consider the push-forward of a Hilbert $C_0(X)$-module $\mathcal{H}$. This gives a Hilbert space $\mathcal{H}_p$. These Hilbert spaces do not form a vector bundle but something more general (not necessarily locally trivial), sometimes called *continuous field of Hilbert spaces* [15, Chap. 10].

12 Homotopy Made Precise and KK-Theory

We first define homotopy and Kasparov's KK-theory in the non-equivariant setting, for pairs of separable C^*-algebras. A first introduction to KK-theory and further references can be found in [20].

Let A be a C^*-algebra and let $\mathcal{H}$ be a Hilbert A-module. Consider $\mathcal{L}(\mathcal{H})$ the bounded operators on $\mathcal{H}$. For each $u, v \in \mathcal{H}$ we have a bounded operator $\theta_{u,v}$ defined as

$$\theta_{u,v}(\xi) = u\langle v, \xi \rangle.$$

It is clear that $\theta_{u,v}^* = \theta_{v,u}$. The $\theta_{u,v}$ are called *rank one operators* on $\mathcal{H}$. A *finite rank operator* on $\mathcal{H}$ is any $T \in \mathcal{L}(\mathcal{H})$ such that T is a finite sum of rank one operators,

$$T = \theta_{u_1,v_1} + \ldots + \theta_{u_n,v_n}.$$

Let $\mathcal{K}(\mathcal{H})$ be the closure (in $\mathcal{L}(\mathcal{H})$) of the set of finite rank operators. $\mathcal{K}(\mathcal{H})$ is an ideal in $\mathcal{L}(\mathcal{H})$. When $A = \mathbb{C}$, $\mathcal{H}$ is a Hilbert space and $\mathcal{K}(\mathcal{H})$ coincides with the usual compact operators on $\mathcal{H}$.

Definition 25. $\mathcal{H}$ is countably generated *if in $\mathcal{H}$ there is a countable (or finite) set such that the A-module generated by this set is dense in $\mathcal{H}$.*

Definition 26. *Let $\mathcal{H}_0$, $\mathcal{H}_1$ be two Hilbert A-modules. We say that $\mathcal{H}_0$ and $\mathcal{H}_1$ are* isomorphic *if there exists an A-module isomorphism $\Phi \colon \mathcal{H}_0 \to \mathcal{H}_1$ with*

$$\langle u,v \rangle_0 = \langle \Phi u, \Phi v \rangle_1 \quad \text{for all } u,v \in \mathcal{H}_0.$$

We want to define non-equivariant KK-theory for pairs of C^*-algebras. Let A and B be C^*-algebras where A is also separable. Define the set

$$\mathcal{E}^1(A,B) = \{(\mathcal{H}, \psi, T)\}$$

such that $\mathcal{H}$ is a countably generated Hilbert B-module, $\psi \colon A \to \mathcal{L}(\mathcal{H})$ is a $*$-homomorphism, $T \in \mathcal{L}(\mathcal{H})$, and the following conditions are satisfied:

- $T = T^*$
- $\psi(a)T - T\psi(a) \in \mathcal{K}(\mathcal{H})$
- $\psi(a)(I - T^2) \in \mathcal{K}(\mathcal{H})$

for all $a \in A$. We call such triples *odd bivariant K-cycles*.

Remark 42. In the Kasparov definition of $KK^1(A,B)$ [24], the conditions of the K-cycles are the same as our conditions except that the requirement $T = T^*$ is replaced by $\psi(a)(T - T^*) \in \mathcal{K}(H)$ for all $a \in A$. The isomorphism of abelian groups from the group defined using these bivariant K-cycles to the group defined using our bivariant K-cycles is obtained by sending a Kasparov cycle (H, ψ, T) to $(H, \psi, \frac{T+T^*}{2})$.

We say that two such triples $(\mathcal{H}_0, \psi_0, T_0)$ and $(\mathcal{H}_1, \psi_1, T_1)$ in $\mathcal{E}^1(A,B)$ are *isomorphic* if there is an isomorphism of Hilbert B-modules $\Phi \colon \mathcal{H}_0 \to \mathcal{H}_1$ with

$$\Phi \psi_0(a) = \psi_1(a)\Phi,$$
$$\Phi T_0 = T_1 \Phi,$$

for all $a \in A$. That is, the following diagrams commute

$$
\begin{array}{ccc}
\mathcal{H}_0 & \xrightarrow{\psi_0(a)} & \mathcal{H}_0 \\
\Phi \downarrow & & \downarrow \Phi \\
\mathcal{H}_1 & \xrightarrow[\psi_1(a)]{} & \mathcal{H}_1
\end{array}
\qquad
\begin{array}{ccc}
\mathcal{H}_0 & \xrightarrow{T_0} & \mathcal{H}_0 \\
\Phi \downarrow & & \downarrow \Phi \\
\mathcal{H}_1 & \xrightarrow[T_1]{} & \mathcal{H}_1
\end{array}
$$

Let A, B, D be C^*-algebras where A is also separable. A $*$-homomorphism $\varphi: B \to D$ induces a map $\varphi_*: \mathscr{E}^1(A,B) \to \mathscr{E}^1(A,D)$ by

$$\varphi_*(\mathscr{H}, \psi, T) = (\mathscr{H} \otimes_B D, \psi \otimes_B I, T \otimes_B I)$$

where I is the identity operator on D, that is, $I(\alpha) = \alpha$ for all $\alpha \in D$.

We can now make the definition of homotopy precise. Consider the C^*-algebra of continuous functions $C([0,1],B)$, and set ρ_0, ρ_1 to be the $*$-homomorphisms

$$C([0,1],B) \underset{\rho_1}{\overset{\rho_0}{\rightrightarrows}} B$$

defined by $\rho_0(f) = f(0)$ and $\rho_1(f) = f(1)$. In particular, we have induced maps

$$(\rho_j)_*: \mathscr{E}^1(A, C([0,1],B)) \longrightarrow \mathscr{E}^1(A,B) \qquad j = 0,1$$

for any separable C^*-algebra A.

Definition 27. *Two triples $(\mathscr{H}_0, \psi_0, T_0)$ and $(\mathscr{H}_1, \psi_1, T_1)$ in $\mathscr{E}^1(A,B)$ are homotopic if there exists $(\mathscr{H}, \psi, T)$ in $\mathscr{E}^1(A, C([0,1],B))$ with*

$$(\rho_j)_*(\mathscr{H}, \psi, T) \cong (\mathscr{H}_j, \psi_j, T_j) \quad j = 0,1.$$

The even case is analogous, removing the self-adjoint condition $T = T^*$.

Remark 43. As above, we do not require the Hilbert B-module $\mathscr{H}$ to be $\mathbb{Z}/2$-graded. The isomorphism between the abelian group we are defining and the group $KK_0(A,B)$ as defined by Kasparov [24] is dealt with as before (see Remark 23).

Hence we have the set of *even bivariant K-cycles*

$$\mathscr{E}^0(A,B) = \{(\mathscr{H}, \psi, T)\}$$

where $\mathscr{H}$ is a countably generated Hilbert B-module, $\psi: A \to \mathscr{L}(\mathscr{H})$ a $*$-homomorphism, $T \in \mathscr{L}(\mathscr{H})$, and the following conditions are satisfied:

- $\psi(a)T - T\psi(a) \in \mathscr{K}(\mathscr{H})$
- $\psi(a)(I - T^*T) \in \mathscr{K}(\mathscr{H})$
- $\psi(a)(I - TT^*) \in \mathscr{K}(\mathscr{H})$

for all $a \in A$. The remaining definitions carry over, in particular the definition of homotopy in $\mathscr{E}^0(A,B)$.

We define the *(non-equivariant) Kasparov KK-theory groups* of the pair (A,B) as

$$KK^1(A,B) = \mathscr{E}^1(A,B)/(\text{homotopy}),$$
$$KK^0(A,B) = \mathscr{E}^0(A,B)/(\text{homotopy}).$$

A key property is that KK-theory incorporates K-theory of C^*-algebras: for any C^*-algebra B, $KK^j(\mathbb{C},B)$ is isomorphic to $K_j(B)$ (see Theorem 25 in [36]).

13 Equivariant *KK*-Theory

We generalize *KK*-theory to the equivariant setting. An alternative definition to ours, by means of a universal property, is described in Sect. 2 of Meyer's notes [36].

All through this section, let A be a *G-C**-algebra.

Definition 28. *A G-Hilbert A-module is a Hilbert A-module $\mathscr{H}$ with a given continuous action* •

$$G \times \mathscr{H} \to \mathscr{H}$$
$$(g, v) \mapsto gv$$

such that

(a) $g(u + v) = gu + gv$
(b) $g(ua) = (gu)(ga)$
(c) $\langle gu, gv \rangle = g\langle u, v \rangle$

for all $g \in G$, $u, v \in \mathscr{H}$, $a \in A$.

Here 'continuous' means that for each $u \in \mathscr{H}$, the map $G \to \mathscr{H}$, $g \mapsto gu$ is continuous.

Example 13. If $A = \mathbb{C}$, a *G*-Hilbert $\mathbb{C}$-module is just a unitary representation of *G* (the action of *G* on $\mathbb{C}$ must be trivial).

Remark 44. Let $\mathscr{H}$ be a *G*-Hilbert *A*-module. For each $g \in G$, denote by L_g the map

$$L_g : \mathscr{H} \to \mathscr{H}, \quad L_g(v) = gv.$$

Note that L_g might not be in $\mathscr{L}(\mathscr{H})$. But if $T \in \mathscr{L}(\mathscr{H})$, then $L_g T L_g^{-1} \in \mathscr{L}(\mathscr{H})$. Hence *G* acts on the *C**-algebra $\mathscr{L}(\mathscr{H})$ by

$$gT = L_g T L_g^{-1}.$$

Example 14. Let A be a *G-C**-algebra. Set $n \geq 1$. Then A^n is a *G*-Hilbert *A*-module (cf. Example 6) with

$$g(a_1, \ldots, a_n) = (ga_1, \ldots, ga_n).$$

Let A and B be *G-C**-algebras, where A is also separable. Define the set

$$\mathscr{E}_G^0(A, B) = \{(\mathscr{H}, \psi, T)\}$$

such that $\mathscr{H}$ is a countably generated *G*-Hilbert *B*-module, $\psi : A \to \mathscr{L}(\mathscr{H})$ is a *-homomorphism with

$$\psi(ga) = g\psi(a) \quad \text{for all } g \in G, a \in A,$$

and $T \in \mathscr{L}(\mathscr{H})$, and so that the following conditions are satisfied:

- $gT - T \in \mathscr{K}(\mathscr{H})$
- $\psi(a)T - T\psi(a) \in \mathscr{K}(\mathscr{H})$
- $\psi(a)(I - T^*T) \in \mathscr{K}(\mathscr{H})$
- $\psi(a)(I - TT^*) \in \mathscr{K}(\mathscr{H})$

for all $g \in G, a \in A$. We define

$$KK_G^0(A, B) = \mathscr{E}_G^0(A, B)/(\text{homotopy}).$$

The definition of *homotopy* in Sect. 12 can be defined in a straightforward way in this setting.

$KK_G^0(A, B)$ is an abelian group with addition and negative

$$(\mathscr{H}, \psi, T) + (\mathscr{H}', \psi', T') = (\mathscr{H} \oplus \mathscr{H}', \psi \oplus \psi', T \oplus T'),$$
$$-(\mathscr{H}, \psi, T) = (\mathscr{H}, \psi, T^*).$$

The odd case is similar, just restricting to self-adjoint operators. Define the set

$$\mathscr{E}_G^1(A, B) = \{(\mathscr{H}, \psi, T)\}$$

such that $\mathscr{H}$ is a countably generated G-Hilbert B-module, $\psi \colon A \to \mathscr{L}(\mathscr{H})$ is a $*$-homomorphism with

$$\psi(ga) = g\psi(a) \quad \text{for all } g \in G, a \in A,$$

and $T \in \mathscr{L}(\mathscr{H})$, and so that the following conditions are satisfied:

- $T = T^*$
- $gT - T \in \mathscr{K}(\mathscr{H})$
- $\psi(a)T - T\psi(a) \in \mathscr{K}(\mathscr{H})$
- $\psi(a)(I - T^2) \in \mathscr{K}(\mathscr{H})$

for all $g \in G, a \in A$.

We define

$$KK_G^1(A, B) = \mathscr{E}_G^1(A, B)/(\text{homotopy}).$$

$KK_G^1(A, B)$ is an abelian group with addition and inverse given by

$$(\mathscr{H}, \psi, T) + (\mathscr{H}', \psi', T') = (\mathscr{H} \oplus \mathscr{H}', \psi \oplus \psi', T \oplus T'),$$
$$-(\mathscr{H}, \psi, T) = (\mathscr{H}, \psi, -T).$$

Remark 45. In the even case we are not requiring a $\mathbb{Z}/2$-grading. The isomorphism to the abelian group defined by Kasparov [25] is given as in Remark 23. Our general principle is that the even and odd cases are identical except that in the odd case the operator T is required to be self-adjoint but not in the even case.

Using equivariant KK-theory, we can introduce coefficients for equivariant K-homology. Let X be a proper G-space with compact quotient. Recall that

$$K_j^G(X) = KK_G^j(C_0(X),\mathbb{C}) \quad \text{and}$$
$$K_j^G(\underline{E}G) = \varinjlim_{\substack{\Delta \subseteq \underline{E}G \\ G\text{-compact}}} K_j^G(\Delta).$$

We define the *equivariant K-homology of* X, respectively *of* $\underline{E}G$, with coefficients in a G-C^*-algebra A as

$$K_j^G(X,A) = KK_G^j(C_0(X),A),$$
$$K_j^G(\underline{E}G,A) = \varinjlim_{\substack{\Delta \subseteq \underline{E}G \\ G\text{-compact}}} K_j^G(\Delta,A).$$

14 The Index Map

Our definition of the index map uses the Kasparov product and the descent map.

14.1 The Kasparov Product

Let A, B, D be (separable) G-C^*-algebras. There is a product

$$KK_G^i(A,B) \otimes_{\mathbb{Z}} KK_G^j(B,D) \longrightarrow KK_G^{i+j}(A,D).$$

The definition is highly non-trivial. Some motivation and examples, in the non-equivariant case, can be found in [20, Sect. 5].

Remark 46. Equivariant KK-theory can be regarded as a category with objects separable G-C^*-algebras and morphisms $\mathrm{mor}(A,B) = KK_G^i(A,B)$ (as a $\mathbb{Z}/2$-graded abelian group), and composition given by the Kasparov product (cf. [36, Thm. 33]).

14.2 The Kasparov Descent Map

Let A and B be (separable) G-C^*-algebras. There is a map between the equivariant KK-theory of (A,B) and the non-equivariant KK-theory of the corresponding reduced crossed-product C^*-algebras,

$$KK_G^j(A,B) \longrightarrow KK^j\left(C_r^*(G,A),C_r^*(G,B)\right) \quad j=0,1.$$

The definition is also highly non-trivial and can be found in [25, Sect. 3]. Alternatively, see Proposition 26 in Meyer's notes [36].

14.3 Definition of the Index Map

We would like to define the index map

$$\mu \colon K_j^G(\underline{E}G) \longrightarrow K_j(C_r^*G).$$

Let X be a proper G-compact G-space. First, we define a map

$$\mu \colon K_j^G(X) = KK_G^j(C_0(X),\mathbb{C}) \longrightarrow K_j(C_r^*G)$$

to be the composition of the Kasparov descent map

$$KK_G^j(C_0(X),\mathbb{C}) \longrightarrow KK^j(C_r^*(G,X),C_r^*(G))$$

(the trivial action of G on $\mathbb{C}$ gives the crossed-product $C_r^*(G,\mathbb{C}) = C_r^*G$) and the Kasparov product with the trivial bundle

$$\mathbb{I} \in K_0(C_r^*(G,X)) = KK^0(\mathbb{C},C_r^*(G,X)),$$

that is, the Kasparov product with the trivial vector bundle $\mathbb{I}$, when $A = \mathbb{C}$, $B = C_r^*(G,X)$, $D = C_r^*G$ and $i = 0$.

Recall that

$$K_j^G(\underline{E}G) = \varinjlim_{\substack{\Delta \subseteq \underline{E}G \\ G\text{-compact}}} KK_G^j(C_0(\Delta),\mathbb{C}).$$

For each G-compact $\Delta \subset \underline{E}G$, we have a map as before

$$\mu \colon KK_G^j(C_0(\Delta),\mathbb{C}) \longrightarrow K_j(C_r^*G).$$

If Δ and Ω are two G-compact subsets of $\underline{E}G$ with $\Delta \subset \Omega$, then by naturality the following diagram commutes:

$$
\begin{array}{ccc}
KK_G^j(C_0(\Delta),\mathbb{C}) & \longrightarrow & KK_G^j(C_0(\Omega),\mathbb{C}) \\
\downarrow & & \downarrow \\
K_jC_r^*G & \xrightarrow{\ =\ } & K_jC_r^*G.
\end{array}
$$

Thus we obtain a well-defined map on the direct limit $\mu \colon K_j^G(\underline{E}G) \to K_jC_r^*G$.

14.4 The Index Map with Coefficients

The coefficients can be introduced in KK-theory at once. Let A be a G-C^*-algebra. We would like to define the index map

$$\mu \colon K_j^G(\underline{E}G;A) \longrightarrow K_jC_r^*(G,A).$$

Let X be a proper G-compact G-space and A a G-C^*-algebra. First, we define a map

$$\mu: KK_G^j(C_0(X),A) \longrightarrow K_j C_r^*(G,A)$$

to be the composition of the Kasparov descent map

$$KK_G^j(C_0(X),A) \longrightarrow KK^j(C_r^*(G,X),C_r^*(G,A))$$

and the Kasparov product with the trivial bundle

$$\mathbb{I} \in K_0 C_r^*(G,X) = KK^0(\mathbb{C},C_r^*(G,X)).$$

For each G-compact $\Delta \subset \underline{E}G$, we have a map as above

$$\mu: KK_G^j(C_0(\Delta),A) \longrightarrow K_j C_r^*(G,A).$$

If Δ and Ω are two G-compact subsets of $\underline{E}G$ with $\Delta \subset \Omega$, then by naturality the following diagram commutes:

$$
\begin{array}{ccc}
KK_G^j(C_0(\Delta),A) & \longrightarrow & KK_G^j(C_0(\Omega),A) \\
\downarrow & & \downarrow \\
K_j C_r^*(G,A) & \xrightarrow{\ =\ } & K_j C_r^*(G,A).
\end{array}
$$

Thus we obtain a well-defined map on the direct limit $\mu: K_j^G(\underline{E}G;A) \to K_j C_r^*(G,A)$.

15 A Brief History of *K*-Theory

15.1 The *K*-Theory Genealogy Tree

Grothendieck invented K-theory to give a conceptual proof of the Hirzebruch–Riemann–Roch theorem. The subject has since then evolved in different directions, as summarized by the following diagram.

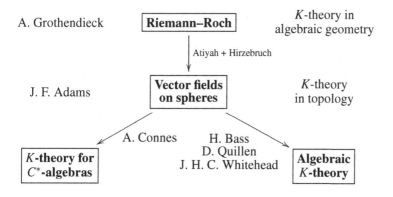

Atiyah and Hirzebruch defined topological K-theory. J. F. Adams then used the Atiyah–Hirzebruch theory to solve the problem of vector fields on spheres. C^*-algebra K-theory developed quite directly out of Atiyah–Hirzebruch topological K-theory. From its inception, C^*-algebra K-theory has been closely linked to problems in geometry-topology (Novikov conjecture, Gromov–Lawson–Rosenberg conjecture, Atiyah–Singer index theorem) and to classification problems within C^*-algebras. More recently, C^*-algebra K-theory has played an essential role in the new subject of non-commutative geometry.

Algebraic K-theory was a little slower to develop [51]; much of the early development in the 1960s was due to H. Bass, who organized the theory on K_0 and K_1 and defined the negative K-groups. J. Milnor introduced K_2. Formulating an appropriate definition for higher algebraic K-theory proved to be a difficult and elusive problem. Definitions were proposed by several authors, including J. Milnor and Karoubi–Villamayor. A remarkable breakthrough was achieved by D. Quillen with his plus-construction. The resulting definition of higher algebraic K-theory (i.e. Quillen's algebraic K-theory) is perhaps the most widely accepted today. Many significant problems and results (e.g. the Lichtenbaum conjecture) have been stated within the context of Quillen algebraic K-theory. In some situations, however, a different definition is relevant. For example, in the recently proved Bloch–Kato conjecture, it is J. Milnor's definition of higher algebraic K-theory which is used.

Since the 1970s, K-theory has grown considerably, and its connections with other parts of mathematics have expanded. For the interested reader, we have included a number of current K-theory textbooks in our reference list [8, 42, 43, 47, 49, 50]. For a taste of the current developments, it is useful to take a look at the *Handbook of K-theory* [16] or at the lectures in this volume. The *Journal of K-theory* (as well as its predecessor, *K-theory*) is dedicated to the subject, as is the website maintained by D. Grayson at http://www.math.uiuc.edu/K-theory. This site, started in 1993, includes a preprint archive which at the moment when this is being written contains 922 preprints. Additionally, see the *Journal of Non-Commutative Geometry* for current results involving C^*-algebra K-theory.

Finally, we have not in these notes emphasized cyclic homology. However, cyclic (co-)homology is an allied theory to K-theory and any state-of-the-art survey of K-theory would have to recognize this central fact.

15.2 The Hirzebruch–Riemann–Roch Theorem

Let M be a non-singular projective algebraic variety over $\mathbb{C}$. Let E be an algebraic vector bundle on M. Write $\underline{E}$ for the sheaf (of germs) of algebraic sections of E. For each $j \geq 0$, consider $H^j(M, \underline{E})$ the j-th cohomology group of M using $\underline{E}$.

Lemma 1. *For all $j \geq 0$, $\dim_{\mathbb{C}} H^j(M, \underline{E}) < \infty$ and for $j > \dim_{\mathbb{C}}(M)$, $H^j(M, \underline{E}) = 0$.*

Define the *Euler characteristic* of M with respect to E as

$$\chi(M, E) = \sum_{j=0}^{n} (-1)^j \dim_{\mathbb{C}} H^j(M, \underline{E}), \quad \text{where } n = \dim_{\mathbb{C}}(M).$$

Theorem 12 (Hirzebruch–Riemann–Roch). *Let M be a non-singular projective algebraic variety over* $\mathbb{C}$ *and let E be an algebraic vector bundle on M. Then*

$$\chi(M,E) = (\mathrm{ch}(E)\cup\mathrm{Td}(M))[M]$$

where $\mathrm{ch}(E)$ *is the Chern character of E,* $\mathrm{Td}(M)$ *is the Todd class of M and* $\cup$ *stands for the cup product.*

15.3 The Unity of *K*-Theory

We explain how *K*-theory for *C**-algebras is a particular case of algebraic *K*-theory of rings.

Let *A* be a *C**-algebra. Consider the inclusion

$$M_n(A) \hookrightarrow M_{n+1}(A)$$

$$\begin{pmatrix} a_{11} & \cdots & a_{1n} \\ \vdots & & \vdots \\ a_{n1} & \cdots & a_{nn} \end{pmatrix} \mapsto \begin{pmatrix} a_{11} & \cdots & a_{1n} & 0 \\ \vdots & & \vdots & \vdots \\ a_{n1} & \cdots & a_{nn} & 0 \\ 0 & \cdots & 0 & 0 \end{pmatrix}. \tag{2}$$

This is a one-to-one *-homomorphism, and it is norm preserving. Define $M_\infty(A)$ as the limit of $M_n(A)$ with respect to these inclusions. That is, $M_\infty(A)$ is the set of infinite matrices where almost all a_{ij} are zero. Finally, define the *stabilization* of *A* (cf. [42, 6.4] or [49, 1.10]) as the closure

$$\dot{A} = \overline{M_\infty(A)}.$$

Here we mean the completion with respect to the norm on $M_\infty(A)$ and the main point is that the inclusions above are all norm-preserving. The result is a *C**-algebra without unit.

Remark 47. There is an equivalent definition of $\dot{A}$ as the tensor product $A \otimes \mathscr{K}$, where $\mathscr{K}$ is the *C**-algebra of all compact operators on a separable infinite-dimensional Hilbert space, and the tensor product is in the sense of *C**-algebras.

Example 15. Let *H* be a separable, infinite-dimensional, Hilbert space. That is, *H* has a countable, but not finite, orthonormal basis. It can be shown that

$$\dot{\mathbb{C}} = \mathscr{K} \subset \mathscr{L}(H),$$

where $\mathscr{K}$ is the subset of compact operators on *H*. We have then

$$K_j(\mathbb{C}) = K_j(\dot{\mathbb{C}}),$$

where $K_j(-)$ is *C**-algebra *K*-theory. This is true in general for any *C**-algebra (Proposition 4 below).

On the other hand, the algebraic K-theory of $\dot{\mathbb{C}}$ is

$$K_j^{\text{alg}}(\dot{\mathbb{C}}) = \begin{cases} \mathbb{Z} & j \text{ even}, \\ 0 & j \text{ odd}, \end{cases}$$

which therefore coincides with the C^*-algebra K-theory of $\mathbb{C}$. This is also true in general (Theorem 13 below). This answer is simple compared with the algebraic K-theory of $\mathbb{C}$, where only some partial results are known.

The stabilization of a C^*-algebra does not change its (C^*-algebra) K-theory.

Proposition 4. *Let A be a C^*-algebra and write $K_j(-)$ for K-theory of C^*-algebras. Then*

$$K_j(A) = K_j(\dot{A}) \quad j \geq 0.$$

The proof is a consequence of the definition of C^*-algebra K-theory: the inclusions (2) induce isomorphisms in K-theory, and the direct limit (in the sense of C^*-algebras) commute with the K-theory functor (cf. [49, 6.2.11 and 7.1.9]).

Remark 48. In the terminology of Cortiñas' notes [12], Proposition 4 says that the functors K_0 and K_1 are $\mathscr{K}$-*stable*.

M. Karoubi conjectured that the algebraic K-theory of $\dot{A}$ is isomorphic to its C^*-algebra K-theory. The conjecture was proved by A. Suslin and M. Wodzicki.

Theorem 13 (A. Suslin and M. Wodzicki [48]). *Let A be a C^*-algebra. Then*

$$K_j(\dot{A}) = K_j^{\text{alg}}(\dot{A}) \quad j \geq 0,$$

where the left-hand side is C^*-algebra K-theory and the right-hand side is (Quillen's) algebraic K-theory of rings.

A proof can be found in Cortiñas' notes [12, Thm. 7.1.3]. In these notes Cortiñas elaborates the isomorphism above into a long exact sequence which involves cyclic homology.

Theorem 13 is the unity of K-theory: It says that C^*-algebra K-theory is a pleasant subdiscipline of algebraic K-theory in which Bott periodicity is valid and certain basic examples are easy to calculate.

References

1. Arveson, W.: An invitation to C^*-algebras. In: Graduate Texts in Mathematics, vol. 39. Springer, New York (1976) MR0512360 (58 #23621)
2. Atiyah, M.F.: Global theory of elliptic operators. In: Proceedings of International Conference on Functional Analysis and Related Topics (Tokyo, 1969), pp. 21–30. University of Tokyo Press, Tokyo (1970) MR0266247 (42 #1154)
3. Atiyah, M.F., Hirzebruch, F.: Riemann-Roch theorems for differentiable manifolds. Bull. Am. Math. Soc. **65**, 276–281 (1959) MR0110106 (22 #989)

4. Baum, P., Connes, A.: Chern character for discrete groups. In: A Fête of Topology, pp. 163–232. Academic Press, Boston, MA, (1988) MR928402 (90e:58149)

5. Baum, P., Connes, A., Higson, N.: Classifying space for proper actions and K-theory of group C^*-algebras, C^*-algebras: 1943–1993 (San Antonio, TX, 1993). In: Contemporary Mathematics, vol. 167, pp. 240–291. American Mathematical Society, Providence, RI (1994) MR1292018 (96c:46070)

6. Baum, P., Higson, N., Schick, T.: A geometric description of equivariant K-homology. In: The Conference Proceedings of the Non Commutative Geometry Conference in honor of Alain Connes, Paris (2007)

7. Biller, H.: Characterizations of proper actions, Math. Proc. Camb. Philol. Soc. **136**(2), 429–439 (2004) MR2040583 (2004k:57043)

8. Blackadar, B.: K-theory for Operator Algebras, second edn., vol. 5. Mathematical Sciences Research Institute Publications, Cambridge University Press, Cambridge (1998) MR1656031 (99g:46104)

9. Bott, R.: The stable homotopy of the classical groups. Ann. Math. **70**(2), 313–337 (1959) MR0110104 (22 #987)

10. Brown, K.S.: Cohomology of groups. In: Graduate Texts in Mathematics, vol. 87. Springer, New York (1982) MR672956 (83k:20002)

11. Chabert, J., Echterhoff, S., Meyer, R.: Deux remarques sur l'application de Baum-Connes. C. R. Acad. Sci. Paris I Math. **332**(7), 607–610 (2001). MR1841893 (2002k:19004)

12. Cortiñas, G.: Algebraic v. topological K-theory: a friendly match (this volume)

13. Davidoff, G., Sarnak, P., Valette, A.: Elementary number theory, group theory, and Ramanujan graphs. In: London Mathematical Society Student Texts, vol. 55. Cambridge University Press, Cambridge (2003) MR1989434 (2004f:11001)

14. Davis, J.F., Lück, W.: Spaces over a category and assembly maps in isomorphism conjectures in K- and L-theory, K-Theory **15**(3), 201–252 (1998) MR1659969 (99m:55004)

15. Dixmier, J.: C^*-algebras. North-Holland Publishing Co., Amsterdam, Translated from the French by Francis Jellett, North-Holland Mathematical Library, vol. 15 (1977) MR0458185 (56 #16388)

16. Friedlander, E.M., Grayson, D.R. (eds.), Handbook of K-theory, vol. 1, 2. Springer, Berlin (2005) MR2182598 (2006e:19001)

17. Guentner, E., Higson, N., Weinberger, S.: The Novikov conjecture for linear groups, Publ. Math. Inst. Hautes Études Sci., **101**, 243–268 (2005) MR2217050 (2007c:19007)

18. Hambleton, I., Pedersen, E.K.: Identifying assembly maps in K- and L-theory. Math. Ann. **328**(1–2), 27–57 (2004) MR2030369 (2004j:19001)

19. Higson, N., Lafforgue, V., Skandalis, G.: Counterexamples to the Baum-Connes conjecture. Geomet. Funct. Anal. **12**(2), 330–354 (2002) MR1911663 (2003g:19007)

20. Higson, N.: A primer on KK-theory, Operator theory: operator algebras and applications, Part 1 (Durham, NH, 1988). In: Proceedings of Symposia in Pure Mathematics, vol. 51, pp. 239–283. American Mathematical Society, Providence, RI (1990) MR1077390 (92g:19005)

21. Higson, N., Kasparov, G.: Operator K-theory for groups which act properly and isometrically on Hilbert space. Electron. Res. Announc. Am. Math. Soc. **3**, 131–142 (1997) (electronic) MR1487204 (99e:46090)

22. Higson, N., Kasparov, G.: E-theory and KK-theory for groups which act properly and isometrically on Hilbert space. Invent. Math. **144**(1), 23–74 (2001) MR1821144 (2002k:19005)

23. Julg, P.: La conjecture de Baum-Connes à coefficients pour le groupe $Sp(n, 1)$. C. R. Math. Acad. Sci. Paris **334**(7), 533–538 (2002) MR1903759 (2003d:19007)

24. Kasparov, G.G.: The operator K-functor and extensions of C^*-algebras. Izv. Akad. Nauk SSSR Ser. Mat. **44**(3), 571–636, 719 (1980) MR582160 (81m:58075)
25. Kasparov, G.G.: Equivariant KK-theory and the Novikov conjecture. Invent. Math. **91**(1), 147–201 (1988) MR918241 (88j:58123)
26. Kasparov, G.G.: K-theory, group C^*-algebras, and higher signatures (conspectus). In: Novikov Conjectures, Index Theorems and Rigidity, vol. 1 (Oberwolfach, 1993). London Mathematical Society Lecture Note Series, vol. 226, pp. 101–146. Cambridge University Press, Cambridge (1995) MR1388299 (97j:58153)
27. Kasparov, G.G., Skandalis, G.: Groups acting on buildings, operator K-theory, and Novikov's conjecture. K-Theory **4**(4), 303–337 (1991) MR1115824 (92h:19009)
28. Lafforgue, V.: K-théorie bivariante pour les algèbres de Banach et conjecture de Baum-Connes. Invent. Math. **149**(1), 1–95 (2002) MR1914617 (2003d:19008)
29. Lance, E.C.: Hilbert C^*-modules.A toolkit for operator algebraists. In: London Mathematical Society Lecture Note Series, vol. 210. Cambridge University Press, Cambridge (1995) MR1325694 (96k:46100)
30. Lück, W.: Transformation groups and algebraic K-theory. In: Lecture Notes in Mathematics, Mathematica Gottingensis, vol. 1408. Springer, Berlin (1989) . MR1027600 (91g:57036)
31. Lück, W.: Chern characters for proper equivariant homology theories and applications to K- and L-theory. J. Reine Angew. Math. **543**, 193–234 (2002)
32. Lück, W.: Survey on classifying spaces for families of subgroups, Infinite groups: geometric, combinatorial and dynamical aspects. In: Progress in Mathematics, vol. 248, pp. 269–322. Birkhäuser, Basel (2005) MR2195456 (2006m:55036)
33. Lück, W., Oliver, B.: The completion theorem in K-theory for proper actions of a discrete group. Topology **40**(3), 585–616 (2001) MR1838997 (2002f:19010)
34. Lück, W., Reich, H.: The Baum-Connes and the Farrell-Jones conjectures in K- and L-theory. In: Handbook of K-theory, vol. 1, 2, pp. 703–842. Springer, Berlin (2005) MR2181833 (2006k:19012)
35. Matthey, M., Oyono-Oyono, H., Pitsch, W.: Homotopy invariance of higher signatures and 3-manifold groups. Bull. Soc. Math. France **136**(1), 1–25 (2008) MR2415334
36. Meyer, R.: Universal Coefficient Theorems and assembly maps in KK-theory (this volume).
37. Mineyev, I., Yu, G.: The Baum-Connes conjecture for hyperbolic groups. Invent. Math. **149**(1), 97–122 (2002) MR1914618 (2003f:20072)
38. Mislin, G.: Equivariant K-homology of the classifying space for proper actions, Proper group actions and the Baum-Connes conjecture. In: Advanced Courses in Mathematics CRM Barcelona, pp. 1–78. Birkhäuser, Basel (2003) MR2027169 (2005e:19008)
39. Murphy, G.J.: C^*-algebras and operator theory. Academic Press Inc., Boston, MA (1990) MR1074574 (91m:46084)
40. H. Oyono-Oyono: Baum-Connes conjecture and group actions on trees. K-Theory **24**(2), 115–134 (2001) MR1869625 (2002m:19004)
41. Pedersen, G.K.: C^*-algebras and their automorphism groups. In: London Mathematical Society Monographs, vol. 14. Academic Press Inc. [Harcourt Brace Jovanovich Publishers], London (1979) MR548006 (81e:46037)
42. Rørdam, M., Larsen, F., Laustsen, N.: An introduction to K-theory for C^*-algebras. In: London Mathematical Society Student Texts, vol. 49. Cambridge University Press, Cambridge (2000) MR1783408 (2001g:46001)
43. Rosenberg, J.: Algebraic K-theory and its applications. In: Graduate Texts in Mathematics, vol. 147. Springer, New York (1994) MR1282290 (95e:19001)

44. Sánchez-García, R.: Bredon homology and equivariant *K*-homology of SL$(3,\mathbb{Z})$. J. Pure Appl. Algebra **212**(5), 1046–1059 (2008) MR2387584
45. Schick, T.: Finite group extensions and the Baum-Connes conjecture. Geom. Topology **11**, 1767–1775 (2007) MR2350467
46. Serre, J.-P.: Trees, Springer Monographs in Mathematics, Springer, Berlin. Translated from the French original by John Stillwell, Corrected 2nd printing of the 1980 English translation, (2003) MR1954121 (2003m:20032)
47. Srinivas, V.: Algebraic *K*-theory, In: Progress in Mathematics, 2nd edn., vol. 90. Birkhäuser, Boston, MA (1996) MR1382659 (97c:19001)
48. Suslin, A.A., Wodzicki, M.: Excision in algebraic *K*-theory. Ann. Math. (2) **136**(1), 51–122 (1992) MR1173926 (93i:19006)
49. Wegge-Olsen, N.E.: *K*-theory and *C**-algebras. A friendly approach. Oxford Science Publications, The Clarendon Press, Oxford University Press, New York (1993) MR1222415 (95c:46116)
50. Weibel, C.A.: The *K*-book: An introduction to algebraic *K*-theory (book in progress), Available at , OPTkey = , OPTmonth = , OPTyear = , OPTannote = .
51. Weibel, C.A.: The development of algebraic *K*-theory before 1980. Algebra, *K*-theory, groups, and education, New York, 1997. In: Contemporary Mathematics, vol. 243, pp. 211–238. American Mathematical Society, Providence, RI (1999) MR1732049 (2000m:19001)

Universal Coefficient Theorems and Assembly Maps in KK-Theory

Ralf Meyer

Mathematisches Institut and Courant Centre "Higher Order Structures",
Georg-August-Universität Göttingen,
Bunsenstraße 3–5,
37073 Göttingen,
Germany,
rameyer@uni-math.gwdg.de

Abstract We introduce equivariant Kasparov theory using its universal property and construct the Baum–Connes assembly map by localising the Kasparov category at a suitable subcategory. Then we explain a general machinery to construct derived functors and spectral sequences in triangulated categories. This produces various generalisations of the Rosenberg–Schochet Universal Coefficient Theorem.

Key words: K-theory, KK-theory, Baum–Connes assembly map, Triangulated category, Universal Coefficient Theorem

1 Introduction

We may view Kasparov theory and its equivariant generalisations as categories. These categories are non-commutative analogues of (equivariant generalisations of) the stable homotopy category of spectra. These equivariant Kasparov categories can be described in two ways:

Abstractly, as the universal split-exact C^*-stable functor on the appropriate category of C^*-algebras – this approach is due to Cuntz and Higson [10, 11, 15, 16].
It is useful for *general* constructions like the descent functor or the adjointness between induction and restriction functors (see Sect. 2.1.6 or [26]).
Concretely, using Fredholm operators on equivariant Hilbert bimodules – this is the original definition of Kasparov [17, 18].
It is useful for *specific* constructions that use, say, geometric properties of a group to construct elements in Kasparov groups.

We mainly treat Kasparov theory as a black box. We define G-equivariant Kasparov theory via its universal property and equip it with a triangulated category structure. This formalises some basic properties of the stable homotopy category that are needed for algebraic topology. We later apply this structure to construct spectral sequences in Kasparov theory.

P.F. Baum et al., *Topics in Algebraic and Topological K-Theory*,
Lecture Notes in Mathematics 2008, DOI 10.1007/978-3-642-15708-0_2,
© Springer-Verlag Berlin Heidelberg 2011

We use the universal property to construct the descent functor and induction and restriction functors for closed subgroups, and to verify that the latter are adjoint for open subgroups.

Then we turn to the Baum–Connes assembly map for a locally compact group G, which we treat as in [26]. Green's Imprimitivity Theorem suggests that we understand crossed products for compactly induced actions much better than general crossed products. We want to construct more general actions out of compactly induced actions by an analogue of the construction of CW-complexes. The notion of *localising subcategory* makes this idea precise.

The orthogonal complement of the compactly induced actions consists of actions that are KK^H-equivalent to 0 for all compact subgroups H of G. We call such actions *weakly contractible*.

The compactly induced and weakly contractible objects together generate the whole Kasparov category. This allows us to compute the localisation of a functor at the weakly contractible objects. The general machinery of localisation yields the Baum–Connes assembly map

$$\mu_* \colon \mathrm{K}^{\mathrm{top}}_*(G,A) \to \mathrm{K}_*(G \ltimes_r A)$$

when we apply it to the functor $A \mapsto \mathrm{K}_*(G \ltimes_r A)$. Roughly speaking, this means that $A \mapsto \mathrm{K}^{\mathrm{top}}_*(G,A)$ is the best possible approximation to $\mathrm{K}_*(G \ltimes_r A)$ that vanishes for weakly contractible objects. The above statements involve functors and the Baum–Connes assembly map with coefficients. The above approach only works if we study this generalisation right away.

The groups $\mathrm{K}^{\mathrm{top}}_*(G,A)$ are supposed to be computable by topological methods. We present one approach to make this precise that works completely within equivariant Kasparov theory and is a special case of a very general machinery for constructing spectral sequences. We carry over notions from homological algebra like exact chain complexes and projective objects to our category and use them to define derived functors (see [27]). The derived functors of $\mathrm{K}_*(G \ltimes_r A)$ and $\mathrm{K}^{\mathrm{top}}_*(G,A)$ agree and form the E^2-term of a spectral sequence that converges towards $\mathrm{K}^{\mathrm{top}}_*(G,A)$. Many other spectral sequences like the Adams spectral sequence in topology can be constructed with the same machinery. In simple special cases, the spectral sequence degenerates to an exact sequence. The Universal Coefficient Theorem by Rosenberg–Schochet in [35] and the Pimsner–Voiculescu exact sequence are special cases of this machinery.

2 Kasparov Theory and Baum–Connes Conjecture

2.1 Kasparov Theory via Its Universal Property

This subsection is mostly taken from [25], where more details can be found.

Let G be a locally compact group.

Definition 1. A G-C*-*algebra* is a C*-algebra with a strongly continuous representation of G by *-automorphisms.

Let $G\text{-}\mathfrak{C}^*\mathfrak{alg}$ be the category of G-C*-algebras; its objects are G-C*-algebras and its morphisms $A \to B$ are the G-equivariant *-homomorphisms $A \to B$; we sometimes denote this morphism set by $\mathrm{Hom}_G(A,B)$.

A C*-algebra is *separable* if it has a countable dense subset. We often restrict attention to the full subcategory $G\text{-}\mathfrak{C}^*\mathfrak{sep} \subseteq G\text{-}\mathfrak{C}^*\mathfrak{alg}$ of separable G-C*-algebras.

Homology theories for C*-algebras are usually required to be homotopy invariant, stable, and exact in a suitable sense. We can characterise G-equivariant Kasparov theory as the *universal functor* on $G\text{-}\mathfrak{C}^*\mathfrak{sep}$ with these properties, in the following sense.

Definition 2. Let P be a property for functors defined on $G\text{-}\mathfrak{C}^*\mathfrak{sep}$. A *universal functor* with P is a functor $u\colon G\text{-}\mathfrak{C}^*\mathfrak{sep} \to \mathfrak{U}_P(G\text{-}\mathfrak{C}^*\mathfrak{sep})$ such that

- $\bar{F} \circ u$ has P for each functor $\bar{F}\colon \mathfrak{U}_P(G\text{-}\mathfrak{C}^*\mathfrak{sep}) \to \mathfrak{C}$
- Any functor $F\colon G\text{-}\mathfrak{C}^*\mathfrak{sep} \to \mathfrak{C}$ with P factors uniquely as $F = \bar{F} \circ u$ for some functor $\bar{F}\colon \mathfrak{U}_P(G\text{-}\mathfrak{C}^*\mathfrak{sep}) \to \mathfrak{C}$
- Let $F,G\colon G\text{-}\mathfrak{C}^*\mathfrak{sep} \rightrightarrows \mathfrak{C}$ be two functors with P, let $\bar{F},\bar{G}\colon \mathfrak{U}_P(G\text{-}\mathfrak{C}^*\mathfrak{sep}) \rightrightarrows \mathfrak{C}$ be their unique extensions; then any natural transformation $\Phi\colon F \Rightarrow G$ remains a natural transformation $\bar{F} \Rightarrow \bar{G}$

Of course, such a functor need not exist. If it does, then it restricts to a bijection between objects of $G\text{-}\mathfrak{C}^*\mathfrak{sep}$ and $\mathfrak{U}_P(G\text{-}\mathfrak{C}^*\mathfrak{sep})$. Hence we can completely describe it by the sets of morphisms $\mathfrak{U}_P(A,B)$ from A to B in $\mathfrak{U}_P(G\text{-}\mathfrak{C}^*\mathfrak{sep})$ and the maps $G\text{-}\mathfrak{C}^*\mathfrak{sep}(A,B) \to \mathfrak{U}_P(A,B)$ for $A,B \in\in G\text{-}\mathfrak{C}^*\mathfrak{sep}$. The universal property means that for any functor $F\colon G\text{-}\mathfrak{C}^*\mathfrak{sep} \to \mathfrak{C}$ with P there is a unique functorial way to extend the maps $\mathrm{Hom}_G(A,B) \to \mathfrak{C}\big(F(A),F(B)\big)$ to $\mathfrak{U}_P(A,B)$. The condition about natural transformations makes precise what it means for this extension to be natural.

2.1.1 Some Basic Homotopy Theory

We define cylinders, cones, and suspensions of objects, and mapping cones and mapping cylinders of morphisms in $G\text{-}\mathfrak{C}^*\mathfrak{alg}$. Then we define homotopy invariance for functors. Mapping cones will be used later to introduce the triangulated category structure on Kasparov theory.

Definition 3. Let A be a G-C*-algebra. We define the *cylinder*, *cone*, and *suspension* over A by

$$\mathrm{Cyl}(A) := \mathrm{C}([0,1],A),$$
$$\mathrm{Cone}(A) := \mathrm{C}_0\big([0,1] \setminus \{0\},A\big),$$
$$\mathrm{Sus}(A) := \mathrm{C}_0\big([0,1] \setminus \{0,1\},A\big) \cong \mathrm{C}_0(\mathbb{S}^1,A).$$

If $A = \mathrm{C}_0(X)$ for a pointed compact space X, then the cylinder, cone, and suspension of A are $\mathrm{C}_0(Y)$ with Y equal to the usual cylinder $[0,1]_+ \wedge X$, cone $[0,1] \wedge X$, or suspension $\mathbb{S}^1 \wedge X$, respectively; here $[0,1]$ has the base point 0.

Definition 4. Let $f\colon A \to B$ be a morphism in $G\text{-}\mathcal{C}^*\mathfrak{alg}$. The *mapping cylinder* $\mathrm{Cyl}(f)$ and the *mapping cone* $\mathrm{Cone}(f)$ of f are the limits of the diagrams

$$A \xrightarrow{f} B \xleftarrow{\mathrm{ev}_1} \mathrm{Cyl}(B), \qquad A \xrightarrow{f} B \xleftarrow{\mathrm{ev}_1} \mathrm{Cone}(B)$$

in $G\text{-}\mathcal{C}^*\mathfrak{alg}$. More concretely,

$$\mathrm{Cone}(f) := \{(a,b) \in A \times C_0((0,1],B) \mid f(a) = b(1)\},$$
$$\mathrm{Cyl}(f) := \{(a,b) \in A \times C([0,1],B) \mid f(a) = b(1)\}.$$

If $f\colon X \to Y$ is a morphism of pointed compact spaces, then the mapping cone and mapping cylinder of the induced *-homomorphism $C_0(f)\colon C_0(Y) \to C_0(X)$ agree with $C_0(\mathrm{Cyl}(f))$ and $C_0(\mathrm{Cone}(f))$, respectively.

The familiar maps relating mapping cones and cylinders to cones and suspensions continue to exist in our case. For any morphism $f\colon A \to B$ in $G\text{-}\mathcal{C}^*\mathfrak{alg}$, we get a morphism of extensions

$$\begin{array}{ccccc}
\mathrm{Sus}(B) & \rightarrowtail & \mathrm{Cone}(f) & \twoheadrightarrow & A \\
\downarrow & & \downarrow & & \| \\
\mathrm{Cone}(B) & \rightarrowtail & \mathrm{Cyl}(f) & \twoheadrightarrow & A
\end{array}$$

The bottom extension splits and the maps $A \leftrightarrow \mathrm{Cyl}(f)$ are inverse to each other up to homotopy. The composite map $\mathrm{Cone}(f) \to A \to B$ factors through $\mathrm{Cone}(\mathrm{id}_B) \cong \mathrm{Cone}(B)$ and hence is homotopic to the zero map.

Definition 5. Let $f_0, f_1\colon A \rightrightarrows B$ be two parallel morphisms in $G\text{-}\mathcal{C}^*\mathfrak{alg}$. We write $f_0 \sim f_1$ and call f_0 and f_1 *homotopic* if there is a morphism $f\colon A \to \mathrm{Cyl}(B)$ with $\mathrm{ev}_t \circ f = f_t$ for $t = 0, 1$.

A functor $F\colon G\text{-}\mathcal{C}^*\mathfrak{alg} \to \mathfrak{C}$ is called *homotopy invariant* if $f_0 \sim f_1$ implies $F(f_0) = F(f_1)$.

It is easy to check that homotopy is an equivalence relation on $\mathrm{Hom}_G(A,B)$. We let $[A,B]$ be the set of equivalence classes. The composition of morphisms in $G\text{-}\mathcal{C}^*\mathfrak{alg}$ descends to maps

$$[B,C] \times [A,B] \to [A,C], \qquad ([f],[g]) \mapsto [f \circ g],$$

that is, $f_1 \sim f_2$ and $g_1 \sim g_2$ implies $f_1 \circ g_1 \sim f_2 \circ g_2$. Thus the sets $[A,B]$ form the morphism sets of a category, called the *homotopy category of G-C^*-algebras*. A functor is homotopy invariant if and only if it descends to the homotopy category. Other characterisations of homotopy invariance are listed in [25, Sect. 3.1].

Of course, our notion of homotopy restricts to the usual one for pointed compact spaces.

2.1.2 Morita–Rieffel Equivalence and Stable Isomorphism

One of the basic ideas of non-commutative geometry is that $G \ltimes C_0(X)$ (or $G \ltimes_r C_0(X)$) should be a substitute for the quotient space $G \backslash X$, which may have bad singularities. In the special case of a free and proper G-space X, we expect that $G \ltimes C_0(X)$ and $C_0(G \backslash X)$ are "equivalent" in a suitable sense. Already the simplest possible case $X = G$ shows that we cannot expect an isomorphism here because

$$G \ltimes C_0(G) \cong G \ltimes_r C_0(G) \cong \mathbb{K}(L^2 G).$$

The right notion of equivalence is a C*-version of Morita equivalence introduced by Marc A. Rieffel [32–34]; therefore, we call it Morita–Rieffel equivalence.

The definition of Morita–Rieffel equivalence involves Hilbert modules over C*-algebras and the C*-algebras of compact operators on them; these notions are crucial for Kasparov theory as well. We refer to [20] for the definition and a discussion of their basic properties.

Definition 6. Two G-C*-algebras A and B are called Morita–Rieffel equivalent if there are a full G-equivariant Hilbert B-module $\mathscr{E}$ and a G-equivariant *-isomorphism $\mathbb{K}(\mathscr{E}) \cong A$.

It is possible (and desirable) to express this definition more symmetrically: $\mathscr{E}$ is an A, B-bimodule with two inner products taking values in A and B, satisfying various conditions (see also [32]). Two Morita–Rieffel equivalent G-C*-algebras have equivalent categories of G-equivariant Hilbert modules via the functor $\mathscr{H} \mapsto \mathscr{H} \otimes_A \mathscr{E}$. The converse is not so clear.

Example 1. The following is a more intricate example of a Morita–Rieffel equivalence. Let Γ and P be two subgroups of a locally compact group G. Then Γ acts on G/P by left translation and P acts on $\Gamma \backslash G$ by right translation. The corresponding orbit space is the double coset space $\Gamma \backslash G / P$. Both $\Gamma \ltimes C_0(G/P)$ and $P \ltimes C_0(\Gamma \backslash G)$ are non-commutative models for this double coset space. They are indeed Morita–Rieffel equivalent; the bimodule that implements the equivalence is a suitable completion of $C_c(G)$.

These examples suggest that Morita–Rieffel equivalent C*-algebras are different ways to describe the *same* non-commutative space. Therefore, we expect that reasonable functors on $\mathfrak{C}^* \mathfrak{alg}$ should not distinguish between Morita–Rieffel equivalent C*-algebras.

Definition 7. Two G-C*-algebras A and B are called *stably isomorphic* if there is a G-equivariant *-isomorphism $A \otimes \mathbb{K}(\mathscr{H}_G) \cong B \otimes \mathbb{K}(\mathscr{H}_G)$, where $\mathscr{H}_G := L^2(G \times \mathbb{N})$ is the direct sum of countably many copies of the regular representation of G; we let G act on $\mathbb{K}(\mathscr{H}_G)$ by conjugation, of course.

The following technical condition is often needed in connection with Morita–Rieffel equivalence.

Definition 8. A C*-algebra is called *σ-unital* if it has a countable approximate identity or, equivalently, contains a strictly positive element.

All separable C*-algebras and all unital C*-algebras are σ-unital; the algebra $\mathbb{K}(\mathcal{H})$ is σ-unital if and only if $\mathcal{H}$ is separable.

Theorem 1 ([7]). *σ-unital G-C*-algebras are G-equivariantly Morita–Rieffel equivalent if and only if they are stably isomorphic.*

In the non-equivariant case, this theorem is due to Brown–Green–Rieffel [7]. A simpler proof that carries over to the equivariant case appeared in [28].

2.1.3 C*-Stable Functors

The definition of C*-stability is more intuitive in the non-equivariant case:

Definition 9. Fix a rank-one projection $p \in \mathbb{K}(\ell^2\mathbb{N})$. The resulting embedding $A \to A \otimes \mathbb{K}(\ell^2\mathbb{N})$, $a \mapsto a \otimes p$, is called a *corner embedding* of A.

A functor $F: \mathfrak{C}^*\mathfrak{alg} \to \mathfrak{C}$ is called C*-*stable* if any corner embedding induces an isomorphism $F(A) \cong F(A \otimes \mathbb{K}(\ell^2\mathbb{N}))$.

The correct equivariant generalisation is the following:

Definition 10. A functor $F: G\text{-}\mathfrak{C}^*\mathfrak{alg} \to \mathfrak{C}$ is called C*-*stable* if the canonical embeddings $\mathcal{H}_1 \to \mathcal{H}_1 \oplus \mathcal{H}_2 \leftarrow \mathcal{H}_2$ induce isomorphisms

$$F\big(A \otimes \mathbb{K}(\mathcal{H}_1)\big) \xrightarrow{\cong} F\big(A \otimes \mathbb{K}(\mathcal{H}_1 \oplus \mathcal{H}_2)\big) \xleftarrow{\cong} F\big(A \otimes \mathbb{K}(\mathcal{H}_2)\big)$$

for all non-zero G-Hilbert spaces $\mathcal{H}_1$ and $\mathcal{H}_2$.

Of course, it suffices to require $F\big(A \otimes \mathbb{K}(\mathcal{H}_1)\big) \xrightarrow{\cong} F\big(A \otimes \mathbb{K}(\mathcal{H}_1 \oplus \mathcal{H}_2)\big)$. It is not hard to check that Definitions 9 and 10 are equivalent for trivial G.

Our next goal is to describe the *universal* C*-stable functor. We abbreviate $A_{\mathbb{K}} := \mathbb{K}(L^2G) \otimes A$.

Definition 11. A *correspondence* from A to B (or $A \dashrightarrow B$) is a G-equivariant Hilbert $B_{\mathbb{K}}$-module $\mathcal{E}$ together with a G-equivariant essential (or non-degenerate) *-homomorphism $f: A_{\mathbb{K}} \to \mathbb{K}(\mathcal{E})$.

Given correspondences $\mathcal{E}$ from A to B and $\mathcal{F}$ from B to C, their *composition* is the correspondence from A to C with underlying Hilbert module $\mathcal{E} \bar{\otimes}_{B_{\mathbb{K}}} \mathcal{F}$ and map $A_{\mathbb{K}} \to \mathbb{K}(\mathcal{E}) \to \mathbb{K}(\mathcal{E} \bar{\otimes}_{B_{\mathbb{K}}} \mathcal{F})$, where the last map sends $T \mapsto T \otimes 1$; this yields compact operators because $B_{\mathbb{K}}$ maps to $\mathbb{K}(\mathcal{F})$. See [20] for the definition of the relevant completed tensor product of Hilbert modules.

The composition of correspondences is only defined up to isomorphism. It is associative and the identity maps $A \to A = \mathbb{K}(A)$ act as unit elements, so that we get a category $\mathfrak{Corr}_G$ whose morphisms are the *isomorphism classes* of correspondences. Any *-homomorphism $\varphi: A \to B$ yields a correspondence: let $\mathcal{E}$ be the right ideal $\varphi(A_{\mathbb{K}}) \cdot B_{\mathbb{K}}$ in $B_{\mathbb{K}}$, viewed as a Hilbert B-module, and let $a \cdot b = \varphi(a) \cdot b$; this restricts to a compact operator on $\mathcal{E}$. This defines a canonical functor $\natural: G\text{-}\mathfrak{C}^*\mathfrak{alg} \to \mathfrak{Corr}_G$.

Proposition 1. *The functor* $\natural\colon G\text{-}\mathfrak{C}^*\mathfrak{alg} \to \mathfrak{Corr}_G$ *is the universal C*-stable functor on* $G\text{-}\mathfrak{C}^*\mathfrak{alg}$*; that is, it is C*-stable, and any other such functor factors uniquely through* $\natural$*.*

Proof. First we sketch the proof in the non-equivariant case. First we must verify that $\natural$ is C*-stable. The Morita–Rieffel equivalence between $\mathbb{K}(\ell^2 \mathbb{N}) \otimes A \cong \mathbb{K}(\ell^2(\mathbb{N},A))$ and A is implemented by the Hilbert module $\ell^2(\mathbb{N},A)$, which yields a correspondence $\big(\mathrm{id}, \ell^2(\mathbb{N},A)\big)$ from $\mathbb{K}(\ell^2 \mathbb{N}) \otimes A$ to A; this is inverse to the correspondence induced by a corner embedding $A \to \mathbb{K}(\ell^2 \mathbb{N}) \otimes A$.

A Hilbert B-module $\mathscr{E}$ with an essential *-homomorphism $A \to \mathbb{K}(\mathscr{E})$ is countably generated because A is assumed σ-unital. Kasparov's Stabilisation Theorem yields an isometric embedding $\mathscr{E} \to \ell^2(\mathbb{N},B)$. Hence we get *-homomorphisms

$$A \to \mathbb{K}(\ell^2 \mathbb{N}) \otimes B \leftarrow B.$$

This diagram induces a map $F(A) \to F(\mathbb{K}(\ell^2 \mathbb{N}) \otimes B) \cong F(B)$ for any stable functor F. Now we should check that this well-defines a functor $\bar{F}\colon \mathfrak{Corr}_G \to \mathfrak{C}$ with $\bar{F} \circ \natural = F$, and that this yields the only such functor. We omit these computations.

The generalisation to the equivariant case uses the crucial property of the left regular representation that $L^2(G) \otimes \mathscr{H} \cong L^2(G \times \mathbb{N})$ for any countably infinite-dimensional G-Hilbert space $\mathscr{H}$. Since we replace A and B by $A_{\mathbb{K}}$ and $B_{\mathbb{K}}$ in the definition of correspondence right away, we can use this to repair a possible lack of G-equivariance; similar ideas appear in [23]. □

Example 2. Let u be a G-invariant multiplier of B. Then the identity map and the *inner automorphism* $B \to B$, $b \mapsto ubu^*$, associated to u define isomorphic correspondences $B \dashrightarrow B$ (via u). Hence inner automorphisms act trivially on C*-stable functors. Actually, this is one of the computations that we have omitted in the proof above; the argument can be found in [12].

2.1.4 Exactness Properties

Definition 12. A diagram $I \to E \to Q$ in $G\text{-}\mathfrak{C}^*\mathfrak{alg}$ is an *extension* if it is isomorphic to the canonical diagram $I \to A \to A/I$ for some G-invariant ideal I in a G-C*-algebra A. We write $I \rightarrowtail E \twoheadrightarrow Q$ to denote extensions. A *section* for an extension

$$I \overset{i}{\rightarrowtail} E \overset{p}{\twoheadrightarrow} Q \tag{1}$$

in $G\text{-}\mathfrak{C}^*\mathfrak{alg}$ is a map (of sets) $Q \to E$ with $p \circ s = \mathrm{id}_Q$. We call (1) *split* if there is a section that is a G-equivariant *-homomorphism. We call (1) *G-equivariantly cp-split* if there is a G-equivariant, completely positive, contractive, linear section.

Sections are also often called *lifts*, *liftings*, or *splittings*.

Definition 13. A functor F on $G\text{-}\mathfrak{C}^*\mathfrak{alg}$ is *split-exact* if, for any split extension $K \overset{i}{\rightarrowtail} E \overset{p}{\twoheadrightarrow} Q$ with section $s\colon Q \to E$, the map $\big(F(i), F(s)\big)\colon F(K) \oplus F(Q) \to F(E)$ is invertible.

Split-exactness is useful because of the following construction of Joachim Cuntz [10].

Let $B \lhd E$ be a G-invariant ideal and let $f_+, f_- : A \rightrightarrows E$ be G-equivariant $*$-homomorphisms with $f_+(a) - f_-(a) \in B$ for all $a \in A$. Equivalently, f_+ and f_- both lift the same morphism $\bar{f} : A \to E/B$. The data (A, f_+, f_-, E, B) is called a *quasi-homomorphism* from A to B.

Pulling back the extension $B \rightarrowtail E \twoheadrightarrow E/B$ along $\bar{f}$, we get an extension $B \rightarrowtail E' \twoheadrightarrow A$ with two sections $f'_+, f'_- : A \rightrightarrows E'$. The split-exactness of F shows that $F(B) \rightarrowtail F(E') \twoheadrightarrow F(A)$ is a split extension in $\mathfrak{C}$. Since both $F(f'_-)$ and $F(f'_+)$ are sections for it, we get a map $F(f'_+) - F(f'_-) : F(A) \to F(B)$. *Thus a quasi-homomorphism induces a map $F(A) \to F(B)$ if F is split-exact.* The formal properties of this construction are summarised in [12].

Given a C*-algebra A, there is a *universal quasi-homomorphism* out of A. Let $Q(A) := A * A$ be the free product of two copies of A and let $\pi_A : Q(A) \to A$ be the *folding homomorphism* that restricts to id_A on both factors. Let $q(A)$ be its kernel. The two canonical embeddings $A \to A * A$ are sections for the folding homomorphism. Hence we get a quasi-homomorphism $A \rightrightarrows Q(A) \rhd q(A)$. The universal property of the free product shows that any quasi-homomorphism yields a G-equivariant $*$-homomorphism $q(A) \to B$.

Theorem 2. *Functors that are C*-stable and split-exact are automatically homotopy invariant.*

This is a deep result of Nigel Higson [16]; a simple proof can be found in [12]. Besides basic properties of quasi-homomorphisms, it only uses that inner endomorphisms act identically on C*-stable functors.

Definition 14. We call F *exact* if $F(K) \to F(E) \to F(Q)$ is exact (at $F(E)$) for any extension $K \rightarrowtail E \twoheadrightarrow Q$ in G-$\mathfrak{C}^*\mathfrak{alg}$. More generally, given a class $\mathscr{E}$ of extensions in G-$\mathfrak{C}^*\mathfrak{alg}$ like, say, the class of equivariantly cp-split extensions, we define exactness for extensions in $\mathscr{E}$.

Most functors we are interested in satisfy homotopy invariance and Bott periodicity, and these two properties prevent a functor from being exact in the stronger sense of being *left* or *right* exact. This explains why our notion of exactness is much weaker than usual in homological algebra.

It is reasonable to require that a functor be part of a *homology* theory, that is, a sequence of functors $(F_n)_{n \in \mathbb{Z}}$ together with natural long exact sequences for all extensions. We do not require this because this additional information tends to be hard to get a priori but often comes for free a posteriori:

Proposition 2. *Suppose that F is homotopy invariant and exact (or exact for equivariantly cp-split extensions). Then F has long exact sequences of the form*

$$\cdots \to F\big(\mathrm{Sus}(K)\big) \to F\big(\mathrm{Sus}(E)\big) \to F\big(\mathrm{Sus}(Q)\big) \to F(K) \to F(E) \to F(Q)$$

for any (equivariantly cp-split) extension $K \rightarrowtail E \twoheadrightarrow Q$. In particular, F is split-exact.

See Sect. 21.4 in [4] for the proof.

Together with Bott periodicity, this yields long exact sequences that extend towards $\pm\infty$ in *both* directions, showing that an exact homotopy invariant functor that satisfies Bott periodicity is part of a homology theory in a canonical way.

2.1.5 Definition of Kasparov Theory

Kasparov theory maps two $\mathbb{Z}/2$-graded C*-algebras to an Abelian group $\mathrm{KK}_0^G(A,B)$; this is a vast generalisation of K-theory and K-homology. The most remarkable feature of this theory is an associative product on KK called *Kasparov product*, which generalises various known product constructions in K-theory and K-homology. We do not discuss KK^G for $\mathbb{Z}/2$-graded G-C*-algebras here because it does not fit so well with the universal property approach.

Fix a locally compact group G. The Kasparov groups $\mathrm{KK}_0^G(A,B)$ for $A,B \in G\text{-}\mathfrak{C}^*\mathfrak{sep}$ form morphisms sets $A \to B$ of a category, which we denote by $\mathfrak{KR}^G$; the composition in $\mathfrak{KR}^G$ is the Kasparov product. The categories $G\text{-}\mathfrak{C}^*\mathfrak{sep}$ and $\mathfrak{KR}^G$ have the same objects. We have a canonical functor

$$\mathrm{KK}: G\text{-}\mathfrak{C}^*\mathfrak{sep} \to \mathfrak{KR}^G$$

that acts identically on objects.

Theorem 3. *The functor* $\mathrm{KK}^G: G\text{-}\mathfrak{C}^*\mathfrak{sep} \to \mathfrak{KR}^G$ *is the universal split-exact C*-stable functor; in particular, $\mathfrak{KR}^G$ is an additive category. In addition, KK^G also has the following properties and is, therefore, universal among functors with some of these extra properties:* KK^G *is*

- *Homotopy invariant*
- *Exact for G-equivariantly cp-split extensions*
- *Satisfies Bott periodicity, that is, in $\mathfrak{KR}^G$ there are natural isomorphisms* $\mathrm{Sus}^2(A) \cong A$ *for all* $A \in \mathfrak{KR}^G$.

Definition 15. *A G-equivariant* *-homomorphism $f: A \to B$ is called a KK^G-equivalence if $\mathrm{KK}(f)$ is invertible in $\mathfrak{KR}^G$.*

Corollary 1. *Let $F: G\text{-}\mathfrak{C}^*\mathfrak{sep} \to \mathfrak{C}$ be split-exact and C*-stable. Then F factors uniquely through KK^G, is homotopy invariant, and satisfies Bott periodicity. A KK^G-equivalence $A \to B$ induces an isomorphism $F(A) \to F(B)$.*

We will take the universal property of Theorem 3 as a definition of $\mathfrak{KR}^G$ and thus of the groups $\mathrm{KK}_0^G(A,B)$. We also let

$$\mathrm{KK}_n^G(A,B) := \mathrm{KK}^G\big(A, \mathrm{Sus}^n(B)\big);$$

since the Bott periodicity isomorphism identifies $\mathrm{KK}_2^G \cong \mathrm{KK}_0^G$, this yields a $\mathbb{Z}/2$-graded theory.

By the universal property, K-theory descends to a functor on $\mathfrak{KK}$, that is, we get canonical maps

$$\mathrm{KK}_0(A,B) \to \mathrm{Hom}\big(\mathrm{K}_*(A), \mathrm{K}_*(B)\big)$$

for all separable C*-algebras A,B, where the right hand side denotes grading-preserving group homomorphisms. For $A = \mathbb{C}$, this yields a map $\mathrm{KK}_0(\mathbb{C},B) \to \mathrm{Hom}\big(\mathbb{Z}, \mathrm{K}_0(B)\big) \cong \mathrm{K}_0(B)$. Using suspensions, we also get a corresponding map $\mathrm{KK}_1(\mathbb{C},B) \to \mathrm{K}_1(B)$.

Theorem 4. *The maps* $\mathrm{KK}_*(\mathbb{C},B) \to \mathrm{K}_*(B)$ *constructed above are isomorphisms for all* $B \in\!\in \mathfrak{C}^*\mathfrak{sep}$.

Thus Kasparov theory is a bivariant generalisation of K-theory.

Roughly speaking, $\mathrm{KK}_*(A,B)$ is the place where maps between K-theory groups live. Most constructions of such maps, say, in index theory can in fact be improved to yield elements of $\mathrm{KK}_*(A,B)$. One reason for this is the *Universal Coefficient Theorem* (UCT) by Rosenberg and Schochet [35], which computes $\mathrm{KK}_*(A,B)$ from $\mathrm{K}_*(A)$ and $\mathrm{K}_*(B)$ for many C*-algebras A,B. If A satisfies the UCT, then any group homomorphism $\mathrm{K}_*(A) \to \mathrm{K}_*(B)$ lifts to an element of $\mathrm{KK}_*(A,B)$ of the same parity.

With our definition, it is not obvious how to construct elements in $\mathrm{KK}_0^G(A,B)$. The only source we know so far are G-equivariant *-homomorphisms. Another important source are *extensions*, more precisely, equivariantly cp-split extensions. Any such extension $I \rightarrowtail E \twoheadrightarrow Q$ yields a class in $\mathrm{KK}_1^G(Q,I) \cong \mathrm{KK}_0^G(\mathrm{Sus}(Q),I) \cong \mathrm{KK}_0^G(Q,\mathrm{Sus}(I))$. Conversely, any element in $\mathrm{KK}_1^G(Q,I)$ comes from an extension in this fashion in a rather transparent way.

Thus it may seem that we can understand all of Kasparov theory from an abstract, category theoretic point of view. But this is not the case. To get a category, we must *compose* extensions; this leads to extensions of higher length. If we allow such higher-length extensions, we can easily construct a category that is isomorphic to Kasparov theory; this generalisation still works for more general algebras than C*-algebras (see [12]) because it does not involve any difficult analysis any more. But such a setup offers no help to *compute* products. Here computing products means identifying them with other simple things like, say, the identity morphism. This is why the more concrete approach to Kasparov theory is still necessary for the interesting applications of the theory.

In connection with the Baum–Connes conjecture, our abstract approach allows us to formulate it and analyse its consequences. But to verify it, say, for amenable groups we must show that a certain morphism in KK^G is invertible. This involves constructing its inverse and checking that the two Kasparov products in both orders are 1. These computations require the concrete description of Kasparov theory that we omit here. We merely refer to [4] for a detailed treatment.

2.1.6 Extending Functors and Identities to $\mathfrak{KK}^G$

We use the universal property to extend functors from G-$\mathfrak{C}^*\mathfrak{alg}$ to $\mathfrak{KK}^G$ and check identities in $\mathfrak{KK}^G$ without computing Kasparov products. As our first example, consider the full and reduced crossed product functors

$$G \ltimes_r \_, G \ltimes \_ : G\text{-}\mathfrak{C}^*\mathfrak{alg} \to \mathfrak{C}^*\mathfrak{alg}.$$

Proposition 3. *These two functors extend to functors*

$$G \ltimes_r \_, G \ltimes \_ \colon \mathfrak{RR}^G \to \mathfrak{RR}$$

called descent functors.

Kasparov constructs these functors directly using the concrete description of Kasparov cycles. This requires a certain amount of work; in particular, checking functoriality involves knowing how to compute Kasparov products.

Proof. We only write down the argument for *reduced* crossed products, the other case is similar. It is well-known that $G \ltimes_r \big(A \otimes \mathbb{K}(\mathcal{H})\big) \cong (G \ltimes_r A) \otimes \mathbb{K}(\mathcal{H})$ for any G-Hilbert space $\mathcal{H}$. Therefore, the composite functor

$$G\text{-}\mathfrak{C}^*\mathfrak{sep} \xrightarrow{G \ltimes_r \_} \mathfrak{C}^*\mathfrak{sep} \xrightarrow{\mathrm{KK}} \mathfrak{RR}$$

is C*-stable. This functor is split-exact as well (we omit the proof). Now the universal property provides an extension to a functor $\mathfrak{RR}^G \to \mathfrak{RR}$.   □

Similarly, we get functors

$$A \otimes_{\min} \_, A \otimes_{\max} \_ \colon \mathfrak{RR}^G \to \mathfrak{RR}^G$$

for any G-C*-algebra A. Since these extensions are natural, we even get bifunctors

$$\otimes_{\min}, \otimes_{\max} \colon \mathfrak{RR}^G \times \mathfrak{RR}^G \to \mathfrak{RR}^G.$$

For the Baum–Connes assembly map, we need the *induction functors*

$$\mathrm{Ind}_H^G \colon \mathfrak{RR}^H \to \mathfrak{RR}^G$$

for closed subgroups $H \subseteq G$. For a finite group H, $\mathrm{Ind}_H^G(A)$ is the H-fixed point algebra of $C_0(G,A)$, where H acts by $h \cdot f(g) = \alpha_h\big(f(gh)\big)$. For infinite H, we have

$$\mathrm{Ind}_H^G(A) = \{f \in C_b(G,A) \mid$$
$$\alpha_h f(gh) = f(g) \text{ for all } g \in G,\, h \in H, \text{ and } gH \mapsto \|f(g)\| \text{ is } C_0\};$$

the group G acts by translations on the left. This construction is clearly functorial for equivariant *-homomorphisms. Furthermore, it commutes with C*-stabilisations and maps split extensions again to split extensions. Therefore, the same argument as above allows us to extend it to a functor

$$\mathrm{Ind}_H^G \colon \mathfrak{RR}^H \to \mathfrak{RR}^G$$

The following examples are more trivial. Let $\tau \colon \mathfrak{C}^*\mathfrak{alg} \to G\text{-}\mathfrak{C}^*\mathfrak{alg}$ equip a C*-algebra with the trivial G-action; it extends to a functor $\tau \colon \mathfrak{RR} \to \mathfrak{RR}^G$. The *restriction functors*

$$\mathrm{Res}_G^H \colon \mathfrak{RR}^G \to \mathfrak{RR}^H$$

for closed subgroups $H \subseteq G$ are defined by forgetting part of the equivariance.

The universal property also allows us to *prove identities* between functors. For instance, Green's Imprimitivity Theorem provides Morita equivalences

$$G \ltimes \mathrm{Ind}_H^G(A) \sim_M H \ltimes A, \qquad G \ltimes_{\mathrm{r}} \mathrm{Ind}_H^G(A) \sim_M H \ltimes_{\mathrm{r}} A \qquad (2)$$

for any H-C^*-algebra A. This is proved by completing $C_0(G,A)$ to an imprimitivity bimodule for both C^*-algebras. This equivalence is clearly natural for H-equivariant *-homomorphisms. Since all functors involved are C^*-stable and split exact, the uniqueness part of the universal property of KK^H shows that the KK-equivalences $G \ltimes \mathrm{Ind}_H^G(A) \cong H \ltimes A$ and $G \ltimes_{\mathrm{r}} \mathrm{Ind}_H^G(A) \cong H \ltimes_{\mathrm{r}} A$ are natural for morphisms in KK^H. That is, the diagram

$$
\begin{array}{ccc}
G \ltimes_{\mathrm{r}} \mathrm{Ind}_H^G(A_1) & \xrightarrow{\ \cong\ } & H \ltimes_{\mathrm{r}} A_1 \\
{\scriptstyle G \ltimes_{\mathrm{r}} \mathrm{Ind}_H^G(f)} \downarrow & & \downarrow {\scriptstyle H \ltimes_{\mathrm{r}} f} \\
G \ltimes_{\mathrm{r}} \mathrm{Ind}_H^G(A_2) & \xrightarrow{\ \cong\ } & H \ltimes_{\mathrm{r}} A_2
\end{array}
$$

in $\mathfrak{K}\mathfrak{K}$ commutes for any $f \in \mathrm{KK}_0^H(A_1,A_2)$. More examples of this kind are discussed in Sect. 4.1 of [25].

We can also prove *adjointness* relations in Kasparov theory in an abstract way by constructing the unit and counit of the adjunction. An important example is the adjointness between induction and restriction functors (see also Sect. 3.2 of [26]).

Proposition 4. *Let $H \subseteq G$ be a closed subgroup. If H is open, then we have natural isomorphisms*

$$\mathrm{KK}^G(\mathrm{Ind}_H^G A, B) \cong \mathrm{KK}^H(A, \mathrm{Res}_G^H B) \qquad (3)$$

for all $A \in\in H$-$\mathfrak{C}^*\mathfrak{alg}$, $B \in\in G$-$\mathfrak{C}^*\mathfrak{alg}$. If $H \subseteq G$ is cocompact, then we have natural isomorphisms

$$\mathrm{KK}^G(A, \mathrm{Ind}_H^G B) \cong \mathrm{KK}^H(\mathrm{Res}_G^H A, B) \qquad (4)$$

for all $A \in\in G$-$\mathfrak{C}^*\mathfrak{alg}$, $B \in\in H$-$\mathfrak{C}^*\mathfrak{alg}$.

Proof. We will not use (4) later and therefore only prove (3). We must construct natural elements

$$\alpha_A \in \mathrm{KK}_0^G(\mathrm{Ind}_H^G \mathrm{Res}_G^H A, A), \qquad \beta_B \in \mathrm{KK}_0^H(B, \mathrm{Res}_G^H \mathrm{Ind}_H^G B)$$

that satisfy the conditions for unit and counit of adjunction [22].

We have a natural G-equivariant *-isomorphism $\mathrm{Ind}_H^G \mathrm{Res}_H^G(A) \cong C_0(G/H) \otimes A$ for any G-C^*-algebra A. Since H is open in G, the homogeneous space G/H is discrete. We represent $C_0(G/H)$ on the Hilbert space $\ell^2(G/H)$ by pointwise multiplication operators. This is G-equivariant for the representation of G on $\ell^2(G/H)$ by left translations. Thus we get a correspondence from $\mathrm{Ind}_H^G \mathrm{Res}_G^H(A)$ to A, which yields $\alpha_A \in \mathrm{KK}_0^G(\mathrm{Ind}_H^G \mathrm{Res}_G^H(A), A)$ because KK^G is C^*-stable.

For any H-C^*-algebra B, we may embed B in $\mathrm{Res}_G^H \mathrm{Ind}_H^G(B)$ as the subalgebra of functions supported on the single coset H. This embedding is H-equivariant and provides $\beta_B \in \mathrm{KK}_0^H(B, \mathrm{Res}_G^H \mathrm{Ind}_H^G B)$.

Now we have to check that the following two composite maps are identity morphisms in $\mathfrak{K}\mathfrak{K}^G$ and $\mathfrak{K}\mathfrak{K}^H$, respectively:

$$\mathrm{Ind}_H^G(B) \xrightarrow{\mathrm{Ind}_H^G(\beta_B)} \mathrm{Ind}_H^G \mathrm{Res}_G^H \mathrm{Ind}_H^G(B) \xrightarrow{\alpha_{\mathrm{Ind}_H^G(B)}} \mathrm{Ind}_H^G(B)$$

$$\mathrm{Res}_G^H A \xrightarrow{\beta_{\mathrm{Res}_G^H A}} \mathrm{Res}_G^H \mathrm{Ind}_H^G \mathrm{Res}_G^H(A) \xrightarrow{\mathrm{Res}_G^H \alpha_A} \mathrm{Res}_G^H A$$

This yields the desired adjointness by a general argument from category theory (see [22]). In fact, both composites are already equal to the identity as *correspondences*. Hence we need no knowledge of Kasparov theory except for its C*-stability to prove (3).   □

The following example is discussed in detail in Sect. 4.1 of [25]. If G is compact, then the trivial action functor $\tau \colon \mathfrak{K}\mathfrak{K} \to \mathfrak{K}\mathfrak{K}^G$ is left adjoint to $G \ltimes \text{\textvisiblespace} = G \ltimes_r \text{\textvisiblespace}$, that is, we have natural isomorphisms

$$\mathrm{KK}_*^G(\tau(A), B) \cong \mathrm{KK}_*(A, G \ltimes B). \tag{5}$$

This is also known as the *Green–Julg Theorem*. For $A = \mathbb{C}$, it specialises to a natural isomorphism $\mathrm{K}_*^G(B) \cong \mathrm{K}_*(G \ltimes B)$.

2.1.7 Triangulated Category Structure

We can turn $\mathfrak{K}\mathfrak{K}^G$ into a triangulated category by extending standard constructions for topological spaces (see [26]). A *triangulated category* is a category $\mathfrak{T}$ with a *suspension automorphism* $\Sigma \colon \mathfrak{T} \to \mathfrak{T}$ and a class of *exact triangles*, subject to various axioms (see [26, 29, 37]). An exact triangle is a diagram in $\mathfrak{T}$ of the form

$$A \to B \to C \to \Sigma A \qquad \text{or} \qquad \begin{array}{ccc} A & \longrightarrow & B \\ & {\scriptstyle [1]}\nwarrow & \swarrow \\ & C, & \end{array}$$

where the [1] in the arrow $C \to A$ warns us that this map has *degree* 1. A *morphism* of triangles is a triple of maps α, β, γ making the obvious diagram commute.

A typical example is the homotopy category $\mathrm{Ho}(\mathfrak{C})$ of chain complexes over an additive category $\mathfrak{C}$. Here the suspension functor is the (signed) *translation* functor

$$\Sigma\big((C_n, d_n)\big) := (C_{n-1}, -d_{n-1}) \qquad \text{on objects,}$$
$$\Sigma\big((f_n)\big) := (f_{n-1}) \qquad \text{on morphisms;}$$

a triangle is exact if it is isomorphic to a *mapping cone triangle*

$$A \xrightarrow{f} B \to \mathrm{Cone}(f) \to \Sigma A$$

for some chain map f; the maps $B \to \mathrm{Cone}(f) \to \Sigma A$ are the canonical ones. It is well-known that this defines a triangulated category.

Another classical example is the stable homotopy category, say, of compactly generated pointed topological spaces (it is not particularly relevant which category of spaces or spectra we use). The suspension is $\Sigma(A) := \mathbb{S}^1 \wedge A$; a triangle is exact if it is isomorphic to a *mapping cone triangle*

$$A \xrightarrow{f} B \to \mathrm{Cone}(f) \to \Sigma A$$

for some map f; the maps $B \to \mathrm{Cone}(f) \to \Sigma A$ are the canonical ones.

We are mainly interested in the categories $\mathfrak{K}\mathfrak{K}$ and $\mathfrak{K}\mathfrak{K}^G$ introduced above. Their triangulated category structure is discussed in detail in [26]. We are facing a notational problem because the functor $X \mapsto C_0(X)$ from pointed compact spaces to C^*-algebras is *contravariant*, so that *mapping cone triangles* now have the form

$$A \xleftarrow{f} B \leftarrow \mathrm{Cone}(f) \leftarrow C_0(\mathbb{R},A)$$

for a *-homomorphism $f \colon B \to A$ as in Definition 4.

It is reasonable to view a *-homomorphism from A to B as a morphism from B to A. Nevertheless, we prefer the convention that an algebra homomorphism $A \to B$ is a morphism $A \to B$. But then the most natural triangulated category structure lives on the opposite category KK^{op}. This creates only notational difficulties because the opposite category of a triangulated category inherits a canonical triangulated category structure, which has "the same" exact triangles. However, the passage to opposite categories exchanges suspensions and desuspensions and modifies some sign conventions. Thus the functor $A \mapsto C_0(\mathbb{R},A)$, which is the suspension functor in KK^{op}, becomes the *desuspension* functor in KK. Fortunately, Bott periodicity implies that $\Sigma^2 \cong \mathrm{id}$, so that Σ and Σ^{-1} agree.

Depending on your definition of a triangulated category, you may want the suspension to be an equivalence or isomorphism of categories. In the latter case, you must replace $\mathfrak{K}\mathfrak{K}^{(G)}$ by an equivalent category (see [26]); since this is not important here, we do not bother about this issue.

Definition 16. A triangle $A \to B \to C \to \Sigma A$ in $\mathfrak{K}\mathfrak{K}^G$ is called *exact* if it is isomorphic as a triangle to the *mapping cone triangle*

$$\mathrm{Sus}(B) \to \mathrm{Cone}(f) \to A \xrightarrow{f} B$$

for some G-equivariant *-homomorphism f.

Alternatively, we can use G-equivariantly cp-split extensions in $G\text{-}\mathfrak{C}^*\mathfrak{sep}$. Any such extension $I \rightarrowtail E \twoheadrightarrow Q$ determines a class in $KK_1^G(Q,I) \cong KK_0^G(\mathrm{Sus}(Q),I)$, so that we get a triangle $\mathrm{Sus}(Q) \to I \to E \to Q$ in $\mathfrak{K}\mathfrak{K}^G$. Such triangles are called *extension triangles*. A triangle in $\mathfrak{K}\mathfrak{K}^G$ is exact if and only if it is isomorphic to the extension triangle of a G-equivariantly cp-split extension.

Theorem 5. *With the suspension automorphism and exact triangles defined above, $\mathfrak{K}\mathfrak{K}^G$ is a triangulated category.*

Proof. This is proved in detail in [26]. □

Triangulated categories clarify the basic bookkeeping with long exact sequences. Mayer–Vietoris exact sequences and inductive limits are discussed from this point of view in [26]. More importantly, this framework sheds light on more advanced constructions like the Baum–Connes assembly map.

The triangulated category axioms are discussed in greater detail in [26, 29, 37]. They encode some standard machinery for manipulating long exact sequences. Most of them amount to formal properties of mapping cones and mapping cylinders, which we can prove as in classical topology. The only axiom that requires more care is that any morphism $f: A \to B$ should be part of an exact triangle.

Unlike in [26], we prefer to construct this triangle as an extension triangle because this works in greater generality; we have learned this idea from Alexander Bonkat [5]. Any element in $KK_0^G(A, B) \cong KK_1^G(A, C_0(\mathbb{R}, B))$ can be represented by an extension $\mathbb{K}(\mathcal{H}) \rightarrowtail E \twoheadrightarrow A$ with an equivariant completely positive contractive section, where $\mathcal{H}$ is a full G-equivariant Hilbert $C_0(\mathbb{R}, B)$-module, so that $\mathbb{K}(\mathcal{H})$ is KK^G-equivalent to $C_0(\mathbb{R}, B)$. Hence the resulting extension triangle in KK^G is isomorphic to one of the form

$$C_0(\mathbb{R}, A) \to C_0(\mathbb{R}, B) \to E \to A;$$

by construction, it contains the suspension of the given class in $KK_0^G(A, B)$; it is easy to remove the suspension.

Definition 17. Let $\mathfrak{T}$ be a triangulated and $\mathfrak{C}$ an Abelian category. A covariant functor $F: \mathfrak{T} \to \mathfrak{C}$ is called *homological* if $F(A) \to F(B) \to F(C)$ is exact at $F(B)$ for all exact triangles $A \to B \to C \to \Sigma A$. A contravariant functor with the analogous exactness property is called *cohomological*.

Let $A \to B \to C \to \Sigma A$ be an exact triangle. Then a homological functor $F: \mathfrak{T} \to \mathfrak{C}$ yields a natural long exact sequence

$$\cdots \to F_{n+1}(C) \to F_n(A) \to F_n(B) \to F_n(C) \to F_{n-1}(A) \to F_{n-1}(B) \to \cdots$$

with $F_n(A) := F(\Sigma^{-n}A)$ for $n \in \mathbb{Z}$, and a cohomological functor $F: \mathfrak{T}^{\mathrm{op}} \to \mathfrak{C}$ yields a natural long exact sequence

$$\cdots \leftarrow F^{n+1}(C) \leftarrow F^n(A) \leftarrow F^n(B) \leftarrow F^n(C) \leftarrow F^{n-1}(A) \leftarrow F^{n-1}(B) \leftarrow \cdots$$

with $F^n(A) := F(\Sigma^{-n}A)$.

Proposition 5. *Let $\mathfrak{T}$ be a triangulated category. The functors*

$$\mathfrak{T}(A, \_): \mathfrak{T} \to \mathfrak{Ab}, \qquad B \mapsto \mathfrak{T}(A, B)$$

are homological for all $A \in \in \mathfrak{T}$. Dually, the functors

$$\mathfrak{T}(\_, B): \mathfrak{T}^{\mathrm{op}} \to \mathfrak{Ab}, \qquad A \mapsto \mathfrak{T}(A, B)$$

are cohomological for all $B \in \in \mathfrak{T}$.

Observe that

$$\mathfrak{T}^n(A,B) = \mathfrak{T}(\Sigma^{-n}A, B) \cong \mathfrak{T}(A, \Sigma^n B) \cong \mathfrak{T}_{-n}(A,B).$$

Definition 18. A functor $F\colon \mathfrak{T} \to \mathfrak{T}'$ between two triangulated categories is called *exact* if it intertwines the suspension automorphisms (up to specified natural isomorphisms) and maps exact triangles in $\mathfrak{T}$ again to exact triangles in $\mathfrak{T}'$.

Example 3. The restriction functor $\mathrm{Res}_G^H\colon \mathrm{KK}^G \to \mathrm{KK}^H$ for a closed subgroup H of a locally compact group G and the crossed product functors $G \ltimes \lrcorner, G \ltimes_r \lrcorner\colon \mathrm{KK}^G \to \mathrm{KK}$ are exact because they preserve mapping cone triangles.

Let $F\colon \mathfrak{T}_1 \to \mathfrak{T}_2$ be an exact functor. If $G\colon \mathfrak{T}_2 \to?$ is exact, homological, or cohomological, then so is $G \circ F$.

2.2 Subcategories in $\mathfrak{KK}^G$

Now we turn to the construction of the Baum–Connes assembly map by studying various subcategories of $\mathfrak{KK}^G$ that are related to it.

2.2.1 Compactly Induced Actions

Definition 19. Let G be a locally compact group. A G-C*-algebra is *compactly induced* if it is of the form $\mathrm{Ind}_H^G(A)$ for some compact subgroup H of G and some H-C*-algebra A. We let $\mathscr{CI}$ be the class of all G-C*-algebras that are KK^G-equivalent to a direct summand of $\bigoplus_{i \in \mathbb{N}} A_i$ with compactly induced G-C*-algebras A_i for $i \in \mathbb{N}$.

Equivalently, $\mathscr{CI}$ is the smallest class of objects in $\mathfrak{KK}^G$ that is closed under direct sums, direct summands and isomorphism and contains all compactly induced G-C*-algebras.

Green's Imprimitivity Theorem (2) tells us that the (reduced) crossed product for a compactly induced action $\mathrm{Ind}_H^G(A)$ is equivalent to the crossed product $H \ltimes_r A$ for the compact group H. Hence we have

$$K_*(G \ltimes \mathrm{Ind}_H^G A) \cong K_*(G \ltimes_r \mathrm{Ind}_H^G A) \cong K_*(H \ltimes A) \cong K_*^H(A)$$

by the Green–Julg Theorem, compare (5).

Since the computation of equivariant K-theory for compact groups is a problem of *classical* topology, operator algebraists can pretend that it is *Somebody Else's Problem*. We are more fascinated by the analytic difficulties created by crosssed products by infinite groups. For instance, it is quite hard to see which Laurent series $\sum_{n \in \mathbb{Z}} a_n z^n$ correspond to an element of $C_{red}^* \mathbb{Z} = C^* \mathbb{Z}$ or, equivalently, which of them are the Fourier series of a continuous function on the unit circle. The Baum–Connes conjecture, when true, implies that such analytic difficulties do not influence the K-theory.

2.2.2 Two Simple Examples

It is best to explain our goals with two examples, namely, the groups $\mathbb{R}$ and $\mathbb{Z}$. The Baum–Connes conjectures for these groups hold and are equivalent to the *Connes–Thom isomorphism* [9] and a *Pimsner–Voiculescu exact sequence* [30]. Although the Baum–Connes conjecture only concerns the K-theory of C_{red}^*G and, more generally, of crossed products $G \ltimes_r A$, we get much stronger statements in this case.

Both $\mathbb{R}$ and $\mathbb{Z}$ are torsion-free, that is, they have no non-trivial compact subgroups. Hence the compactly induced actions are of the form $C_0(G,A)$ with $G \in \{\mathbb{R},\mathbb{Z}\}$ acting by translation. If A carries another action of G, then it makes no difference whether we let G act on $C_0(G,A)$ by $t \cdot f(x) := f(t^{-1}x)$ or $t \cdot f(x) := \alpha_t\big(f(t^{-1}x)\big)$: both definitions yield isomorphic G-C^*-algebras.

Theorem 6. *Any $\mathbb{R}$-C^*-algebra is $KK^{\mathbb{R}}$-equivalent to a compactly induced one. More briefly, $\mathscr{CI} = KK^{\mathbb{R}}$.*

Proof. Let A be an $\mathbb{R}$-C^*-algebra. Let $\mathbb{R}$ act on $\mathbb{R}$ by translation and extend this to an action on $X = (-\infty,\infty]$ by $t \cdot \infty := \infty$ for all $t \in \mathbb{R}$. Then we get an extension of $\mathbb{R}$-C^*-algebras

$$C_0(\mathbb{R},A) \rightarrowtail C_0(X,A) \twoheadrightarrow A,$$

where we let $\mathbb{R}$ act diagonally. It does not yet have an $\mathbb{R}$-*equivariant* completely positive section, but it becomes equivariantly cp-split if we tensor with $\mathbb{K}(L^2\mathbb{R})$. Therefore, it yields an extension triangle in $KK^{\mathbb{R}}$.

The Dirac operator for the standard Riemannian metric on $\mathbb{R}$ defines a class in $KK_1^{\mathbb{R}}(C_0(\mathbb{R}),\mathbb{C})$ which we may then map to $KK_1^{\mathbb{R}}(C_0(\mathbb{R},A),A)$ by exterior product. This yields another cp-split extension

$$\mathbb{K}(L^2\mathbb{R}) \otimes A \rightarrowtail \mathscr{T} \otimes A \twoheadrightarrow C_0(\mathbb{R},A).$$

The main work in the proof is to check that the resulting classes in $KK_1^{\mathbb{R}}\big(A,C_0(\mathbb{R},A)\big)$ and $KK_1^{\mathbb{R}}(C_0(\mathbb{R},A),A)$ are inverse to each other, so that A is $KK^{\mathbb{R}}$-equivalent to the induced $\mathbb{R}$-C^*-algebra $C_0(\mathbb{R},A)$.

This involves computing their Kasparov products in both orders and writing down homotopies to the identity transformation. The coefficient algebra A is irrelevant for these computations, that is, we may reduce everything to computations in $KK_0^{\mathbb{R}}(\mathbb{C},\mathbb{C})$ and $KK_0^{\mathbb{R}}\big(C_0(\mathbb{R}),C_0(\mathbb{R})\big)$. Since we did not describe the definition of the Kasparov product, we cannot say much about what is involved in this computation. We may, however, remark that we are dealing with an equivariant form of Bott periodicity, and that we merely have to follow a particular proof of Bott periodicity involving Kasparov theory and check that all the homotopies involved are sufficiently $\mathbb{R}$-equivariant to lift from KK to $KK^{\mathbb{R}}$.   □

Since the crossed product is functorial on Kasparov categories, this implies

$$\mathbb{R} \ltimes A = \mathbb{R} \ltimes_r A \sim \mathbb{R} \ltimes_r \mathrm{Sus}\big(C_0(\mathbb{R},A)\big) \cong \mathrm{Sus}(\mathbb{K}(L^2\mathbb{R}) \otimes A) \sim \mathrm{Sus}(A),$$

where $\sim$ denotes KK-equivalence. Taking K-theory, we get the Connes–Thom Isomorphism $K_*(\mathbb{R} \ltimes A) \cong K_{*+1}(A)$.

For most groups, we have $\mathscr{CI} \neq KK^G$. We now study the simplest case where this happens, namely, $G = \mathbb{Z}$.

We have seen above that $C_0(\mathbb{R})$ with the translation action of $\mathbb{R}$ is $KK^{\mathbb{R}}$-equivalent to $C_0(\mathbb{R})$ with trivial action. This equivalence persists if we restrict the action from $\mathbb{R}$ to the subgroup $\mathbb{Z} \subseteq \mathbb{R}$. Hence we get a $KK^{\mathbb{Z}}$-equivalence

$$A \sim \mathrm{Sus}\big(C_0(\mathbb{R},A)\big),$$

where $n \in \mathbb{Z}$ acts on $C_0(\mathbb{R},A) \cong C_0(\mathbb{R}) \otimes A$ by $(\alpha_n f)(x) := \alpha_n\big(f(x-n)\big)$. Although the $\mathbb{Z}$-action on $\mathbb{R}$ is free and proper, the action of $\mathbb{Z}$ on $C_0(\mathbb{R},A)$ need not be induced from the trivial subgroup.

Theorem 7. *For any $\mathbb{Z}$-C^*-algebra A, there is an exact triangle*

$$P_1 \to P_0 \to A \to \Sigma P_1$$

in $\mathfrak{RR}^{\mathbb{Z}}$ with compactly induced P_0 and P_1; more explicitly, $P_0 = P_1 = C_0(\mathbb{Z},A)$.

Proof. Restriction to $\mathbb{Z} \subseteq \mathbb{R}$ provides a surjection $C_0(\mathbb{R},A) \twoheadrightarrow C_0(\mathbb{Z},A)$, whose kernel may be identified with $C_0\big((0,1)\big) \otimes C_0(\mathbb{Z},A)$. The resulting extension

$$C_0\big((0,1)\big) \otimes C_0(\mathbb{Z},A) \rightarrowtail C_0(\mathbb{R},A) \twoheadrightarrow C_0(\mathbb{Z},A)$$

is $\mathbb{Z}$-equivariantly cp-split and hence provides an extension triangle in $KK^{\mathbb{Z}}$. Since $C_0(\mathbb{R},A)$ is $KK^{\mathbb{Z}}$-equivalent to the suspension of A, we get an exact triangle of the desired form. $\qquad\square$

When we apply a homological functor $\mathfrak{RR}^G \to \mathfrak{C}$ such as $K_*(\mathbb{Z} \ltimes \_)$ to the exact triangle in Theorem 7, then we get the *Pimsner–Voiculescu exact sequence*

$$
\begin{array}{ccccc}
K_1(A) & \longrightarrow & K_0(\mathbb{Z} \ltimes A) & \longrightarrow & K_0(A) \\
{\scriptstyle \alpha_* - 1}\big\uparrow & & & & \big\downarrow{\scriptstyle \alpha_* - 1} \\
K_1(A) & \longleftarrow & K_1(\mathbb{Z} \ltimes A) & \longleftarrow & K_0(A).
\end{array}
$$

Here $\alpha_* \colon K_*(A) \to K_*(A)$ is the map induced by the automorphism $\alpha(1)$ of A. It is not hard to identify the boundary map for the above extension with this map. Our approach yields such exact sequences for any homological functor.

Now we formulate some structural results for $\mathbb{R}$ and $\mathbb{Z}$ that have a chance to generalise to other groups.

Theorem 8. *Let G be $\mathbb{R}$ or $\mathbb{Z}$. Let A_1 and A_2 be G-C^*-algebras and let $f \in KK^G(A_1,A_2)$. If $\mathrm{Res}_G(f) \in KK(A_1,A_2)$ is invertible, then so is f itself. In particular, if $\mathrm{Res}_G(A_1) \cong 0$ in KK, then already $A \cong 0$ in KK^G.*

Proof. We only write down the proof for $G = \mathbb{Z}$; the case $G = \mathbb{R}$ is similar but simpler. If f were an equivariant *-homomorphism, then it would induce a morphism of extensions

$$
\begin{array}{ccccc}
C_0((0,1) \times \mathbb{Z}, A_1) & \longrightarrow & C_0(\mathbb{R}, A_1) & \longrightarrow & C_0(\mathbb{Z}, A_1) \\
\downarrow {\scriptstyle f_*} & & \downarrow {\scriptstyle f_*} & & \downarrow {\scriptstyle f_*} \\
C_0((0,1) \times \mathbb{Z}, A_2) & \longrightarrow & C_0(\mathbb{R}, A_2) & \longrightarrow & C_0(\mathbb{Z}, A_2)
\end{array}
$$

and hence a morphism of triangles between the resulting extension triangles. The latter morphism still exists even if f is merely a morphism in $KK^{\mathbb{Z}}$. This can be checked directly or deduced in a routine fashion from the uniqueness part of the universal property of $KK^{\mathbb{Z}}$. If $\mathrm{Res}_G(f)$ is invertible, then so are the induced maps $C_0((0,1) \times \mathbb{Z}, A_1) \to C_0((0,1) \times \mathbb{Z}, A_2)$ and $C_0(\mathbb{Z}, A_1) \to C_0(\mathbb{Z}, A_2)$ because $C_0(\mathbb{Z}, A) \cong \mathrm{Ind}^{\mathbb{Z}} \mathrm{Res}_{\mathbb{Z}}(A)$. Hence the Five Lemma in triangulated categories shows that f itself is invertible. To get the second statement, apply the first one to the zero maps $0 \to A_1 \to 0$. □

Definition 20. A path of G-actions $(\alpha_t)_{t \in [0,1]}$ is *continuous* if its pointwise application defines a strongly continuous action of G on $\mathrm{Cyl}(A) := C([0,1], A)$.

Corollary 2. *Let $G = \mathbb{R}$ or $\mathbb{Z}$. If $(\alpha_t)_{t \in [0,1]}$ is a continuous path of G-actions on A, then there is a canonical KK^G-equivalence $(A, \alpha_0) \sim (A, \alpha_1)$. As a consequence, the crossed products for both actions are KK-equivalent.*

Proof. Equip $\mathrm{Cyl}(A)$ with the automorphism α. Evaluation at 0 and 1 provides elements in $KK^{\mathbb{Z}}(\mathrm{Cyl}(A), (A, \alpha_t))$ that are non-equivariantly invertible because KK is homotopy invariant. Hence they are invertible in KK^G by Theorem 8. Their composition yields the desired KK^G-equivalence $(A, \alpha_0) \sim (A, \alpha_1)$. □

It is not hard to extend Theorem 8 and hence Corollary 2 to the groups $\mathbb{R}^n$ and $\mathbb{Z}^n$ for any $n \in \mathbb{Z}$. With a bit more work, we could also treat solvable Lie groups. But Theorem 8 as stated above fails for finite groups: there exists a space X and two homotopic actions α_0, α_1 of $\mathbb{Z}/2$ on X for which $K^*_{\mathbb{Z}/2}(X, \alpha_t)$ are different for $t = 0, 1$. Reversing the argument in the proof of Corollary 2, this provides the desired counterexample. Less complicated counterexamples can be constructed where A is a UHF C^*-algebra.

Such counterexamples force us to amend our question:

Suppose $\mathrm{Res}^H_G(A) \cong 0$ for all compact subgroups $H \subseteq G$. Does it follow that $A \cong 0$ in KK^G? Or at least that $K_*(G \ltimes_r A) \cong 0$?

It is shown in [26] that the second question has a positive answer if and only if the *Baum–Connes conjecture* holds for G *with arbitrary coefficients*. For many groups for which we know the Baum–Connes conjecture with coefficients, we also know that the first question has a positive answer. But the first question can only have a positive answer if the group is K-amenable, that is, if reduced and full crossed products have the same K-theory. The Lie group $Sp(n,1)$ and its cocompact subgroups are examples where we know the Baum–Connes conjecture with coefficients although the group is not K-amenable.

Definition 21. A G-C^*-algebra A is called *weakly contractible* if $\mathrm{Res}_G^H(A) \cong 0$ for all compact subgroups $H \subseteq G$. Let $\mathscr{CC}$ be the class of weakly contractible objects.

A morphism $f \in \mathrm{KK}^G(A_1, A_2)$ is called a *weak equivalence* if $\mathrm{Res}_G^H(f)$ is invertible for all compact subgroups $H \subseteq G$.

Recall that any $f \in \mathrm{KK}^G(A_1, A_2)$ is part of an exact triangle $A_1 \to A_2 \to C \to \Sigma A_1$ in KK^G. We have $C \in \mathscr{CC}$ if and only if f is a weak equivalence. Hence our two questions above are equivalent to:

Are all weak equivalences invertible in KK^G? Do they at least act invertibly on $K_*(G \ltimes_r \_)$?

The second question is equivalent to the Baum–Connes conjecture.

Suppose now that G is discrete. Then any subgroup is open, so that the adjointness isomorphism (3) always applies. It asserts that the subcategories $\mathscr{CI}$ and $\mathscr{CC}$ are orthogonal, that is, $\mathrm{KK}^G(A,B) = 0$ if $A \in\in \mathscr{CI}$, $B \in\in \mathscr{CC}$. Even more, if $\mathrm{KK}^G(A,B) = 0$ for all $A \in\in \mathscr{CI}$, then it follows that $B \in\in \mathscr{CC}$. A more involved argument in [26] extends these observations to all locally compact groups G.

Definition 22. Let $\langle \mathscr{CI} \rangle$ be the smallest full subcategory of KK^G that contains $\mathscr{CI}$ and is closed under suspensions, (countable) direct sums, and exact triangles.

We may think of objects of $\langle \mathscr{CI} \rangle$ as generalised CW-complexes that are built out of the cells in $\mathscr{CI}$.

Theorem 9. *The pair of subcategories* $(\langle \mathscr{CI} \rangle, \mathscr{CC})$ *is* complementary *in the following sense (see [26]):*

- $\mathrm{KK}^G(P,N) = 0$ *if* $P \in\in \langle \mathscr{CI} \rangle$, $N \in\in \mathscr{CC}$
- *For any* $A \in\in \mathrm{KK}^G$, *there is an exact triangle* $P \to A \to N \to \Sigma P$ *with* $P \in\in \langle \mathscr{CI} \rangle$, $N \in\in \mathscr{CC}$

Moreover, the exact triangle $P \to A \to N \to \Sigma P$ *above is unique up to a canonical isomorphism and depends functorially on A, and the ensuing functors $A \mapsto P(A)$, $A \mapsto N(A)$ are exact functors on $\mathfrak{KK}^G$.*

Proof. The orthogonality of $\langle \mathscr{CI} \rangle$ and $\mathscr{CC}$ follows easily from the orthogonality of $\mathscr{CI}$ and $\mathscr{CC}$. The existence of an exact triangle decomposition is more difficult. The proof in [26] reduces this to the special case $A = \mathbb{C}$. A more elementary construction of this exact triangle is explained in [12]. $\square$

Theorem 9 asserts that $\mathscr{CI}$ and $\mathscr{CC}$ together generate all of $\mathfrak{KK}^G$. This is why the vanishing of $K_*(G \ltimes_r A)$ for $A \in\in \mathscr{CC}$ is so useful: it allows us to replace an arbitrary object by one in $\langle \mathscr{CI} \rangle$. The latter is built out of objects in $\mathscr{CI}$. We have already agreed that the computation of $K_*(G \ltimes_r A)$ for $A \in\in \mathscr{CI}$ is Somebody Else's

Problem. Once we understand a mechanism for decomposing objects of $\langle \mathscr{C}\mathscr{I} \rangle$ into objects of $\mathscr{C}\mathscr{I}$, the computation of $\mathrm{K}_*(G \ltimes_{\mathrm{r}} A)$ for $A \in \langle \mathscr{C}\mathscr{I} \rangle$ becomes a purely topological affair and hence Somebody Else's Problem as well.

For the groups $\mathbb{Z}^n$ and $\mathbb{R}^n$, the subcategory $\mathscr{C}\mathscr{C}$ is trivial, so that Theorem 9 simply asserts that $\mathfrak{RR}^G = \langle \mathscr{C}\mathscr{I} \rangle$ is *generated* by the compactly induced actions. More generally, this is the case for all amenable groups; the proof of the Baum–Connes conjecture by Higson and Kasparov for such groups also yields this stronger assertion (see [26]).

Definition 23. Let $F \colon \mathfrak{RR}^G \to \mathfrak{C}$ be a functor. Its *localisation* at $\mathscr{C}\mathscr{C}$ (or at the weak equivalences) is the functor

$$\mathbb{L}F := F \circ P \colon \mathfrak{RR}^G \to \langle \mathscr{C}\mathscr{I} \rangle \subseteq \mathfrak{RR}^G \to \mathfrak{C},$$

where we use the functors $P \colon \mathfrak{RR}^G \to \langle \mathscr{C}\mathscr{I} \rangle$ and $N \colon \mathfrak{RR}^G \to \mathscr{C}\mathscr{C}$ that are part of a natural exact triangle $P(A) \to A \to N(A) \to \Sigma P(A)$.

The natural transformation $P(A) \to A$ furnishes a natural transformation $\mathbb{L}F(A) \to F(A)$. If F is homological or exact, then $F \circ N(A)$ is the *obstruction* to invertibility of this map.

The localisation $\mathbb{L}F$ can be characterised by a universal property. First of all, it vanishes on $\mathscr{C}\mathscr{C}$ because $P(A) \cong 0$ whenever $A \in \mathscr{C}\mathscr{C}$. If $\tilde{F}$ is another functor with this property, then any natural transformation $\tilde{F} \to F$ factors uniquely through $\mathbb{L}F \to F$. This universal property characterises $\mathbb{L}F$ uniquely up to natural isomorphism of functors.

Theorem 10. *The natural transformation $\mathbb{L}F(A) \to F(A)$ for $F(A) := \mathrm{K}_*(G \ltimes_{\mathrm{r}} A)$ is equivalent to the Baum–Connes assembly map. That is, there is a natural isomorphism $\mathbb{L}F(A) \cong \mathrm{K}_*^{\mathrm{top}}(G, A)$ compatible with the maps to $F(A)$.*

Proof. It is known (but not obvious) that $\mathrm{K}_*^{\mathrm{top}}(G, A)$ vanishes for $\mathscr{C}\mathscr{C}$ and that the Baum–Connes assembly map is an isomorphism for coefficients in $\mathscr{C}\mathscr{I}$. These two facts together imply the result.   □

The Baum–Connes *conjecture* asserts that the assembly map $\mathbb{L}F(A) \to F(A)$ is invertible for all A if $F(A) := \mathrm{K}_*(G \ltimes_{\mathrm{r}} A)$. This follows if $\mathscr{C}\mathscr{I} = \mathfrak{RR}^G$, of course. In particular, the Baum–Connes *conjecture* is trivial if G itself is compact.

3 Homological Algebra

It is well-known that many basic constructions from homotopy theory extend to categories of C*-algebras. As we argued in [26], the framework of *triangulated categories* is ideal for this purpose. The notion of triangulated category was introduced by Jean-Louis Verdier to formalise the properties of the derived category of an Abelian category. Stable homotopy theory provides further classical examples of triangulated categories. The triangulated category structure encodes basic information about manipulations with long exact sequences and (total) derived functors. The main point of [26] is that the domain of the Baum–Connes assembly map is the total left derived functor of the functor that maps a G-C^*-algebra A to $K_*(G \ltimes_r A)$.

Projective resolutions are among the most fundamental concepts in homological algebra; several others like derived functors are based on it. Projective resolutions seem to live in the underlying Abelian category and not in its derived category. This is why *total* derived functors make more sense in triangulated categories than the derived functors themselves. Nevertheless, we can define derived functors in triangulated categories and far more general categories. This goes back to Samuel Eilenberg and John C. Moore [13]. We learned about this theory in articles by Apostolos Beligiannis [3] and J. Daniel Christensen [8].

Homological algebra in non-Abelian categories is always *relative*, that is, we need additional structure to get started. This is useful because we may fit the additional data to our needs. In a triangulated category $\mathfrak{T}$, there are several kinds of additional data that yield equivalent theories; following [8], we use an *ideal* in $\mathfrak{T}$. We only consider ideals $\mathfrak{I} \subseteq \mathfrak{T}$ of the form

$$\mathfrak{I}(A,B) := \{x \in \mathfrak{T}(A,B) \mid F(x) = 0\}$$

for a stable homological functor $F \colon \mathfrak{T} \to \mathfrak{C}$ into a stable Abelian category $\mathfrak{C}$. Here *stable* means that $\mathfrak{C}$ carries a suspension automorphism and that F intertwines the suspension automorphisms on $\mathfrak{T}$ and $\mathfrak{C}$, and homological means that exact triangles yield exact sequences. Ideals of this form are called *homological ideals*.

A basic example is the ideal in the Kasparov category KK defined by

$$\mathfrak{I}_K(A,B) := \{f \in KK(A,B) \mid 0 = K_*(f) \colon K_*(A) \to K_*(B)\}. \tag{6}$$

For a locally compact group G and a (suitable) family of subgroups $\mathscr{F}$, we define the homological ideal

$$\mathscr{VC}_{\mathscr{F}}(A,B) := \{f \in KK^G(A,B) \mid \mathrm{Res}_G^H(f) = 0 \text{ in } KK^H(A,B) \text{ for all } H \in \mathscr{F}\}. \tag{7}$$

If $\mathscr{F}$ is the family of compact subgroups, then $\mathscr{VC}_{\mathscr{F}}$ is related to the Baum–Connes assembly map [26]. Of course, there are analogous ideals in more classical categories of (spectra of) G-CW-complexes.

All these examples can be analysed using the machinery we explain; but we only carry this out in some cases.

We use an ideal $\mathfrak{I}$ to carry over various notions from homological algebra to our triangulated category $\mathfrak{T}$. In order to see what they mean in examples, we characterise them using a stable homological functor $F \colon \mathfrak{T} \to \mathfrak{C}$ with $\ker F = \mathfrak{I}$. This is often easy. For instance, a chain complex with entries in $\mathfrak{T}$ is $\mathfrak{I}$-*exact* if and only if F maps it to an exact chain complex in the Abelian category $\mathfrak{C}$, and a morphism in $\mathfrak{T}$ is an $\mathfrak{I}$-*epimorphism* if and only if F maps it to an epimorphism. Here we may take any functor F with $\ker F = \mathfrak{I}$.

But the most crucial notions like projective objects and resolutions require a more careful choice of the functor F. Here we need the *universal $\mathfrak{I}$-exact functor*, which is a stable homological functor F with $\ker F = \mathfrak{I}$ such that any other such functor factors uniquely through F (up to natural equivalence). The universal $\mathfrak{I}$-exact functor and its applications are due to Apostolos Beligiannis [3].

If $F \colon \mathfrak{T} \to \mathfrak{C}$ is universal, then F detects $\mathfrak{I}$-projective objects, and it identifies $\mathfrak{I}$-derived functors with derived functors in the Abelian category $\mathfrak{C}$. Thus all our homological notions reduce to their counterparts in the Abelian category $\mathfrak{C}$.

In order to apply this, we need to know when a functor F with $\ker F = \mathfrak{I}$ is the universal one. We are going to develop a useful criterion for this purpose that uses partially defined adjoint functors.

Our criterion shows that the universal $\mathfrak{I}_K$-exact functor for the ideal $\mathfrak{I}_K \subseteq KK$ in (6) is the K-theory functor K_*, considered as a functor from KK to the category $\mathfrak{Ab}_c^{\mathbb{Z}/2}$ of countable $\mathbb{Z}/2$-graded Abelian groups. Hence the derived functors for $\mathfrak{I}_K$ only involve Ext and Tor for Abelian groups.

The derived functors that we have discussed above appear in a spectral sequence which – in favourable cases – computes morphism spaces in $\mathfrak{T}$ (like $KK^G(A,B)$) and other homological functors. This spectral sequence is a generalisation of the *Adams spectral sequence* in stable homotopy theory and is the main motivation for [8]. Much earlier, such spectral sequences were studied by Hans-Berndt Brinkmann in [6]. Here we concentrate on the much easier case where this spectral sequence degenerates to an exact sequence. This generalises the familiar Universal Coefficient Theorem for $KK_*(A,B)$.

3.1 Homological Ideals in Triangulated Categories

After fixing some basic notation, we introduce several interesting ideals in bivariant Kasparov categories; we are going to discuss these ideals throughout this article. Then we briefly recall what a triangulated category is and introduce homological ideals. Before we begin, we should point out that the choice of ideal is important because all our homological notions depend on it. It seems to be a matter of experimentation and experience to find the right ideal for a given purpose.

3.1.1 Generalities About Ideals in Additive Categories

All categories we consider will be *additive*, that is, they have a zero object and finite direct products and coproducts which agree, and the morphism spaces carry Abelian group structures such that the composition is additive in each variable [22].

Definition 24. Let $\mathfrak{C}$ be an additive category. We write $\mathfrak{C}(A,B)$ for the group of morphisms $A \to B$ in $\mathfrak{C}$, and $A \in\in \mathfrak{C}$ to denote that A is an object of the category $\mathfrak{C}$.

Definition 25. An *ideal* $\mathfrak{I}$ in $\mathfrak{C}$ is a family of subgroups $\mathfrak{I}(A,B) \subseteq \mathfrak{C}(A,B)$ for all $A, B \in\in \mathfrak{C}$ such that

$$\mathfrak{C}(C,D) \circ \mathfrak{I}(B,C) \circ \mathfrak{C}(A,B) \subseteq \mathfrak{I}(A,D) \qquad \text{for all } A, B, C, D \in\in \mathfrak{C}.$$

We write $\mathfrak{I}_1 \subseteq \mathfrak{I}_2$ if $\mathfrak{I}_1(A,B) \subseteq \mathfrak{I}_2(A,B)$ for all $A, B \in\in \mathfrak{C}$. Clearly, the ideals in $\mathfrak{C}$ form a complete lattice. The largest ideal $\mathfrak{C}$ consists of all morphisms in $\mathfrak{C}$; the smallest ideal 0 contains only zero morphisms.

Definition 26. Let $\mathfrak{C}$ and $\mathfrak{C}'$ be additive categories and let $F\colon \mathfrak{C} \to \mathfrak{C}'$ be an additive functor. Its *kernel* $\ker F$ is the ideal in $\mathfrak{C}$ defined by

$$\ker F(A,B) := \{f \in \mathfrak{C}(A,B) \mid F(f) = 0\}.$$

This should be distinguished from the *kernel on objects*, consisting of all objects with $F(A) \cong 0$, which is used much more frequently. This agrees with the class of $\ker F$-contractible objects that we introduce below.

Definition 27. Let $\mathfrak{I} \subseteq \mathfrak{C}$ be an ideal. Its *quotient category* $\mathfrak{C}/\mathfrak{I}$ has the same objects as $\mathfrak{C}$ and morphism groups $\mathfrak{C}(A,B)/\mathfrak{I}(A,B)$.

The quotient category is again additive, and the obvious functor $F\colon \mathfrak{C} \to \mathfrak{C}/\mathfrak{I}$ is additive and satisfies $\ker F = \mathfrak{I}$. Thus any ideal $\mathfrak{I} \subseteq \mathfrak{C}$ is of the form $\ker F$ for a canonical additive functor F.

The additivity of $\mathfrak{C}/\mathfrak{I}$ and F depends on the fact that any ideal $\mathfrak{I}$ is compatible with *finite* products in the following sense: the natural isomorphisms

$$\mathfrak{C}(A, B_1 \times B_2) \xrightarrow{\cong} \mathfrak{C}(A, B_1) \times \mathfrak{C}(A, B_2), \quad \mathfrak{C}(A_1 \times A_2, B) \xrightarrow{\cong} \mathfrak{C}(A_1, B) \times \mathfrak{C}(A_2, B)$$

restrict to isomorphisms

$$\mathfrak{I}(A, B_1 \times B_2) \xrightarrow{\cong} \mathfrak{I}(A, B_1) \times \mathfrak{I}(A, B_2), \quad \mathfrak{I}(A_1 \times A_2, B) \xrightarrow{\cong} \mathfrak{I}(A_1, B) \times \mathfrak{I}(A_2, B).$$

3.1.2 Examples of Ideals

Example 4. Let KK be the *Kasparov category*, whose objects are the separable C^*-algebras and whose morphism spaces are the Kasparov groups $KK_0(A,B)$, with the Kasparov product as composition. Let $\mathfrak{Ab}^{\mathbb{Z}/2}$ be the category of $\mathbb{Z}/2$-graded Abelian groups. Both categories are evidently additive.

K-theory is an additive functor $K_*\colon KK \to \mathfrak{Ab}^{\mathbb{Z}/2}$. We let $\mathfrak{I}_K := \ker K_*$ (as in (6)). Thus $\mathfrak{I}_K(A,B) \subseteq KK(A,B)$ is the kernel of the natural map

$$\gamma\colon KK(A,B) \to \mathrm{Hom}\big(K_*(A), K_*(B)\big) := \prod_{n \in \mathbb{Z}/2} \mathrm{Hom}\big(K_n(A), K_n(B)\big).$$

There is another interesting ideal in KK, namely, the kernel of a natural map

$$\kappa\colon \mathfrak{I}_K(A,B) \to \mathrm{Ext}\big(K_*(A), K_{*+1}(B)\big) := \prod_{n \in \mathbb{Z}/2} \mathrm{Ext}\big(K_n(A), K_{n+1}(B)\big)$$

due to Lawrence Brown (see [35]), whose definition we now recall. We represent $f \in KK(A,B) \cong \mathrm{Ext}\big(A, C_0(\mathbb{R}, B)\big)$ by a C^*-algebra extension $C_0(\mathbb{R}, B) \otimes \mathbb{K} \rightarrowtail E \twoheadrightarrow A$. This yields an exact sequence

$$
\begin{array}{ccc}
K_1(B) & \longrightarrow K_0(E) \longrightarrow & K_0(A) \\
\big\uparrow{\scriptstyle f_*} & & \big\downarrow{\scriptstyle f_*} \\
K_1(A) & \longleftarrow K_1(E) \longleftarrow & K_0(B).
\end{array}
\tag{8}
$$

The vertical maps in (8) are the two components of $\gamma(f)$. If $f \in \mathfrak{I}_K(A,B)$, then (8) splits into two extensions of Abelian groups, which yield an element $\kappa(f)$ in $\mathrm{Ext}\big(K_*(A), K_{*+1}(B)\big)$.

Example 5. Let G be a second countable, locally compact group. Let KK^G be the associated *equivariant Kasparov category*; its objects are the separable G-C^*-algebras and its morphism spaces are the groups $KK^G(A,B)$, with the Kasparov product as composition. If $H \subseteq G$ is a closed subgroup, then there is a *restriction functor* $\mathrm{Res}^H_G \colon KK^G \to KK^H$, which simply forgets part of the equivariance.

If $\mathscr{F}$ is a set of closed subgroups of G, we define an ideal $\mathscr{VC}_{\mathscr{F}} \subseteq KK^G$ by

$$
\mathscr{VC}_{\mathscr{F}}(A,B) := \{f \in KK^G(A,B) \mid \mathrm{Res}^H_G(f) = 0 \text{ for all } H \in \mathscr{F}\}
$$

as in (7). Of course, the condition $\mathrm{Res}^H_G(f) = 0$ is supposed to hold in $KK^H(A,B)$. We are mainly interested in the case where $\mathscr{F}$ is the family of all compact subgroups of G and simply denote the ideal by $\mathscr{VC}$ in this case.

This ideal arises if we try to compute G-equivariant homology theories in terms of H-equivariant homology theories for $H \in \mathscr{F}$. The ideal $\mathscr{VC}$ is closely related to the approach to the Baum–Connes assembly map in [26].

Many readers will prefer to work in categories of spectra of, say, G-CW-complexes. We do not introduce these categories here because the author happens to be more familiar with Kasparov theory than with spectra; but it shoud be clear enough that spectra of G-CW-complexes support similar restriction functors, which provide analogues of the ideals $\mathscr{VC}_{\mathscr{F}}$.

Finally, we consider a classical example from homological algebra.

Example 6. Let $\mathfrak{C}$ be an Abelian category. Let $\mathrm{Ho}(\mathfrak{C})$ be the homotopy category of unbounded chain complexes

$$
\cdots \to C_n \xrightarrow{\delta_n} C_{n-1} \xrightarrow{\delta_{n-1}} C_{n-2} \xrightarrow{\delta_{n-2}} C_{n-3} \to \cdots
$$

over $\mathfrak{C}$. The space of morphisms $A \to B$ in $\mathrm{Ho}(\mathfrak{C})$ is the space $[A,B]$ of *homotopy classes* of chain maps $A \to B$.

Taking homology defines functors $H_n \colon \mathrm{Ho}(\mathfrak{C}) \to \mathfrak{C}$ for $n \in \mathbb{Z}$, which we combine to a single functor $H_* \colon \mathrm{Ho}(\mathfrak{C}) \to \mathfrak{C}^{\mathbb{Z}}$. We let $\mathfrak{I}_H \subseteq \mathrm{Ho}(\mathfrak{C})$ be its kernel:

$$
\mathfrak{I}_H(A,B) := \{f \in [A,B] \mid H_*(f) = 0\}.
\tag{9}
$$

We also consider the category $\mathrm{Ho}(\mathfrak{C}; \mathbb{Z}/p)$ of *p-periodic* chain complexes over $\mathfrak{C}$ for $p \in \mathbb{N}_{\geq 1}$; its objects satisfy $C_n = C_{n+p}$ and $\delta_n = \delta_{n+p}$ for all $n \in \mathbb{Z}$, and chain maps and homotopies are required p-periodic as well. The category $\mathrm{Ho}(\mathfrak{C};\mathbb{Z}/2)$ plays a role in connection with cyclic cohomology, especially with local cyclic cohomology [24, 31]. The category $\mathrm{Ho}(\mathfrak{C};\mathbb{Z}/1)$ is isomorphic to the category of chain complexes without grading. By convention, we let $\mathbb{Z}/0 = \mathbb{Z}$, so that $\mathrm{Ho}(\mathfrak{C};\mathbb{Z}/0) = \mathrm{Ho}(\mathfrak{C})$.

The homology of a periodic chain complex is, of course, periodic, so that we get a homological functor $\mathrm{H}_* \colon \mathrm{Ho}(\mathfrak{C};\mathbb{Z}/p) \to \mathfrak{C}^{\mathbb{Z}/p}$; here $\mathfrak{C}^{\mathbb{Z}/p}$ denotes the category of $\mathbb{Z}/p$-graded objects of $\mathfrak{C}$. We let $\mathfrak{I}_\mathrm{H} \subseteq \mathrm{Ho}(\mathfrak{C};\mathbb{Z}/p)$ be the kernel of H_* as in (9).

Definition 28. A *stable additive category* is an additive category equipped with an (additive) automorphism $\Sigma \colon \mathfrak{C} \to \mathfrak{C}$, called *suspension*.

A *stable homological functor* is a homological functor $F \colon \mathfrak{T} \to \mathfrak{C}$ into a stable Abelian category $\mathfrak{C}$ together with natural isomorphisms $F\bigl(\Sigma_{\mathfrak{T}}(A)\bigr) \cong \Sigma_{\mathfrak{C}}\bigl(F(A)\bigr)$ for all $A \in \mathfrak{T}$.

Example 7. The category $\mathfrak{C}^{\mathbb{Z}/p}$ of $\mathbb{Z}/p$-graded objects of an Abelian category $\mathfrak{C}$ is stable for any $p \in \mathbb{N}$; the suspension automorphism merely shifts the grading. The functors $\mathrm{K}_* \colon \mathrm{KK} \to \mathfrak{Ab}^{\mathbb{Z}/2}$ and $\mathrm{H}_* \colon \mathrm{Ho}(\mathfrak{C};\mathbb{Z}/p) \to \mathfrak{C}^{\mathbb{Z}/p}$ introduced in Examples 4 and 6 are stable homological functors.

If $F \colon \mathfrak{T} \to \mathfrak{C}$ is any homological functor, then

$$F_* \colon \mathfrak{T} \to \mathfrak{C}^{\mathbb{Z}}, \qquad A \mapsto \bigl(F_n(A)\bigr)_{n\in\mathbb{Z}}$$

is a stable homological functor. Many of our examples satisfy *Bott periodicity*, that is, there is a natural isomorphism $F_2(A) \cong F(A)$. Then we get a stable homological functor $F_* \colon \mathfrak{T} \to \mathfrak{C}^{\mathbb{Z}/2}$. A typical example for this is the functor K_*.

Examples 7 and 3 show that the functors that define the ideals $\ker \gamma$ in Example 4, $\mathscr{VC}_{\mathscr{F}}$ in Example 5, and $\mathfrak{I}_\mathrm{H}$ in Example 6 are all stable and either homological or exact.

3.1.3 The Universal Homological Functor

The following general construction of Peter Freyd [14] plays an important role in [3]. For an additive category $\mathfrak{C}$, let $\mathfrak{Fun}(\mathfrak{C}^{\mathrm{op}}, \mathfrak{Ab})$ be the category of contravariant additive functors $\mathfrak{C}^{\mathrm{op}} \to \mathfrak{Ab}$, with natural transformations as morphisms. Unless $\mathfrak{C}$ is essentially small, this is not quite a category because the morphisms may form classes instead of sets. We may ignore this set-theoretic problem because the bivariant Kasparov categories that we are interested in are essentially small, and the subcategory $\mathrm{Coh}(\mathfrak{C})$ of $\mathfrak{Fun}(\mathfrak{C}^{\mathrm{op}}, \mathfrak{Ab})$ defined below is an honest category for any $\mathfrak{C}$.

The category $\mathfrak{Fun}(\mathfrak{C}^{\mathrm{op}}, \mathfrak{Ab})$ is Abelian: if $f \colon F_1 \to F_2$ is a natural transformation, then its kernel, cokernel, image, and co-image are computed pointwise on the objects of $\mathfrak{C}$, so that they boil down to the corresponding constructions with Abelian groups.

The *Yoneda embedding* is an additive functor

$$\mathbb{Y} \colon \mathfrak{C} \to \mathfrak{Fun}(\mathfrak{C}^{\mathrm{op}}, \mathfrak{Ab}), \qquad B \mapsto \mathfrak{C}(\_, B).$$

This functor is fully faithful, and there are natural isomorphisms

$$\mathrm{Hom}(\mathbb{Y}(B), F) \cong F(B) \qquad \text{for all } F \in\in \mathfrak{Fun}(\mathfrak{C}^{\mathrm{op}}, \mathfrak{Ab}), B \in\in \mathfrak{C}$$

by the *Yoneda Lemma*. A functor $F \in\in \mathfrak{Fun}(\mathfrak{C}^{\mathrm{op}}, \mathfrak{Ab})$ is called *representable* if it is isomorphic to $\mathbb{Y}(B)$ for some $B \in\in \mathfrak{C}$. Hence $\mathbb{Y}$ yields an equivalence of categories between $\mathfrak{C}$ and the subcategory of representable functors in $\mathfrak{Fun}(\mathfrak{C}^{\mathrm{op}}, \mathfrak{Ab})$.

A functor $F \in\in \mathfrak{Fun}(\mathfrak{C}^{\mathrm{op}}, \mathfrak{Ab})$ is called *finitely presented* if there is an exact sequence $\mathbb{Y}(B_1) \to \mathbb{Y}(B_2) \to F \to 0$ with $B_1, B_2 \in\in \mathfrak{C}$. Since $\mathbb{Y}$ is fully faithful, this means that F is the cokernel of $\mathbb{Y}(f)$ for a morphism f in $\mathfrak{C}$.

Definition 29. We let $\mathrm{Coh}(\mathfrak{C})$ be the full subcategory of finitely presented functors in $\mathfrak{Fun}(\mathfrak{C}^{\mathrm{op}}, \mathfrak{Ab})$.

Since representable functors belong to $\mathfrak{Coh}(\mathfrak{C})$, we still have a Yoneda embedding $\mathbb{Y} \colon \mathfrak{C} \to \mathfrak{Coh}(\mathfrak{C})$. Although the category $\mathfrak{Coh}(\mathfrak{C})$ tends to be very big and therefore unwieldy, it plays an important theoretical role.

Theorem 11 (Freyd's Theorem). *Let $\mathfrak{T}$ be a triangulated category.*

Then $\mathfrak{Coh}(\mathfrak{T})$ is a stable Abelian category that has enough projective and enough injective objects, and the projective and injective objects coincide.

The functor $\mathbb{Y} \colon \mathfrak{T} \to \mathfrak{Coh}(\mathfrak{T})$ is fully faithful, stable, and homological. Its essential range $\mathbb{Y}(\mathfrak{T})$ consists of projective-injective objects. Conversely, an object of $\mathfrak{Coh}(\mathfrak{T})$ is projective-injective if and only if it is a retract of an object of $\mathbb{Y}(\mathfrak{T})$.

The functor $\mathbb{Y}$ is the universal (stable) homological functor in the following sense: any (stable) homological functor $F \colon \mathfrak{T} \to \mathfrak{C}'$ to a (stable) Abelian category $\mathfrak{C}'$ factors uniquely as $F = \bar{F} \circ \mathbb{Y}$ for a (stable) exact functor $F \colon \mathfrak{Coh}(\mathfrak{T}) \to \mathfrak{C}'$.

If idempotents in $\mathfrak{T}$ split – as in all our examples – then $\mathbb{Y}(\mathfrak{T})$ is closed under retracts, so that $\mathbb{Y}(\mathfrak{T})$ is equal to the class of projective-injective objects in $\mathfrak{Coh}(\mathfrak{T})$.

3.1.4 Homological Ideals in Triangulated Categories

Let $\mathfrak{T}$ be a triangulated category, let $\mathfrak{C}$ be a stable additive category, and let $F \colon \mathfrak{T} \to \mathfrak{C}$ be a stable homological functor. Then $\ker F$ is a stable ideal in the following sense:

Definition 30. An ideal $\mathfrak{I} \subseteq \mathfrak{T}$ is called *stable* if the suspension isomorphisms $\Sigma \colon \mathfrak{T}(A, B) \xrightarrow{\cong} \mathfrak{T}(\Sigma A, \Sigma B)$ for $A, B \in\in \mathfrak{T}$ restrict to isomorphisms

$$\Sigma \colon \mathfrak{I}(A, B) \xrightarrow{\cong} \mathfrak{I}(\Sigma A, \Sigma B).$$

If $\mathfrak{I}$ is stable, then there is a unique suspension automorphism on $\mathfrak{T}/\mathfrak{I}$ for which the canonical functor $\mathfrak{T} \to \mathfrak{T}/\mathfrak{I}$ is stable. Thus the stable ideals are exactly the kernels of stable additive functors.

Definition 31. An ideal $\mathfrak{I} \subseteq \mathfrak{T}$ in a triangulated category is called *homological* if it is the kernel of a stable homological functor.

Remark 1. Freyd's Theorem shows that $\mathbb{Y}$ induces a bijection between (stable) exact functors $\mathfrak{Coh}(\mathfrak{T}) \to \mathfrak{C}'$ and (stable) homological functors $\mathfrak{T} \to \mathfrak{C}'$ because $\bar{F} \circ \mathbb{Y}$ is homological if $\bar{F} \colon \mathfrak{Coh}(\mathfrak{T}) \to \mathfrak{C}'$ is exact. Hence the notion of homological functor is independent of the triangulated category structure on $\mathfrak{T}$ because the Yoneda embedding $\mathbb{Y} \colon \mathfrak{T} \to \mathfrak{Coh}(\mathfrak{T})$ does not involve any additional structure. Hence the notion of homological ideal only uses the suspension automorphism, not the class of exact triangles.

All the ideals considered in Sect. 3.1.2 except for $\ker \kappa$ in Example 4 are kernels of stable homological functors or exact functors. Those of the first kind are homological by definition. If $F \colon \mathfrak{T} \to \mathfrak{T}'$ is an exact functor between two triangulated categories, then $\mathbb{Y} \circ F \colon \mathfrak{T} \to \mathfrak{Coh}(\mathfrak{T}')$ is a stable homological functor with $\ker \mathbb{Y} \circ F = \ker F$ by Freyd's Theorem 11. Hence kernels of exact functors are homological as well.

Is any homological ideal the kernel of an exact functor? This is *not* the case:

Proposition 6. *Let $\mathfrak{Der}(\mathfrak{Ab})$ be the derived category of the category $\mathfrak{Ab}$ of Abelian groups. Define the ideal $\mathfrak{I}_H \subseteq \mathfrak{Der}(\mathfrak{Ab})$ as in Example 6. This ideal is not the kernel of an exact functor.*

We postpone the proof to the end of Sect. 3.2.1 because it uses the machinery of Sect. 3.2.1.

It takes some effort to characterise homological ideals because $\mathfrak{T}/\mathfrak{I}$ is almost never Abelian. The results in [3, Sect. 2–3] show that an ideal is homological if and only if it is *saturated* in the notation of [3]. We do not discuss this notion here because most ideals that we consider are obviously homological. The only example where we could profit from an abstract characterisation is the ideal $\ker \kappa$ in Example 4.

There is no obvious homological functor whose kernel is $\ker \kappa$ because κ is not a functor on KK. Nevertheless, $\ker \kappa$ is the kernel of an exact functor; the relevant functor is the functor KK $\to$ UCT, where UCT is the variant of KK that satisfies the Universal Coefficient Theorem in complete generality. This functor can be constructed as a localisation of KK (see [26]). The Universal Coefficient Theorem implies that its kernel is exactly $\ker \kappa$.

3.2 From Homological Ideals to Derived Functors

Various notions from homological algebra still make sense in the context of homological ideals in triangulated categories. Our discussion mostly follows [1, 3, 8, 13]. Once we have a stable homological functor $F \colon \mathfrak{T} \to \mathfrak{C}$, it is not surprising that we can do a certain amount of homological algebra in $\mathfrak{T}$. For instance, we may call a chain complex of objects of $\mathfrak{T}$ *F-exact* if F maps it to an exact chain complex in $\mathfrak{C}$; and we may call an object *F-projective* if F maps it to a projective object in $\mathfrak{C}$. But are these definitions reasonable?

We propose that a reasonable homological notion should depend only on the ideal $\ker F$. We will see that the notion of F-exact chain complex is reasonable and only depends on $\ker F$. In contrast, the notion of projectivity above depends on F and is only reasonable in special cases. There is another, more useful, notion of projective object that depends only on the ideal ker F.

All our definitions involve only the ideal, not a stable homological functor that defines it. We reformulate them in terms of an exact or a stable homological functor defining the ideal in order to understand what they mean in concrete cases. Following [13], we construct projective objects using adjoint functors.

The most sophisticated concept in this subsection is the *universal $\mathfrak{I}$-exact functor*, which gives us complete control over projective resolutions and derived functors. We can usually describe such functors very concretely.

3.2.1 Basic Notions

We introduce some useful terminology related to an ideal:

Definition 32. Let $\mathfrak{I}$ be a homological ideal in a triangulated category $\mathfrak{T}$.

- Let $f: A \to B$ be a morphism in $\mathfrak{T}$; embed it in an exact triangle $A \xrightarrow{f} B \xrightarrow{g} C \xrightarrow{h} \Sigma A$. We call f
  - $\mathfrak{I}$-*monic* if $h \in \mathfrak{I}$
  - $\mathfrak{I}$-*epic* if $g \in \mathfrak{I}$
  - An $\mathfrak{I}$-*equivalence* if it is both $\mathfrak{I}$-monic and $\mathfrak{I}$-epic, that is, $g, h \in \mathfrak{I}$
  - An $\mathfrak{I}$-*phantom map* if $f \in \mathfrak{I}$
- An object $A \in \in \mathfrak{T}$ is called $\mathfrak{I}$-*contractible* if $\mathrm{id}_A \in \mathfrak{I}(A,A)$
- An exact triangle $A \xrightarrow{f} B \xrightarrow{g} C \xrightarrow{h} \Sigma A$ in $\mathfrak{T}$ is called $\mathfrak{I}$-*exact* if $h \in \mathfrak{I}$

The notions of monomorphism (or monic morphism) and epimorphism (or epic morphism) – which can be found in any book on category theory such as [22] – are categorical ways to express injectivity or surjectivity of maps of sets. A morphism in an Abelian category that is both monic and epic is invertible.

The classes of $\mathfrak{I}$-phantom maps, $\mathfrak{I}$-monomorphisms, $\mathfrak{I}$-epimorphisms, and of $\mathfrak{I}$-exact triangles determine each other uniquely because we can embed any morphism in an exact triangle in any position. It is a matter of taste which of these is considered most fundamental. Following Daniel Christensen [8], we favour the phantom maps. Other authors prefer exact triangles instead [1, 3, 13]. Of course, the notion of an $\mathfrak{I}$-phantom map is redundant; it becomes more relevant if we consider, say, the class of $\mathfrak{I}$-exact triangles as our basic notion.

Notice that f is $\mathfrak{I}$-epic or $\mathfrak{I}$-monic if and only if $-f$ is. If f is $\mathfrak{I}$-epic or $\mathfrak{I}$-monic, then so are $\Sigma^n(f)$ for all $n \in \mathbb{Z}$ because $\mathfrak{I}$ is stable. Similarly, (signed) suspensions of $\mathfrak{I}$-exact triangles remain $\mathfrak{I}$-exact triangles.

Lemma 1. *Let $F: \mathfrak{T} \to \mathfrak{C}$ be a stable homological functor into a stable Abelian category $\mathfrak{C}$.*

- *A morphism f in $\mathfrak{T}$ is*
  - *A $\ker F$-phantom map if and only if $F(f) = 0$*
  - *$\ker F$-monic if and only if $F(f)$ is monic*

  – ker F-*epic if and only if* $F(f)$ *is epic*
    – *A* ker F-*equivalence if and only if* $F(f)$ *is invertible*
  • *An object* $A \in\in \mathfrak{T}$ *is* ker F-*contractible if and only if* $F(A) = 0$
  • *An exact triangle* $A \to B \to C \to \Sigma A$ *is* ker F-*exact if and only if*

$$0 \to F(A) \to F(B) \to F(C) \to 0$$

is a short exact sequence in $\mathfrak{C}$

Proof. Sequences in $\mathfrak{C}$ of the form $X \xrightarrow{0} Y \xrightarrow{f} Z$ or $X \xrightarrow{f} Y \xrightarrow{0} Z$ are exact at Y if and only if f is monic or epic, respectively. Moreover, a sequence of the form $X \xrightarrow{0} Y \to Z \to U \xrightarrow{0} W$ is exact if and only if $0 \to Y \to Z \to U \to 0$ is exact.

Combined with the long exact homology sequences for F and suitable exact triangles, these observations yield the assertions about monomorphisms, epimorphisms, and exact triangles. The description of equivalences and contractible objects follows.   □

Now we specialise these notions to the ideal $\mathfrak{J}_K \subseteq$ KK of Example 4, replacing $\mathfrak{J}_K$ by K in our notation to avoid clutter.

  • Let $f \in$ KK(A,B) and let $K_*(f) \colon K_*(A) \to K_*(B)$ be the induced map. Then f is
      – A K-*phantom map if and only if* $K_*(f) = 0$
      – K-*epic if and only if* $K_*(f)$ *is surjective*
      – K-*monic if and only if* $K_*(f)$ *is injective*
      – A K-*equivalence if and only if* $K_*(f)$ *is invertible*
  • A C*-algebra $A \in\in$ KK is K-*contractible if and only if* $K_*(A) = 0$
  • An exact triangle $A \to B \to C \to \Sigma A$ in KK is K-*exact if and only if*

$$0 \to K_*(A) \to K_*(B) \to K_*(C) \to 0$$

is a short exact sequence (of $\mathbb{Z}/2$-graded Abelian groups)

Similar things happen for the other ideals in Sect. 3.1.2 that are *naturally* defined as kernels of stable homological functors.

Remark 2. It is crucial for the above theory that we consider functors that are both *stable* and *homological*. Everything fails if we drop either assumption and consider functors such as $K_0(A)$ or $\mathrm{Hom}\bigl(\mathbb{Z}/4, K_*(A)\bigr)$.

Lemma 2. *An object* $A \in\in \mathfrak{T}$ *is* $\mathfrak{J}$-*contractible if and only if* $0 \colon 0 \to A$ *is an* $\mathfrak{J}$-*equivalence. A morphism* f *in* $\mathfrak{T}$ *is an* $\mathfrak{J}$-*equivalence if and only if its cone is* $\mathfrak{J}$-*contractible.*

Thus the classes of $\mathfrak{J}$-equivalences and of $\mathfrak{J}$-contractible objects determine each other. But they do not allow us to recover the ideal itself. For instance, the ideals $\mathfrak{J}_K$ and ker κ in Example 4 have the same contractible objects and equivalences.

Proof. Recall that the cone of f is the object C that fits in an exact triangle $A \xrightarrow{f} B \to C \to \Sigma A$. The long exact sequence for this triangle yields that $F(f)$ is invertible if and only if $F(C) = 0$, where F is some stable homological functor F with $\ker F = \mathfrak{I}$. Now the second assertion follows from Lemma 1. Since the cone of $0 \to A$ is A, the first assertion is a special case of the second one. $\square$

Many ideals are defined as $\ker F$ for an exact functor $F \colon \mathfrak{T} \to \mathfrak{T}'$ between triangulated categories. We can also use such a functor to describe the above notions:

Lemma 3. *Let $\mathfrak{T}$ and $\mathfrak{T}'$ be triangulated categories and let $F \colon \mathfrak{T} \to \mathfrak{T}'$ be an exact functor.*

- *A morphism $f \in \mathfrak{T}(A, B)$ is*
    - *A $\ker F$-phantom map if and only if $F(f) = 0$*
    - *$\ker F$-monic if and only if $F(f)$ is (split) monic*
    - *$\ker F$-epic if and only if $F(f)$ is (split) epic*
    - *A $\ker F$-equivalence if and only if $F(f)$ is invertible*
- *An object $A \in\in \mathfrak{T}$ is $\ker F$-contractible if and only if $F(A) = 0$*
- *An exact triangle $A \to B \to C \to \Sigma A$ is $\ker F$-exact if and only if the exact triangle $F(A) \to F(B) \to F(C) \to F(\Sigma A)$ in $\mathfrak{T}'$ splits*

We will explain the notation during the proof.

Proof. A morphism $f \colon X \to Y$ in $\mathfrak{T}'$ is called *split epic* (*split monic*) if there is $g \colon Y \to X$ with $f \circ g = \mathrm{id}_Y$ ($g \circ f = \mathrm{id}_X$). An exact triangle $X \xrightarrow{f} Y \xrightarrow{g} Z \xrightarrow{h} \Sigma X$ is said to *split* if $h = 0$. This immediately yields the characterisation of $\ker F$-exact triangles. Any split triangle is isomorphic to a direct sum triangle, so that f is split monic and g is split epic [29, Corollary 1.2.7]. Conversely, either of these conditions implies that the triangle splits.

Since the $\ker F$-exact triangles determine the $\ker F$-epimorphisms and $\ker F$-monomorphisms, the latter are detected by $F(f)$ being split epic or split monic, respectively. It is clear that split epimorphisms and split monomorphisms are epimorphisms and monomorphisms, respectively. The converse holds in a triangulated category because if we embed a monomorphism or epimorphism in an exact triangle, then one of the maps is forced to vanish, so that the exact triangle splits.

Finally, a morphism is invertible if and only if it is both split monic and split epic, and the zero map $F(A) \to F(A)$ is invertible if and only if $F(A) = 0$. $\square$

We may also prove Lemma 1 using the Yoneda embedding $\mathbb{Y} \colon \mathfrak{T}' \to \mathfrak{Coh}(\mathfrak{T}')$. The assertions about phantom maps, equivalences, and contractibility boil down to the observation that $\mathbb{Y}$ is fully faithful. The assertions about monomorphisms and epimorphisms follow because a map $f \colon A \to B$ in $\mathfrak{T}'$ becomes epic (monic) in $\mathfrak{Coh}(\mathfrak{T}')$ if and only if it is split epic (monic) in $\mathfrak{T}'$.

There is a similar description for $\bigcap \ker F_i$ for a set $\{F_i\}$ of exact functors. This applies to the ideal $\mathcal{VC}_{\mathscr{F}}$ for a family of subgroups $\mathscr{F}$ in a locally compact group G (Example 5). Replacing $\mathcal{VC}_{\mathscr{F}}$ by $\mathscr{F}$ in our notation to avoid clutter, we get:

- A morphism $f \in KK^G(A, B)$ is
  - An $\mathscr{F}$-phantom map if and only if $\text{Res}_G^H(f) = 0$ in KK^H for all $H \in \mathscr{F}$
  - $\mathscr{F}$-epic if and only if $\text{Res}_G^H(f)$ is (split) epic in KK^H for all $H \in \mathscr{F}$
  - $\mathscr{F}$-monic if and only if $\text{Res}_G^H(f)$ is (split) monic in KK^H for all $H \in \mathscr{F}$
  - An $\mathscr{F}$-equivalence if and only if $\text{Res}_G^H(f)$ is a KK^H-equivalence for all $H \in \mathscr{F}$
- A G-C^*-algebra $A \in\in KK^G$ is $\mathscr{F}$-contractible if and only if $\text{Res}_G^H(A) \cong 0$ in KK^H for all $H \in \mathscr{F}$
- An exact triangle $A \to B \to C \to \Sigma A$ in KK^G is $\mathscr{F}$-exact if and only if

$$\text{Res}_G^H(A) \to \text{Res}_G^H(B) \to \text{Res}_G^H(C) \to \Sigma \text{Res}_G^H(A)$$

is a split exact triangle in KK^H for all $H \in \mathscr{F}$

Lemma 3 allows us to prove that the ideal $\mathfrak{I}_H$ in $\mathfrak{Der}(\mathfrak{Ab})$ cannot be the kernel of an exact functor:

Proof (of Proposition 6). We embed $\mathfrak{Ab} \to \mathfrak{Der}(\mathfrak{Ab})$ as chain complexes concentrated in degree 0. The generator $\tau \in \text{Ext}(\mathbb{Z}/2, \mathbb{Z}/2)$ corresponds to the extension of Abelian groups $\mathbb{Z}/2 \rightarrowtail \mathbb{Z}/4 \twoheadrightarrow \mathbb{Z}/2$, where the first map is multiplication by 2 and the second map is the natural projection. We get an exact triangle

$$\mathbb{Z}/2 \to \mathbb{Z}/4 \to \mathbb{Z}/2 \xrightarrow{\tau} \mathbb{Z}/2[1]$$

in $\mathfrak{Der}(\mathfrak{Ab})$. This triangle is $\mathfrak{I}_H$-exact because the map $\mathbb{Z}/2 \to \mathbb{Z}/4$ is injective as a group homomorphism and hence $\mathfrak{I}_H$-monic in $\mathfrak{Der}(\mathfrak{Ab})$.

Assume there were an exact functor $F \colon \mathfrak{Der}(\mathfrak{Ab}) \to \mathfrak{T}'$ with $\ker F = \mathfrak{I}_H$. Then $F(\tau) = 0$, so that F maps our triangle to a split triangle and $F(\mathbb{Z}/4) \cong F(\mathbb{Z}/2) \oplus F(\mathbb{Z}/2)$ by Lemma 3. It follows that $F(2 \cdot \text{id}_{\mathbb{Z}/4}) = 2 \cdot \text{id}_{F(\mathbb{Z}/4)} = 0$ because $2 \cdot \text{id}_{F(\mathbb{Z}/2)} = F(2 \cdot \text{id}_{\mathbb{Z}/2}) = 0$. Hence $2 \cdot \text{id}_{\mathbb{Z}/4} \in \ker F = \mathfrak{I}_H$, which is false. This contradiction shows that there is no exact functor F with $\ker F = \mathfrak{I}_H$.   □

One of the most interesting questions about an ideal is whether all $\mathfrak{I}$-contractible objects vanish or, equivalently, whether all $\mathfrak{I}$-equivalences are invertible. These two questions are equivalent by Lemma 2. The answer is negative for the ideal $\mathfrak{I}_K \subseteq KK$ because the Universal Coefficient Theorem does not hold for arbitrary separable C^*-algebras. If G is an amenable group, then $\mathscr{VC}$-equivalences in KK^G are invertible; this follows from the proof of the Baum–Connes conjecture for these groups by Nigel Higson and Gennadi Kasparov (see [26]). These examples show that this question is subtle and may involve difficult analysis.

3.2.2 Exact Chain Complexes

We are going to extend to chain complexes the notion of $\mathfrak{I}$-exactness, which we have only defined for exact triangles so far. Our definition differs from Beligiannis' one [1, 3], which we recall first.

Let $\mathfrak{T}$ be a triangulated category and let $\mathfrak{I}$ be a homological ideal in $\mathfrak{T}$.

Definition 33. A chain complex

$$C_\bullet := (\cdots \to C_{n+1} \xrightarrow{d_{n+1}} C_n \xrightarrow{d_n} C_{n-1} \xrightarrow{d_{n-1}} C_{n-2} \to \cdots)$$

in $\mathfrak{T}$ is called $\mathfrak{J}$-*decomposable* if there is a sequence of $\mathfrak{J}$-exact triangles

$$K_{n+1} \xrightarrow{g_n} C_n \xrightarrow{f_n} K_n \xrightarrow{h_n} \Sigma K_{n+1}$$

with $d_n = g_{n-1} \circ f_n \colon C_n \to C_{n-1}$.

Such complexes are called $\mathfrak{J}$-exact in [1, 3]. This definition is inspired by the following well-known fact: a chain complex over an Abelian category is exact if and only if it splits into short exact sequences of the form $K_n \rightarrowtail C_n \twoheadrightarrow K_{n-1}$ as in Definition 33.

We prefer another definition of exactness because we have not found a general explicit criterion for a chain complex to be $\mathfrak{J}$-decomposable.

Definition 34. Let $C_\bullet = (C_n, d_n)$ be a chain complex over $\mathfrak{T}$. For each $n \in \mathbb{N}$, embed d_n in an exact triangle

$$C_n \xrightarrow{d_n} C_{n-1} \xrightarrow{f_n} X_n \xrightarrow{g_n} \Sigma C_n. \tag{10}$$

We call $C_\bullet$ $\mathfrak{J}$-*exact in degree* n if the map $X_n \xrightarrow{g_n} \Sigma C_n \xrightarrow{\Sigma f_{n+1}} \Sigma X_{n+1}$ belongs to $\mathfrak{J}(X_n, \Sigma X_{n+1})$. This does not depend on auxiliary choices because the exact triangles in (10) are unique up to (non-canonical) isomorphism.

We call $C_\bullet$ $\mathfrak{J}$-*exact* if it is $\mathfrak{J}$-exact in degree n for all $n \in \mathbb{Z}$.

This definition is designed to make the following lemma true:

Lemma 4. *Let* $F \colon \mathfrak{T} \to \mathfrak{C}$ *be a stable homological functor into a stable Abelian category* $\mathfrak{C}$ *with* $\ker F = \mathfrak{J}$. *A chain complex* $C_\bullet$ *over* $\mathfrak{T}$ *is* $\mathfrak{J}$-*exact in degree n if and only if*

$$F(C_{n+1}) \xrightarrow{F(d_{n+1})} F(C_n) \xrightarrow{F(d_n)} F(C_{n-1})$$

is exact at $F(C_n)$.

Proof. The complex $C_\bullet$ is $\mathfrak{J}$-exact in degree n if and only if the map

$$\Sigma^{-1} F(X_n) \xrightarrow{\Sigma^{-1} F(g_n)} F(C_n) \xrightarrow{F(f_{n+1})} F(X_{n+1})$$

vanishes. Equivalently, the range of $\Sigma^{-1} F(g_n)$ is contained in the kernel of $F(f_{n+1})$. The long exact sequences

$$\cdots \to \Sigma^{-1} F(X_n) \xrightarrow{\Sigma^{-1} F(g_n)} F(C_n) \xrightarrow{F(d_n)} F(C_{n-1}) \to \cdots,$$

$$\cdots \to F(C_{n+1}) \xrightarrow{F(d_{n+1})} F(C_n) \xrightarrow{F(f_{n+1})} F(X_{n+1}) \to \cdots$$

show that the range of $\Sigma^{-1} F(g_n)$ and the kernel of $F(f_{n+1})$ are equal to the kernel of $F(d_n)$ and the range of $F(d_{n+1})$, respectively. Hence $C_\bullet$ is $\mathfrak{J}$-exact in degree n if and only if $\ker F(d_n) \subseteq \operatorname{range} F(d_{n+1})$. Since $d_n \circ d_{n+1} = 0$, this is equivalent to $\ker F(d_n) = \operatorname{range} F(d_{n+1})$. $\qquad\square$

Corollary 3. $\mathfrak{J}$-*decomposable chain complexes are* $\mathfrak{J}$-*exact.*

Proof. Let $F\colon \mathfrak{T} \to \mathfrak{C}$ be a stable homological functor with $\ker F = \mathfrak{J}$. If $C_\bullet$ is $\mathfrak{J}$-decomposable, then $F(C_\bullet)$ is obtained by splicing short exact sequences in $\mathfrak{C}$. This implies that $F(C_\bullet)$ is exact, so that $C_\bullet$ is $\mathfrak{J}$-exact by Lemma 4. $\qquad\square$

Example 8. For the ideal $\mathfrak{J}_K \subseteq KK$, Lemma 4 yields that a chain complex $C_\bullet$ over KK is K-*exact* (in degree n) if and only if the chain complex

$$\cdots \to K_*(C_{n+1}) \to K_*(C_n) \to K_*(C_{n-1}) \to \cdots$$

of $\mathbb{Z}/2$-graded Abelian groups is exact (in degree n). Similar remarks apply to the other ideals in Sect. 3.1.2 that are defined as kernels of stable homological functors.

As a trivial example, we consider the largest possible ideal $\mathfrak{J} = \mathfrak{T}$. This ideal is defined by the zero functor. Lemma 4 or the definition yield that *all* chain complexes are $\mathfrak{T}$-exact. In contrast, it seems hard to characterise the $\mathfrak{J}$-decomposable chain complexes, already for $\mathfrak{J} = \mathfrak{T}$.

Lemma 5. *A chain complex of length* 3

$$\cdots \to 0 \to A \xrightarrow{f} B \xrightarrow{g} C \to 0 \to \cdots$$

is $\mathfrak{J}$-*exact if and only if there are an* $\mathfrak{J}$-*exact, exact triangle* $A' \xrightarrow{f'} B' \xrightarrow{g'} C' \to \Sigma A'$ *and a commuting diagram*

$$
\begin{array}{ccccc}
A' & \xrightarrow{\;f'\;} & B' & \xrightarrow{\;g'\;} & C' \\
{\scriptstyle\sim}\big\downarrow{\scriptstyle\alpha} & & {\scriptstyle\sim}\big\downarrow{\scriptstyle\beta} & & {\scriptstyle\sim}\big\downarrow{\scriptstyle\gamma} \\
A & \xrightarrow[\;f\;]{} & B & \xrightarrow[\;g\;]{} & C
\end{array}
\tag{11}
$$

where the vertical maps α, β, γ *are* $\mathfrak{J}$-*equivalences. Furthermore, we can achieve that* α *and* β *are identity maps.*

Proof. Let F be a stable homological functor with $\mathfrak{J} = \ker F$.

Suppose first that we are in the situation of (11). Lemma 1 yields that $F(\alpha)$, $F(\beta)$, and $F(\gamma)$ are invertible and that $0 \to F(A') \to F(B') \to F(C') \to 0$ is a short exact sequence. Hence so is $0 \to F(A) \to F(B) \to F(C) \to 0$. Now Lemma 4 yields that our given chain complex is $\mathfrak{J}$-exact.

Conversely, suppose that we have an $\mathfrak{J}$-exact chain complex. By Lemma 4, this means that $0 \to F(A) \to F(B) \to F(C) \to 0$ is a short exact sequence. Hence $f\colon A \to B$ is $\mathfrak{J}$-monic. Embed f in an exact triangle $A \to B \to C' \to \Sigma A$. Since f is $\mathfrak{J}$-monic, this triangle is $\mathfrak{J}$-exact. Let $\alpha = \mathrm{id}_A$ and $\beta = \mathrm{id}_B$. Since the functor $\mathfrak{T}(\_,C)$ is cohomological and $g \circ f = 0$, we can find a map $\gamma\colon C' \to C$ making (11) commute. The functor F maps the rows of (11) to short exact sequences by Lemmas 4 and 1. Now the Five Lemma yields that $F(\gamma)$ is invertible, so that γ is an $\mathfrak{J}$-equivalence. $\qquad\square$

Remark 3. Lemma 5 implies that $\mathfrak{I}$-exact chain complexes of length 3 are $\mathfrak{I}$-decomposable. We do not expect this for chain complexes of length 4. But we have not searched for a counterexample.

Which chain complexes over $\mathfrak{T}$ are $\mathfrak{I}$-exact for $\mathfrak{I} = 0$ and hence for any homological ideal? The next definition provides the answer.

Definition 35. A chain complex $C_\bullet$ over a triangulated category is called *homologically exact* if $F(C_\bullet)$ is exact for any homological functor $F: \mathfrak{T} \to \mathfrak{C}$.

Example 9. If $A \to B \to C \to \Sigma A$ is an exact triangle, then the chain complex

$$\cdots \to \Sigma^{-1}A \to \Sigma^{-1}B \to \Sigma^{-1}C \to A \to B \to C \to \Sigma A \to \Sigma B \to \Sigma C \to \cdots$$

is homologically exact by the definition of a homological functor.

Lemma 6. *Let* $F: \mathfrak{T} \to \mathfrak{T}'$ *be an exact functor between two triangulated categories. Let* $C_\bullet$ *be a chain complex over* $\mathfrak{T}$. *The following are equivalent:*

(1) $C_\bullet$ *is* $\ker F$-*exact in degree n.*
(2) $F(C_\bullet)$ *is* $\mathfrak{I}$-*exact in degree n with respect to the zero ideal* $\mathfrak{I} = 0$.
(3) *The chain complex* $\mathbb{Y} \circ F(C_\bullet)$ *in* $\mathfrak{Coh}(\mathfrak{T}')$ *is exact in degree n.*
(4) $F(C_\bullet)$ *is homologically exact in degree n.*
(5) *The chain complexes of Abelian groups* $\mathfrak{T}'(A, F(C_\bullet))$ *are exact in degree n for all* $A \in\in \mathfrak{T}'$.

Proof. By Freyd's Theorem 11, $\mathbb{Y} \circ F: \mathfrak{T} \to \mathfrak{Coh}(\mathfrak{T}')$ is a stable homological functor with $\ker F = \ker(\mathbb{Y} \circ F)$. Hence Lemma 4 yields (1) $\Longleftrightarrow$ (3). Similarly, we have (2) $\Longleftrightarrow$ (3) because $\mathbb{Y}: \mathfrak{T}' \to \mathfrak{Coh}(\mathfrak{T}')$ is a stable homological functor with $\ker \mathbb{Y} = 0$. Freyd's Theorem 11 also asserts that any homological functor $F: \mathfrak{T}' \to \mathfrak{C}'$ factors as $\bar{F} \circ \mathbb{Y}$ for an exact functor $\bar{F}$. Hence (3)$\Longrightarrow$(4). Proposition 5 yields (4)$\Longrightarrow$(5). Finally, (5) $\Longleftrightarrow$ (3) because kernels and cokernels in $\mathfrak{Coh}(\mathfrak{T}')$ are computed pointwise on objects of $\mathfrak{T}'$. $\square$

Remark 4. More generally, consider a set of exact functors $F_i: \mathfrak{T} \to \mathfrak{T}'_i$. As in the proof of the equivalence (1) $\Longleftrightarrow$ (2) in Lemma 6, we see that a chain complex $C_\bullet$ is $\bigcap \ker F_i$-exact (in degree n) if and only if the chain complexes $F_i(C_\bullet)$ are exact (in degree n) for all i.

As a consequence, a chain complex $C_\bullet$ over KK^G for a locally compact group G is $\mathscr{F}$-exact if and only if $\mathrm{Res}^H_G(C_\bullet)$ is homologically exact for all $H \in \mathscr{F}$.

Example 10. We exhibit an $\mathfrak{I}$-exact chain complex that is not $\mathfrak{I}$-decomposable for the ideal $\mathfrak{I} = 0$. By Lemma 3, any 0-exact triangle is split. Therefore, a chain complex is 0-decomposable if and only if it is a direct sum of chain complexes of the form $0 \to K_n \xrightarrow{\mathrm{id}} K_n \to 0$. Hence any decomposable chain complex is contractible and therefore mapped by any homological functor to a contractible chain complex. By the way, if idempotents in $\mathfrak{T}$ split then a chain complex is 0-decomposable if and only if it is contractible.

As we have remarked in Example 9, the chain complex

$$\cdots \to \Sigma^{-1}C \to A \to B \to C \to \Sigma A \to \Sigma B \to \Sigma C \to \Sigma^2 A \to \cdots$$

is homologically exact for any exact triangle $A \to B \to C \to \Sigma A$. But such chain complexes need not be contractible. A counterexample is the exact triangle $\mathbb{Z}/2 \to \mathbb{Z}/4 \to \mathbb{Z}/2 \to \Sigma\mathbb{Z}/2$ in $\mathfrak{Der}(\mathfrak{Ab})$, which we have already used in the proof of Proposition 6. The resulting chain complex over $\mathfrak{Der}(\mathfrak{Ab})$ cannot be contractible because H_* maps it to a non-contractible chain complex.

3.2.3 More Homological Algebra with Chain Complexes

Using our notion of exactness for chain complexes, we can do homological algebra in the homotopy category $\mathrm{Ho}(\mathfrak{T})$. We briefly sketch some results in this direction, assuming some familiarity with more advanced notions from homological algebra. We will not use this later.

The $\mathfrak{I}$-exact chain complexes form a thick subcategory of $\mathrm{Ho}(\mathfrak{T})$ because of Lemma 4. We let $\mathfrak{Der} := \mathfrak{Der}(\mathfrak{T}, \mathfrak{I})$ be the localisation of $\mathrm{Ho}(\mathfrak{T})$ at this subcategory and call it the *derived category of $\mathfrak{T}$ with respect to $\mathfrak{I}$*.

We let $\mathfrak{Der}^{\geq n}$ and $\mathfrak{Der}^{\leq n}$ be the full subcategories of $\mathfrak{Der}$ consisting of chain complexes that are $\mathfrak{I}$-exact in degrees $< n$ and $> n$, respectively.

Theorem 12. *The pair of subcategories $\mathfrak{Der}^{\geq 0}$, $\mathfrak{Der}^{\leq 0}$ forms a* truncation structure *(t-structure) on $\mathfrak{Der}$ in the sense of* [2].

Proof. The main issue here is the truncation of chain complexes. Let $C_\bullet$ be a chain complex over $\mathfrak{T}$. We embed the map d_0 in an exact triangle $C_0 \to C_{-1} \to X \to \Sigma C_0$ and let $C_\bullet^{\geq 0}$ be the chain complex

$$\cdots \to C_2 \to C_1 \to C_0 \to C_{-1} \to X \to \Sigma C_0 \to \Sigma C_{-1} \to \Sigma X \to \Sigma^2 C_0 \to \cdots.$$

This chain complex is $\mathfrak{I}$-exact – even homologically exact – in negative degrees, that is, $C_\bullet^{\geq 0} \in \mathfrak{Der}^{\geq 0}$. The triangulated category structure allows us to construct a chain map $C_\bullet^{\geq 0} \to C_\bullet$ that is an isomorphism on C_n for $n \geq -1$. Hence its mapping cone $C_\bullet^{\leq -1}$ is $\mathfrak{I}$-exact – even contractible – in degrees ≥ 0, that is, $C_\bullet^{\leq -1} \in\in \mathfrak{Der}^{\leq -1}$. By construction, we have an exact triangle

$$C_\bullet^{\geq 0} \to C_\bullet \to C_\bullet^{\leq -1} \to \Sigma C_\bullet^{\geq 0}$$

in $\mathfrak{Der}$.

We also have to check that there is no non-zero morphism $C_\bullet \to D_\bullet$ in $\mathfrak{Der}$ if $C_\bullet \in\in \mathfrak{Der}^{\geq 0}$ and $D_\bullet \in\in \mathfrak{Der}^{\leq -1}$. Recall that morphisms in $\mathfrak{Der}$ are represented by diagrams $C_\bullet \xleftarrow{\sim} \tilde{C}_\bullet \to D_\bullet$ in $\mathrm{Ho}(\mathfrak{T})$, where the first map is an $\mathfrak{I}$-equivalence. Hence $\tilde{C}_\bullet \in\in \mathfrak{Der}^{\geq 0}$ as well. We claim that any chain map $f\colon \tilde{C}_\bullet^{\geq 0} \to D_\bullet^{\leq -1}$ is homotopic to 0. Since the maps $\tilde{C}_\bullet^{\geq 0} \to C_\bullet$ and $D_\bullet \to D_\bullet^{\leq -1}$ are $\mathfrak{I}$-equivalences, any morphism $C_\bullet \to D_\bullet$ vanishes in $\mathfrak{Der}$.

It remains to prove the claim. In a first step, we use that $D_\bullet^{\leq-1}$ is contractible in degrees ≥ 0 to replace f by a homotopic chain map supported in degrees < 0. In a second step, we use that $\tilde{C}_\bullet^{\geq 0}$ is homologically exact in the relevant degrees to recursively construct a chain homotopy between f and 0.   □

Any truncation structure gives rise to an Abelian category, its *core*. In our case, we get the full subcategory $\mathfrak{C} \subseteq \mathfrak{Der}$ of all chain complexes that are $\mathfrak{I}$-exact except in degree 0. This is a stable Abelian category, and the standard embedding $\mathfrak{T} \to \mathrm{Ho}(\mathfrak{T})$ yields a stable homological functor $F: \mathfrak{T} \to \mathfrak{C}$ with $\ker F = \mathfrak{I}$.

This functor is characterised uniquely by the following universal property: any (stable) homological functor $H: \mathfrak{T} \to \mathfrak{C}'$ with $\mathfrak{I} \subseteq \ker H$ factors uniquely as $H = \bar{H} \circ F$ for an exact functor $\bar{H}: \mathfrak{C} \to \mathfrak{C}'$. We construct $\bar{H}$ in three steps.

First, we lift H to an exact functor $\mathrm{Ho}(H): \mathrm{Ho}(\mathfrak{T},\mathfrak{I}) \to \mathrm{Ho}(\mathfrak{C}')$. Secondly, $\mathrm{Ho}(H)$ descends to a functor $\mathfrak{Der}(H): \mathfrak{Der}(\mathfrak{T},\mathfrak{I}) \to \mathfrak{Der}(\mathfrak{C}')$. Finally, $\mathfrak{Der}(H)$ restricts to a functor $\bar{H}: \mathfrak{C} \to \mathfrak{C}'$ between the cores. Since $\mathfrak{I} \subseteq \ker H$, an $\mathfrak{I}$-exact chain complex is also $\ker H$-exact. Hence $\mathrm{Ho}(H)$ preserves exactness of chain complexes by Lemma 4. This allows us to construct $\mathfrak{Der}(H)$ and shows that $\mathfrak{Der}(H)$ is compatible with truncation structures. This allows us to restrict it to an exact functor between the cores. Finally, we use that the core of the standard truncation structure on $\mathfrak{Der}(\mathfrak{C})$ is $\mathfrak{C}$. It is easy to see that we have $\bar{H} \circ F = H$.

Especially, we get an exact functor $\mathfrak{Der}(F): \mathfrak{Der}(\mathfrak{T},\mathfrak{I}) \to \mathfrak{Der}(\mathfrak{C})$, which restricts to the identity functor $\mathrm{id}_\mathfrak{C}$ between the cores. Hence $\mathfrak{Der}(F)$ is fully faithful on the thick subcategory generated by $\mathfrak{C} \subseteq \mathfrak{Der}(\mathfrak{T},\mathfrak{I})$. It seems plausible that $\mathfrak{Der}(F)$ should be an equivalence of categories under some mild conditions on $\mathfrak{I}$ and $\mathfrak{T}$.

We will continue our study of the functor $\mathfrak{T} \to \mathfrak{C}$ in Sect. 3.2.8. The universal property determines it uniquely. Beligiannis [3] has another, simpler construction.

3.2.4 Projective Objects

Let $\mathfrak{I}$ be a homological ideal in a triangulated category $\mathfrak{T}$.

Definition 36. A homological functor $F: \mathfrak{T} \to \mathfrak{C}$ is called $\mathfrak{I}$-*exact* if $F(f) = 0$ for all $\mathfrak{I}$-phantom maps f or, equivalently, $\mathfrak{I} \subseteq \ker F$. An object $A \in\in \mathfrak{T}$ is called $\mathfrak{I}$-*projective* if the functor $\mathfrak{T}(A,\_): \mathfrak{T} \to \mathfrak{Ab}$ is $\mathfrak{I}$-exact. Dually, an object $B \in\in \mathfrak{T}$ is called $\mathfrak{I}$-*injective* if the functor $\mathfrak{T}(\_,B): \mathfrak{T} \to \mathfrak{Ab}^{\mathrm{op}}$ is $\mathfrak{I}$-exact.

We write $\mathfrak{P}_\mathfrak{I}$ for the class of $\mathfrak{I}$-projective objects in $\mathfrak{T}$.

The notions of projective and injective object are dual to each other: if we pass to the opposite category $\mathfrak{T}^{\mathrm{op}}$ with the canonical triangulated category structure and use the same ideal $\mathfrak{I}^{\mathrm{op}}$, then this exchanges the roles of projective and injective objects. Therefore, it suffices to discuss one of these two notions in the following. We will only treat projective objects because all the ideals in Sect. 3.1.2 have enough projective objects, but most of them do not have enough injective objects.

Notice that the functor F is $\mathfrak{I}$-exact if and only if the associated stable functor $F_*: \mathfrak{T} \to \mathfrak{C}^\mathbb{Z}$ is $\mathfrak{I}$-exact because $\mathfrak{I}$ is stable.

Since we require F to be homological, the long exact homology sequence and Lemma 4 yield that the following conditions are all equivalent to F being $\mathfrak{I}$-exact:

- F maps $\mathfrak{I}$-epimorphisms to epimorphisms in $\mathfrak{C}$
- F maps $\mathfrak{I}$-monomorphisms to monomorphisms in $\mathfrak{C}$
- $0 \to F(A) \to F(B) \to F(C) \to 0$ is a short exact sequence in $\mathfrak{C}$ for any $\mathfrak{I}$-exact triangle $A \to B \to C \to \Sigma A$
- F maps $\mathfrak{I}$-exact chain complexes to exact chain complexes in $\mathfrak{C}$

This specialises to equivalent definitions of $\mathfrak{I}$-projective objects.

Lemma 7. *An object $A \in\in \mathfrak{T}$ is $\mathfrak{I}$-projective if and only if $\mathfrak{I}(A,B) = 0$ for all $B \in\in \mathfrak{T}$.*

Proof. If $f \in \mathfrak{I}(A,B)$, then $f = f_*(\mathrm{id}_A)$. This has to vanish if A is $\mathfrak{I}$-projective. Suppose, conversely, that $\mathfrak{I}(A,B) = 0$ for all $B \in\in \mathfrak{T}$. If $f \in \mathfrak{I}(B,B')$, then $\mathfrak{T}(A,f)$ maps $\mathfrak{T}(A,B)$ to $\mathfrak{I}(A,B') = 0$, so that $\mathfrak{T}(A,f) = 0$. Hence A is $\mathfrak{I}$-projective. □

An $\mathfrak{I}$-exact functor also has the following properties (which are strictly weaker than being $\mathfrak{I}$-exact):

- F maps $\mathfrak{I}$-equivalences to isomorphisms in $\mathfrak{C}$
- F maps $\mathfrak{I}$-contractible objects to 0 in $\mathfrak{C}$

Again we may specialise this to $\mathfrak{I}$-projective objects.

Lemma 8. *The class of $\mathfrak{I}$-exact homological functors $\mathfrak{T} \to \mathfrak{Ab}$ or $\mathfrak{T} \to \mathfrak{Ab}^{\mathrm{op}}$ is closed under composition with $\Sigma^{\pm 1} : \mathfrak{T} \to \mathfrak{T}$, retracts, direct sums, and direct products. The class $\mathfrak{P}_\mathfrak{I}$ of $\mathfrak{I}$-projective objects is closed under (de)suspensions, retracts, and possibly infinite direct sums (as far as they exist in $\mathfrak{T}$).*

Proof. The first assertion follows because direct sums and products of Abelian groups are exact; the second one is a special case. □

Definition 37. Let $\mathfrak{P} \subseteq \mathfrak{T}$ be a set of objects. We let $(\mathfrak{P})_\oplus$ be the smallest class of objects of $\mathfrak{T}$ that contains $\mathfrak{P}$ and is closed under retracts and direct sums (as far as they exist in $\mathfrak{T}$).

By Lemma 8, $(\mathfrak{P})_\oplus$ consists of $\mathfrak{I}$-projective objects if $\mathfrak{P}$ does. We say that $\mathfrak{P}$ *generates all $\mathfrak{I}$-projective objects* if $(\mathfrak{P})_\oplus = \mathfrak{P}_\mathfrak{I}$. In examples, it is usually easier to describe a class of generators in this sense.

Example 11. Suppose that G is discrete. Then the adjointness between induction and restriction functors implies that all compactly induced objects are projective for the ideal $\mathscr{V}\mathscr{C}$. Even more, the techniques that we develop below show that $\mathfrak{P}_{\mathscr{V}\mathscr{C}} = \mathscr{C}\mathscr{I}$.

3.2.5 Projective Resolutions

Definition 38. Let $\mathfrak{I} \subseteq \mathfrak{T}$ be a homological ideal in a triangulated category and let $A \in\!\!\in \mathfrak{T}$. A *one-step $\mathfrak{I}$-projective resolution* is an $\mathfrak{I}$-epimorphism $\pi\colon P \to A$ with $P \in\!\!\in \mathfrak{P}_{\mathfrak{I}}$. An *$\mathfrak{I}$-projective resolution of A* is an $\mathfrak{I}$-exact chain complex

$$\cdots \xrightarrow{\delta_{n+1}} P_n \xrightarrow{\delta_n} P_{n-1} \xrightarrow{\delta_{n-1}} \cdots \xrightarrow{\delta_1} P_0 \xrightarrow{\delta_0} A$$

with $P_n \in\!\!\in \mathfrak{P}_{\mathfrak{I}}$ for all $n \in \mathbb{N}$.

We say that $\mathfrak{I}$ has *enough projective objects* if each $A \in\!\!\in \mathfrak{T}$ has a one-step $\mathfrak{I}$-projective resolution.

The following proposition contains the basic properties of projective resolutions, which are familiar from the similar situation for Abelian categories.

Proposition 7. *If $\mathfrak{I}$ has enough projective objects, then any object of $\mathfrak{T}$ has an $\mathfrak{I}$-projective resolution (and vice versa).*

Let $P_\bullet \to A$ and $P'_\bullet \to A'$ be $\mathfrak{I}$-projective resolutions. Then any map $A \to A'$ may be lifted to a chain map $P_\bullet \to P'_\bullet$, and this lifting is unique up to chain homotopy. Two $\mathfrak{I}$-projective resolutions of the same object are chain homotopy equivalent. As a result, the construction of projective resolutions provides a functor

$$P\colon \mathfrak{T} \to \mathrm{Ho}(\mathfrak{T}).$$

Let $A \xrightarrow{f} B \xrightarrow{g} C \xrightarrow{h} \Sigma A$ be an $\mathfrak{I}$-exact triangle. Then there exists a canonical map $\eta\colon P(C) \to P(A)[1]$ in $\mathrm{Ho}(\mathfrak{T})$ such that the triangle

$$P(A) \xrightarrow{P(f)} P(B) \xrightarrow{P(g)} P(C) \xrightarrow{\eta} P(A)[1]$$

in $\mathrm{Ho}(\mathfrak{T})$ is exact; here $[1]$ denotes the translation functor in $\mathrm{Ho}(\mathfrak{T})$, which has nothing to do with the suspension in $\mathfrak{T}$.

Proof. Let $A \in\!\!\in \mathfrak{T}$. By assumption, there is a one-step $\mathfrak{I}$-projective resolution $\delta_0\colon P_0 \to A$, which we embed in an exact triangle $A_1 \to P_0 \to A \to \Sigma A_1$. Since δ_0 is $\mathfrak{I}$-epic, this triangle is $\mathfrak{I}$-exact. By induction, we construct a sequence of such $\mathfrak{I}$-exact triangles $A_{n+1} \to P_n \to A_n \to \Sigma A_{n+1}$ for $n \in \mathbb{N}$ with $P_n \in\!\!\in \mathfrak{P}$ and $A_0 = A$. By composition, we obtain maps $\delta_n\colon P_n \to P_{n-1}$ for $n \geq 1$, which satisfy $\delta_n \circ \delta_{n+1} = 0$ for all $n \geq 0$. The resulting chain complex

$$\cdots \to P_n \xrightarrow{\delta_n} P_{n-1} \xrightarrow{\delta_{n-1}} P_{n-2} \to \cdots \to P_1 \xrightarrow{\delta_1} P_0 \xrightarrow{\delta_0} A \to 0$$

is $\mathfrak{I}$-decomposable by construction and therefore $\mathfrak{I}$-exact by Corollary 3.

The remaining assertions are proved exactly as their classical counterparts in homological algebra. We briefly sketch the arguments. Let $P_\bullet \to A$ and $P'_\bullet \to A'$ be $\mathfrak{I}$-projective resolutions and let $f \in \mathfrak{T}(A, A')$. We construct $f_n \in \mathfrak{T}(P_n, P'_n)$ by induction on n such that the diagrams

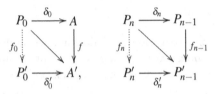

for $n \geq 1$ commute. We must check that this is possible. Since the chain complex $P'_\bullet \to A$ is $\mathfrak{J}$-exact and P_n is $\mathfrak{J}$-projective for all $n \geq 0$, the chain complexes

$$\cdots \to \mathfrak{T}(P_n, P'_m) \xrightarrow{(\delta'_m)_*} \mathfrak{T}(P_n, P'_{m-1}) \to \cdots \to \mathfrak{T}(P_n, P'_0) \xrightarrow{(\delta'_0)_*} \mathfrak{T}(P_n, A) \to 0$$

are exact for all $n \in \mathbb{N}$. This allows us to find the needed maps f_n. By construction, these maps form a chain map lifting $f \colon A \to A'$. Its uniqueness up to chain homotopy is proved similarly. If we apply this unique lifting result to two $\mathfrak{J}$-projective resolutions of the same object, we get the uniqueness of $\mathfrak{J}$-projective resolutions up to chain homotopy equivalence. Hence we get a well-defined functor $P \colon \mathfrak{T} \to \mathrm{Ho}(\mathfrak{T})$.

Now consider an $\mathfrak{J}$-exact triangle $A \to B \to C \to \Sigma A$ as in the third paragraph of the lemma. Let $X_\bullet$ be the mapping cone of some chain map $P(A) \to P(B)$ in the homotopy class $P(f)$. This chain complex is supported in degrees ≥ 0 and has $\mathfrak{J}$-projective entries because $X_n = P(A)_{n-1} \oplus P(B)_n$. The map $X_0 = 0 \oplus P(B)_0 \to B \to C$ yields a chain map $X_\bullet \to C$, that is, the composite map $X_1 \to X_0 \to C$ vanishes. By construction, this chain map lifts the given map $B \to C$ and we have an exact triangle $P(A) \to P(B) \to X_\bullet \to P(A)[1]$ in $\mathrm{Ho}(\mathfrak{T})$. It remains to observe that $X_\bullet \to C$ is $\mathfrak{J}$-exact. Then $X_\bullet$ is an $\mathfrak{J}$-projective resolution of C. Since such resolutions are unique up to chain homotopy equivalence, we get a canonical isomorphism $X_\bullet \cong P(C)$ in $\mathrm{Ho}(\mathfrak{T})$ and hence the assertion in the third paragraph.

Let F be a stable homological functor with $\mathfrak{J} = \ker F$. We have to check that $F(X_\bullet) \to F(C)$ is a resolution. This reduces to a well-known diagram chase in Abelian categories, using that $F(P(A)) \to F(A)$ and $F(P(B)) \to F(B)$ are resolutions and that $F(A) \rightarrowtail F(B) \twoheadrightarrow F(C)$ is exact. □

3.2.6 Derived Functors

We only define derived functors if there are enough projective objects because this case is rather easy and suffices for our applications. The general case can be reduced to the familiar case of Abelian categories using the results of Sect. 3.2.3.

Definition 39. Let $\mathfrak{J}$ be a homological ideal in a triangulated category $\mathfrak{T}$ with enough projective objects. Let $F \colon \mathfrak{T} \to \mathfrak{C}$ be an additive functor with values in an Abelian category $\mathfrak{C}$. It induces a functor $\mathrm{Ho}(F) \colon \mathrm{Ho}(\mathfrak{T}) \to \mathrm{Ho}(\mathfrak{C})$, applying F pointwise to chain complexes. Let $P \colon \mathfrak{T} \to \mathrm{Ho}(\mathfrak{T})$ be the projective resolution functor constructed in Proposition 7. Let $\mathrm{H}_n \colon \mathrm{Ho}(\mathfrak{C}) \to \mathfrak{C}$ be the nth homology functor for some $n \in \mathbb{N}$. The composite functor

$$\mathbb{L}_n F \colon \mathfrak{T} \xrightarrow{P} \mathrm{Ho}(\mathfrak{T}) \xrightarrow{\mathrm{Ho}(F)} \mathrm{Ho}(\mathfrak{C}) \xrightarrow{\mathrm{H}_n} \mathfrak{C}$$

is called the nth *left derived functor* of F. If $F: \mathfrak{T}^{op} \to \mathfrak{C}$ is a contravariant additive functor, then the corresponding functor $H^n \circ \mathrm{Ho}(F) \circ P: \mathfrak{T}^{op} \to \mathfrak{C}$ is denoted by $\mathbb{R}^n F$ and called the nth *right derived functor* of F.

More concretely, let $A \in\in \mathfrak{T}$ and let $(P_\bullet, \delta_\bullet)$ be an $\mathfrak{J}$-projective resolution of A. If F is covariant, then $\mathbb{L}_n F(A)$ is the homology at $F(P_n)$ of the chain complex

$$\cdots \to F(P_{n+1}) \xrightarrow{F(\delta_{n+1})} F(P_n) \xrightarrow{F(\delta_n)} F(P_{n-1}) \to \cdots \to F(P_0) \to 0.$$

If F is contravariant, then $\mathbb{R}^n F(A)$ is the cohomology at $F(P_n)$ of the cochain complex

$$\cdots \leftarrow F(P_{n+1}) \xleftarrow{F(\delta_{n+1})} F(P_n) \xleftarrow{F(\delta_n)} F(P_{n-1}) \leftarrow \cdots \leftarrow F(P_0) \leftarrow 0.$$

Lemma 9. *Let $A \to B \to C \to \Sigma A$ be an $\mathfrak{J}$-exact triangle. If $F: \mathfrak{T} \to \mathfrak{C}$ is a covariant additive functor, then there is a long exact sequence*

$$\cdots \to \mathbb{L}_n F(A) \to \mathbb{L}_n F(B) \to \mathbb{L}_n F(C) \to \mathbb{L}_{n-1} F(A)$$
$$\to \cdots \to \mathbb{L}_1 F(C) \to \mathbb{L}_0 F(A) \to \mathbb{L}_0 F(B) \to \mathbb{L}_0 F(C) \to 0.$$

If $F: \mathfrak{T}^{op} \to \mathfrak{C}$ is contravariant instead, then there is a long exact sequence

$$\cdots \leftarrow \mathbb{R}^n F(A) \leftarrow \mathbb{R}^n F(B) \leftarrow \mathbb{R}^n F(C) \leftarrow \mathbb{R}^{n-1} F(A)$$
$$\leftarrow \cdots \leftarrow \mathbb{R}^1 F(C) \leftarrow \mathbb{R}^0 F(A) \leftarrow \mathbb{R}^0 F(B) \leftarrow \mathbb{R}^0 F(C) \leftarrow 0.$$

Proof. This follows from the third assertion of Proposition 7 together with the well-known long exact homology sequence for exact triangles in $\mathrm{Ho}(\mathfrak{C})$. □

Lemma 10. *Let $F: \mathfrak{T} \to \mathfrak{C}$ be a homological functor. The following assertions are equivalent:*

(1) *F is $\mathfrak{J}$-exact*
(2) *$\mathbb{L}_0 F(A) \cong F(A)$ and $\mathbb{L}_p F(A) = 0$ for all $p > 0$, $A \in\in \mathfrak{T}$*
(3) *$\mathbb{L}_0 F(A) \cong F(A)$ for all $A \in\in \mathfrak{T}$*

The analogous assertions for contravariant functors are equivalent as well.

Proof. If F is $\mathfrak{J}$-exact, then F maps $\mathfrak{J}$-exact chain complexes in $\mathfrak{T}$ to exact chain complexes in $\mathfrak{C}$. This applies to $\mathfrak{J}$-projective resolutions, so that $(1)\Longrightarrow(2)\Longrightarrow(3)$. It follows from (3) and Lemma 9 that F maps $\mathfrak{J}$-epimorphisms to epimorphisms. Since this characterises $\mathfrak{J}$-exact functors, we get $(3)\Longrightarrow(1)$. □

It can happen that $\mathbb{L}_p F = 0$ for all $p > 0$ although F is not $\mathfrak{J}$-exact.

We have a natural transformation $\mathbb{L}_0 F(A) \to F(A)$ (or $F(A) \to \mathbb{R}^0 F(A)$), which is induced by the augmentation map $P_\bullet \to A$ for an $\mathfrak{J}$-projective resolution. Lemma 10 shows that these maps are usually not bijective, although this happens frequently for derived functors on Abelian categories.

Definition 40. We let $\mathrm{Ext}^n_{\mathfrak{T},\mathfrak{I}}(A,B)$ be the nth right derived functor with respect to $\mathfrak{I}$ of the contravariant functor $A \mapsto \mathfrak{T}(A,B)$.

We have natural maps $\mathfrak{T}(A,B) \to \mathrm{Ext}^0_{\mathfrak{T},\mathfrak{I}}(A,B)$, which usually are not invertible. Lemma 9 yields long exact sequences

$$\cdots \leftarrow \mathrm{Ext}^n_{\mathfrak{T},\mathfrak{I}}(A,D) \leftarrow \mathrm{Ext}^n_{\mathfrak{T},\mathfrak{I}}(B,D) \leftarrow \mathrm{Ext}^n_{\mathfrak{T},\mathfrak{I}}(C,D) \leftarrow \mathrm{Ext}^{n-1}_{\mathfrak{T},\mathfrak{I}}(A,D) \leftarrow$$
$$\cdots \leftarrow \mathrm{Ext}^1_{\mathfrak{T},\mathfrak{I}}(C,D) \leftarrow \mathrm{Ext}^0_{\mathfrak{T},\mathfrak{I}}(A,D) \leftarrow \mathrm{Ext}^0_{\mathfrak{T},\mathfrak{I}}(B,D) \leftarrow \mathrm{Ext}^0_{\mathfrak{T},\mathfrak{I}}(C,D) \leftarrow 0$$

for any $\mathfrak{I}$-exact, exact triangle $A \to B \to C \to \Sigma A$ and any $D \in\in \mathfrak{T}$.

We claim that there are similar long exact sequences

$$0 \to \mathrm{Ext}^0_{\mathfrak{T},\mathfrak{I}}(D,A) \to \mathrm{Ext}^0_{\mathfrak{T},\mathfrak{I}}(D,B) \to \mathrm{Ext}^0_{\mathfrak{T},\mathfrak{I}}(D,C) \to \mathrm{Ext}^1_{\mathfrak{T},\mathfrak{I}}(D,A) \to \cdots$$
$$\to \mathrm{Ext}^{n-1}_{\mathfrak{T},\mathfrak{I}}(D,C) \to \mathrm{Ext}^n_{\mathfrak{T},\mathfrak{I}}(D,A) \to \mathrm{Ext}^n_{\mathfrak{T},\mathfrak{I}}(D,B) \to \mathrm{Ext}^n_{\mathfrak{T},\mathfrak{I}}(D,C) \to \cdots$$

in the second variable. Since $P(D)_n$ is $\mathfrak{I}$-projective, the sequences

$$0 \to \mathfrak{T}(P(D)_n,A) \to \mathfrak{T}(P(D)_n,B) \to \mathfrak{T}(P(D)_n,C) \to 0$$

are exact for all $n \in \mathbb{N}$. This extension of chain complexes yields the desired long exact sequence.

We list a few more elementary properties of derived functors. We only spell things out for the left derived functors $\mathbb{L}_n F : \mathfrak{T} \to \mathfrak{C}$ of a covariant functor $F : \mathfrak{T} \to \mathfrak{C}$. Similar assertions hold for right derived functors of contravariant functors.

The derived functors $\mathbb{L}_n F$ satisfy $\mathfrak{I} \subseteq \ker \mathbb{L}_n F$ and hence descend to functors $\mathbb{L}_n F : \mathfrak{T}/\mathfrak{I} \to \mathfrak{C}$ because the zero map $P(A) \to P(B)$ is a chain map lifting of f if $f \in \mathfrak{I}(A,B)$. As a consequence, $\mathbb{L}_n F(A) \cong 0$ if A is $\mathfrak{I}$-contractible. The long exact homology sequences of Lemma 9 show that $\mathbb{L}_n F(f) : \mathbb{L}_n F(A) \to \mathbb{L}_n F(B)$ is invertible if $f \in \mathfrak{T}(A,B)$ is an $\mathfrak{I}$-equivalence.

Remark 5. The derived functors $\mathbb{L}_n F$ are *not homological* and therefore do not deserve to be called $\mathfrak{I}$-exact even though they vanish on $\mathfrak{I}$-phantom maps. Lemma 9 shows that these functors are only half-exact on $\mathfrak{I}$-exact triangles. Thus $\mathbb{L}_n F(f)$ need not be monic (or epic) if f is $\mathfrak{I}$-monic (or $\mathfrak{I}$-epic). The problem is that the $\mathfrak{I}$-projective resolution functor $P : \mathfrak{T} \to \mathrm{Ho}(\mathfrak{T})$ is not exact; it even fails to be stable.

The following remarks require a more advanced background in homological algebra and are not going to be used in the sequel.

Remark 6. The derived functors introduced above, especially the Ext functors, can be interpreted in terms of *derived categories*.

We have already observed in Sect. 3.2.3 that the $\mathfrak{I}$-exact chain complexes form a thick subcategory of $\mathrm{Ho}(\mathfrak{T})$. The augmentation map $P(A) \to A$ of an $\mathfrak{I}$-projective resolution of $A \in\in \mathfrak{T}$ is a quasi-isomorphism with respect to this thick subcategory. The chain complex $P(A)$ is projective (see [19]), that is, for any chain complex $C_\bullet$, the space of morphisms $A \to C_\bullet$ in the derived category $\mathfrak{Der}(\mathfrak{T},\mathfrak{I})$ agrees with $[P(A),C_\bullet]$. Especially, $\mathrm{Ext}^n_{\mathfrak{T},\mathfrak{I}}(A,B)$ is the space of morphisms $A \to B[n]$ in $\mathfrak{Der}(\mathfrak{T},\mathfrak{I})$.

Now let $F: \mathfrak{T} \to \mathfrak{C}$ be an additive covariant functor. Extend it to an exact functor $\bar{F}: \text{Ho}(\mathfrak{T}) \to \text{Ho}(\mathfrak{C})$. It has a total left derived functor

$$\mathbb{L}\bar{F}: \mathfrak{Der}(\mathfrak{T}, \mathfrak{I}) \to \mathfrak{Der}(\mathfrak{C}), \qquad A \mapsto \bar{F}(P(A)).$$

By definition, we have $\mathbb{L}_n F(A) := \text{H}_n(\mathbb{L}\bar{F}(A))$.

Remark 7. In classical Abelian categories, the Ext groups form a graded ring, and the derived functors form graded modules over this graded ring. The same happens in our context. The most conceptual construction of these products uses the description of derived functors sketched in Remark 6.

Recall that we may view elements of $\text{Ext}^n_{\mathfrak{T},\mathfrak{I}}(A, B)$ as morphisms $A \to B[n]$ in the derived category $\mathfrak{Der}(\mathfrak{T}, \mathfrak{I})$. Taking translations, we can also view them as morphisms $A[m] \to B[n+m]$ for any $m \in \mathbb{Z}$. The usual composition in the category $\mathfrak{Der}(\mathfrak{T}, \mathfrak{I})$ therefore yields an associative product

$$\text{Ext}^n_{\mathfrak{T},\mathfrak{I}}(B, C) \otimes \text{Ext}^m_{\mathfrak{T},\mathfrak{I}}(A, B) \to \text{Ext}^{n+m}_{\mathfrak{T},\mathfrak{I}}(A, C).$$

Thus we get a graded additive category with morphism spaces $\left(\text{Ext}^n_{\mathfrak{T},\mathfrak{I}}(A, B)\right)_{n \in \mathbb{N}}$.

Similarly, if $F: \mathfrak{T} \to \mathfrak{C}$ is an additive functor and $\mathbb{L}\bar{F}: \mathfrak{Der}(\mathfrak{T}, \mathfrak{I}) \to \mathfrak{Der}(\mathfrak{C})$ is as in Remark 7, then a morphism $A \to B[n]$ in $\mathfrak{Der}(\mathfrak{T}, \mathfrak{I})$ induces a morphism $\mathbb{L}\bar{F}(A) \to \mathbb{L}\bar{F}(B)[n]$ in $\mathfrak{Der}(\mathfrak{C})$. Passing to homology, we get canonical maps

$$\text{Ext}^n_{\mathfrak{T},\mathfrak{I}}(A, B) \to \text{Hom}_{\mathfrak{C}}\left(\mathbb{L}F_m(A), \mathbb{L}F_{m-n}(B)\right) \qquad \forall m \geq n,$$

which satisfy an appropriate associativity condition. For a contravariant functor, we get canonical maps

$$\text{Ext}^n_{\mathfrak{T},\mathfrak{I}}(A, B) \to \text{Hom}_{\mathfrak{C}}\left(\mathbb{R}F^m(B), \mathbb{R}F^{m+n}(A)\right) \qquad \forall m \geq 0.$$

3.2.7 Projective Objects via Adjointness

We develop a method for constructing enough projective objects. Let $\mathfrak{T}$ and $\mathfrak{C}$ be stable additive categories, let $F: \mathfrak{T} \to \mathfrak{C}$ be a stable additive functor, and let $\mathfrak{I} := \ker F$. In our applications, $\mathfrak{T}$ is triangulated and the functor F is either exact or stable and homological.

Recall that a covariant functor $R: \mathfrak{T} \to \mathfrak{Ab}$ is *(co)representable* if it is naturally isomorphic to $\mathfrak{T}(A, \_)$ for some $A \in\in \mathfrak{T}$, which is then unique. If the functor $B \mapsto \mathfrak{C}(A, F(B))$ on $\mathfrak{T}$ is representable, we write $F^\dagger(A)$ for the representing object. By construction, we have natural isomorphisms

$$\mathfrak{T}(F^\dagger(A), B) \cong \mathfrak{C}(A, F(B))$$

for all $B \in \mathfrak{T}$. Let $\mathfrak{C}' \subseteq \mathfrak{C}$ be the full subcategory of all objects $A \in\in \mathfrak{C}$ for which $F^\dagger(A)$ is defined. Then $F^\dagger$ is a functor $\mathfrak{C}' \to \mathfrak{T}$, which we call the *(partially defined) left adjoint* of F. Although one usually assumes $\mathfrak{C} = \mathfrak{C}'$, we shall also need $F^\dagger$ in cases where it is not defined everywhere.

The functor $B \mapsto \mathfrak{C}(A, F(B))$ for $A \in\in \mathfrak{C}'$ vanishes on $\mathfrak{I} = \ker F$ for trivial reasons. Hence $F^\dagger(A) \in\in \mathfrak{T}$ is $\mathfrak{I}$-projective. This simple observation is surprisingly powerful: as we shall see, it often yields all $\mathfrak{I}$-projective objects.

Remark 8. We have $F^\dagger(\Sigma A) \cong \Sigma F^\dagger(A)$ for all $A \in\in \mathfrak{C}'$, so that $\Sigma(\mathfrak{C}') = \mathfrak{C}'$. Moreover, $F^\dagger$ commutes with infinite direct sums (as far as they exist in $\mathfrak{T}$) because

$$\mathfrak{T}\left(\bigoplus F^\dagger(A_i), B\right) \cong \prod \mathfrak{T}(F^\dagger(A_i), B) \cong \prod \mathfrak{C}(A_i, F(B)) \cong \mathfrak{C}\left(\bigoplus A_i, F(B)\right).$$

Example 12. Consider the functor $K_* \colon KK \to \mathfrak{Ab}^{\mathbb{Z}/2}$. Let $\mathbb{Z} \in\in \mathfrak{Ab}^{\mathbb{Z}/2}$ denote the trivially graded Abelian group $\mathbb{Z}$. Notice that

$$\mathrm{Hom}\big(\mathbb{Z}, K_*(A)\big) \cong K_0(A) \cong KK(\mathbb{C}, A),$$
$$\mathrm{Hom}\big(\mathbb{Z}[1], K_*(A)\big) \cong K_1(A) \cong KK(C_0(\mathbb{R}), A),$$

where $\mathbb{Z}[1]$ means $\mathbb{Z}$ in odd degree. Hence $K_*^\dagger(\mathbb{Z}) = \mathbb{C}$ and $K_*^\dagger(\mathbb{Z}[1]) = C_0(\mathbb{R})$. More generally, Remark 8 shows that $K_*^\dagger(A)$ is defined if both the even and odd parts of $A \in\in \mathfrak{Ab}^{\mathbb{Z}/2}$ are countable free Abelian groups: it is a direct sum of at most countably many copies of $\mathbb{C}$ and $C_0(\mathbb{R})$. Hence all such countable direct sums are $\mathfrak{I}_K$-projective (we briefly say K-*projective*). As we shall see, $K_*^\dagger$ is not defined on all of $\mathfrak{Ab}^{\mathbb{Z}/2}$; this is typical of homological functors.

Example 13. Consider the functor $H_* \colon \mathrm{Ho}(\mathfrak{C}; \mathbb{Z}/p) \to \mathfrak{C}^{\mathbb{Z}/p}$ of Example 6. Let $j \colon \mathfrak{C}^{\mathbb{Z}/p} \to \mathrm{Ho}(\mathfrak{C}; \mathbb{Z}/p)$ be the functor that views an object of $\mathfrak{C}^{\mathbb{Z}/p}$ as a p-periodic chain complex whose boundary map vanishes.

A chain map $j(A) \to B_\bullet$ for $A \in\in \mathfrak{C}^{\mathbb{Z}/p}$ and $B_\bullet \in\in \mathrm{Ho}(\mathfrak{C}; \mathbb{Z}/p)$ is a family of maps $\varphi_n \colon A_n \to \ker(d_n \colon B_n \to B_{n-1})$. Such a family is chain homotopic to 0 if and only if each φ_n lifts to a map $A_n \to B_{n+1}$. Suppose that A_n is projective for all $n \in \mathbb{Z}/p$. Then such a lifting exists if and only if $\varphi_n(A_n) \subseteq d_{n+1}(B_{n+1})$. Hence

$$[j(A), B_\bullet] \cong \prod_{n \in \mathbb{Z}/p} \mathfrak{C}(A_n, H_n(B_\bullet)) \cong \mathfrak{C}^{\mathbb{Z}/p}(A, H_*(B_\bullet)).$$

As a result, the left adjoint of H_* is defined on the subcategory of projective objects $\mathfrak{P}(\mathfrak{C})^{\mathbb{Z}/p} \subseteq \mathfrak{C}^{\mathbb{Z}/p}$ and agrees there with the restriction of j. We will show in Sect. 3.2.9 that $\mathfrak{P}(\mathfrak{C})^{\mathbb{Z}/p}$ is equal to the domain of definition of $H_*^\dagger$ and that all $\mathfrak{I}_H$-projective objects are of the form $H_*^\dagger(A)$ (provided $\mathfrak{C}$ has enough projective objects).

These examples show that $F^\dagger$ yields many $\ker F$-projective objects. We want to get *enough* $\ker F$-projective objects in this fashion, assuming that $F^\dagger$ is defined on enough of $\mathfrak{C}$. In order to treat ideals of the form $\bigcap F_i$, we now consider a more complicated setup. Let $\{\mathfrak{C}_i \mid i \in I\}$ be a set of stable homological or triangulated categories together with full subcategories $\mathfrak{P}\mathfrak{C}_i \subseteq \mathfrak{C}_i$ and stable homological or exact functors $F_i \colon \mathfrak{T} \to \mathfrak{C}_i$ for all $i \in I$. Assume that

- The left adjoint $F_i^\dagger$ is defined on $\mathfrak{P}\mathfrak{C}_i$ for all $i \in I$
- There is an epimorphism $P \to F_i(A)$ in $\mathfrak{C}_i$ with $P \in\in \mathfrak{P}\mathfrak{C}_i$ for any $i \in I$, $A \in\in \mathfrak{T}$
- The set of functors $F_i^\dagger \colon \mathfrak{P}\mathfrak{C}_i \to \mathfrak{T}$ is *cointegrable*, that is, $\bigoplus_{i \in I} F_i^\dagger(B_i)$ exists for all families of objects $B_i \in \mathfrak{P}\mathfrak{C}_i$, $i \in I$

The reason for the notation $\mathfrak{PC}_i$ is that for a homological functor F_i we usually take $\mathfrak{PC}_i$ to be the class of projective objects of $\mathfrak{C}_i$; if F_i is exact, then we often take $\mathfrak{PC}_i = \mathfrak{C}_i$. But it may be useful to choose a smaller category, as long as it satisfies the second condition above.

Proposition 8. *In this situation, there are enough $\mathfrak{I}$-projective objects, and $\mathfrak{P}_\mathfrak{I}$ is generated by $\bigcup_{i \in I} \{F_i^\dagger(B) \mid B \in \mathfrak{PC}_i\}$. More precisely, an object of $\mathfrak{T}$ is $\mathfrak{I}$-projective if and only if it is a retract of $\bigoplus_{i \in I} F_i^\dagger(B_i)$ for a family of objects $B_i \in \mathfrak{PC}_i$.*

Proof. Let $\tilde{\mathfrak{P}}_0 := \bigcup_{i \in I} \{F_i^\dagger(B) \mid B \in \mathfrak{PC}_i\}$ and $\mathfrak{P}_0 := (\tilde{\mathfrak{P}}_0)_\oplus$. To begin with, we observe that any object of the form $F_i^\dagger(B)$ with $B \in\in \mathfrak{PC}_i$ is $\ker F_i$-projective and hence $\mathfrak{I}$-projective because $\mathfrak{I} \subseteq \ker F_i$. Hence $\mathfrak{P}_0$ consists of $\mathfrak{I}$-projective objects.

Let $A \in\in \mathfrak{T}$. For each $i \in I$, there is an epimorphism $p_i \colon B_i \to F_i(A)$ with $B_i \in \mathfrak{PC}_i$. The direct sum $B := \bigoplus_{i \in I} F_i^\dagger(B_i)$ exists. We have $B \in\in \mathfrak{P}_0$ by construction. We are going to construct an $\mathfrak{I}$-epimorphism $p \colon B \to A$. This shows that there are enough $\mathfrak{I}$-projective objects.

The maps $p_i \colon B_i \to F_i(A)$ provide maps $\hat{p}_i \colon F_i^\dagger(B_i) \to A$ via the adjointness isomorphisms $\mathfrak{T}(F_i^\dagger(B_i), A) \cong \mathfrak{C}_i(B_i, F_i(A))$. We let $p := \sum \hat{p}_i \colon \bigoplus F_i^\dagger(B_i) \to A$. We must check that p is an $\mathfrak{I}$-epimorphism. Equivalently, p is $\ker F_i$-epic for all $i \in I$; this is, in turn equivalent to $F_i(p)$ being an epimorphism in $\mathfrak{C}_i$ for all $i \in I$, because of Lemma 1 or 3. This is what we are going to prove.

The identity map on $F_i^\dagger(B_i)$ yields a map $\alpha_i \colon B_i \to F_i F_i^\dagger(B_i)$ via the adjointness isomorphism $\mathfrak{T}(F_i^\dagger(B_i), F_i^\dagger(B_i)) \cong \mathfrak{C}_i(B_i, F_i F_i^\dagger(B_i))$. Composing with the map

$$F_i F_i^\dagger(B_i) \to F_i\left(\bigoplus F_i^\dagger(B_i)\right) = F_i(B)$$

induced by the coordinate embedding $F_i^\dagger(B_i) \to B$, we get a map $\alpha_i' \colon B_i \to F_i(B)$. The naturality of the adjointness isomorphisms yields $F_i(\hat{p}_i) \circ \alpha_i = p_i$ and hence $F_i(p) \circ \alpha_i' = p_i$. The map p_i is an epimorphism by assumption. Now we use a cancellation result for epimorphisms: if $f \circ g$ is an epimorphism, then so is f. Thus $F_i(p)$ is an epimorphism as desired.

If A is $\mathfrak{I}$-projective, then the $\mathfrak{I}$-epimorphism $p \colon B \to A$ splits; to see this, embed p in an exact triangle $N \to B \to A \to \Sigma N$ and observe that the map $A \to \Sigma N$ belongs to $\mathfrak{I}(A, \Sigma N) = 0$. Therefore, A is a retract of B. Since $\mathfrak{P}_0$ is closed under retracts and $B \in\in \mathfrak{P}_0$, we get $A \in\in \mathfrak{P}_0$. Hence $\tilde{\mathfrak{P}}_0$ generates all $\mathfrak{I}$-projective objects. □

3.2.8 The Universal Exact Homological Functor

For the following results, it is essential to define an ideal by a single functor F instead of a family of functors as in Proposition 8.

Definition 41. Let $\mathfrak{I} \subseteq \mathfrak{T}$ be a homological ideal. An $\mathfrak{I}$-exact stable homological functor $F \colon \mathfrak{T} \to \mathfrak{C}$ is called *universal* if any other $\mathfrak{I}$-exact stable homological functor $G \colon \mathfrak{T} \to \mathfrak{C}'$ factors as $\bar{G} = G \circ F$ for a stable exact functor $\bar{G} \colon \mathfrak{C} \to \mathfrak{C}'$ that is unique up to natural isomorphism.

This universal property characterises F uniquely up to natural isomorphism. We have constructed such a functor in Sect. 3.2.3. Beligiannis constructs it in [3, Sect. 3] using a localisation of the Abelian category $\mathfrak{Coh}(\mathfrak{T})$ at a suitable Serre subcategory; he calls this functor *projectivisation functor* and its target category *Steenrod category*. This notation is motivated by the special case of the Adams spectral sequence. The following theorem allows us to check whether a given functor is universal:

Theorem 13. *Let $\mathfrak{T}$ be a triangulated category, let $\mathfrak{J} \subseteq \mathfrak{T}$ be a homological ideal, and let $F \colon \mathfrak{T} \to \mathfrak{C}$ be an $\mathfrak{J}$-exact stable homological functor into a stable Abelian category $\mathfrak{C}$; let $\mathfrak{PC}$ be the class of projective objects in $\mathfrak{C}$. Suppose that idempotent morphisms in $\mathfrak{T}$ split.*

The functor F is the universal $\mathfrak{J}$-exact stable homological functor and there are enough $\mathfrak{J}$-projective objects in $\mathfrak{T}$ if and only if

- *$\mathfrak{C}$ has enough projective objects*
- *The adjoint functor $F^\dagger$ is defined on $\mathfrak{PC}$*
- *$F \circ F^\dagger(A) \cong A$ for all $A \in \mathfrak{PC}$.*

Proof. Suppose first that F is universal and that there are enough $\mathfrak{J}$-projective objects. Then F is equivalent to the projectivisation functor of [3]. The various properties of this functor listed in [3, Proposition 4.19] include the following:

- There are enough projective objects in $\mathfrak{C}$
- F induces an equivalence of categories $\mathfrak{P}_\mathfrak{J} \cong \mathfrak{PC}$ ($\mathfrak{P}_\mathfrak{J}$ is the class of projective objects in $\mathfrak{T}$)
- $\mathfrak{C}(F(A), F(B)) \cong \mathfrak{T}(A, B)$ for all $A \in \mathfrak{P}_\mathfrak{J}, B \in \mathfrak{T}$

Here we use the assumption that idempotents in $\mathfrak{T}$ split. The last property is equivalent to $F^\dagger \circ F(A) \cong A$ for all $A \in \mathfrak{P}_\mathfrak{J}$. Since $\mathfrak{P}_\mathfrak{J} \cong \mathfrak{PC}$ via F, this implies that $F^\dagger$ is defined on $\mathfrak{PC}$ and that $F \circ F^\dagger(A) \cong A$ for all $A \in \mathfrak{PC}$. Thus F has the properties listed in the statement of the theorem.

Now suppose conversely that F has these properties. Let $\mathfrak{P}'_\mathfrak{J} \subseteq \mathfrak{T}$ be the essential range of $F^\dagger \colon \mathfrak{PC} \to \mathfrak{T}$. We claim that $\mathfrak{P}'_\mathfrak{J}$ is the class of all $\mathfrak{J}$-projective objects in $\mathfrak{T}$. Since $F \circ F^\dagger$ is equivalent to the identity functor on $\mathfrak{PC}$ by assumption, $F|_{\mathfrak{P}'_\mathfrak{J}}$ and $F^\dagger$ provide an equivalence of categories $\mathfrak{P}'_\mathfrak{J} \cong \mathfrak{PC}$. Since $\mathfrak{C}$ is assumed to have enough projectives, the hypotheses of Proposition 8 are satisfied. Hence there are enough $\mathfrak{J}$-projective objects in $\mathfrak{T}$, and any object of $\mathfrak{P}_\mathfrak{J}$ is a retract of an object of $\mathfrak{P}'_\mathfrak{J}$. Idempotent morphisms in the category $\mathfrak{P}'_\mathfrak{J} \cong \mathfrak{PC}$ split because $\mathfrak{C}$ is Abelian and retracts of projective objects are again projective. Hence $\mathfrak{P}'_\mathfrak{J}$ is closed under retracts in $\mathfrak{T}$, so that $\mathfrak{P}'_\mathfrak{J} = \mathfrak{P}_\mathfrak{J}$. It also follows that F and $F^\dagger$ provide an equivalence of categories $\mathfrak{P}_\mathfrak{J} \cong \mathfrak{PC}$. Hence $F^\dagger \circ F(A) \cong A$ for all $A \in \mathfrak{P}_\mathfrak{J}$, so that we get $\mathfrak{C}(F(A), F(B)) \cong \mathfrak{T}(F^\dagger \circ F(A), B) \cong \mathfrak{T}(A, B)$ for all $A \in \mathfrak{P}_\mathfrak{J}, B \in \mathfrak{T}$.

Now let $G \colon \mathfrak{T} \to \mathfrak{C}'$ be a stable homological functor. We will later assume G to be $\mathfrak{J}$-exact, but the first part of the following argument works in general. Since F provides an equivalence of categories $\mathfrak{P}_\mathfrak{J} \cong \mathfrak{PC}$, the rule $\bar{G}(F(P)) := G(P)$ defines a functor $\bar{G}$ on $\mathfrak{PC}$. This yields a functor $\mathrm{Ho}(\bar{G}) \colon \mathrm{Ho}(\mathfrak{PC}) \to \mathrm{Ho}(\mathfrak{C}')$. Since $\mathfrak{C}$

has enough projective objects, the construction of projective resolutions provides a functor $P\colon \mathfrak{C} \to \mathrm{Ho}(\mathfrak{PC})$. We let $\bar{G}$ be the composite functor

$$\bar{G}\colon \mathfrak{C} \xrightarrow{P} \mathrm{Ho}(\mathfrak{PC}) \xrightarrow{\mathrm{Ho}(\tilde{G})} \mathrm{Ho}(\mathfrak{C}') \xrightarrow{\mathrm{H}_0} \mathfrak{C}'.$$

This functor is right-exact and satisfies $\bar{G} \circ F = G$ on $\mathfrak{I}$-projective objects of $\mathfrak{T}$.

Now suppose that G is $\mathfrak{I}$-exact. Then we get $\bar{G} \circ F = G$ for all objects of $\mathfrak{T}$ because this holds for $\mathfrak{I}$-projective objects. We claim that $\bar{G}$ is exact. Let $A \in\in \mathfrak{C}$. Since $\mathfrak{C}$ has enough projective objects, we can find a projective resolution of A. We may assume this resolution to have the form $F(P_\bullet)$ with $P_\bullet \in\in \mathrm{Ho}(\mathfrak{P}_\mathfrak{I})$ because $F(\mathfrak{P}_\mathfrak{I}) \cong \mathfrak{PC}$. Lemma 4 yields that $P_\bullet$ is $\mathfrak{I}$-exact except in degree 0. Since $\mathfrak{I} \subseteq \ker G$, the chain complex $P_\bullet$ is $\ker G$-exact in positive degrees as well, so that $G(P_\bullet)$ is exact except in degree 0 by Lemma 4. As a consequence, $\mathbb{L}_p \bar{G}(A) = 0$ for all $p > 0$. We also have $\mathbb{L}_0 \bar{G}(A) = \bar{G}(A)$ by construction. Thus $\bar{G}$ is exact.

As a result, G factors as $G = \bar{G} \circ F$ for an exact functor $\bar{G}\colon \mathfrak{C} \to \mathfrak{C}'$. It is clear that $\bar{G}$ is stable. Finally, since $\mathfrak{C}$ has enough projective objects, a functor on $\mathfrak{C}$ is determined up to natural equivalence by its restriction to projective objects. Therefore, our factorisation of G is unique up to natural equivalence. Thus F is the universal $\mathfrak{I}$-exact functor. □

Remark 9. Let $\mathfrak{P}'\mathfrak{C} \subseteq \mathfrak{PC}$ be some subcategory such that any object of $\mathfrak{C}$ is a quotient of a direct sum of objects of $\mathfrak{P}'\mathfrak{C}$. Equivalently, $(\mathfrak{P}'\mathfrak{C})_\oplus = \mathfrak{PC}$. Theorem 13 remains valid if we only assume that $F^\dagger$ is defined on $\mathfrak{P}'\mathfrak{C}$ and that $F \circ F^\dagger(A) \cong A$ holds for $A \in\in \mathfrak{P}'\mathfrak{C}$ because both conditions are evidently hereditary for direct sums and retracts.

Theorem 14. *In the situation of Theorem 13, the functors F and $F^\dagger$ restrict to equivalences of categories $\mathfrak{P}_\mathfrak{I} \cong \mathfrak{PC}$ inverse to each other.*

An object $A \in\in \mathfrak{T}$ is $\mathfrak{I}$-projective if and only if $F(A)$ is projective and

$$\mathfrak{C}\big(F(A), F(B)\big) \cong \mathfrak{T}(A, B)$$

for all $B \in\in \mathfrak{T}$; following Ross Street [36], we call such objects F-projective. $F(A)$ is projective if and only if there is an $\mathfrak{I}$-equivalence $P \to A$ with $P \in\in \mathfrak{P}_\mathfrak{I}$.

The functors F and $F^\dagger$ induce bijections between isomorphism classes of projective resolutions of $F(A)$ in $\mathfrak{C}$ and isomorphism classes of $\mathfrak{I}$-projective resolutions of $A \in\in \mathfrak{T}$ in $\mathfrak{T}$.

If $G\colon \mathfrak{T} \to \mathfrak{C}'$ is any (stable) homological functor, then there is a unique right-exact (stable) functor $\bar{G}\colon \mathfrak{C} \to \mathfrak{C}'$ such that $\bar{G} \circ F(P) = G(P)$ for all $P \in\in \mathfrak{P}_\mathfrak{I}$.

The left derived functors of G with respect to $\mathfrak{I}$ and of $\bar{G}$ are related by natural isomorphisms $\mathbb{L}_n \bar{G} \circ F(A) = \mathbb{L}_n G(A)$ for all $A \in\in \mathfrak{T}$, $n \in \mathbb{N}$. There is a similar statement for cohomological functors, which specialises to natural isomorphisms

$$\mathrm{Ext}^n_{\mathfrak{T},\mathfrak{I}}(A, B) \cong \mathrm{Ext}^n_\mathfrak{C}\big(F(A), F(B)\big).$$

Proof. We have already seen during the proof of Theorem 13 that F restricts to an equivalence of categories $\mathfrak{P}_\mathfrak{I} \xrightarrow{\cong} \mathfrak{PC}$, whose inverse is the restriction of $F^\dagger$, and that $\mathfrak{C}\big(F(A), F(B)\big) \cong \mathfrak{T}(A, B)$ for all $A \in\in \mathfrak{P}_\mathfrak{I}$, $B \in\in \mathfrak{PC}$.

Conversely, if A is F-projective in the sense of Street, then A is $\mathfrak{J}$-projective because already $\mathfrak{T}(A,B) \cong \mathfrak{C}\big(F(A),F(B)\big)$ for all $B \in\in \mathfrak{T}$ yields $A \cong F^{\dagger} \circ F(A)$, so that A is $\mathfrak{J}$-projective; notice that the projectivity of $F(A)$ is automatic.

Since F maps $\mathfrak{J}$-equivalences to isomorphisms, $F(A)$ is projective whenever there is an $\mathfrak{J}$-equivalence $P \to A$ with $\mathfrak{J}$-projective P. Conversely, suppose that $F(A)$ is $\mathfrak{J}$-projective. Let $P_0 \to A$ be a one-step $\mathfrak{J}$-projective resolution. Since $F(A)$ is projective, the epimorphism $F(P_0) \to F(A)$ splits by some map $F(A) \to F(P_0)$. The resulting map $F(P_0) \to F(A) \to F(P_0)$ is idempotent and comes from an idempotent endomorphism of P_0 because F is fully faithful on $\mathfrak{P}_{\mathfrak{J}}$. Its range object P exists because we require idempotent morphisms in $\mathfrak{C}$ to split. It belongs again to $\mathfrak{P}_{\mathfrak{J}}$, and the induced map $F(P) \to F(A)$ is invertible by construction. Hence we get an $\mathfrak{J}$-equivalence $P \to A$.

If $C_\bullet$ is a chain complex over $\mathfrak{T}$, then we know already from Lemma 4 that $C_\bullet$ is $\mathfrak{J}$-exact if and only if $F(C_\bullet)$ is exact. Hence F maps an $\mathfrak{J}$-projective resolution of A to a projective resolution of $F(A)$. Conversely, if $P_\bullet \to F(A)$ is any projective resolution in $\mathfrak{C}$, then it is of the form $F(\hat{P}_\bullet) \to F(A)$ where $\hat{P}_\bullet := F^{\dagger}(P_\bullet)$ and where we get the map $\hat{P}_0 \to A$ by adjointness from the given map $P_0 \to F(A)$. This shows that F induces a bijection between isomorphism classes of $\mathfrak{J}$-projective resolutions of A and projective resolutions of $F(A)$.

We have seen during the proof of Theorem 13 how a stable homological functor $G \colon \mathfrak{T} \to \mathfrak{C}'$ gives rise to a unique right-exact functor $\bar{G} \colon \mathfrak{C} \to \mathfrak{C}'$ that satisfies $\bar{G} \circ F(P) = G(P)$ for all $P \in\in \mathfrak{P}_{\mathfrak{J}}$. The derived functors $\mathbb{L}_n \bar{G}\big(F(A)\big)$ for $A \in\in \mathfrak{T}$ are computed by applying $\bar{G}$ to a projective resolution of $F(A)$. Since such a projective resolution is of the form $F(P_\bullet)$ for an $\mathfrak{J}$-projective resolution $P_\bullet \to A$ and since $\bar{G} \circ F = G$ on $\mathfrak{J}$-projective objects, the derived functors $\mathbb{L}_n G(A)$ and $\mathbb{L}_n \bar{G}(F(A))$ are computed by the same chain complex and agree. The same reasoning applies to cohomological functors and yields the assertion about Ext.   □

Remark 10. The assumption that idempotents split is only needed to check that the universal $\mathfrak{J}$-exact functor has the properties listed in Theorem 13. The converse directions of Theorem 13 and Theorem 14 do not need this assumption.

If $\mathfrak{T}$ has countable direct sums or countable direct products, then idempotents in $\mathfrak{T}$ automatically split by [29, Sect. 1.3]. This covers categories such as KK^G because they have countable direct sums.

3.2.9 Derived Functors in Homological Algebra

Now we study the kernel $\mathfrak{J}_H$ of the homology functor $H_* \colon \mathrm{Ho}(\mathfrak{C}; \mathbb{Z}/p) \to \mathfrak{C}^{\mathbb{Z}/p}$ introduced Example 6. We get exactly the same statements if we replace the homotopy category by its derived category and study the kernel of $H_* \colon \mathfrak{Der}(\mathfrak{C}; \mathbb{Z}/p) \to \mathfrak{C}^{\mathbb{Z}/p}$. We often abbreviate $\mathfrak{J}_H$ to H and speak of H-epimorphisms, H-exact chain complexes, H-projective resolutions, and so on. We denote the full subcategory of H-projective objects in $\mathrm{Ho}(\mathfrak{C}; \mathbb{Z}/p)$ by $\mathfrak{P}_H$.

We assume that the underlying Abelian category $\mathfrak{C}$ has enough projective objects. Then the same holds for $\mathfrak{C}^{\mathbb{Z}/p}$, and we have $\mathfrak{P}(\mathfrak{C}^{\mathbb{Z}/p}) \cong (\mathfrak{P}\mathfrak{C})^{\mathbb{Z}/p}$. That is, an object of $\mathfrak{C}^{\mathbb{Z}/p}$ is projective if and only if its homogeneous pieces are.

Theorem 15. *The category* $\mathrm{Ho}(\mathfrak{C};\mathbb{Z}/p)$ *has enough* H-*projective objects, and the functor* $\mathrm{H}_*\colon \mathrm{Ho}(\mathfrak{C};\mathbb{Z}/p) \to \mathfrak{C}^{\mathbb{Z}/p}$ *is the universal* H-*exact stable homological functor. Its restriction to* $\mathfrak{P}_{\mathrm{H}}$ *provides an equivalence of categories* $\mathfrak{P}_{\mathrm{H}} \cong \mathfrak{P}\mathfrak{C}^{\mathbb{Z}/p}$. *More concretely, a chain complex in* $\mathrm{Ho}(\mathfrak{C};\mathbb{Z}/p)$ *is* H-*projective if and only if it is homotopy equivalent to one with vanishing boundary map and projective entries.*

The functor H_* *maps isomorphism classes of* H-*projective resolutions of* A *in* $\mathrm{Ho}(\mathfrak{C};\mathbb{Z}/p)$ *bijectively to isomorphism classes of projective resolutions of* $\mathrm{H}_*(A)$ *in* $\mathfrak{C}^{\mathbb{Z}/p}$. *We have*

$$\mathrm{Ext}^n_{\mathrm{Ho}(\mathfrak{C};\mathbb{Z}/p),\mathfrak{I}_{\mathrm{H}}}(A,B) \cong \mathrm{Ext}^n_{\mathfrak{C}}\bigl(\mathrm{H}_*(A),\mathrm{H}_*(B)\bigr).$$

Let $F\colon \mathfrak{C} \to \mathfrak{C}'$ *be some covariant additive functor and define*

$$\bar{F}\colon \mathrm{Ho}(\mathfrak{C};\mathbb{Z}/p) \to \mathrm{Ho}(\mathfrak{C}';\mathbb{Z}/p)$$

by applying F *entrywise. Then* $\mathbb{L}_n\bar{F}(A) \cong \mathbb{L}_n F\bigl(\mathrm{H}_*(A)\bigr)$ *for all* $n \in \mathbb{N}$. *Similarly, we have* $\mathbb{R}^n\bar{F}(A) \cong \mathbb{R}^n F\bigl(\mathrm{H}_*(A)\bigr)$ *if* F *is a contravariant functor.*

Proof. The category $\mathfrak{C}^{\mathbb{Z}/p}$ has enough projective objects by assumption. We have already seen in Example 13 that $\mathrm{H}_*^\dagger$ is defined on $\mathfrak{P}\mathfrak{C}^{\mathbb{Z}/p}$; this functor is denoted by j in Example 13. It is clear that $\mathrm{H}_* \circ j(A) \cong A$ for all $A \in\in \mathfrak{C}^{\mathbb{Z}/p}$. Now Theorem 13 shows that H_* is universal. We do not need idempotent morphisms in $\mathrm{Ho}(\mathfrak{C};\mathbb{Z}/p)$ to split by Remark 10. □

Remark 11. Since the universal $\mathfrak{I}$-exact functor is essentially unique, the universality of $\mathrm{H}_*\colon \mathfrak{Der}(\mathfrak{C};\mathbb{Z}/p) \to \mathfrak{C}^{\mathbb{Z}/p}$ means that we can recover this functor and hence the stable Abelian category $\mathfrak{C}^{\mathbb{Z}/p}$ from the ideal $\mathfrak{I}_{\mathrm{H}} \subseteq \mathfrak{Der}(\mathfrak{C};\mathbb{Z}/p)$. That is, the ideal $\mathfrak{I}_{\mathrm{H}}$ and the functor $\mathrm{H}_*\colon \mathfrak{Der}(\mathfrak{C};\mathbb{Z}/p) \to \mathfrak{C}^{\mathbb{Z}/p}$ contain exactly the same amount of information.

For instance, if we forget the precise category $\mathfrak{C}$ by composing H_* with some *faithful* functor $\mathfrak{C} \to \mathfrak{C}'$, then the resulting homology functor $\mathrm{Ho}(\mathfrak{C};\mathbb{Z}/p) \to \mathfrak{C}'$ still has kernel $\mathfrak{I}_{\mathrm{H}}$. We can recover $\mathfrak{C}^{\mathbb{Z}/p}$ by passing to the universal $\mathfrak{I}$-exact functor.

We compare this with the situation for truncation structures [2]. These cannot exist for periodic categories such as $\mathfrak{Der}(\mathfrak{C};\mathbb{Z}/p)$ for $p \geq 1$. Given the standard truncation structure on $\mathfrak{Der}(\mathfrak{C})$, we can recover the Abelian category $\mathfrak{C}$ as its core; we also get back the homology functors $\mathrm{H}_n\colon \mathfrak{Der}(\mathfrak{C}) \to \mathfrak{C}$ for all $n \in \mathbb{Z}$. Conversely, the functor $\mathrm{H}_*\colon \mathfrak{Der}(\mathfrak{C}) \to \mathfrak{C}^{\mathbb{Z}}$ together with the grading on $\mathfrak{C}^{\mathbb{Z}}$ tells us what it means for a chain complex to be exact in degrees ≥ 0 or ≤ 0 and thus determines the truncation structure. Hence the standard truncation structure on $\mathfrak{Der}(\mathfrak{C})$ contains the same amount of information as the functor $\mathrm{H}_*\colon \mathfrak{Der}(\mathfrak{C}) \to \mathfrak{C}^{\mathbb{Z}}$ together with the grading on $\mathfrak{C}^{\mathbb{Z}}$.

3.3 Universal Coefficient Theorems

First we study the ideal $\mathfrak{I}_{\mathrm{K}} := \ker \mathrm{K}_* \subseteq \mathrm{KK}$ of Example 4. We complete our analysis of this example and explain the Universal Coefficient Theorem for KK in our framework. We call $\mathfrak{I}_{\mathrm{K}}$-projective objects and $\mathfrak{I}_{\mathrm{K}}$-exact functors briefly K-*projective* and K-*exact* and let $\mathfrak{P}_{\mathrm{K}} \subseteq \mathrm{KK}$ be the class of K-projective objects.

Let $\mathfrak{Ab}_c^{\mathbb{Z}/2} \subseteq \mathfrak{Ab}^{\mathbb{Z}/2}$ be the full subcategory of *countable* $\mathbb{Z}/2$-graded Abelian groups. Since the K-theory of a separable C*-algebra is countable, we may view K_* as a stable homological functor $K_* \colon KK \to \mathfrak{Ab}_c^{\mathbb{Z}/2}$.

Theorem 16. *There are enough K-projective objects in* KK, *and the universal K-exact functor is* $K_* \colon KK \to \mathfrak{Ab}_c^{\mathbb{Z}/2}$. *It restricts to an equivalence of categories between* $\mathfrak{P}_K$ *and the full subcategory* $\mathfrak{Ab}_{fc}^{\mathbb{Z}/2} \subseteq \mathfrak{Ab}_c^{\mathbb{Z}/2}$ *of* $\mathbb{Z}/2$-graded countable free Abelian groups. A separable C*-algebra belongs to $\mathfrak{P}_K$ if and only if it is KK-equivalent to $\bigoplus_{i \in I_0} \mathbb{C} \oplus \bigoplus_{i \in I_1} C_0(\mathbb{R})$ where the sets I_0, I_1 are at most countable.*

If $A \in\in KK$, then K_* maps isomorphism classes of K-projective resolutions of A in KK bijectively to isomorphism classes of free resolutions of $K_*(A)$. We have

$$\mathrm{Ext}_{KK,\mathfrak{J}_K}^n(A,B) \cong \begin{cases} \mathrm{Hom}_{\mathfrak{Ab}^{\mathbb{Z}/2}}\big(K_*(A), K_*(B)\big) & \text{for } n = 0; \\ \mathrm{Ext}_{\mathfrak{Ab}^{\mathbb{Z}/2}}^1\big(K_*(A), K_*(B)\big) & \text{for } n = 1; \\ 0 & \text{for } n \geq 2. \end{cases}$$

Let $F \colon KK \to \mathfrak{C}$ be some covariant additive functor; then there is a unique right-exact functor $\bar{F} \colon \mathfrak{Ab}_c^{\mathbb{Z}/2} \to \mathfrak{C}$ with $\bar{F} \circ K_* = F$. We have $\mathbb{L}_n F = (\mathbb{L}_n \bar{F}) \circ K_*$ for all $n \in \mathbb{N}$; this vanishes for $n \geq 2$. Similar assertions hold for contravariant functors.

Proof. Notice that $\mathfrak{Ab}_c^{\mathbb{Z}/2} \subseteq \mathfrak{Ab}^{\mathbb{Z}/2}$ is an Abelian category. We shall denote objects of $\mathfrak{Ab}^{\mathbb{Z}/2}$ by pairs (A_0, A_1) of Abelian groups. By definition, $(A_0, A_1) \in\in \mathfrak{Ab}_{fc}^{\mathbb{Z}/2}$ if and only if A_0 and A_1 are countable free Abelian groups, that is, they are of the form $A_0 = \mathbb{Z}[I_0]$ and $A_1 = \mathbb{Z}[I_1]$ for at most countable sets I_0, I_1. It is well-known that any Abelian group is a quotient of a free Abelian group and that subgroups of free Abelian groups are again free. Moreover, free Abelian groups are projective. Hence $\mathfrak{Ab}_{fc}^{\mathbb{Z}/2}$ is the subcategory of projective objects in $\mathfrak{Ab}_c^{\mathbb{Z}/2}$ and any object $G \in\in \mathfrak{Ab}_c^{\mathbb{Z}/2}$ has a projective resolution of the form $0 \to F_1 \to F_0 \to G$ with $F_0, F_1 \in\in \mathfrak{Ab}_{fc}^{\mathbb{Z}/2}$. This implies that derived functors on $\mathfrak{Ab}_c^{\mathbb{Z}/2}$ only occur in dimensions 1 and 0.

As in Example 12, we see that $K_*^\dagger$ is defined on $\mathfrak{Ab}_{fc}^{\mathbb{Z}/2}$ and satisfies

$$K_*^\dagger\big(\mathbb{Z}[I_0], \mathbb{Z}[I_1]\big) \cong \bigoplus_{i \in I_0} \mathbb{C} \oplus \bigoplus_{i \in I_1} C_0(\mathbb{R})$$

if I_0, I_1 are countable. We also have $K_* \circ K_*^\dagger\big(\mathbb{Z}[I_0], \mathbb{Z}[I_1]\big) \cong \big(\mathbb{Z}[I_0], \mathbb{Z}[I_1]\big)$, so that the hypotheses of Theorem 13 are satisfied. Hence there are enough K-projective objects and K_* is universal. The remaining assertions follow from Theorem 14 and our detailed knowledge of the homological algebra in $\mathfrak{Ab}_c^{\mathbb{Z}/2}$. $\qquad\square$

Example 14. Consider the stable homological functor

$$F \colon KK \to \mathfrak{Ab}_c^{\mathbb{Z}/2}, \qquad A \mapsto K_*(A \otimes B)$$

for some $B \in\in KK$, where $\otimes$ denotes, say, the spatial C*-tensor product. We claim that the associated right-exact functor $\mathfrak{Ab}_c^{\mathbb{Z}/2} \to \mathfrak{Ab}_c^{\mathbb{Z}/2}$ is

$$\bar{F}\colon \mathfrak{Ab}_c^{\mathbb{Z}/2} \to \mathfrak{Ab}_c^{\mathbb{Z}/2}, \qquad G \mapsto G \otimes \mathrm{K}_*(B).$$

It is easy to check $F \circ \mathrm{K}_*^{\dagger}(G) \cong G \otimes \mathrm{K}_*(B) \cong \bar{F}(G)$ for $G \in\!\in \mathfrak{Ab}_{\mathrm{fc}}^{\mathbb{Z}/2}$. Since the functor $G \mapsto G \otimes \mathrm{K}_*(B)$ is right-exact and agrees with $\bar{F}$ on projective objects, we get $\bar{F}(G) = G \otimes \mathrm{K}_*(B)$ for all $G \in\!\in \mathfrak{Ab}_c^{\mathbb{Z}/2}$. Hence the derived functors of F are

$$\mathbb{L}_n F(A) \cong \begin{cases} \mathrm{K}_*(A) \otimes \mathrm{K}_*(B) & \text{for } n = 0; \\ \mathrm{Tor}^1\big(\mathrm{K}_*(A), \mathrm{K}_*(B)\big) & \text{for } n = 1; \\ 0 & \text{for } n \geq 2. \end{cases}$$

Here we use the same graded version of Tor as in the Künneth Theorem [4].

Example 15. Consider the stable homological functor

$$F\colon \mathrm{KK} \to \mathfrak{Ab}^{\mathbb{Z}/2}, \qquad B \mapsto \mathrm{KK}_*(A, B)$$

for some $A \in\!\in \mathrm{KK}$. We suppose that A is a *compact* object of KK, that is, the functor F commutes with direct sums. Then $\mathrm{KK}_*\big(A, \mathrm{K}_*^{\dagger}(G)\big) \cong \mathrm{KK}_*(A, \mathbb{C}) \otimes G$ for all $G \in\!\in \mathfrak{Ab}_{\mathrm{fc}}^{\mathbb{Z}/2}$ because this holds for $G = (\mathbb{Z}, 0)$ and is inherited by suspensions and direct sums. Now we get $\bar{F}(G) \cong \mathrm{KK}_*(A, \mathbb{C}) \otimes G$ for all $G \in\!\in \mathfrak{Ab}_c^{\mathbb{Z}/2}$ as in Example 14. Therefore,

$$\mathbb{L}_n F(B) \cong \begin{cases} \mathrm{KK}_*(A, \mathbb{C}) \otimes \mathrm{K}_*(B) & \text{for } n = 0; \\ \mathrm{Tor}^1\big(\mathrm{KK}_*(A, \mathbb{C}), \mathrm{K}_*(B)\big) & \text{for } n = 1; \\ 0 & \text{for } n \geq 2. \end{cases}$$

Generalising Examples 14 and 15, we have $\bar{F}(G) \cong F(\mathbb{C}) \otimes G$ and hence

$$\mathbb{L}_n F(B) \cong \begin{cases} F(\mathbb{C}) \otimes \mathrm{K}_*(B) & \text{for } n = 0, \\ \mathrm{Tor}^1\big(F(\mathbb{C}), \mathrm{K}_*(B)\big) & \text{for } n = 1, \end{cases}$$

for any covariant functor $F\colon \mathrm{KK} \to \mathfrak{C}$ that commutes with direct sums.

Similarly, if $F\colon \mathrm{KK}^{\mathrm{op}} \to \mathfrak{C}$ is contravariant and maps direct sums to direct products, then $\bar{F}(G) \cong \mathrm{Hom}(G, F(\mathbb{C}))$ and

$$\mathbb{R}^n F(B) \cong \begin{cases} \mathrm{Hom}\big(\mathrm{K}_*(B), F(\mathbb{C})\big) & \text{for } n = 0, \\ \mathrm{Ext}^1\big(\mathrm{K}_*(B), F(\mathbb{C})\big) & \text{for } n = 1. \end{cases}$$

The description of $\mathrm{Ext}_{\mathrm{KK}, \mathfrak{I}_\mathrm{K}}^n$ in Theorem 16 is a special case of this.

3.3.1 Universal Coefficient Theorem in the Hereditary Case

In general, we need spectral sequences in order to relate the derived functors $\mathbb{L}_n F$ back to F. Here we concentrate on the simple case where we have projective resolutions of length 1, so that the spectral sequence degenerates to a short exact sequence. The following universal coefficient theorem is very similar to but slightly more general than [3, Theorem 4.27] because we do not require *all* $\mathfrak{I}$-equivalences to be invertible.

Theorem 17. *Let $\mathfrak{T}$ be a triangulated category and let $\mathfrak{I} \subseteq \mathfrak{T}$ be a homological ideal. Let $A \in\!\!\in \mathfrak{T}$ have an $\mathfrak{I}$-projective resolution of length 1. Suppose also that $\mathfrak{T}(A,B) = 0$ for all $\mathfrak{I}$-contractible B. Let $F\colon \mathfrak{T} \to \mathfrak{C}$ be a homological functor, $\tilde{F}\colon \mathfrak{T}^{op} \to \mathfrak{C}$ a cohomological functor, and $B \in\!\!\in \mathfrak{T}$. Then there are natural short exact sequences*

$$0 \to \mathbb{L}_0 F_*(A) \to F_*(A) \to \mathbb{L}_1 F_{*-1}(A) \to 0,$$
$$0 \to \mathbb{R}^1 \tilde{F}^{*-1}(A) \to \tilde{F}^*(A) \to \mathbb{R}^0 \tilde{F}^*(A) \to 0,$$
$$0 \to \mathrm{Ext}^1_{\mathfrak{T},\mathfrak{I}}(\Sigma A, B) \to \mathfrak{T}(A,B) \to \mathrm{Ext}^0_{\mathfrak{T},\mathfrak{I}}(A,B) \to 0.$$

Example 16. For the ideal $\mathfrak{I}_K \subseteq KK$, any object has a K-projective resolution of length 1 by Theorem 16. The other hypothesis of Theorem 17 holds if and only if A satisfies the Universal Coefficient Theorem (UCT). The UCT for $KK(A,B)$ predicts $KK(A,B) = 0$ if $K_*(B) = 0$. Conversely, if this is the case, then Theorem 17 applies, and our description of $\mathrm{Ext}_{KK,\mathfrak{I}_K}$ in Theorem 16 yields the UCT for $KK(A,B)$ for all B. This yields our claim.

Thus the UCT for $KK(A,B)$ is a special of Theorem 17. In the situations of Examples 14 and 15, we get the familiar Künneth Theorems for $K_*(A \otimes B)$ and $KK_*(A,B)$. These arguments are very similar to the original proofs (see [4]). Our machinery allows us to treat other situations in a similar fashion.

Proof (of Theorem 17). We only write down the proof for homological functors. The cohomological case is dual and contains $\mathfrak{T}(\lrcorner, B)$ as a special case.

Let $0 \to P_1 \xrightarrow{\delta_1} P_0 \xrightarrow{\delta_0} A$ be an $\mathfrak{I}$-projective resolution of length 1 and view it as an $\mathfrak{I}$-exact chain complex of length 3. Lemma 5 yields a commuting diagram

$$
\begin{array}{ccccc}
P_1 & \xrightarrow{\delta_1} & P_0 & \xrightarrow{\tilde{\delta}_0} & \tilde{A} \\
\| & & \| & & \downarrow{\scriptstyle\alpha} \\
P_1 & \xrightarrow{\delta_1} & P_0 & \xrightarrow{\delta_0} & A,
\end{array}
$$

such that the top row is part of an $\mathfrak{I}$-exact, exact triangle $P_1 \to P_0 \to \tilde{A} \to \Sigma P_1$ and α is an $\mathfrak{I}$-equivalence. We claim that α is an isomorphism in $\mathfrak{T}$.

We embed α in an exact triangle $\Sigma^{-1} B \to \tilde{A} \xrightarrow{\alpha} A \xrightarrow{\beta} B$. Lemma 2 shows that B is $\mathfrak{I}$-contractible because α is an $\mathfrak{I}$-equivalence. Hence $\mathfrak{T}(A,B) = 0$ by our assumption on A. This forces $\beta = 0$, so that our exact triangle splits: $A \cong \tilde{A} \oplus B$. Then $\mathfrak{T}(B,B) \subseteq \mathfrak{T}(A,B)$ vanishes as well, so that $B \cong 0$. Thus α is invertible.

We get an exact triangle in $\mathfrak{T}$ of the form $P_1 \xrightarrow{\delta_1} P_0 \xrightarrow{\delta_0} A \to \Sigma P_1$ because any triangle isomorphic to an exact one is itself exact.

Now we apply F. Since F is homological, we get a long exact sequence

$$\cdots \to F_*(P_1) \xrightarrow{F_*(\delta_1)} F_*(P_0) \to F_*(A) \to F_{*-1}(P_1) \xrightarrow{F_{*-1}(\delta_1)} F_{*-1}(P_0) \to \cdots.$$

We cut this into short exact sequences of the form

$$\operatorname{coker}\bigl(F_*(\delta_1)\bigr) \rightarrowtail F_*(A) \twoheadrightarrow \ker\bigl(F_{*-1}(\delta_1)\bigr).$$

Since $\operatorname{coker} F_*(\delta_1) = \mathbb{L}_0 F_*(A)$ and $\ker F_*(\delta_1) = \mathbb{L}_1 F_*(A)$, we get the desired exact sequence. The map $\mathbb{L}_0 F_*(A) \to F_*(A)$ is the canonical map induced by δ_0. The other map $F_*(A) \to \mathbb{L}_1 F_{*-1}(A)$ is natural for all morphisms between objects with an $\mathfrak{J}$-projective resolution of length 1 by Proposition 7.   □

The proof shows that – in the situation of Theorem 17 – we have

$$\operatorname{Ext}^0_{\mathfrak{T},\mathfrak{J}}(A,B) \cong \mathfrak{T}/\mathfrak{J}(A,B), \qquad \operatorname{Ext}^1_{\mathfrak{T},\mathfrak{J}}(A,B) \cong \mathfrak{J}(A,\Sigma B).$$

More generally, we can construct a natural map $\mathfrak{J}(A,\Sigma B) \to \operatorname{Ext}^1_{\mathfrak{T},\mathfrak{J}}(A,B)$ for any homological ideal, using the $\mathfrak{J}$-universal homological functor $F\colon \mathfrak{T} \to \mathfrak{C}$. We embed $f \in \mathfrak{J}(A,\Sigma B)$ in an exact triangle $B \to C \to A \to \Sigma B$. We get an extension

$$\bigl[F(B) \rightarrowtail F(C) \twoheadrightarrow F(A)\bigr] \in\in \operatorname{Ext}^1_{\mathfrak{C}}\bigl(F(A),F(B)\bigr)$$

because this triangle is $\mathfrak{J}$-exact. This class $\kappa(f)$ in $\operatorname{Ext}^1_{\mathfrak{C}}\bigl(F(A),F(B)\bigr)$ does not depend on auxiliary choices because the exact triangle $B \to C \to A \to \Sigma B$ is unique up to isomorphism. Theorem 14 yields $\operatorname{Ext}^1_{\mathfrak{T},\mathfrak{J}}(A,B) \cong \operatorname{Ext}^1_{\mathfrak{C}}\bigl(F(A),F(B)\bigr)$ because F is universal. Hence we get a natural map

$$\kappa\colon \mathfrak{J}(A,\Sigma B) \to \operatorname{Ext}^1_{\mathfrak{T},\mathfrak{J}}(A,B).$$

We may view κ as a secondary invariant generated by the canonical map

$$\mathfrak{T}(A,B) \to \operatorname{Ext}^0_{\mathfrak{T},\mathfrak{J}}(A,B).$$

For the ideal $\mathfrak{J}_K$, we get the same map κ as in Example 4.

An Abelian category with enough projective objects is called *hereditary* if any subobject of a projective object is again projective. Equivalently, any object has a projective resolution of length 1. This motivates the following definition:

Definition 42. A homological ideal $\mathfrak{J}$ in a triangulated category $\mathfrak{T}$ is called *hereditary* if any object of $\mathfrak{T}$ has a projective resolution of length 1.

If $\mathfrak{J}$ is hereditary and if $\mathfrak{J}$-equivalences are invertible, then Theorem 17 applies to all $A \in\in \mathfrak{T}$ (and vice versa).

Example 17. As another example, consider the ideal $\mathscr{V}\mathscr{C} \subseteq \operatorname{KK}^{\mathbb{Z}}$ for the group $\mathbb{Z}$. Here the family of subgroups only contains the trivial one. Theorem 7 shows that Theorem 17 applies to all objects of $\operatorname{KK}^{\mathbb{Z}}$. The resulting extensions are equivalent to the Pimsner–Voiculescu exact sequence. To see this, first cut the latter into two short exact sequences involving the kernel and cokernel of $\alpha_* - 1$. Then notice that the latter coincide with the group homology of the induced action of $\mathbb{Z}$ on $K_*(A)$.

3.3.2 The Adams Resolution

Let $\mathfrak{J}$ be an ideal in a triangulated category and let $\mathfrak{P}$ be its class of projective objects. We assume that $\mathfrak{J}$ has enough projective objects. Let $A \in\in \mathfrak{T}$. Write $A = B_0$ and let $\Sigma B_1 \to P_0 \to B_0 \to B_1$ be a one-step $\mathfrak{J}$-projective resolution of $A = B_0$. Similarly, let $\Sigma B_2 \to P_1 \to B_1 \to B_2$ be a one-step $\mathfrak{J}$-projective resolution of B_1. Repeating this process we obtain objects $B_n \in\in \mathfrak{T}$, $P_n \in \mathfrak{P}$ for $n \in \mathbb{N}$ with $B_0 = A$ and morphisms $\beta_n^{n+1} \in \mathfrak{J}(B_n, B_{n+1})$, $\pi_n \in \mathfrak{T}(P_n, B_n)$, $\alpha_n \in \mathfrak{T}_1(B_{n+1}, P_n)$ that are part of distinguished triangles

$$\Sigma B_{n+1} \xrightarrow{\alpha_n} P_n \xrightarrow{\pi_n} B_n \xrightarrow{\beta_n^{n+1}} B_{n+1} \tag{12}$$

for all $n \in \mathbb{N}$. Thus the maps π_n are $\mathfrak{J}$-epic for all $n \in \mathbb{N}$. We can assemble these data in a diagram

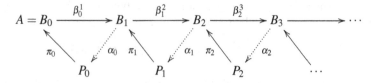

called an *Adams resolution of A*. We also let

$$\beta_m^n := \beta_{n-1}^n \circ \cdots \circ \beta_m^{m+1} : B_m \to B_n$$

for all $n \geq m$ (by convention, $\beta_m^m = \mathrm{id}$). We have $\beta_m^n \in \mathfrak{J}^{n-m}(B_m, B_n)$, that is, β_m^n is a product of $n - m$ factors in $\mathfrak{J}$.

We are particularly interested in the maps $\beta^n := \beta_0^n : A \to B_n$ for $n \in \mathbb{N}$. Taking a mapping cone of β^n we obtain a distinguished triangle

$$\Sigma B_n \xrightarrow{\sigma_n} C_n \xrightarrow{\rho_n} A \xrightarrow{\beta^n} B_n \tag{13}$$

for each n, which is determined uniquely by β^n up to non-canonical isomorphism. Applying the octahedral axiom (TR4) of [37] or, equivalently, [29, Proposition 1.4.12], we get maps $\gamma_n^{n+1} : C_n \to C_{n+1}$ and $\nu_n : C_{n+1} \to P_n$ that are part of morphisms of distinguished triangles

$$
\begin{array}{ccccccc}
\Sigma B_n & \xrightarrow{\sigma_n} & C_n & \xrightarrow{\rho_n} & A & \xrightarrow{\beta^n} & B_n \\
\downarrow{\scriptstyle \Sigma\beta_n^{n+1}} & & \downarrow{\scriptstyle \gamma_n^{n+1}} & & \| & & \downarrow{\scriptstyle \beta_n^{n+1}} \\
\Sigma B_{n+1} & \xrightarrow{\sigma_{n+1}} & C_{n+1} & \xrightarrow{\rho_{n+1}} & A & \xrightarrow{\beta^{n+1}} & B_{n+1} \\
\| & & \downarrow{\scriptstyle \nu_n} & & \downarrow{\scriptstyle \beta^n} & & \| \\
\Sigma B_{n+1} & \xrightarrow{\alpha_{n+1}} & P_n & \xrightarrow{\pi_n} & B_n & \xrightarrow{\beta_n^{n+1}} & B_{n+1}
\end{array}
\tag{14}
$$

and of a distinguished triangle

$$\Sigma P_n \xrightarrow{\sigma_n \circ \Sigma \pi_n} C_n \xrightarrow{\gamma_n^{n+1}} C_{n+1} \xrightarrow{\nu_n} P_n. \tag{15}$$

It follows by induction on n that $C_n \in \mathfrak{P}_n$ for all $n \in \mathbb{N}$. Since $\beta^n \in \mathfrak{I}^n$ by construction, the distinguished triangle (13) shows that $\rho_n \colon C_n \to A$ is a one-step $\mathfrak{I}^n$-projective resolution.

3.3.3 Spectral Sequences from the Adams Resolution

The Adams resolution gives rise to an exact couple and thus to a spectral sequence in a canonical way (our reference for exact couples and spectral sequences is [21]). We let $F \colon \mathfrak{T} \to \mathfrak{Ab}$ be a contravariant cohomological functor and define $F^n(A) := F(\Sigma^n A)$ for $n \in \mathbb{Z}$. We define $\mathbb{Z} \times \mathbb{N}$-graded Abelian groups

$$D_1^{pq} := F^{p+q}(B_p), \qquad E_1^{pq} := F^{p+q}(P_p),$$

and homomorphisms

$$
\begin{aligned}
i_1^{pq} &:= F^*(\beta_{p-1}^p) \colon D_1^{p,q} \to D_1^{p-1,q+1}, \\
j_1^{pq} &:= F^*(\pi_p) \colon D_1^{p,q} \to E_1^{p,q}, \\
k_1^{pq} &:= F^*(\alpha_p) \colon E_1^{p,q} \to D_1^{p+1,q}
\end{aligned}
$$

of bidegree

$$\deg i_1 = (-1,1), \qquad \deg j_1 = (0,0), \qquad \deg k_1 = (1,0).$$

Since F is cohomological, we get long exact sequences for the distinguished triangles (12). This means that $(D_1, E_1, i_1, j_1, k_1)$ is an exact couple. As in [21, Sect. XI.5] we form the derived exact couples $(D_r, E_r, i_r, j_r, k_r)$ for $r \in \mathbb{N}_{\geq 2}$ and let $d_r = j_r k_r \colon E_r \to E_r$. The map d_r has bidegree $(r, 1 - r)$ and the data (E_r, d_r) define a cohomological spectral sequence.

Now consider instead a covariant homological functor $F \colon \mathfrak{T} \to \mathfrak{Ab}$.

Let $F_n(A) := F(\Sigma^n A)$ for $n \in \mathbb{Z}$ and define $\mathbb{Z} \times \mathbb{N}$-graded Abelian groups

$$D_{pq}^1 := F_{p+q}(B_p), \qquad E_{pq}^1 := F_{p+q}(P_p)$$

and homomorphisms

$$
\begin{aligned}
i_{pq}^1 &:= F_*(\beta_p^{p+1}) \colon D_{p,q}^1 \to D_{p+1,q-1}^1, \\
j_{pq}^1 &:= F_*(\alpha_p) \colon D_{p,q}^1 \to E_{p-1,q}^1. \\
k_{pq}^1 &:= F_*(\pi_p) \colon E_{p,q}^1 \to D_{p,q}^1
\end{aligned}
$$

of bidegree

$$\deg i^1 = (1,-1), \qquad \deg j^1 = (-1,0), \qquad \deg k^1 = (0,0).$$

This is an exact couple because F is homological. We form derived exact couples $(D^r, E^r, i^r, j^r, k^r)$ for $r \in \mathbb{N}_{\geq 2}$ and let $d^r = j^r k^r$. This map has bidegree $(-r, r - 1)$, so that (E^r, d^r) is a homological spectral sequence.

The boundary maps d^1 and d_1 in the above spectral sequences are induced by the composition

$$\delta_n := \alpha_n \circ \Sigma \pi_{n+1} \colon \Sigma P_{n+1} \to P_n.$$

Letting $\delta_{-1} := \pi_0 \colon P_0 \to A$, we obtain a chain complex

$$A \xleftarrow{\delta_0} P_0 \xleftarrow{\delta_1} \Sigma P_1 \xleftarrow{\delta_2} \Sigma^2 P_2 \xleftarrow{\delta_3} \Sigma^3 P_3 \xleftarrow{\delta_4} \cdots$$

in $\mathfrak{T}$. This is an $\mathfrak{J}$-projective resolution of A.

Let $F \colon \mathfrak{T} \to \mathfrak{Ab}$ be a covariant functor. By construction, we have

$$E^2_{pq} \cong H_p\big(F_q(\Sigma^\bullet P_\bullet, \delta_\bullet)\big) \cong \mathbb{L}_p F_{p+q}(A).$$

Thus the second tableau of our spectral sequence comprises the derived functors of suspensions of F.

We do not analyse the convergence of the above spectral sequence here in detail. In general, we cannot hope for convergence towards $F(A)$ itself because the derived functors vanish if A is $\mathfrak{J}$-contractible, but $F(A)$ need not vanish. Thus we should replace F by $\mathbb{L}F$ right away. Under mild conditions, the spectral sequence converges towards $\mathbb{L}F$.

References

1. Asadollahi, J., Salarian, S.: Gorenstein objects in triangulated categories. J. Algebra **281**(1), 264–286 (2004) MR 2091971 (2006b:18011)
2. Beĭlinson, A.A., Bernstein, J., Deligne, P.: Faisceaux pervers, Analysis and topology on singular spaces, I (Luminy, 1981), Astérisque, vol. 100, pp. 5–171. Soc. Math. France, Paris, (1982) MR 751966 (86g:32015)
3. Beligiannis, A.: Relative homological algebra and purity in triangulated categories. J. Algebra **227**(1), 268–361 (2000) MR 1754234 (2001e:18012)
4. Blackadar, B.: K-theory for operator algebras. In: Mathematical Sciences Research Institute Publications, 2nd edn., vol. 5. Cambridge University Press, Cambridge (1998) MR 1656031 (99g:46104)
5. Bonkat, A.: Bivariante K-Theorie für Kategorien projektiver Systeme von C^*-Algebren. Ph.D. thesis, Westf. Wilhelms-Universität Münster (2002) (German). Electronically available at the Deutsche Nationalbibliothek at dokserv?idn=967387191
6. Brinkmann, H.-B.: Relative homological algebra and the Adams spectral sequence. Arch. Math. **19**, 137–155 (1968) MR 0230788 (37 #6348)
7. Brown, L.G., Green, P., Rieffel, M.A.: Stable isomorphism and strong Morita equivalence of C^*-algebras, Pac. J. Math. **71**(2), 349–363 (1977) MR 0463928 (57 #3866)
8. Christensen, J.D.: Ideals in triangulated categories: phantoms, ghosts and skeleta. Adv. Math. **136**(2), 284–339 (1998) MR 1626856 (99g:18007)
9. Connes, A.: An analogue of the Thom isomorphism for crossed products of a C^*-algebra by an action of **R**. Adv. Math. **39**(1), 31–55 (1981) MR 605351 (82j:46084)
10. Cuntz, J.: Generalized homomorphisms between C^*-algebras and KK-theory. In: Dynamics and processes (Bielefeld, 1981). Lecture Notes in Mathematics, vol. 1031, pp. 31–45. Springer, Berlin (1983) MR 733641 (85j:46126)

11. Cuntz, J.: A new look at KK-theory. K-Theory **1**(1), 31–51 (1987) MR 899916 (89a:46142)
12. Cuntz, J., Meyer, R., Rosenberg, J.M.: Topological and bivariant K-theory. In: Oberwolfach Seminars, vol. 36. Birkhäuser, Basel (2007) MR 2340673 (2008j:19001)
13. Eilenberg, S., Moore, J.C.: Foundations of relative homological algebra. Mem. Am. Math. Soc. **55**, 39 (1965) MR 0178036 (31 #2294)
14. Freyd, P.: Representations in abelian categories. In: Proceedings of the Conference on Categorical Algebra (La Jolla, Calif., 1965), pp. 95–120. Springer, New York (1966) MR 0209333 (35 #231)
15. Higson, N.: A characterization of KK-theory. Pac. J. Math. **126**(2), 253–276 (1987) MR 869779
16. Higson, N.: Algebraic K-theory of stable C^*-algebras. Adv. Math. **67**(1), 140 (1988) MR 922140 (89g:46110)
17. Kasparov, G.G.: The operator K-functor and extensions of C^*-algebras. Izv. Akad. Nauk SSSR Ser. Mat. **44**(3), 571–636, 719 (1980) MR 582160 (81m:58075)
18. Kasparov, G.G.: Equivariant KK-theory and the Novikov conjecture. Invent. Math. **91**(1), 147–201 (1988) MR 918241 (88j:58123)
19. Keller, B.: Derived categories and their uses. In: Handbook of algebra, vol. 1, pp. 671–701. North-Holland, Amsterdam (1996) MR 1421815 (98h:18013)
20. Lance, E.C.: Hilbert C^*-modules. A toolkit for operator algebraists. In: London Mathematical Society Lecture Note Series, vol. 210. Cambridge University Press, Cambridge (1995) MR 1325694 (96k:46100)
21. Mac Lane, S.: Homology. In: Classics in Mathematics. Springer, Berlin (1995) Reprint of the 1975 edition. MR 1344215 (96d:18001)
22. MacLane, S.: Categories for the working mathematician. In: Graduate Texts in Mathematics, vol. 5. Springer, New York (1971) MR 0354798 (50 #7275)
23. Meyer, R.: Equivariant Kasparov theory and generalized homomorphisms. K-Theory **21**(3), 201–228 (2000) MR 1803228 (2001m:19013)
24. Meyer, R.: Local and analytic cyclic homology. In: EMS Tracts in Mathematics, vol. 3. European Mathematical Society (EMS), Zürich (2007) MR 2337277 (2009g:46138)
25. Meyer, R.: Categorical aspects of bivariant K-theory, K-theory and noncommutative geometry. In: EMS Series of Congress Reports, pp. 1–39. European Mathematical Society, Zürich, (2008) MR 2513331
26. Meyer, R., Nest, R.: The Baum-Connes conjecture via localisation of categories. Topology **45**(2), 209–259 (2006) MR 2193334 (2006k:19013)
27. Meyer, R., Nest, R.: Homological algebra in bivariant K-theory and other triangulated categories. I, eprint (2007) arXiv: math.KT/0702146
28. Mingo, J.A., Phillips, W.J.: Equivariant triviality theorems for Hilbert C^*-modules. Proc. Am. Math. Soc. **91**(2), 225–230 (1984) MR 740176 (85f:46111)
29. Neeman, A.: Triangulated categories. In: Annals of Mathematics Studies, vol. 148. Princeton University Press, Princeton, NJ (2001)
30. Pimsner, M., Voiculescu, D.: Exact sequences for K-groups and Ext-groups of certain cross-product C^*-algebras, J. Oper. Theory **4**(1), 93–118 (1980) MR 587369 (82c:46074)
31. Puschnigg, M.: Diffeotopy functors of ind-algebras and local cyclic cohomology. Doc. Math. **8**, 143–245 (2003) (electronic) MR 2029166 (2004k:46128)
32. Rieffel, M.A.: Induced representations of C^*-algebras. Adv. Math. **13**, 176–257 (1974) MR 0353003 (50 #5489)
33. Rieffel, M.A.: Morita equivalence for C^*-algebras and W^*-algebras. J. Pure Appl. Algebra **5**, 51–96 (1974) MR 0367670 (51 #3912)

34. Rieffel, M.A.: Strong Morita equivalence of certain transformation group C^*-algebras. Math. Ann. **222**(1), 7–22 (1976) MR 0419677 (54 #7695)

35. Rosenberg, J., Schochet, C.: The Künneth theorem and the universal coefficient theorem for Kasparov's generalized K-functor. Duke Math. J. **55**(2), 431–474 (1987) MR 894590 (88i:46091)

36. Street, R.: Homotopy classification of filtered complexes. J. Aust. Math. Soc. **15**, 298–318 (1973) MR 0340380 (49 #5135)

37. Verdier, J.-L.: Des catégories dérivées des catégories abéliennes, Astérisque (1996), vol. 239, xii+253 pp. (1997), With a preface by Luc Illusie, Edited and with a note by Georges Maltsiniotis. MR 1453167 (98c:18007)

Algebraic v. Topological K-Theory: A Friendly Match

Guillermo Cortiñas

Dep. Matemática,
Facultad de Ciencias Exactas y Naturales,
Universidad de Buenos Aires,
Ciudad Universitaria Pab 1,
(1428) Buenos Aires, Argentina,
gcorti@dm.uba.ar

1 Introduction

These notes evolved from the lecture notes of a minicourse given in Swisk, the Sedano Winter School on K-theory held in Sedano, Spain, during the week January 22–27 of 2007, and from those of a longer course given in the University of Buenos Aires, during the second half of 2006. They intend to be an introduction to K-theory, with emphasis in the comparison between its algebraic and topological variants. We have tried to keep as elementary as possible. Section 2 introduces K_n for $n \leq 1$. Elementary properties such as matrix stability and excision are discussed. Section 3 is concerned with topological K-theory of Banach algebras; its excision property is derived from the excision sequence for algebraic K_0 and K_1. Cuntz' proof of Bott periodicity for C^*-algebras, via the C^*-Toeplitz extension, is sketched. In the next section we review Karoubi–Villamayor K-theory, which is an algebraic version of K^{top}, and has some formally similar properties, such as (algebraic) homotopy invariance, but does not satisfy excision in general. Section 5 discusses KH, Weibel's homotopy K-theory, which is introduced in a purely algebraic, spectrum-free manner. Several of its properties, including excision, homotopy invariance and the fundamental theorem, are proved. The parallelism between Bott periodicity and the fundamental theorem for KH is emphasized by the use of the algebraic Toeplitz extension in the proof of the latter. Quillen's higher K-theory is introduced in Sect. 6, via the plus construction of the classifying space of the general linear group. This is the first place where some algebraic topology is needed. The "décalage" formula $K_n \Sigma R = K_{n-1} R$ via Karoubi's suspension is proved, and some of the deep results of Suslin and Wodzicki on excision are discussed. Then the fundamental theorem for K-theory is reviewed, and its formal connection to Bott periodicity via the algebraic Toeplitz extension is established. The next section is the first of three devoted to the comparison between algebraic and topological K-theory of topological algebras. Using Higson's homotopy invariance theorem, and the excision results of Suslin and Wodzicki, we give proofs of the C^*- and Banach variants of Karoubi's conjecture, that algebraic and

P.F. Baum et al., *Topics in Algebraic and Topological K-Theory*,
Lecture Notes in Mathematics 2008, DOI 10.1007/978-3-642-15708-0_3,
© Springer-Verlag Berlin Heidelberg 2011

topological K-theory become isomorphic after stabilizing with respect to the ideal of compact operators (theorems of Suslin–Wodzicki and Wodzicki, respectively). Section 8 defines two variants of topological K-theory for locally convex algebras: KV^{dif} and KD which are formally analogue to KV and KH. Some of their basic properties are similar and derived with essentially the same arguments as their algebraic counterparts. We also give a proof of Bott periodicity for KD of locally convex algebras stabilized by the algebra of smooth compact operators. The proof uses the locally convex Toeplitz extension, and is modelled on Cuntz' proof of Bott periodicity for his bivariant K-theory of locally convex algebras. In Sect. 9 we review some of the results of [12]. Using the homotopy invariance theorem of Cuntz and Thom, we show that KH and KD agree on locally convex algebras stabilized by Fréchet operator ideals. The spectra for Quillen's and Weibel's K-theory, and the space for Karoubi–Villamayor K-theory are introduced in Sect. 10, where also the primary and secondary characters going from K-theory to cyclic homology are reviewed. The technical results of this section are used in the next, where we again deal with the comparison between algebraic and topological K-theory of locally convex algebras. We give proofs of the Fréchet variant of Karoubi's conjecture (due to Wodzicki), and of the 6-term exact sequence of [12], which relates algebraic K-theory and cyclic homology to topological K-theory of a stable locally convex algebra.

2 The Groups K_n for $n \leq 1$

Notation. Throughout these notes, A, B, C will be rings and R, S, T will be rings with unit.

2.1 Definition and Basic Properties of K_j for $j = 0, 1$

Let R be a ring with unit. Write $M_n R$ for the matrix ring. Regard $M_n R \subset M_{n+1} R$ via

$$a \mapsto \begin{bmatrix} a & 0 \\ 0 & 0 \end{bmatrix} \tag{1}$$

Put

$$M_\infty R = \bigcup_{n=1}^{\infty} M_n R$$

Note $M_\infty R$ is a ring (without unit). We write $\mathrm{Idem}_n R$ and $\mathrm{Idem}_\infty R$ for the set of idempotent elements of $M_n R$ and $M_\infty R$. Thus

$$M_\infty R \supset \mathrm{Idem}_\infty R = \bigcup_{n=1}^{\infty} \mathrm{Idem}_n R.$$

We write $\mathrm{GL}_n R = (M_n R)^*$ for the group of invertible matrices. Regard $\mathrm{GL}_n R \subset \mathrm{GL}_{n+1} R$ via

$$g \mapsto \begin{bmatrix} g & 0 \\ 0 & 1 \end{bmatrix}$$

Put

$$GLR := \bigcup_{n=1}^{\infty} GL_n R.$$

Note GLR acts by conjugation on $M_\infty R$, Idem$_\infty R$ and, of course, GLR.

For $a, b \in M_\infty R$ there is defined a *direct sum* operation

$$a \oplus b := \begin{bmatrix} a_{1,1} & 0 & a_{1,2} & 0 & a_{1,3} & 0 & \cdots \\ 0 & b_{1,1} & 0 & b_{1,2} & 0 & b_{1,3} & \cdots \\ a_{2,1} & 0 & a_{2,2} & 0 & a_{2,3} & 0 & \cdots \\ \vdots & \vdots & \vdots & \vdots & \vdots & \vdots & \ddots \end{bmatrix}. \tag{2}$$

We remark that if $a, b \in M_p R$ then $a \oplus b \in M_{2p} R$ and is conjugate, by a permutation matrix, to the usual direct sum

$$\begin{bmatrix} a & 0 \\ 0 & b \end{bmatrix}.$$

One checks that $\oplus$ is associative and commutative up to conjugation. Thus the coinvariants under the conjugation action

$$I(R) := ((\text{Idem}_\infty R)_{GLR}, \oplus)$$

form an abelian monoid.

Exercise 2.1.1. The operation (2) can be described as follows. Consider the decomposition $\mathbb{N} = \mathbb{N}_0 \sqcup \mathbb{N}_1$ into even and odd positive integers; write ϕ_i for the bijection $\phi_i : \mathbb{N} \to \mathbb{N}_i$, $\phi_i(n) = 2n - i$ $i = 0, 1$. The map ϕ_i induces an *R*-module monomorphism

$$\phi_i : R^{(\mathbb{N})} := \bigoplus_{n=1}^{\infty} R \to R^{(\mathbb{N}_i)} \subset R^{(\mathbb{N})}, \qquad e_n \mapsto e_{\phi_i(n)}.$$

We abuse notation and also write ϕ_i for the matrix of this homomorphism with respect to the canonical basis and ϕ_i^t for its transpose. Check the formula

$$a \oplus b = \phi_0 a \phi_0^t + \phi_1 a \phi_1^t.$$

Observe that the same procedure can be applied to any decomposition $\mathbb{N} = \mathbb{N}_0' \sqcup \mathbb{N}_1'$ into two infinite disjoint subsets and any choice of bijections $\phi_i' : \mathbb{N} \to \mathbb{N}_i'$, to obtain an operation $\oplus_{\phi'} : M_\infty R \times M_\infty R \to M_\infty R$. Verify that the operation so obtained defines the same monoid structure on the coinvariants $(M_\infty R)_{GLR}$, and thus also on $I(R)$.

Lemma 2.1.2. *Let M be an abelian monoid. Then there exist an abelian group M^+ and a monoid homomorphism $M \to M^+$ such that if $M \to G$ is any other such homomorphism, then there exists a unique group homomorphism $M^+ \to G$ such that*

commutes.

Proof. Let $F = \mathbb{Z}^{(M)}$ be the free abelian group on one generator e_m for each $m \in M$, and let $S \subset F$ be the subgroup generated by all elements of the form $e_{m_1} + e_{m_2} - e_{m_1+m_2}$. One checks that $M^+ = F/S$ satisfies the desired properties.   □

Definition 2.1.3.

$$K_0(R) := I(R)^+$$

$$K_1(R) := \frac{GLR}{[GLR, GLR]} = (GLR)_{ab}.$$

Here $[,]$ denotes the commutator subgroup, and the subscript $_{ab}$ indicates abelianization.

Proposition 2.1.4. *(see [42, Sect. 2.1])*

- $[GLR, GLR] = ER := <1 + ae_{i,j} : a \in R, i \neq j>$, *the subgroup of GLR generated by elementary matrices.*
- *If* $\alpha \in GL_n R$ *then*

$$\begin{bmatrix} \alpha & 0 \\ 0 & \alpha^{-1} \end{bmatrix} \in E_{2n}R \qquad \qquad (Whitehead's\ Lemma).$$

(Here $E_{2n}R = ER \cap GL_{2n}R$*).*

As a consequence of Whitehead's lemma above, if $\beta \in GL_n R$, then

$$\alpha\beta = \begin{bmatrix} \alpha\beta & 0 \\ 0 & 1_{n \times n} \end{bmatrix}$$

$$= \begin{bmatrix} \alpha & 0 \\ 0 & \beta \end{bmatrix} \begin{bmatrix} \beta & 0 \\ 0 & \beta^{-1} \end{bmatrix} \qquad (3)$$

$$\equiv \alpha \oplus \beta \quad -\text{mod}\, ER.$$

Exercise 2.1.5. Let R be a unital ring, and let ϕ' and $\oplus_{\phi'}$ be as in Exercise 2.1.1. Prove that $\oplus_{\phi'}$ and $\oplus$ define the same operation in $K_1(R)$, which coincides with the product of matrices.

Let $r \geq 1$. Then

$$p_r = 1_{r \times r} \in \text{Idem}_\infty R.$$

Because $p_r \oplus p_s = p_{r+s}$, the assignment $r \mapsto p_r$ defines a monoid homomorphism $\mathbb{N} \to I(R)$. Applying the group completion functor we obtain a group homomorphism

$$\mathbb{Z} = \mathbb{N}^+ \to I(R)^+ = K_0 R. \qquad (4)$$

Similarly, the inclusion $R^* = GL_1 R \subset GLR$ induces a homomorphism

$$R^*_{ab} \to K_1 R. \qquad (5)$$

Example 2.1.6. If F is a field, and $e \in \mathrm{Idem}_\infty F$ is of rank r, then e is conjugate to p_r; moreover p_r and p_s are conjugate $\iff r = s$. Thus (4) is an isomorphism in this case. Assume more generally that R is commutative. Then (4) is a split monomorphism. Indeed, there exists a surjective unital homomorphism $R \twoheadrightarrow F$ onto a field F; the induced map $K_0(R) \to K_0(F) = \mathbb{Z}$ is a left inverse of (4). Similarly, for commutative R, the homomorphism (5) is injective, since it is split by the map $det : K_1 R \to R^*$ induced by the determinant.

Example 2.1.7. The following are examples of rings for which the maps (4) and (5) are isomorphisms (see [42, Chap. 1, Sect. 3; Chap. 2, Sect. 2, Sect. 3]): fields, division rings, principal ideal domains and local rings. Recall that a ring R is a *local ring* if the subset $R \backslash R^*$ of noninvertible elements is an ideal of R. For instance if k is a field, then the k-algebra $k[\varepsilon] := k \oplus k\varepsilon$ with $\varepsilon^2 = 0$ is a local ring. Indeed $k[\varepsilon]^* = k^* + k\varepsilon$ and $k[\varepsilon] \backslash k[\varepsilon]^* = k\varepsilon \lhd k[\varepsilon]$.

Example 2.1.8. Here is an example of a local ring involving operator theory. Let H be a separable Hilbert space over $\mathbb{C}$; put $\mathscr{B} = \mathscr{B}(H)$ for the algebra of bounded operators. Write $\mathscr{K} \subset \mathscr{B}$ for the ideal of compact operators, and $\mathscr{F}$ for that of finite rank operators. The Riesz–Schauder theorem from elementary operator theory implies that if $\lambda \in \mathbb{C}^*$ and $T \in \mathscr{K}$ then there exists an $f \in \mathscr{F}$ such that $\lambda + T + f$ is invertible in $\mathscr{B}$. In fact one checks that if $\mathscr{F} \subset I \subset \mathscr{K}$ is an ideal of $\mathscr{B}$ such that $T \in I$ then the inverse of $\lambda + T + f$ is again in $\mathbb{C} \oplus I$. Hence the ring

$$R_I := \mathbb{C} \oplus I / \mathscr{F}$$

is local, and thus $K_0(R_I) = \mathbb{Z}$.

Remark 2.1.9. (K_0 *from projective modules*) In the literature, K_0 of a unital ring is often defined in terms of finitely generated projective modules. This approach is equivalent to ours, as we shall see presently. If R is a unital ring and $e \in \mathrm{Idem}_n R$, then left multiplication by e defines a right module homomorphism $R^n = R^{n \times 1} \to R^n$ with image eR^n. Similarly $(1 - e)R^n \subset R^n$ is a submodule, and we have a direct sum decomposition

$$R^n = eR^n \oplus (1 - e)R^n.$$

Hence eR^n is a finitely generated projective module, as it is a direct summand of a finitely generated free R-module. Note every finitely generated projective right R-module arises in this way for some n and some $e \in \mathrm{Idem}_n R$. Moreover, one checks that if $e \in \mathrm{Idem}_n R$ and $f \in \mathrm{Idem}_m R$, then the modules eR^n and fR^m are isomorphic if and only if the images of e and f in $\mathrm{Idem}_\infty R$ define the same class in $I(R)$ (see [42, Lemma 1.2.1]). Thus we have a natural bijection from the monoid $I(R)$ to the set $P(R)$ of isomorphism classes of finitely generated projective modules; further, one checks that the direct sum of idempotents corresponds to the direct sum of modules. Hence the monoids $I(R)$ and $P(R)$ are isomorphic, and therefore they have the same group completion:

$$K_0(R) = I(R)^+ = P(R)^+.$$

Additivity

If R_1 and R_2 are unital rings, then $M_\infty(R_1 \times R_2) \to M_\infty R_1 \times M_\infty R_2$ is an isomorphism. It follows from this that the natural map induced by the projections $R_1 \times R_2 \to R_i$ is an isomorphism:

$$K_j(R_1 \times R_2) \to K_j R_1 \oplus K_j R_2 \qquad (j = 0, 1).$$

Application: Extension to Nonunital Rings

If A is any (not necessarily unital) ring, then the abelian group $\tilde{A} = A \oplus \mathbb{Z}$ equipped with the multiplication

$$(a+n)(b+m) := ab + nm \qquad (a, b \in A, \ n, m \in \mathbb{Z}) \tag{6}$$

is a unital ring, with unit element $1 \in \mathbb{Z}$, and $\tilde{A} \to \mathbb{Z}$, $a + n \mapsto n$, is a unital homomorphism. Put

$$K_j(A) := \ker(K_j \tilde{A} \to K_j \mathbb{Z}) \qquad (j = 0, 1).$$

If A happens to have a unit, we have two definitions for $K_j A$. To check that they are the same, one observes that the map

$$\tilde{A} \to A \times \mathbb{Z}, \ a + n \mapsto (a + n \cdot 1, n) \tag{7}$$

is a unital isomorphism. One verifies that, under this isomorphism, $\tilde{A} \to \mathbb{Z}$ identifies with the projection $A \times \mathbb{Z} \to \mathbb{Z}$, and $\ker(K_j(\tilde{A}) \to K_j \mathbb{Z})$ with $\ker(K_j A \oplus K_j \mathbb{Z} \to K_j \mathbb{Z}) = K_j A$. Note that the same procedure works to extend any additive functor of unital rings unambiguously to all rings.

Notation. We write $\mathfrak{Ass}$ for the category of rings and ring homomorphisms, and $\mathfrak{Ass}_1$ for the subcategory of unital rings and unit preserving ring homomorphisms.

Remark 2.1.10. The functor GL : $\mathfrak{Ass}_1 \to \mathfrak{Grp}$ preserves products. Hence it extends to all rings by

$$\mathrm{GL}(A) := \ker(\mathrm{GL}(\tilde{A}) \to \mathrm{GL}\mathbb{Z})$$

It is a straightforward exercise to show that, with this definition, GL becomes a left exact functor in $\mathfrak{Ass}$; thus if $A \triangleleft B$ is an ideal embedding, then $\mathrm{GL}(A) = \ker(\mathrm{GL}(B) \to \mathrm{GL}(B/A))$. It is straightforward from this that the group $K_1 A$ defined above can be described as

$$K_1 A = \mathrm{GL}(A) / E(\tilde{A}) \cap \mathrm{GL}(A) \tag{8}$$

A little more work shows that $E(\tilde{A}) \cap \mathrm{GL}(A)$ is the smallest normal subgroup of $E(\tilde{A})$ which contains the elementary matrices $1 + ae_{i,j}$ with $a \in A$ (see [42, Sect. 2.5]).

Matrix Stability

Let R be a unital ring and $n \geq 2$. A choice of bijection $\phi : \mathbb{N} \times \mathbb{N}_{\leq n} \cong \mathbb{N}$ gives a ring isomorphism $\phi : M_\infty(M_n R) \cong M_\infty(R)$ which induces, for $j = 0, 1$, a group isomorphism $\phi_j : K_j(M_n R) \cong K_j R$. Next, consider the decomposition $\mathbb{N} = \mathbb{N}'_0 \sqcup \mathbb{N}'_1$, $\mathbb{N}'_0 = \phi(\mathbb{N} \times \{1\})$, $\mathbb{N}'_1 = \phi(\mathbb{N} \times \mathbb{N}_{<n} \setminus \mathbb{N} \times \{1\})$. Setting $\psi_0 : \mathbb{N} \to \mathbb{N}'_0$, $\psi_0(m) = \phi(m, 1)$ and choosing any bijection $\psi_1 : \mathbb{N} \to \mathbb{N}'_1$, we obtain, as in Exercise 2.1.1, a direct sum operation $\oplus_\psi : M_\infty R \times M_\infty R \to M_\infty R$. Set $\iota : R \mapsto M_n R$, $r \mapsto r e_{11}$. The composite of $M_\infty \iota$ followed by the isomorphism induced by ϕ is the map sending

$$e_{i,j}(r) \mapsto e_{\phi(i,1),\phi(j,1)}(r) = e_{i,j}(r) \oplus_\psi 0. \qquad (9)$$

By Exercise 2.1.1 the latter map induces the identity in K_0. Moreover, one checks that (9) induces the map $\mathrm{GL}(M_n R) \to \mathrm{GL}(M_n R)$, $g \mapsto g \oplus_\psi 1$, whence it also gives the identity in K_1, by Exercise 2.1.5. It follows that, for $j = 0, 1$, the map

$$K_j(\iota) : K_j(R) \to K_j(M_n R)$$

is an isomorphism, inverse to ϕ_j. Starting with a bijection $\phi : \mathbb{N} \times \mathbb{N} \to \mathbb{N}$ and using the same argument as above, one shows that also

$$K_j(\iota) : K_j(R) \to K_j(M_\infty R)$$

is an isomorphism.

Nilinvariance for K_0

If $I \lhd R$ is a nilpotent ideal, then $K_0(R) \to K_0(R/I)$ is an isomorphism. This property is a consequence of the well-known fact that nilpotent extensions admit idempotent liftings, and that any two liftings of the same idempotent are conjugate (see for example [3, Theorem 1.7.3]). Note that K_1 does not have the same property, as the following example shows.

Example 2.1.11. Let k be a field. Then by Example 2.1.7, $K_1(k[\varepsilon]) = k^* + k\varepsilon$ and $K_1(k) = k^*$. Thus $k[\varepsilon] \to k[\varepsilon]/\varepsilon k[\varepsilon] = k$ does not become an isomorphism under K_1.

Example 2.1.12. Let A be an abelian group; make it into a ring with the trivial product: $ab = 0 \ \forall a, b \in A$. The map $A \to \mathrm{GL}_1 A$, $a \mapsto 1 + a$ is an isomorphism of groups, and thus induces a group homomorphism $A \to K_1 A$. We are going to show that the latter map is an isomorphism. First of all, it is injective, since $\mathrm{GL}_1(\tilde{A}) \to K_1(\tilde{A})$ is (by Example 2.1.6) and since by definition, $K_1 A \subset K_1(\tilde{A})$. Second, note that if $\varepsilon = 1 + a e_{ij}$ is an elementary matrix with $a \in A$ and $g \in \mathrm{GL} A$, then $(\varepsilon g)_{ij} = g_{ij} + a$, and $(\varepsilon g)_{p,q} = g_{p,q}$ for $(p, q) \neq (i, j)$. Thus g is congruent to its diagonal in $K_1 A$. But by Whitehead's lemma, any diagonal matrix in $\mathrm{GL}(\tilde{A})$ is K_1-equivalent to its determinant (see (3)). This shows that $A \to K_1 A$ is surjective, whence an isomorphism.

Remark 2.1.13. The example above shows that K_1 is no longer matrix stable when extended to general nonunital rings. In addition, it gives another example of the failure of nilinvariance for K_1 of unital rings. It follows from Examples 2.1.11 and 2.1.12 that if k and ε are as in Example 2.1.11, then $K_1(k\varepsilon) = \ker(K_1(k[\varepsilon]) \to K_1(k))$. In Example 2.4.4 below, we give an example of a unital ring T such that $k\varepsilon$ is an ideal in T, and such that $\ker(T \to T/k\varepsilon) = 0$.

Exercise 2.1.14. Prove that K_0 and K_1 commute with filtering colimits; that is, show that if I is a small filtering category and $A : I \to \mathfrak{Ass}$ is a functor, then for $j = 0, 1$, the map $\mathrm{colim}_I K_j A_i \to K_j(\mathrm{colim}_I A_i)$ is an isomorphism.

2.2 Matrix-Stable Functors

Definition 2.2.1. *Let* $\mathfrak{C} \subset \mathfrak{Ass}$ *be a subcategory of the category of rings,* $S : \mathfrak{C} \to \mathfrak{C}$ *a functor, and* $\gamma : 1_{\mathfrak{C}} \to S$ *a natural transformation. If* $\mathfrak{D}$ *is any category,* $F : \mathfrak{C} \to \mathfrak{D}$ *a functor and* $A \in \mathfrak{C}$, *then we say that* F *is stable on* A *with respect to* (S, γ) *(or S-stable on A, for short) if the map* $F(\gamma_A) : F(A) \to F(S(A))$ *is an isomorphism. We say that* F *is S-stable if it is stable on every* $A \in \mathfrak{C}$.

Example 2.2.2. We showed in Sect. 2.1 that K_j is M_n and even M_∞-stable on unital rings; in both cases, the natural transformation of the definition above is $r \mapsto re_{11}$.

Exercise 2.2.3. Let $F : \mathfrak{Ass} \to \mathfrak{Ab}$ be a functor and A a ring. Prove:
(i) The following are equivalent:

- For all $n, p \in \mathbb{N}$, F is M_p-stable on $M_n A$
- For all $n \in \mathbb{N}$, F is M_2-stable on $M_n A$

In particular, an M_2-stable functor is M_n-stable, for all n.
(ii) If F is M_∞-stable on both A and $M_n A$, then F is M_n-stable on A. In particular, if F is M_∞-stable, then it is M_n-stable for all n.

Lemma 2.2.4. *Let* $F : \mathfrak{Ass} \to \mathfrak{D}$ *be a functor, and* $A \in \mathfrak{Ass}$. *Assume* F *is* M_2-stable *on both* A *and* $M_2 A$. *Then the inclusions* $\iota_0, \iota_1 : A \to M_2 A$

$$\iota_0(a) = ae_{11}, \qquad \iota_1(a) = ae_{22}$$

induce the same isomorphism $FA \to FM_2 A$.

Proof. Consider the composites $j_0 = \iota_0 M_2 \circ \iota_0$ and $j_1 = \iota_0 M_2 \circ \iota_1$, and the matrices

$$J_2 = \begin{bmatrix} 0 & 1 & 0 & 0 \\ 1 & 0 & 0 & 0 \\ 0 & 0 & 1 & 0 \\ 0 & 0 & 0 & 1 \end{bmatrix}, \quad J_3 = \begin{bmatrix} 0 & 0 & 1 & 0 \\ 1 & 0 & 0 & 0 \\ 0 & 1 & 0 & 0 \\ 0 & 0 & 0 & 1 \end{bmatrix} \in \mathrm{GL}_4 \mathbb{Z}.$$

Conjugation by J_i induces an automorphism σ_i of $M_4 A = M_2 M_2 A$ of order i such that

$$\sigma_i j_0 = j_1 \qquad (i = 2, 3).$$

Since $F(j_0)$ is an isomorphism, and the orders of σ_2 and σ_3 are relatively prime, it follows that $F(\sigma_2) = F(\sigma_3) = 1_{F(M_4A)}$ and hence that $F(j_0) = F(j_1)$ and $F(\iota_0) = F(\iota_1)$.   □

Exercise 2.2.5. Let F and A be as in Lemma 2.2.4. Assume in addition that $\mathcal{D}$ and F are additive. Consider the map

$$\text{diag} : A \times A \to M_2A, \quad \text{diag}(a,b) = \begin{bmatrix} a & 0 \\ 0 & b \end{bmatrix}.$$

Prove that the composite

$$F(A) \oplus F(A) = F(A \times A) \xrightarrow{F(\text{diag})} F(M_2A) \xrightarrow{F(i_0)^{-1}} F(A)$$

is the codiagonal map (i.e. it restricts to the identity on each copy of $F(A)$).

Proposition 2.2.6. *Let F and A be as in Lemma 2.2.4, $A \subset B$ an overring, and $V, W \in B$ elements such that*

$$WA, AV \subset A, \quad aVWa' = aa' \quad (a, a' \in A).$$

Then

$$\phi^{V,W} : A \to A, \quad a \mapsto WaV$$

is a ring homomorphism, and

$$F(\phi^{V,W}) = 1_{F(A)}.$$

Proof. We may assume that B is unital. Consider the elements $V \oplus 1$ and $W \oplus 1 \in M_2B$. The hypothesis guarantee that both $\phi := \phi^{V,W}$ and $\phi' := \phi^{V \oplus 1, W \oplus 1} : M_2A \to M_2A$ are well-defined ring homomorphisms. Moreover, $\phi'\iota_1 = \iota_1$ and $\phi'\iota_0 = \iota_0\phi$. It follows that $F(\phi')$ and $F(\phi)$ are the identity maps, by Lemma 2.2.4.   □

Exercise 2.2.7.
(i) Let R be a unital ring and L a free, finitely generated R-module of rank n. A choice of basis $\mathfrak{B}$ of L gives an isomorphism $\phi = \phi_{\mathfrak{B}} : M_nR \to \text{End}_R L$. Use Proposition 2.2.6 to show that $K_j(\phi)$ is independent of the choice of $\mathfrak{B}$ $(j = 0, 1)$.

(ii) Assume R is a field. If $e \in \text{End}_R L$ is idempotent, then $\iota_e : R \to \text{End}_R L$, $x \mapsto xe$ is a ring monomorphism. Show that if $e \in \text{End}_R L$ is of rank 1, then $K_j(\iota_e) = K_j(\phi\iota)$. In particular, $K_j(\iota_e)$ is independent of the choice of the rank-one idempotent e.

(iii) Let H and $\mathscr{F}$ be as in Example 2.1.8. If $V \subset W \subset H$ are finite dimensional subspaces and $U = V^\perp \cap W$ then the decomposition $W = V \oplus U$ induces an inclusion $\text{End}_\mathbb{C}(V) \subset \text{End}_\mathbb{C}(W)$. Show that

$$\mathscr{F} = \bigcup_{\dim V < \infty} \text{End}_\mathbb{C}(V)$$

(iv) Prove that if $e \in \mathscr{F}$ is any self-adjoint, rank-one idempotent, then the inclusion $\mathbb{C} \to \mathscr{F}, x \mapsto xe$, induces an isomorphism $K_j(\mathbb{C}) \xrightarrow{\cong} K_j(\mathscr{F})$. Show moreover that this isomorphism is independent of the choice of e.

2.3 Sum Rings and Infinite Sum Rings

Recall from [51] that a *sum ring* is a unital ring R together with elements α_i, β_i, $i = 0, 1$ such that the following identities hold

$$\alpha_0 \beta_0 = \alpha_1 \beta_1 = 1$$
$$\beta_0 \alpha_0 + \beta_1 \alpha_1 = 1 \tag{10}$$

If R is a sum ring, then

$$\boxplus : R \times R \to R, \tag{11}$$
$$(a, b) \mapsto a \boxplus b = \beta_0 a \alpha_0 + \beta_1 b \alpha_1$$

is a unital ring homomorphism. An *infinite sum ring* is a sum ring R together with a unit preserving ring homomorphism $\infty : R \to R$, $a \mapsto a^\infty$ such that

$$a \boxplus a^\infty = a^\infty \qquad (a \in R). \tag{12}$$

Proposition 2.3.1. *Let $\mathfrak{D}$ be an additive category, $F : \mathfrak{Ass} \to \mathfrak{D}$ a functor, and R a sum ring. Assume that the sum of projection maps $\gamma = F\pi_0 + F\pi_1 : F(R \times R) \to FR \oplus FR$ is an isomorphism, and that F is M_2-stable on both R and $M_2 R$. Then the composite*

$$F(R) \oplus F(R) \xrightarrow{\gamma^{-1}} F(R \times R) \xrightarrow{F(\boxplus)} F(R)$$

is the codiagonal map; that is, it restricts to the identity on each copy of $F(R)$. If moreover R is an infinite sum ring, then $F(R) = 0$.

Proof. Let $j_0, j_1 : R \to R \times R$, $j_0(x) = (x, 0)$, $j_1(x) = (0, x)$. Note that $\gamma^{-1} = F j_0 + F j_1$. Because F is M_2-stable on both R and $M_2 R$, $F(\boxplus) F(j_i) = 1_{F(R)}$, by Proposition 2.2.6. Thus $F(\boxplus) \circ \gamma^{-1}$ is the codiagonal map, as claimed. It follows that if $\alpha, \beta : R \to R$ are homomorphisms, then $F\alpha + F\beta = F(\boxplus(\alpha, \beta))$. In particular, if R is an infinite sum ring, then

$$F(\infty) + 1_{F(R)} = F(\infty) + F(1_R) = F(\boxplus(\infty, 1_R)) = F(\infty).$$

Thus $1_{F(R)} = 0$, whence $F(R) = 0$. $\qquad\square$

Example 2.3.2. Let A be a ring. Write ΓA for the ring of all $\mathbb{N} \times \mathbb{N}$ matrices $a = (a_{i,j})_{i,j \geq 1}$ which satisfy the following two conditions:

(i) The set $\{a_{ij} : i, j \in \mathbb{N}\}$ is finite.
(ii) There exists a natural number $N \in \mathbb{N}$ such that each row and each column has at most N nonzero entries.

It is an exercise to show that ΓA is indeed a ring and that $M_\infty A \subset \Gamma A$ is an ideal. The ring ΓA is called (Karoubi's) *cone ring*; the quotient $\Sigma A := \Gamma A / M_\infty A$ is the *suspension* of A. A useful fact about Γ and Σ is that the well-known isomorphism $M_\infty \mathbb{Z} \otimes A \cong M_\infty A$ extends to Γ, so that there are isomorphisms (see [11, Lemma 4.7.1])

$$\Gamma \mathbb{Z} \otimes A \overset{\cong}{\to} \Gamma A \text{ and } \Sigma \mathbb{Z} \otimes A \overset{\cong}{\to} \Sigma A. \tag{13}$$

Let R be a unital ring. One checks that the following elements of ΓR satisfy the identities (10):

$$\alpha_0 = \sum_{i=1}^{\infty} e_{i,2i}, \quad \beta_0 = \sum_{i=1}^{\infty} e_{2i,i}, \quad \alpha_1 = \sum_{i=1}^{\infty} e_{i,2i-1}, \quad \text{and} \quad \beta_1 = \sum_{i=1}^{\infty} e_{2i-1,i}.$$

Let $a \in \Gamma R$. Because the map $\mathbb{N} \times \mathbb{N} \to \mathbb{N}$, $(k,i) \mapsto 2^{k+1}i + 2^k - 1$, is injective, the following assignment gives a well-defined, $\mathbb{N} \times \mathbb{N}$-matrix

$$\phi^{\infty}(a) = \sum_{k=0}^{\infty} \beta_1^k \beta_0 a \alpha_0 \alpha_1^k = \sum_{k,i,j} e_{2^{k+1}i+2^k-1, 2^{k+1}j+2^k-1} \otimes a_{i,j}. \tag{14}$$

One checks that $\alpha_1 \beta_0 = \alpha_0 \beta_1 = 0$ and $\alpha_0 \alpha_1^i \beta_1^j \beta_0 = \delta_{ij}$. It follows from this that ϕ^{∞} is a ring endomorphism of ΓR; it is straightforward that (12) is satisfied too. In particular $K_n \Gamma R = 0$ for $n = 0, 1$.

Exercise 2.3.3. Let A be a ring. If $m = (m_{i,j})$ is an $\mathbb{N} \times \mathbb{N}$-matrix with coefficients in A, and $x \in M_{\infty}\tilde{A}$, then both $m \cdot x$ and $x \cdot m$ are well-defined $\mathbb{N} \times \mathbb{N}$-matrices. Put

$$\Gamma^{\ell} A := \{ m \in M_{\mathbb{N} \times \mathbb{N}} A : m \cdot M_{\infty}\tilde{A} \subset M_{\infty} A \supset M_{\infty}\tilde{A} \cdot m \}. \tag{15}$$

Prove
(i) $\Gamma^{\ell} A$ consists of those matrices in $M_{\mathbb{N} \times \mathbb{N}} A$ having finitely many nonzero elements in each row and column. In particular, $\Gamma^{\ell} A \supset \Gamma A$.
(ii) The usual matrix sum and product operations make $\Gamma^{\ell} A$ into a ring.
(iii) If R is a unital ring then $\Gamma^{\ell} R$ is an infinite sum ring.

Remark 2.3.4. The ring $\Gamma^{\ell} A$ is the cone ring considered by Wagoner in [51], where it was denoted ℓA. The notion of infinite sum ring was introduced in *loc. cit.*, where it was also shown that if R is unital, then $\Gamma^{\ell} R$ is an example of such a ring.

Exercise 2.3.5. Let $F : \mathfrak{Ass} \to \mathfrak{Ab}$ be a functor. Assume F is both additive and M_2-stable for unital rings and for rings of the form $M_{\infty} R$, with R unital. Show that if R is a unital ring, then the direct sum operation (2), induces the group operation in $F(M_{\infty} R)$, and that the same is true of any of the other direct sum operations of Exercise 2.1.1.

Exercise 2.3.6. Let $\mathscr{B}$ and H be as in Example 2.1.8. Choose a Hilbert basis $\{e_i\}_{i \geq 1}$ of H, and regard $\mathscr{B}$ as a ring of $\mathbb{N} \times \mathbb{N}$ matrices. With these identifications, show that $\mathscr{B} \supset \Gamma \mathbb{C}$. Deduce from this that $\mathscr{B}$ is a sum ring. Further show that (14) extends to $\mathscr{B}$, so that the latter is in fact an infinite sum ring.

2.4 The Excision Sequence for K_0 and K_1

A reason for considering K_0 and K_1 as part of the same theory is that they are connected by a long exact sequence, as shown in Theorem 2.4.1 below. We need some notation. Let

$$0 \to A \to B \to C \to 0 \tag{16}$$

be an exact sequence of rings. If $\hat{g} \in M_n B$ maps to an invertible matrix $g \in GL_n C$ and $\hat{g}^*$ maps to g^{-1}, then

$$h = h(\hat{g}, \hat{g}^*) := \begin{bmatrix} 2\hat{g} - \hat{g}\hat{g}^*\hat{g} & \hat{g}\hat{g}^* - 1 \\ 1 - \hat{g}^*\hat{g} & \hat{g}^* \end{bmatrix} \tag{17}$$

$$= \begin{bmatrix} 1 & \hat{g} \\ 0 & 1 \end{bmatrix} \cdot \begin{bmatrix} 1 & 0 \\ -\hat{g}^* & 1 \end{bmatrix} \cdot \begin{bmatrix} 1 & \hat{g} \\ 0 & 1 \end{bmatrix} \cdot \begin{bmatrix} 0 & -1 \\ 1 & 0 \end{bmatrix} \in E_{2n}(\tilde{B}) \subset GL_{2n}(\tilde{B})$$

Note that h maps to $\text{diag}(g, g^{-1}) \in GL_{2n}(C)$. Thus $h p_n h^{-1}$ maps to p_n, whence $h p_n h^{-1} - p_n \in M_{2n} A$ and $h p_n h^{-1} \in M_{2n} \tilde{A}$. Put

$$\partial(\hat{g}, \hat{g}^*) := [h p_n h^{-1}] - [p_n] \in \ker(K_0(\tilde{A}) \to K_0 \mathbb{Z}) = K_0 A \tag{18}$$

Theorem 2.4.1. *If* (16) *is an exact sequence of rings, then there is a long exact sequence*

$$\begin{array}{ccc}
K_1 A \longrightarrow K_1 B \longrightarrow & K_1 C \\
& & \downarrow \partial \\
K_0 C \longleftarrow K_0 B \longleftarrow & K_0 A
\end{array}$$

The map ∂ sends the class of an element $g \in GL_n C$ to the class of the element (18); *in particular the latter depends only on the K_1-class of g.*

Proof. (Sketch) The exactness of the top row of the sequence of the theorem is straightforward. Putting together [42, Thms. 1.5.5, 1.5.9 and 2.5.4] we obtain the theorem for those sequences (16) in which $B \to C$ is a unital homomorphism. It follows that we have a map of exact sequences

$$\begin{array}{ccccccccc}
K_1 \tilde{B} & \longrightarrow & K_1 \tilde{C} & \longrightarrow & K_0 A & \longrightarrow & K_0 \tilde{B} & \longrightarrow & K_0 \tilde{C} \\
\downarrow & & \downarrow & & \downarrow & & \downarrow & & \downarrow \\
K_1 \mathbb{Z} & = & K_1 \mathbb{Z} & \longrightarrow & 0 & \longrightarrow & K_0 \mathbb{Z} & = & K_0 \mathbb{Z}
\end{array}$$

Taking kernels of the vertical maps, we obtain an exact sequence

$$K_1 B \longrightarrow K_1 C \longrightarrow K_0 A \longrightarrow K_0 B \longrightarrow K_0 C$$

It remains to show that the map $K_1 C \to K_0 A$ of this sequence is given by the formula of the theorem. This is done by tracking down the maps and identifications of the proofs of [42, Thms. 1.5.5, 1.5.9 and 2.5.4] (see also [36, Sect. 3, Sect. 4]), and computing the idempotent matrices to which the projective modules appearing there correspond, taking into account that $B \to C$ sends the matrix $h \in GL_{2n} \tilde{B}$ of (17) to the diagonal matrix $\text{diag}(g, g^{-1}) \in GL_{2n} C$. □

Remark 2.4.2. In [42, Theorem 2.5.4], a sequence similar to that of the theorem above is obtained, in which $K_1 A$ is replaced by a relative K_1-group $K_1(B : A)$, depending on both A and B. For example if $B \to B/A$ is a split surjection, then [42, Exer. 2.5.19]

$$K_1(B : A) = \ker(K_1 B \to K_1(B/A))$$

The groups $K_1(B : A)$ and $K_1 A$ are not isomorphic in general (see Example 2.4.4 below); however their images in $K_1 B$ coincide. We point out also that the theorem above can be deduced directly from Milnor's Mayer-Vietoris sequence for a Milnor square [36, Sect. 4].

The following corollary is immediate from the theorem.

Corollary 2.4.3. *Assume* (16) *is split by a ring homomorphism $C \to B$. Then $K_0 A \to K_0 B$ is injective, and induces an isomorphism*

$$K_0 A = \ker(K_0 B \to K_0 C)$$

Because of this we say that K_0 is split exact.

Example 2.4.4. (Swan's example [48]) We shall give an example which shows that K_1 is not split exact. Let k be a field with at least three elements (i.e. $k \neq \mathbb{F}_2$). Consider the ring of upper triangular matrices

$$T := \begin{bmatrix} k & k \\ 0 & k \end{bmatrix}$$

with coefficients in k. The set I of strictly upper triangular matrices forms an ideal of T, isomorphic as a ring, to the ideal $k\varepsilon \lhd k[\varepsilon]$, via the identification $\varepsilon = e_{12}$. By Examples 2.1.11 and 2.1.12, $\ker(K_1(k[\varepsilon]) \to K_1(k)) = K_1(k\varepsilon) \cong k\varepsilon$, the additive group underlying k. If K_1 were split exact, then also

$$K_1(T : I) = \ker(K_1 T \to K_1(k \times k)) \tag{19}$$

should be isomorphic to $k\varepsilon$. However we shall see presently that $K_1(T : I) = 0$. Note that $T \to k \times k$ is split by the natural inclusion $\mathrm{diag} : k \times k \to T$. Thus any element of $K_1(T : I)$ is the class of an element in $\mathrm{GL}(k\varepsilon)$, and by Example 2.1.11 it is congruent to the class of an element in $\mathrm{GL}_1(k\varepsilon) = 1 + k\varepsilon$. We shall show that if $\lambda \in k$, then $1 + \lambda\varepsilon \in [\mathrm{GL}_1 T, \mathrm{GL}_1 T]$. Because we are assuming that $k \neq \mathbb{F}_2$, there exists $\mu \in k - \{0, 1\}$; one checks that

$$1 + \lambda\varepsilon = \begin{bmatrix} 1 & \lambda \\ 0 & 1 \end{bmatrix} = \left[\begin{bmatrix} \mu & 0 \\ 0 & 1 \end{bmatrix}, \begin{bmatrix} 1 & \frac{\lambda}{\mu-1} \\ 0 & 1 \end{bmatrix} \right] \in [\mathrm{GL}_1 T, \mathrm{GL}_1 T].$$

Example 2.4.5. Let R be a unital ring. Applying the theorem above to the cone sequence

$$0 \to M_\infty R \to \Gamma R \to \Sigma R \to 0 \tag{20}$$

we obtain an isomorphism

$$K_1 \Sigma R = K_0 R. \tag{21}$$

Exercise 2.4.6. Use Corollary 2.4.3 to prove that all the properties of K_0 stated in Sect. 2.1 for unital rings, remain valid for all rings. Further, show that $K_0(\Gamma A) = 0$ for all rings A, and thus that for any ring A, the boundary map gives a surjection

$$K_1 \Sigma A \twoheadrightarrow K_0 A.$$

2.5 Negative K-Theory

Definition 2.5.1. *Let A be a ring and $n \geq 0$. Put*

$$K_{-n}A := K_0 \Sigma^n A.$$

Proposition 2.5.2.
(i) For $n \leq 0$, the functors $K_n : \mathfrak{Ass} \to \mathfrak{Ab}$ are additive, nilinvariant and M_∞-stable.
(ii) The exact sequence of Theorem 2.4.1 extends to negative K-theory. Thus if

$$0 \to A \to B \to C \to 0$$

is a short exact sequence of rings, then for $n \leq 0$ we have a long exact sequence

$$
\begin{array}{ccc}
K_n A & \longrightarrow K_n B & \longrightarrow K_n C \\
& & \Big\downarrow \partial \\
K_{n-1} C \longleftarrow & K_{n-1} B \longleftarrow & K_{n-1} A
\end{array}
$$

Proof.
(i) By (13), we have $\Sigma A = \Sigma \mathbb{Z} \otimes A$. Thus Σ commutes with finite products and with M_∞, and sends nilpotent rings to nilpotent rings. Moreover, Σ is exact, because both M_∞ and Γ are. Hence the general case of (i) follows from the case $n = 0$, which is proved in Sect. 2.1.

(ii) Consider the sequence
$$0 \to A \to \tilde{B} \to \tilde{C} \to 0$$

Applying Σ, we obtain
$$0 \to \Sigma A \to \Sigma \tilde{B} \to \Sigma \tilde{C} \to 0$$

By (21), if D is any ring, then $K_0 \tilde{D} = K_1 \Sigma \tilde{D}$. Thus by Theorem 2.4.1 and Corollary 2.4.3, we get an exact sequence

$$
\begin{array}{ccc}
K_0 A & \longrightarrow K_0 B \oplus K_0 \mathbb{Z} & \longrightarrow K_0 C \oplus K_0 \mathbb{Z} \\
& & \Big\downarrow \partial \\
K_{-1} C \oplus K_{-1} \mathbb{Z} \longleftarrow & K_{-1} B \oplus K_{-1} \mathbb{Z} \longleftarrow & K_{-1} A
\end{array}
$$

Splitting off the $K_j \mathbb{Z}$ summands, we obtain

$$
\begin{array}{ccc}
K_0 A & \longrightarrow K_0 B & \longrightarrow K_0 C \\
& & \Big\downarrow \partial \\
K_{-1} C \longleftarrow & K_{-1} B \longleftarrow & K_{-1} A
\end{array}
$$

This proves the case $n = 0$ of the proposition. The general case follows from this. $\square$

Example 2.5.3. Let $\mathscr{B}$ be as in Example 2.1.8 and $I \lhd \mathscr{B}$ a proper ideal. It is classical that $\mathscr{F} \subset I \subset \mathscr{K}$ for any such ideal. We shall show that the map

$$K_0\mathscr{F} \to K_0 I \qquad (22)$$

is an isomorphism; thus $K_0 I = K_0 \mathscr{F} = \mathbb{Z}$, by Exercise 2.2.7. As in Example 2.1.8 we consider the local ring $R_I = \mathbb{C} \oplus I/\mathscr{F}$. We have a commutative diagram with exact rows and split exact columns

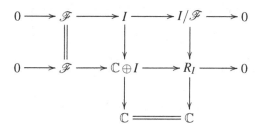

By Example 2.1.8 and split exactness, $K_0(I/\mathscr{F}) = 0$. Thus the map (22) is onto. From the diagram above, $K_1(I/\mathscr{F}) \to K_0\mathscr{F}$ factors through $K_1(R_I) \to K_0\mathscr{F}$. But it follows from the discussion of Example 2.1.8 that the map $K_1(\mathbb{C} \oplus I) \to K_1(R_I)$ is onto, whence $K_1(R_I) \to K_0\mathscr{F}$ and thus also $K_1(I/\mathscr{F}) \to K_0\mathscr{F}$, are zero. Thus (22) is an isomorphism.

Exercise 2.5.4. A theorem of Karoubi asserts that $K_{-1}(\mathscr{K}) = 0$ [30]. Use this and excision to show that $K_{-1}(I) = 0$ for any operator ideal I.

Exercise 2.5.5. Prove that if $n < 0$, then the functor K_n commutes with filtering colimits.

Remark 2.5.6. The definition of negative K-theory used here is taken from Karoubi–Villamayor's paper [33], where cohomological notation is used. Thus what we call $K_n A$ here is denoted $K^{-n}A$ in *loc. cit.* ($n \leq 0$). There is also another definition, due to Bass [2]. A proof that the two definitions agree is given in [33, Théorème 7.9].

3 Topological K-Theory

We saw in the previous section (Example 2.4.4) that K_1 is not split exact. It follows from this that there is no way of defining higher K-groups such that the long exact sequence of Theorem 2.4.1 can be extended to higher K-theory. This motivates the question of whether this problem could be fixed if we replaced K_1 by some other functor. This is successfully done in topological K-theory of Banach algebras.

3.1 Topological K-Theory of Banach Algebras

A *Banach* ($\mathbb{C}$-) algebra is a $\mathbb{C}$-algebra together with a norm $|| \; ||$ which makes it into a Banach space and is such that $||xy|| \leq ||x|| \cdot ||y||$ for all $x, y \in A$. If A is a Banach algebra then its $\mathbb{C}$-unitalization is the unital Banach algebra

$$\tilde{A}_{\mathbb{C}} = A \oplus \mathbb{C}$$

equipped with the product (6) and the norm $||a + \lambda|| := ||a|| + |\lambda|$. An algebra homomorphism is a morphism of Banach algebras if it is continuous. If X is a compact Hausdorff space and V is any topological vectorspace, we write $\mathscr{C}(X, V)$ for the topological vectorspace of all continuous maps $X \to V$. If A is a Banach algebra, then $\mathscr{C}(X, A)$ is again a Banach algebra with norm $||f||_\infty := \sup_x ||f(x)||$. If X is locally compact, X^+ its one point compactification, and V a topological vectorspace, we write

$$V(X) = \mathscr{C}_0(X, V) = \{ f \in \mathscr{C}(X^+, V) : f(\infty) = 0 \}.$$

Note that if X is compact, $V(X) = \mathscr{C}(X, V)$. If A is a Banach algebra then $A(X)$ is again a Banach algebra, as it is the kernel of the homomorphism $\mathscr{C}(X^+, A) \to A$, $f \mapsto f(\infty)$. For example, $A[0, 1]$ is the algebra of continous functions $[0, 1] \to A$, and $A(0, 1]$ and $A(0, 1)$ are identified with the ideals of $A[0, 1]$ consisting of those functions which vanish respectively at 0 and at both endpoints. Two homomorphisms $f_0, f_1 : A \to B$ of Banach algebras are called *homotopic* if there exists a homomorphism $H : A \to B[0, 1]$ such that the following diagram commutes.

$$
\begin{array}{ccc}
 & & B[0,1] \\
 & \overset{H}{\nearrow} & \big\downarrow {\scriptstyle (\mathrm{ev}_0, \mathrm{ev}_1)} \\
A & \underset{(f_0, f_1)}{\longrightarrow} & B \times B
\end{array}
$$

A functor G from Banach algebras to a category $\mathfrak{D}$ is called *homotopy invariant* if it maps homotopic maps to equal maps.

Exercise 3.1.1. Prove that G is homotopy invariant if and only if for every Banach algebra A the map $G(A) \to G(A[0, 1])$ induced by the natural inclusion $A \subset A[0, 1]$ is an isomorphism.

Theorem 3.1.2. *([42, Corollary 1.6.11]) The functor* $K_0 : ((\text{Banach Algebras})) \to \mathfrak{Ab}$ *is homotopy invariant.*

Example 3.1.3. K_1 *is not homotopy invariant.* The algebra $A := \mathbb{C}[\varepsilon]$ is a Banach algebra with norm $||a + b\varepsilon|| = |a| + |b|$. Both the inclusion $\iota : \mathbb{C} \to A$ and the projection $\pi : A \to \mathbb{C}$ are homomorphisms of Banach algebras; they satisfy $\pi\iota = 1$. Moreover the map $H : A \to A[0, 1]$, $H(a + b\varepsilon)(t) = a + tb\varepsilon$ is also a Banach algebra homomorphism, and satisfies $\mathrm{ev}_0 H = \iota\pi$, $\mathrm{ev}_1 H = 1$. Thus any homotopy invariant functor G sends ι and π to inverse homomorphisms; since K_1 does not do so by Example 2.1.11, it is not homotopy invariant.

Next we consider a variant of K_1 which is homotopy invariant.

Definition 3.1.4. *Let R be a unital Banach algebra. Put*

$$\mathrm{GL}R_0 := \{g \in \mathrm{GL}R : \exists h \in \mathrm{GL}(R[0,1]) : h(0) = 1, h(1) = g\}.$$

Note $\mathrm{GL}(R)_0 \lhd \mathrm{GL}R$. *The* topological K_1 *of R is*

$$K_1^{\mathrm{top}}R = \mathrm{GL}R / \mathrm{GL}(R)_0.$$

Exercise 3.1.5. Show that if we regard $\mathrm{GL}R = \mathrm{colim}_n \mathrm{GL}_n R$ with the weak topology inherited from the topology of R, then $K_1^{\mathrm{top}}R = \pi_0(\mathrm{GL}R) = \mathrm{colim}_n \pi_0(\mathrm{GL}_n R)$. Then show that K_1^{top} is homotopy invariant.

Note that if R is a unital Banach algebra, $a \in R$ and $i \neq j$, then $1 + tae_{i,j} \in E(R[0,1])$ is a path connecting 1 to the elementary matrix $1 + ae_{i,j}$. Thus $ER = [\mathrm{GL}(R), \mathrm{GL}(R)] \subset \mathrm{GL}(R)_0$, whence $\mathrm{GL}(R)_0$ is normal, and we have a surjection

$$K_1 R \twoheadrightarrow K_1^{\mathrm{top}}R. \tag{23}$$

In particular, $K_1^{\mathrm{top}}R$ is an abelian group.

Example 3.1.6. Because $\mathbb{C}$ is a field, $K_1\mathbb{C} = \mathbb{C}^*$. Since on the other hand $\mathbb{C}^*$ is path connected, we have $K_1^{\mathrm{top}}\mathbb{C} = 0$.

Note that K_1^{top} is additive. Thus we can extend K_1^{top} to nonunital Banach algebras in the usual way, i.e.

$$K_1^{\mathrm{top}}A := \ker(K_1^{\mathrm{top}}(\tilde{A}_{\mathbb{C}}) \to K_1^{\mathrm{top}}\mathbb{C}) = K_1^{\mathrm{top}}(\tilde{A}_{\mathbb{C}})$$

Exercise 3.1.7. Show that if A is a (not necessarily unital) Banach algebra, then

$$K_1^{\mathrm{top}}A = \mathrm{GL}A / \mathrm{GL}(A)_0.$$

Proposition 3.1.8. *([5, Corollary 3.4.4]) If $R \twoheadrightarrow S$ is a surjective unital homomorphism of unital Banach algebras, then $\mathrm{GL}(R)_0 \to \mathrm{GL}(S)_0$ is surjective.*

Let

$$0 \to A \to B \to C \to 0 \tag{24}$$

be an exact sequence of Banach algebras. Then

$$0 \to A \to \tilde{B}_{\mathbb{C}} \to \tilde{C}_{\mathbb{C}} \to 0$$

is again exact. By (23) and Proposition 3.1.8, the connecting map $\partial : K_1(\tilde{C}_{\mathbb{C}}) \to K_0 A$ of Theorem 2.4.1 sends $\ker(K_1(\tilde{C}_{\mathbb{C}}) \to K_1^{\mathrm{top}}C)$ to zero, and thus induces a homomorphism

$$\partial : K_1^{\mathrm{top}}C \to K_0 A.$$

Theorem 3.1.9. *The sequence*

$$
\begin{array}{ccccc}
K_1^{\text{top}}A & \longrightarrow & K_1^{\text{top}}B & \longrightarrow & K_1^{\text{top}}C \\
 & & & & \downarrow{\scriptstyle \partial} \\
K_0C & \longleftarrow & K_0B & \longleftarrow & K_0A
\end{array}
$$

is exact.

Proof. Straightforward from Theorem 2.4.1 and Proposition 3.1.8.   □

Consider the exact sequences

$$0 \to A(0,1] \to A[0,1] \to A \to 0$$
$$0 \to A(0,1) \to A(0,1] \to A \to 0$$

Note moreover that the first of these sequences is split exact. Because K_0 is homotopy invariant and split exact, and because $A(0,1]$ is contractible, we get an isomorphism

$$K_1^{\text{top}}A = K_0(A(0,1)) \tag{25}$$

Since also K_1^{top} is homotopy invariant, we put

$$K_2^{\text{top}}A = K_1^{\text{top}}(A(0,1)). \tag{26}$$

Lemma 3.1.10. *If* (24) *is exact, then*

$$0 \to A(0,1) \to B(0,1) \to C(0,1) \to 0$$

is exact too.

Proof. See [41, Proposition 10.1.2] for a proof in the C^*-algebra case; a similar argument works for Banach algebras.   □

Taking into account the lemma above, as well as (25) and (26), we obtain the following corollary of Theorem 3.1.9.

Corollary 3.1.11. *There is an exact sequence*

$$
\begin{array}{ccccc}
K_2^{\text{top}}A & \longrightarrow & K_2^{\text{top}}B & \longrightarrow & K_2^{\text{top}}C \\
 & & & & \downarrow{\scriptstyle \partial} \\
K_1^{\text{top}}C & \longleftarrow & K_1^{\text{top}}B & \longleftarrow & K_1^{\text{top}}A
\end{array}
$$

The sequence above can be extended further by defining inductively

$$K_{n+1}^{\text{top}}A := K_n^{\text{top}}(A(0,1)).$$

3.2 Bott Periodicity

Let R be a unital Banach algebra. Consider the map $\beta : \mathrm{Idem}_n R \to \mathrm{GL}_n \mathscr{C}_0(S^1, R)$,

$$\beta(e)(z) = ze + 1 - e \tag{27}$$

This map induces a group homomorphism $K_0 R \to K_1^{\mathrm{top}} \mathscr{C}_0(S^1, R)$ (see [5, Sect. 9.1]). If A is any Banach algebra, we write β for the composite

$$
\begin{aligned}
K_0 A \to K_0(\tilde{A}_{\mathbb{C}}) &\xrightarrow{\beta} K_1^{\mathrm{top}}(\mathscr{C}_0(S^1, \tilde{A}_{\mathbb{C}})) \\
&= K_1^{\mathrm{top}} \mathbb{C}(0,1) \oplus K_1^{\mathrm{top}} A(0,1) \twoheadrightarrow K_1^{\mathrm{top}} A(0,1) = K_2^{\mathrm{top}} A \quad (28)
\end{aligned}
$$

One checks that for unital A this definition agrees with that given above.

Theorem 3.2.1. *(Bott periodicity) ([5, Theorem 9.2.1]) The composite map (28) is an isomorphism.*

Let (24) be an exact sequence of Banach algebras. By Corollary 3.1.11 we have a map $\partial : K_2^{\mathrm{top}} C \to K_1^{\mathrm{top}} A$. Composing with the Bott map, we obtain a homomorphism

$$\partial \beta : K_0 C \to K_1^{\mathrm{top}} A.$$

Theorem 3.2.2. *If (24) is an exact sequence of Banach algebras, then the sequence*

$$
\begin{array}{ccccc}
K_1^{\mathrm{top}} A & \longrightarrow & K_1^{\mathrm{top}} B & \longrightarrow & K_1^{\mathrm{top}} C \\
{\scriptstyle \partial\beta}\big\uparrow & & & & \big\downarrow{\scriptstyle \partial} \\
K_0 C & \longleftarrow & K_0 B & \longleftarrow & K_0 A
\end{array}
$$

is exact.

Proof. Immediate from Theorem 3.1.9, Corollary 3.1.11 and Theorem 3.2.1.   □

3.2.1 Sketch of Cuntz' Proof of Bott Periodicity for C^*-Algebras

A C^*-algebra is a Banach algebra A equipped with additive map $* : A \to A$ such that $(a^*)^* = a$, $(\lambda a)^* = \bar{\lambda} a^*$, $(ab)^* = b^* a^*$ and $\|aa^*\| = \|a\|^2$ ($\lambda \in \mathbb{C}$, $a, b \in A$). The *Toeplitz C^*-algebra* is the free unital C^*-algebra $\mathscr{T}^{\mathrm{top}}$ on a generator α subject to $\alpha\alpha^* = 1$. Since the shift $s : \ell^2(\mathbb{N}) \to \ell^2(\mathbb{N})$, $s(e_1) = 0$, $s(e_{i+1}) = e_i$ satisfies $ss^* = 1$, there is a homomorphism $\mathscr{T}^{\mathrm{top}} \to \mathscr{B} = \mathscr{B}(\ell^2(\mathbb{N}))$. It turns out that this is a monomorphism, that its image contains the ideal $\mathscr{K}$, and that the latter is the kernel of the $*$-homomorphism $\mathscr{T}^{\mathrm{top}} \to \mathbb{C}(S^1)$ which sends α to the identity function $S^1 \to S^1$. We have a commutative diagram with exact rows and split exact columns:

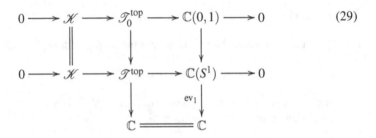

(29)

Here we have used the identification $\mathscr{C}_0(S^1, \mathbb{C}) = \mathbb{C}(0,1)$, via the exponential map; $\mathscr{T}_0^{\text{top}}$ is defined so that the middle column be exact. Write $\tilde{\otimes} = \otimes_{\min}$ for the C^*-algebra tensor product. If now A is any C^*-algebra, and we apply the functor $A\tilde{\otimes}$ to the diagram (29), we obtain a commutative diagram whose columns are split exact and whose rows are still exact (by nuclearity, see [52, Appendix T]).

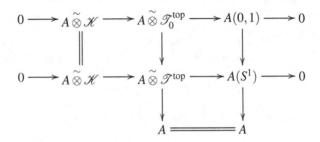

The inclusion $\mathbb{C} \subset M_\infty \mathbb{C} \subset \mathscr{K} = \mathscr{K}(\ell^2(\mathbb{N}))$, $\lambda \mapsto \lambda e_{1,1}$ induces a natural transformation $1 \to \mathscr{K}\tilde{\otimes}-$; a functor G from C^*-algebras to abelian groups is $\mathscr{K}$-*stable* if it is stable with respect to this data in the sense of Definition 2.2.1. We say that G is *half exact* if for every exact sequence (24), the sequence

$$GA \to GB \to GC$$

is exact.

Remark 3.2.3. In general, there is no precedence between the notions of split exact and half exact. However a functor of C^*-algebras which is homotopy invariant, additive and half exact is automatically split exact (see [5, Sect. 21.4]).

The following theorem of Cuntz is stated in the literature for half exact rather than split exact functors. However the proof uses only split exactness.

Theorem 3.2.4. *([14, Theorem 4.4]) Let G be a functor from C^*-algebras to abelian groups. Assume that*

- *G is homotopy invariant.*
- *G is $\mathscr{K}$-stable.*
- *G is split exact.*

Then for every C^*-algebra A,

$$G(A \overset{\sim}{\otimes} \mathscr{T}_0^{\text{top}}) = 0.$$

Proposition 3.2.5. *([41, Proposition 6.4.1]) K_0 is $\mathscr{K}$-stable.*

It follows from the proposition above, (25), Cuntz' theorem and excision, that the connecting map $\partial : K_1^{\text{top}}(A(0,1)) \rightarrow K_0(A \overset{\sim}{\otimes} \mathscr{K})$ is an isomorphism. Further, one checks, using the explicit formulas for β and ∂ ((27), (18)), that the following diagram commutes

$$K_1^{\text{top}} A(0,1) \xrightarrow{\ \partial\ } K_0(A \overset{\sim}{\otimes} \mathscr{K})$$

$$\beta \searrow \qquad \uparrow \wr$$

$$K_0 A$$

This proves that β is an isomorphism.

4 Polynomial Homotopy and Karoubi–Villamayor K-Theory

In this section we analyze to what extent the results of the previous section on topological K-theory of Banach algebras have algebraic analogues valid for all rings. We shall not consider continuous homotopies for general rings, among other reasons, because in general they do not carry any interesting topologies. Instead, we shall consider polynomial homotopies. Two ring homomorphisms $f_0, f_1 : A \rightarrow B$ are called *elementary homotopic* if there exists a ring homomorphism $H : A \rightarrow B[t]$ such that the following diagram commutes

$$B[t]$$

$$H \nearrow \qquad \downarrow (\text{ev}_0, \text{ev}_1)$$

$$A \xrightarrow[(f_0, f_1)]{} B$$

Two homomorphisms $f, g : A \rightarrow B$ are *homotopic* if there is a finite sequence $(f_i)_{0 \leq i \leq n}$ of homomophisms such that $f = f_0$, $f_n = g$, and such that for all i, f_i is homotopic to f_{i+1}. We write $f \sim g$ to indicate that f is homotopic to g. We say that a functor G from rings to a category $\mathfrak{D}$ is *homotopy invariant* if it maps the inclusion $A \rightarrow A[t]$ ($A \in \mathfrak{Ass}$) to an isomorphism. In other words, G is homotopy invariant if it is stable (in the sense of Definition 2.2.1) with respect to the natural inclusion $A \rightarrow A[t]$. One checks that G is homotopy invariant if and only if it preserves the homotopy relation between homomorphisms. If G is any functor, we call a ring A *G-regular* if $GA \rightarrow G(A[t_1, \ldots, t_n])$ is an isomorphism for all n.

Example 4.1. Noetherian regular rings are K_0-regular [42, Corollary 3.2.13] (the same is true for all Quillen's K-groups, by a result of Quillen [39]; see Schlichting's lecture notes [44]) and moreover for $n < 0$, K_n vanishes on such rings (by [42, Definition 3.3.1] and Remark 2.5.6). If k is any field, then the ring $R = k[x,y]/\langle y^2 - x^3 \rangle$ is not K_0-regular (this follows from [53, Theorem I.3.11 and Corollary II.2.3.2]). By Examples 2.1.6 and 2.1.11, the ring $k[\varepsilon]$ is not K_1-regular; indeed the K_1-class of the element $1 + \varepsilon t \in k[\varepsilon][t]^*$ is a nontrivial element of $\ker(K_1(k[\varepsilon][t]) \to K_1(k[\varepsilon])) = \mathrm{coker}(K_1(k[\varepsilon]) \to K_1(k[\varepsilon][t]))$.

The Banach algebras of paths and loops have the following algebraic analogues. Let A be a ring; let $\mathrm{ev}_i : A[t] \to A$ be the evaluation homomorphism ($i = 0, 1$). Put

$$PA := \ker(A[t] \overset{\mathrm{ev}_0}{\to} A) \qquad (30)$$

$$\Omega A := \ker(PA \overset{\mathrm{ev}_1}{\to} A) \qquad (31)$$

The groups $GL(\ )_0$ and K_1^{top} have the following algebraic analogues. Let A be a unital ring. Put

$$GL(A)_0' = \mathrm{Im}(GLPA \to GLA\}$$
$$= \{g \in GLA : \exists h \in GL(A[t]) : h(0) = 1, h(1) = g\}.$$

Set

$$KV_1 A := GLA/GL(A)_0'.$$

The group KV_1 is the K_1-group of Karoubi–Villamayor [33]. It is abelian, since as we shall see in Proposition 4.2 below, there is a natural surjection $K_1 A \twoheadrightarrow KV_1 A$. Unlike what happens with its topological analogue, the functor $GL(\ )_0'$ does not preserve surjections (see Exercise 5.3.7 below). As a consequence, the KV-analogue of Theorem 2.4.1 does not hold for general short exact sequences of rings, but only for those sequences (16) such that $GL(B)_0' \to GL(C)_0'$ is onto, such as, for example, split exact sequences. Next we list some of the basic properties of KV_1; all except nilinvariance (due independently to Weibel [57] and Pirashvili [38]) were proved by Karoubi and Villamayor in [33].

Proposition 4.2.
(i) There is a natural surjective map $K_1 A \twoheadrightarrow KV_1 A$ ($A \in \mathfrak{Ass}$).
(ii) The rule $A \mapsto KV_1 A$ defines a split-exact functor $\mathfrak{Ass} \to \mathfrak{Ab}$.
(iii) If (16) is an exact sequence such that the map $GL(B)_0' \to GL(C)_0'$ is onto, then the map $K_1 C \to K_0 A$ of Theorem 2.4.1 factors through $KV_1 C$, and the resulting sequence

$$
\begin{array}{ccc}
KV_1 A \longrightarrow KV_1 B \longrightarrow KV_1 C \\
\downarrow \partial \\
K_0 C \longleftarrow K_0 B \longleftarrow K_0 A
\end{array}
$$

is exact.
(iv) KV_1 is additive, homotopy invariant, nilinvariant and M_∞-stable.

Proof. If (16) is exact and $GL(B)'_0 \to GL(C)'_0$ is onto, then it is clear that

$$KV_1A \to KV_1B \to KV_1C \tag{32}$$

is exact, and we have a commutative diagram with exact rows and columns

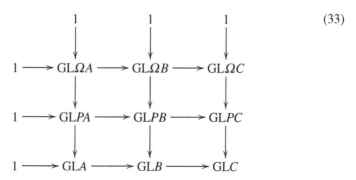

$$(33)$$

If moreover (16) is split exact, then each row in the diagram above is, and one checks, by looking at this diagram, that $GL(A)'_0 = GL(B)'_0 \cap GL(A)$, whence $KV_1A \to KV_1B$ is injective. Thus

$$0 \to KV_1A \to KV_1B \to KV_1C \to 0 \tag{34}$$

is exact. In particular

$$KV_1A = \ker(KV_1\tilde{A} \to KV_1\mathbb{Z}). \tag{35}$$

If R is unital, then $GL(R)'_0 \supset E(R)$, by the same argument as in the Banach algebra case (23). In particular KV_1R is abelian. This proves the unital case of (i); the general case follows from the unital one using (35). Thus the right split exact sequence (34) is in fact a split exact sequence of abelian groups. It follows that KV_1 is split exact, proving (ii). Part (iii) follows from part (i) and (32). The proof that KV_1 is M_∞-stable on unital rings is the same as the proof that K_1 is. By split exactness, it follows that KV_1 is M_∞-stable on all rings. To prove that KV_1 is homotopy invariant, we must show that the split surjection $ev_0 : KV_1(A[t]) \to KV_1A$ is injective. By split exactness, its kernel is $KV_1PA = GLPA/GL(PA)'_0$, so we must prove that $GLPA \subset GL(PA)'_0$. But if $\alpha(s) \in GLPA$, then $\beta(s,t) := \alpha(st) \in GLPPA$ and $ev_{t=1}(\beta) = \alpha$. Thus homotopy invariance is proved. If (16) is exact and A is nilpotent, then PA and ΩA are nilpotent too, whence all those maps displayed in diagram (33) which are induced by $B \to C$ are surjective. Diagram chasing shows that $GL(B)'_0 \to GL(C)'_0$ is surjective, whence by (iii) we have an exact sequence

$$KV_1A \to KV_1B \to KV_1C \to 0$$

Thus to prove KV_1 is nilinvariant, it suffices to show that if $A^2 = 0$ then $KV_1A = 0$. But if $A^2 = 0$, then the map $H : A \to A[t]$, $H(a) = at$ is a ring homomorphism, and satisfies $ev_0H = 0$, $ev_1H = 1_A$. Hence $KV_1A = 0$, by homotopy invariance. $\qquad\square$

Consider the exact sequence

$$0 \to \Omega A \to PA \xrightarrow{ev_1} A \to 0 \tag{36}$$

By definition, $\mathrm{GL}(A)'_0 = \mathrm{Im}(\mathrm{GL}(PA) \to \mathrm{GL}(A))$. But in the course of the proof of Proposition 4.2 above, we have shown that $\mathrm{GL}(PA) = \mathrm{GL}(PA)'_0$, so by Proposition 4.2 (iii), we have a natural map

$$KV_1(A) \xrightarrow{\partial} K_0(\Omega A). \tag{37}$$

Moreover, (37) is injective, by Proposition 4.2 (iii) and (iv). This map will be of use in the next section.

Higher KV-groups are defined by iterating the loop functor Ω:

$$KV_{n+1}(A) = KV_1(\Omega^n A).$$

Higher KV-theory satisfies excision for those exact sequences (16) such that for every n, the map

$$\{\alpha \in \mathrm{GL}(B[t_1,\ldots,t_n]) : \alpha(0,\ldots,0) = 1\} \to \{\alpha \in \mathrm{GL}(C[t_1,\ldots,t_n]) : \alpha(0,\ldots,0) = 1\}$$

is onto. Such sequences are usually called GL-fibration sequences (a different notation was used in [33]). Note that if (16) is a GL-fibration, then $\mathrm{GL}(B)'_0 \to \mathrm{GL}(C)'_0$ is surjective, and thus Proposition 4.2 (iii) applies. Moreover it is proved in [33] that if (16) is a GL-fibration, then there is a long exact sequence $(n \geq 1)$

$$KV_{n+1}B \longrightarrow KV_{n+1}C \longrightarrow KV_n(A) \longrightarrow KV_n(B) \longrightarrow KV_n(C).$$

5 Homotopy K-Theory

5.1 Definition and Basic Properties of KH

Let A be a ring. Consider the natural map

$$\partial : K_0 A \to K_{-1} \Omega A \tag{38}$$

associated with the exact sequence (31). Since $K_{-p} = K_0 \Sigma^p$, we may iterate the construction and form the colimit

$$KH_0 A := \mathrm{colim}_p K_{-p} \Omega^p A. \tag{39}$$

Put

$$KH_n A := \mathrm{colim}_p K_{-p} \Omega^{n+p} A = \begin{cases} KH_0 \Omega^n A & (n \geq 0) \\ KH_0 \Sigma^n A & (n \leq 0) \end{cases} \tag{40}$$

The groups $KH_* A$ are Weibel's *homotopy K-theory* groups of A ([55], [11, Proposition 8.1.1]). One can also express KH in terms of KV, as we shall see presently. We need some preliminaries first. We know that $K_1(S) = 0$ for every infinite sum ring S; in particular $KV_1(\Gamma R) = 0$ for unital R, by Proposition 4.2 (i). Using split exactness of KV_1, it follows that $KV_1 \Gamma A = 0$ for every ring A. Thus the dotted arrow in the commutative diagram with exact row below exists

$$K_1(\Gamma A) \longrightarrow K_1(\Sigma A) \longrightarrow K_0(A) \longrightarrow 0$$

$$\downarrow$$

$$KV_1(\Sigma A)$$

The map $K_1(\Sigma A) \to KV_1(\Sigma A)$ is surjective by Proposition 4.2 (i). Thus the dotted arrow above is a surjective map

$$K_0(A) \twoheadrightarrow KV_1(\Sigma A). \tag{41}$$

On the other hand, the map (37) applied to ΣA gives

$$KV_1(\Sigma A) \xrightarrow{\partial} K_0(\Omega \Sigma A) = K_0(\Sigma \Omega A) = K_{-1}(\Omega A) \tag{42}$$

One checks, by tracking down boundary maps, (see the proof of [11, Proposition 8.1.1]) that the composite of (41) with (42) is the map (38):

$$K_0(A) \xrightarrow{\quad(38)\quad} K_{-1}(\Omega A) \tag{43}$$

$$\begin{array}{cc} (41) \searrow & \nearrow (42) \\ & KV_1(\Sigma A) \end{array}$$

On the other hand, (42) followed by (41) applied to $\Sigma \Omega A$ yields a map

$$KV_1(\Sigma A) \to KV_1(\Sigma^2 \Omega A) = KV_2(\Sigma^2 A).$$

Iterating this map one obtains an inductive system; by (43), we get

$$KH_0(A) = \operatorname*{colim}_r KV_1(\Sigma^{r+1} \Omega^r A) = \operatorname*{colim}_r KV_r(\Sigma^r A)$$

and in general,

$$KH_n(A) = \operatorname*{colim}_r KV_1(\Sigma^{r+1} \Omega^{n+r} A) = \operatorname*{colim}_r KV_{n+r}(\Sigma^r A). \tag{44}$$

Next we list some of the basic properties of KH, proved by Weibel in [55].

Theorem 5.1.1. *([55]) Homotopy K-theory has the following properties.*

(i) It is homotopy invariant, nilinvariant and M_∞-stable.
(ii) It satisfies excision: to the sequence (16) there corresponds a long exact sequence
$(n \in \mathbb{Z})$

$$KH_{n+1}C \to KH_n A \to KH_n B \to KH_n C \to KH_{n-1}A.$$

Proof. From (44) and the fact that, by Proposition 4.2, KV is homotopy invariant, it follows that KH is homotopy invariant. Nilinvariance, M_∞-stability and excision for KH follow from the fact that (by Proposition 2.5.2) these hold for nonpositive K-theory, using the formulas (39), (40). $\qquad\square$

Exercise 5.1.2. Note that in the proof of Theorem 5.1.1, the formula (44) is used only for homotopy invariance. Prove that KH is homotopy invariant without using (44), but using excision instead. Hint: show that the excision map $KH_*(A) \to KH_{*-1}(\Omega A)$ coming from the sequence (36) is an isomorphism.

Exercise 5.1.3. Show that KH commutes with filtering colimits.

5.2 KH for K_0-Regular Rings

Lemma 5.2.1. *Let A be a ring. Assume that A is K_n-regular for all $n \leq 0$. Then $KV_1(A) \to K_0(\Omega A)$ is an isomorphism, and for $n \leq 0$, $K_n(PA) = 0$, PA and ΩA are K_n-regular, and $K_nA \to K_{n-1}\Omega A$ an isomorphism.*

Proof. Consider the split exact sequence

$$0 \longrightarrow PA[t_1,\ldots,t_r] \longrightarrow A[s,t_1,\ldots,t_r] \longrightarrow A[t_1,\ldots,t_r] \longrightarrow 0$$

Applying K_n ($n \leq 0$) and using that K_n is split exact and that, by hypothesis, A is K_n-regular, we get that $K_n(PA[t_1,\ldots,t_r]) = 0$. As this happens for all $r \geq 0$, PA is K_n-regular. Hence the map of exact sequences

$$
\begin{array}{ccccccccc}
0 & \longrightarrow & \Omega A & \longrightarrow & PA & \longrightarrow & A & \longrightarrow & 0 \\
 & & \downarrow & & \downarrow & & \downarrow & & \\
0 & \longrightarrow & \Omega A[t_1,\ldots,t_r] & \longrightarrow & PA[t_1,\ldots,t_r] & \longrightarrow & A[t_1,\ldots,t_r] & \longrightarrow & 0
\end{array}
$$

induces commutative squares with exact rows

$$
\begin{array}{ccccccc}
0 & \longrightarrow & KV_1(A) & \longrightarrow & K_0(\Omega A) & \longrightarrow & 0 \\
 & & \downarrow & & \downarrow & & \\
0 & \longrightarrow & KV_1(A[t_1,\ldots,t_r]) & \longrightarrow & K_0(\Omega A[t_1,\ldots,t_r]) & \longrightarrow & 0
\end{array}
$$

and

$$
\begin{array}{cccccccc}
0 & \longrightarrow & K_n(A) & \longrightarrow & K_{n-1}(\Omega A) & \longrightarrow & 0 & \quad (n \leq 0) \\
 & & \downarrow & & \downarrow & & & \\
0 & \longrightarrow & K_n(A[t_1,\ldots,t_r]) & \longrightarrow & K_{n-1}(\Omega A[t_1,\ldots,t_r]) & \longrightarrow & 0 &
\end{array}
$$

By Proposition 4.2 and our hypothesis, the first vertical map in each diagram is an isomorphism; it follows that the second is also an isomorphism.   □

Remark 5.2.2. A theorem of Vorst [50] implies that if A is K_0-regular then it is K_n-regular for all $n \leq 0$. Thus the lemma above holds whenever A is K_0-regular. The statement of Vorst's theorem is that, for Quillen's K-theory, and $n \in \mathbb{Z}$, a K_n-regular unital ring is also K_{n-1}-regular. (In his paper, Vorst states this only for commutative rings, but his proof works in general). For $n \leq 0$, Vorst's theorem extends to all, not necessarily unital rings. To see this, one shows first, using the fact that $\mathbb{Z}$ is K_n-regular (since it is noetherian regular), and split exactness, that A is K_n-regular if and only if $\tilde{A}$ is. Now Vorst's theorem applied to $\tilde{A}$ implies that if A is K_n-regular then it is K_{n-1}-regular $(n \leq 0)$.

Proposition 5.2.3. *If A satisfies the hypothesis of Lemma 5.2.1, then*

$$KH_n(A) = \begin{cases} KV_n(A) & n \geq 1 \\ K_n(A) & n \leq 0 \end{cases}$$

Proof. By the lemma, $KV_{n+1}(A) = KV_1(\Omega^n A) \to K_0(\Omega^{n+1}A)$ and $K_{-n}(\Omega^p A) \to K_{-n-1}(\Omega^{p+1}A)$ are isomorphisms for all $n, p \geq 0$. $\qquad\square$

5.3 Toeplitz Ring and the Fundamental Theorem for KH

Write $\mathscr{T}$ for the free unital ring on two generators α, α^* subject to $\alpha\alpha^* = 1$. Mapping α to $\sum_i e_{i,i+1}$ and α^* to $\sum_i e_{i+1,i}$ yields a homomorphism $\mathscr{T} \to \Gamma := \Gamma\mathbb{Z}$ which is injective [11, Proof of 4.10.1]; we identify $\mathscr{T}$ with its image in Γ. Note

$$\alpha^{*p-1}\alpha^{q-1} - \alpha^{*p}\alpha^q = e_{p,q} \qquad (p,q \geq 1). \tag{45}$$

Thus $\mathscr{T}$ contains the ideal $M_\infty := M_\infty\mathbb{Z}$. There is a commutative diagram with exact rows and split exact columns:

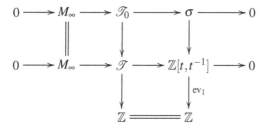

Here the rings $\mathscr{T}_0$ and σ of the top row are defined so that the columns be exact. Note moreover that the rows are split as sequences of abelian groups. Thus tensoring with any ring A yields an exact diagram

Here we have omitted tensor products from our notation; thus $\mathscr{T}A = \mathscr{T} \otimes A$, $\sigma A = \sigma \otimes A$, and $\mathscr{T}_0 A = \mathscr{T}_0 \otimes A$. We have the following algebraic analogue of Cuntz' theorem.

Theorem 5.3.1. *([11, Lemma 7.3.2]) Let G be a functor from rings to abelian groups. Assume that:*

- *G is homotopy invariant.*
- *G is split exact.*
- *G is M_∞-stable.*

Then for any ring A, $G(\mathscr{T}_0 A) = 0$.

Theorem 5.3.2. *Let A be a ring and $n \in \mathbb{Z}$. Then*

$$KH_n(\sigma A) = KH_{n-1}(A).$$

Proof. By Theorem 5.1.1, KH satisfies excision. Apply this to the top row of diagram (46) and use Theorem 5.3.1.   □

The following result of Weibel [55, Theorem 1.2(iii)] is immediate from the theorem above.

Corollary 5.3.3. *(Fundamental theorem for KH,   [55, Theorem 1.2(iii)]).* $KH_n(A[t,t^{-1}]) = KH_n(A) \oplus KH_{n-1}(A)$.

Remark 5.3.4. We regard Theorem 5.3.2 as an algebraic analogue of Bott periodicity for KH. What is missing in the algebraic case is an analogue of the exponential map; there is no isomorphism $\Omega A \to \sigma A$.

Remark 5.3.5. We have shown in Proposition 4.2 that KV_1 satisfies the hypothesis of Theorem 5.3.1. Thus $KV_1(\mathscr{T}_0 A) = 0$ for every ring A. However, there is no natural isomorphism $KV_1(\sigma A) = K_0(A)$. Indeed, since $KV_1(\sigma PA) = KV_1(P\sigma A) = 0$ the existence of such an isomorphism would imply that $K_0(PA) = 0$, which in turn, given the fact that K_0 is split exact, would imply that $K_0(A[t]) = K_0(A)$, a formula which does not hold for general A (see Example 4.1).

We finish the section with a technical result which will be used in Sect. 6.7.

Proposition 5.3.6. *Let $\mathfrak{D}$ be an additive category, $F : \mathfrak{Ass} \to \mathfrak{D}$ an additive functor, and A a ring. Assume F is M_∞-stable on A and M_2-stable on both $\mathscr{T}A$ and $M_2(\mathscr{T}A)$. Then $F(M_\infty A \to \mathscr{T}A)$ is the zero map.*

Proof. Because F is M_∞-stable on A, it suffices to show that F sends the inclusion $\jmath : A \to \mathscr{T}A$, $a \mapsto ae_{11}$, to the zero map. Note that $e_{11} = 1 - \alpha^*\alpha$, by (45). Consider the inclusion $\jmath^\infty : A \to \mathscr{T}A$, $\jmath^\infty(a) = a \cdot 1 = \mathrm{diag}(a,a,a,\dots)$. One checks that the following matrix

$$Q = \begin{bmatrix} 1 - \alpha^*\alpha & \alpha^* \\ \alpha & 0 \end{bmatrix} \in \mathrm{GL}_2 \mathscr{T}(\tilde{A})$$

satisfies $Q^2 = 1$ and

$$Q \begin{bmatrix} \jmath(a) & 0 \\ 0 & \jmath^\infty(a) \end{bmatrix} Q = \begin{bmatrix} \jmath^\infty(a) & 0 \\ 0 & 0 \end{bmatrix}.$$

Since we are assuming that F is additive and M_2-stable on both $\mathscr{T}A$ and $\mathscr{M}_2 \mathscr{T}A$, we may now apply Exercise 2.2.5 and Proposition 2.2.6 to deduce the following identity between elements of the group $\hom_{\mathfrak{D}}(F(A), F(\mathscr{T}A))$:

$$\jmath + \jmath^\infty = \jmath^\infty.$$

It follows that $\jmath = 0$, as we had to prove.   □

Exercise 5.3.7. Deduce from Remark 5.3.5 and Propositions 5.3.6 and 4.2(iii) that the canonical maps $\mathrm{GL}(\mathscr{T}_0 A)'_0 \to \mathrm{GL}(\sigma A)'_0$ and $\mathrm{GL}(\mathscr{T}A)'_0 \to \mathrm{GL}(A[t,t^{-1}])'_0$ are not surjective in general.

6 Quillen's Higher K-Theory

Quillen's higher K-groups of a unital ring R are defined as the homotopy groups of a certain CW-complex; the plus construction of the classifying space of the group $\mathrm{GL}(R)$ [40]. The latter construction is defined more generally for CW-complexes, but we shall not go into this general version; for this and other matters connected with the plus construction approach to K-theory, the interested reader should consult standard references such as Jon Berrick's book [4], or the papers of Loday [35], and Wagoner [42]. We shall need a number of basic facts from algebraic topology, which we shall presently review. First of all we recall that if X and Y are CW-complexes, then the cartesian product $X \times Y$, equipped with the product topology, is not a CW-complex in general. That is, the category of CW-complexes is not closed under finite products in Top. On the other hand, any CW-complex is a compactly generated –or Kelley– space, and the categorical product of two CW-complexes in the category Ke of compactly generated spaces is again CW, and also has the cartesian product $X \times Y$ as underlying set. Moreover, in case the product topology in $X \times Y$ happens to be compactly generated, then it agrees with that of the product in Ke. In these notes, we write $X \times Y$ for the cartesian product equipped with its compactly generated topology. (For a more detailed treatment of the categorical properties of Ke see [21]).

6.1 Classifying Spaces

The *classifying space* of a (discrete) group G is a pointed connected CW-complex BG such that

$$\pi_n BG = \begin{cases} G & n = 1 \\ 0 & n \neq 1 \end{cases}$$

This property characterizes BG and makes it functorial up to homotopy. Further there are various strictly functorial models for BG ([42, Chap. 5 Sect. 1], [24, Example I.1.5]). We choose the model coming from the realization of the simplicial nerve of G [24], and write BG for that model. Here are some basic properties of BG which we shall use.

Properties 6.1.1.

(i) If
$$1 \to G_1 \to G_2 \to G_3 \to 1$$
is an exact sequence of groups, then $BG_2 \to BG_3$ is a fibration with fiber BG_1.
(ii) If G_1 and G_2 are groups, then the map $B(G_1 \times G_2) \to BG_1 \times BG_2$ is a homeomorphism.
(iii) The homology of BG is the same as the group homology of G; if M is $\pi_1 BG = G$-module, then
$$H_n(BG, M) = H_n(G, M) := \mathrm{Tor}_n^{\mathbb{Z}G}(\mathbb{Z}, M)$$

6.2 Perfect Groups and the Plus Construction for BG

A group P is called *perfect* if its abelianization is trivial, or equivalently, if $P = [P, P]$. Note that a group P is perfect if and only if the functor $\hom_{\mathfrak{Grp}}(P, -) : \mathfrak{Ab} \to \mathfrak{Ab}$ is zero. Thus the full subcategory $\subset \mathfrak{Grp}$ of all perfect groups is closed both under colimits and under homomorphic images. In particular, if G is group, then the directed set of all perfect subgroups of G is filtering, and its union is again a perfect subgroup N, the maximal perfect subgroup of G. Since the conjugate of a perfect subgroup is again perfect, it follows that N is normal in G. Note that $N \subset [G, G]$; if moreover the equality holds, then we say that G is *quasi-perfect*. For example, if R is a unital ring, then GLR is quasi-perfect, and ER is its maximal perfect subgroup [42, Proposition 2.1.4]. Quillen's *plus construction* applied to the group G yields a cellular map of CW-complexes $\iota : BG \to (BG)^+$ with the following properties (see [35, Theorem 1.1.1 and Proposition 1.1.2]).

(i) At the level of π_1, ι induces the projection $G \twoheadrightarrow G/N$.

(ii) At the level of homology, ι induces an isomorphism $H_*(G, M) \to H_*((BG)^+, M)$ for each G/N-module M.

(iii) If $BG \to X$ is any continuous function which at the level of π_1 maps $N \to 1$, then the dotted arrow in the following diagram exists and is unique up to homotopy

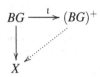

(iv) Properties (i) and (iii) above characterize $\iota : BG \to BG^+$ up to homotopy.

From the universal property, it follows that if $f : BG_1 \to BG_2$ is a continuous map, then there is a (continuous) map $BG_1^+ \to BG_2^+$, unique up to homotopy, which makes the following diagram commute

Properties 6.2.1. *([35, Proposition 1.1.4])*

(i) If G_1 and G_2 are groups, and $\pi_i : B(G_1 \times G_2) \to BG_i$ is the projection, then the map $(\pi_1^+, \pi_2^+) : B(G_1 \times G_2)^+ \to BG_1^+ \times BG_2^+$ is a homotopy equivalence.
(ii) The map $BN^+ \to BG^+$ is the universal covering space of BG^+.

If

$$1 \to G_1 \to G_2 \xrightarrow{\pi} G_3 \to 1 \qquad (47)$$

is an exact sequence of groups, then we can always choose π^+ to be a fibration; write *F* for its fiber. If the induced map $G_1 \to \pi_1 F$ kills the maximal perfect subgroup N_1 of G_1, then $BG_1 \to F$ factors through a map

$$BG_1^+ \to F \qquad (48)$$

Proposition 6.2.2. *Let (47) be an exact sequence of groups. Assume that*

(i) G_1 is quasi-perfect and G_2 is perfect
(ii) G_3 acts trivially on $H_*(G_1, \mathbb{Z})$
(iii) $\pi_1 F$ acts trivially on $H_*(F, \mathbb{Z})$

Then the map (48) is a homotopy equivalence.

Proof. Consider the map of fibration sequences

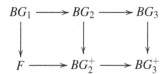

By the second property of the plus construction listed above, the maps $BG_i \to BG_i^+$ are homology equivalences. For $i \geq 2$, we have, in addition, that G_i is perfect, so BG_i^+ is simply connected and *F* is connected with abelian π_1, isomorphic to $\mathrm{coker}(\pi_2 BG_2^+ \to \pi_2 BG_3^+)$. Hence $\pi_1 F \to H_1 F$ is an isomorphism, by Poincaré's theorem. All this together with the Comparison Theorem [61], imply that $BG_1 \to F$ and thus also (48), are homology equivalences. Moreover, because G_1 is quasi-perfect by hypothesis, the Hurewicz map $\pi_1 BG_1^+ \to H_1 BG_1^+$ is an isomorphism, again by Poincaré's theorem. Summing up, $BG_1^+ \to F$ is a homology isomorphism which induces an isomorphism of fundamental groups; since $\pi_1 F$ acts trivially on $H_* F$ by hypothesis, this implies that (48) is a weak equivalence [8, Lemma 4.6.2]. □

Lemma 6.2.3. *Let (47) be an exact sequence of groups. Assume that for every $g \in G_2$ and every finite set $h_1, \ldots, h_k$ of elements of G_1, there exists an $h \in G_1$ such that for all i, $g h_i g^{-1} = h h_i h^{-1}$. Then G_3 acts trivially on $H_*(G_1, \mathbb{Z})$.*

Proof. If $g \in G_2$ maps to $\bar{g} \in G_3$, then the action of $\bar{g}$ on $H_*(G_1, \mathbb{Z})$ is that induced by conjugation by *g*. The hypothesis implies that the action of *g* on any fixed cycle of the standard bar complex which computes $H_*(BG_1, \mathbb{Z})$ [54, Application 6.5.4] coincides with the conjugation action by an element of G_1, whence it is trivial [54, Theorem 6.7.8]. □

Quillen's higher K-groups of a unital ring R are defined as the homotopy groups of $(BGLR)^+$; we put

$$K(R) := (BGLR)^+$$
$$K_n R := \pi_n K(R) \qquad (n \geq 1).$$

In general, for a not necessarily unital ring A, we put

$$K(A) := \text{fiber}(K(\tilde{A}) \to K(\mathbb{Z})), \qquad K_n(A) = \pi_n K(A) \qquad (n \geq 1)$$

One checks, using Property 6.2.1 (i), that when A is unital, these definitions agree with the previous ones.

Remark 6.2.4. As defined, K is a functor from $\mathfrak{Ass}$ to the homotopy category of topological spaces HoTop. Further note that for $n = 1$ we recover the definition of K_1 given in Definition 2.1.3.

We shall see below that the main basic properties of Sect. 2.1 which hold for K_1 hold also for higher K_n [35]. First we need some preliminaries. If $W : \mathbb{N} \to \mathbb{N}$ is an injection, we shall identify W with the endomorphism $\mathbb{Z}^{(\mathbb{N})} \to \mathbb{Z}^{(\mathbb{N})}$, $W(e_i) = e_{w(i)}$ and also with the matrix of the latter in the canonical basis, given by $W_{ij} = \delta_{i,w(j)}$. Let $V = W^t$ be the transpose matrix; then $VW = 1$. If now R is a unital ring, then the endomorphism $\psi^{V,W} : M_\infty R \to M_\infty R$ of Proposition 2.2.6 induces a group endomorphism $GL(R) \to GL(R)$, which in turn yields homotopy classes of maps

$$\psi : K(R) \to K(R), \qquad \psi' : BE(R)^+ \to BE(R)^+. \qquad (49)$$

Lemma 6.2.5. *([35, Proposition 1.2.9]) The maps (49) are homotopic to the identity.*

A proof of the previous lemma for the case of ψ can be found in *loc. cit.*; a similar argument works for ψ'.

Proposition 6.2.6. *Let $n \geq 1$ and let R be a unital ring.*
(i) The functor $K_n : \mathfrak{Ass}_1 \to \mathfrak{Ab}$ is additive.
(ii) The direct sum $\oplus : GLR \times GLR \to GLR$ of (2) induces a map $K(R) \times K(R) \to K(R)$ which makes $K(R)$ into an H-group, that is, into a group up to homotopy. Similarly, $BE(R)^+$ also has an H-group structure induced by $\oplus$.
(iii) The functors $K : \mathfrak{Ass}_1 \to \text{HoTop}$ and $K_n : \mathfrak{Ass}_1 \to \mathfrak{Ab}$ are M_∞-stable.

Proof. Part (i) is immediate from Property 6.2.1 (i). The map of (ii) is the composite of the homotopy inverse of the map of Property 6.2.1 (i) and the map $BGL(\oplus)^+$. One checks that, up to endomorphisms of the form $\psi^{V,W}$ induced by injections $\mathbb{N} \to \mathbb{N}$, the map $\oplus : GL(R) \times GL(R)$ is associative and commutative and the identity matrix is a neutral element for $\oplus$. Hence by Lemma 6.2.5 it follows that $BGL(R)^+$ is a commutative and associative H-space. Since it is connected, this implies that it is an H-group, by [58, X.4.17]. The same argument shows that $BE(R)^+$ is also an H-group. Thus (ii) is proved. Let $\iota : R \to M_\infty R$ be the canonical inclusion. To prove (iii), one observes that a choice of bijection $\mathbb{N} \times \mathbb{N} \to \mathbb{N}$ gives an isomorphism $\phi : M_\infty M_\infty R \overset{\cong}{\to} M_\infty R$ such that the composite with $M_\infty \iota$ is a homomorphism the form $\psi^{V,W}$ for some injection $W : \mathbb{N} \to \mathbb{N}$, whence the induced map $K(R) \to K(R)$ is homotopic to the identity, by Lemma 6.2.5. This proves that $K(\iota)$ is a homotopy equivalence. $\square$

Corollary 6.2.7. *If S is an infinite sum ring, then $K(S)$ is contractible.*

Proof. It follows from the theorem above, using Exercise 2.2.3 (ii) and Proposition 2.3.1. $\square$

Proposition 6.2.8. *Let R be a unital ring, ΣR the suspension, $\Omega K(\Sigma R)$ the loops-pace, and $\Omega_0 K(\Sigma R) \subset \Omega K(\Sigma R)$ the connected component of the trivial loop. There is a homotopy equivalence*

$$K(R) \overset{\sim}{\to} \Omega_0 K(\Sigma R).$$

Proof. Consider the exact sequence of rings

$$0 \to M_\infty R \to \Gamma R \to \Sigma R \to 0$$

Since $K_1(\Gamma R) = 0$, we have

$$GL(\Gamma R) = E(\Gamma R),$$

which applies onto $E\Sigma R$. Thus we have an exact sequence of groups

$$1 \to GLM_\infty R \to GL\Gamma R \to E\Sigma R \to 1$$

One checks that the inclusion $GL(M_\infty R) \to GL(\Gamma R)$ satisfies the hypothesis of Lemma 6.2.3 (see [51, bottom of page 357] for details). Thus the perfect group $E(\Sigma R)$ acts trivially on $H_*(GLM_\infty R, \mathbb{Z})$. On the other hand, by Proposition 6.2.6 (ii), both $K(\Gamma R)$ and $BE(\Sigma R)^+$ are H-groups, and moreover since $\pi : GL(\Gamma R) \to GL(\Sigma R)$ is compatible with $\oplus$, the map $\pi^+ : K(\Gamma R) \to BE(\Sigma R)^+$ can be chosen to be compatible with the induced operation. This implies that the fiber of π^+ is a connected H-space (whence an H-group) and so its fundamental group acts trivially on its homology. Hence by Propositions 6.2.6 (iii) and 6.2.2, we have a homotopy fibration

$$K(R) \to K(\Gamma R) \to BE(\Sigma R)^+$$

By Corollary 6.2.7, the map

$$\Omega BE(\Sigma R)^+ \to K(R) \tag{50}$$

is a homotopy equivalence. Finally, by Property 6.2.1 (ii),

$$\Omega BE(\Sigma R)^+ \overset{\sim}{\to} \Omega_0 K(\Sigma R). \tag{51}$$

Now compose (51) with a homotopy inverse of (50) to obtain the theorem. $\square$

Corollary 6.2.9. *For all $n \in \mathbb{Z}$, $K_n(\Sigma R) = K_{n-1}(R)$.*

Proof. For $n \leq 0$, the statement of the proposition is immediate from the definition of K_n. For $n = 1$, it is (21). If $n \geq 2$, then

$$K_n(\Sigma R) = \pi_n(K(\Sigma R)) = \pi_{n-1}(\Omega K(\Sigma R)) = \pi_{n-1}(\Omega_0 K(\Sigma R)) = \pi_{n-1} K(R) = K_{n-1} R. \ \Box$$

Remark 6.2.10. The homotopy equivalence of Proposition 6.2.8 is the basis for the construction of the nonconnective K-theory spectrum; we will come back to this in Sect. 10.

6.3 Functoriality Issues

As defined, the rule $K : R \mapsto BGL(R)^+$ is only functorial up to homotopy. Actually its possible to choose a functorial model for KR; this can be done in different ways (see for example [34, Sects. 11.2.4 and 11.2.11], [6, Chap. VII, Sect. 3]). However, in the constructions and arguments we have made (notably in the proof of Proposition 6.2.8) we have often used Whitehead's theorem that a map of CW-complexes which induces an isomorphism at the level of homotopy groups (a *weak equivalence*) always has a homotopy inverse. Now, there is in principle no reason why a natural weak equivalence between functorial CW-complexes will admit a homotopy inverse which is also natural; thus for example, the weak equivalence of Proposition 6.2.8 need not be natural for an arbitrarily chosen functorial version of KR. What we need is to be able to choose functorial models so that any natural weak equivalence between them automatically has a natural homotopy inverse. In fact we can actually do this, as we shall now see. First of all, as a technical restriction we have to choose a small full subcategory I of the category $\mathfrak{Ass}$, and look at K-theory as a functor on I. This is no real restriction, as in practice we always start with a set of rings (ofter with only one element) and then all the arguments and constructions we perform take place in a set (possibly larger than the one we started with, but still a set). Next we invoke the fact that the category Top_*^I of functors from I to pointed spaces is a closed model category where fibrations and weak equivalences are defined objectwise (by [28, Theorem 11.6.1] this is true of the category of functors to any cofibrantly generated model category; by [28, Example 11.1.9], Top_* is such a category). This implies, among other things, that there is a full subcategory $(\mathrm{Top}_*^I)_c \to \mathrm{Top}_*^I$, the subcategory of cofibrant objects (among which any natural weak equivalence has a natural homotopy inverse), a functor $\mathrm{Top}_*^I \to (\mathrm{Top}_*^I)_c$, $X \mapsto \hat{X}$ and a natural transformation $\hat{X} \to X$ such that $\hat{X}(R) \to X(R)$ is a fibration and weak equivalence for all R. Thus we can replace our given functorial model for $BGL^+(R)$ by $\widehat{BGL(R)}^+$, and redefine $K(R) = \widehat{BGL(R)}^+$.

6.4 Relative K-Groups and Excision

Let R be a unital ring, $I \triangleleft R$ an ideal, and $S = R/I$. Put

$$\overline{GLS} := \mathrm{Im}(GLR \to GLS)$$

The inclusion $\overline{\mathrm{GL}}S \subset \mathrm{GL}S$ induces a map

$$(B\overline{\mathrm{GL}}S)^+ \to K(S) \tag{52}$$

By Property 6.2.1 (ii), (52) induces an isomorphism

$$\pi_n(B\overline{\mathrm{GL}}S)^+ = K_n S \qquad (n \geq 2).$$

On the other hand,

$$\pi_1(B\overline{\mathrm{GL}}S^+) = \overline{\mathrm{GL}}S/ES = \mathrm{Im}(K_1 R \to K_1 S).$$

Consider the homotopy fiber

$$K(R:I) := \mathrm{fiber}((BGLR)^+ \to (B\overline{\mathrm{GL}}S)^+).$$

The *relative K-groups* of I with respect to the ideal embedding $I \triangleleft R$ are defined by

$$K_n(R:I) := \begin{cases} \pi_n K(R:I) & n \geq 1 \\ K_n(I) & n \leq 0 \end{cases}$$

The long exact sequence of homotopy groups of the fibration which defines $K(R:I)$, spliced together with the exact sequences of Theorem 2.4.1 and Proposition 2.5.2, yields a long exact sequence

$$K_{n+1}R \to K_{n+1}S \to K_n(R:I) \to K_n R \to K_n(S) \qquad (n \in \mathbb{Z}) \tag{53}$$

The canonical map $\tilde{I} \to R$ induces a map

$$K_n(I) \to K_n(R:I) \tag{54}$$

This map is an isomorphism for $n \leq 0$, but not in general (see Remark 2.4.4). The rings I so that this map is an isomorphism for all n and R are called K-excisive. Suslin and Wodzicki have completely characterized K-excisive rings [46, 47, 59]. We have

Theorem 6.4.1. *([46]) The map (54) is an isomorphism for all n and R* $\iff$ $\mathrm{Tor}_n^{\tilde{I}}(\mathbb{Z}, I) = 0\ \forall n$.

Note that

$$\mathrm{Tor}_0^{\tilde{I}}(\mathbb{Z}, I) = I/I^2$$
$$\mathrm{Tor}_n^{\tilde{I}}(\mathbb{Z}, I) = \mathrm{Tor}_{n+1}^{\tilde{I}}(\mathbb{Z}, \mathbb{Z}).$$

Example 6.4.2. Let G be a group, $IG \triangleleft \mathbb{Z}G$ the augmentation ideal. Then $\mathbb{Z}G = \widetilde{IG}$ is the unitalization of IG. Hence

$$\mathrm{Tor}_n^{\widetilde{IG}}(\mathbb{Z}, IG) = H_{n+1}(G, \mathbb{Z}).$$

In particular

$$\mathrm{Tor}_0^{\tilde{I}G}(\mathbb{Z}, I) = G_{ab}$$

So if IG is K-excisive, then G must be a perfect group. Thus, for example, IG is not K-excisive if G is a nontrivial abelian group. In particular, the ring σ is not K-excisive, as it coincides with the augmentation ideal of $\mathbb{Z}[\mathbb{Z}] = \mathbb{Z}[t, t^{-1}]$. As another example, if S is an infinite sum ring, then

$$H_n(\mathrm{GL}(S), \mathbb{Z}) = H_n(K(S), \mathbb{Z}) = H_n(pt, \mathbb{Z}) = 0 \qquad (n \geq 1).$$

Thus the ring $IGL(S)$ is K-excisive.

Remark 6.4.3. We shall introduce a functorial complex $\bar{L}(A)$ which computes $\mathrm{Tor}_*^{\tilde{A}}(\mathbb{Z}, A)$ and use it to show that the functor $\widetilde{\mathrm{Tor}_*^{(-)}}(\mathbb{Z}, -)$ commutes with filtering colimits. Consider the functor $\perp: \tilde{A} - mod \to \tilde{A} - mod$,

$$\perp M = \bigoplus_{m \in M} \tilde{A}.$$

The functor $\perp$ is the free $\tilde{A}$-module cotriple [54, Paradigm 8.6.6]. Let $L(A) \to A$ be the canonical free resolution associated to $\perp$ [54, Example 8.7.2]; by definition, its n-th term is $L_n(A) = \perp^n A$. Put $\bar{L}(A) = \mathbb{Z} \otimes_{\tilde{A}} L(A)$. Then $\bar{L}(A)$ is a functorial chain complex which satisfies $H_*(\bar{L}(A)) = \mathrm{Tor}_*^{\tilde{A}}(\mathbb{Z}, A)$. Because $\perp$ commutes with filtering colimits, it follows that the same is true of L and $\bar{L}$, and therefore also of $\widetilde{\mathrm{Tor}_*^{(-)}}(\mathbb{Z}, -) = H_* \bar{L}(-)$.

Exercise 6.4.4.
(i) Prove that any unital ring is K-excisive.
(ii) Prove that if R is a unital ring, then $M_\infty R$ is K-excisive. (Hint: $M_\infty R = \mathrm{colim}_n M_n R$).

Remark 6.4.5. If A is flat over k (e.g. if k is a field) then the canonical resolution $L^k(A) \xrightarrow{\sim} A$ associated with the induced module cotriple $\tilde{A}_k \otimes_k (-)$, is flat. Thus $\bar{L}^k(A) := L^k(A)/AL^k(A)$ computes $\mathrm{Tor}_*^{\tilde{A}_k}(k, A)$. Modding out by degenerates, we obtain a homotopy equivalent complex [54] $C^{bar}(A/k)$, with $C_n^{bar}(A/k) = A^{\otimes_k n+1}$. The complex C^{bar} is the bar complex considered by Wodzicki in [59]; its homology is the bar homology of A relative to k, $H_*^{bar}(A/k)$. If A is a $\mathbb{Q}$-algebra, then A is flat as a $\mathbb{Z}$-module, and thus $H_*^{bar}(A/\mathbb{Z}) = \mathrm{Tor}_*^{\tilde{A}}(\mathbb{Z}, A)$. Moreover, as $A^{\otimes_{\mathbb{Z}} n} = A^{\otimes_{\mathbb{Q}} n}$, we have $C^{bar}(A/\mathbb{Z}) = C^{bar}(A/\mathbb{Q})$, whence

$$\mathrm{Tor}_*^{\tilde{A}}(\mathbb{Z}, A) = \mathrm{Tor}_*^{\tilde{A}_{\mathbb{Q}}}(\mathbb{Q}, A) = H_*^{bar}(A/\mathbb{Q})$$

6.5 Locally Convex Algebras

A *locally convex* algebra is a complete topological $\mathbb{C}$-algebra L with a locally convex topology. Such a topology is defined by a family of seminorms $\{\rho_\alpha\}$; continuity of the product means that for every α there exists a β such that

$$\rho_\alpha(xy) \le \rho_\beta(x)\rho_\beta(y) \qquad (x, y \in L). \tag{55}$$

If in addition the topology is determined by a countable family of seminorms, we say that L is a *Fréchet* algebra.

Let L be a locally convex algebra. Consider the following two factorization properties:

(a) Cohen–Hewitt factorization.

$$\forall n \ge 1, a = (a_1, \ldots, a_n) \in L^{\oplus n} = \bigoplus_{i=1}^n L \quad \exists z \in L, \quad x \in L^n \text{ such that} \tag{56}$$

$$z \cdot x = a \text{ and } x \in \overline{L \cdot a}$$

Here the bar denotes topological closure in $L^{\oplus n}$.

(b) Triple factorization.

$$\forall a \in L^{\oplus n}, \exists b \in L^{\oplus n}, \quad c, d \in L, \tag{57}$$

$$\text{such that } a = cdb \text{ and } (0:d)_l := \{v \in L : dv = 0\} = (0:cd)_l \tag{58}$$

The right ideal $(0:d)_l$ is called the *left annihilator* of d. Note that property (b) makes sense for an arbitrary ring L.

Lemma 6.5.1. *Cohen–Hewitt factorization implies triple factorization. That is, if L is a locally convex algebra which satisfies property (a) above, then it also satisfies property (b).*

Proof. Let $a \in L^{\oplus n}$. By (a), there exist $b \in L^{\oplus n}$ and $z \in L$ such that $a = zb$. Applying (a) again, we get that $z = cd$ with $d \in \overline{L \cdot z}$; this implies that $(0:d)_l = (0:z)_l$. □

Theorem 6.5.2. *([47, Corollary 3.12 and Remark 3.13(a)]) Let L be a ring. Assume that either L or L^{op} satisfy (57). Then A is K-excisive.*

6.6 Fréchet m-Algebras with Approximate Units

A *uniformly bounded left approximate unit* (ublau) in a locally convex algebra L is a net $\{e_\lambda\}$ of elements of L such that $e_\lambda a \mapsto a$ for all a and $\sup_\alpha \rho_\alpha(a) < \infty$. Right ubau's are defined analogously. If L is a locally convex algebra such that a defining family of seminorms can be chosen so that condition (55) is satisfied with $\alpha = \beta$ (i.e. the seminorms are *submultiplicative*) we say that L is an *m-algebra*. An m-algebra which is also Fréchet will be called a *Fréchet m-algebra*.

Example 6.6.1. Every C^*-algebra has a two-sided ubau [20, Theorem I.4.8]. If G is a locally compact group, then the group algebra $L^1(G)$ is a Banach algebra with two sided ubau [59, Example 8.4]. If L_1 and L_2 are locally convex algebras with ublaus $\{e_\lambda\}$ and $\{f_\mu\}$, then $\{e_\lambda \otimes f_\mu\}$ is a ublau for the projective tensor product $L_1 \hat{\otimes} L_2$, which is a (Fréchet) m-algebra if both L_1 and L_2 are.

Remark 6.6.2. In a Banach algebra, any bounded approximate unit is uniformly bounded. Thus for example, the unit of a unital Banach algebra is an ublau. However, the unit of a general unital locally convex algebra (or even of a Fréchet m-algebra) need not be uniformly bounded.

Let L be an m-Fréchet algebra. A left *Fréchet L-module* is a Fréchet space V equipped with a left L-module structure such that the multiplication map $L \times V \to V$ is continuous. If L is equipped with an ublau e_λ such that $e_\lambda \cdot v \to v$ for all $v \in V$, then we say that V is *essential*.

Example 6.6.3. If L is an m-Fréchet algebra with ublau e_λ and $x \in L^{\oplus n}$ $(n \geq 1)$ then $e_\lambda x \to x$. Thus $L^{\oplus n}$ is an essential Fréchet L-module. The next exercise generalizes this example.

Exercise 6.6.4. Let L be an m-Fréchet algebra with ublau e_λ, M a unital m-Fréchet algebra, and $n \geq 1$. Prove that for every $x \in (L\hat{\otimes}M)^{\oplus n}$, $(e_\lambda \otimes 1)x \to x$. Conclude that $(L\hat{\otimes}M)^{\oplus n}$ is an essential $L\hat{\otimes}M$-module.

The following Fréchet version of Cohen–Hewitt's factorization theorem (originally proved in the Banach setting) is due to M. Summers.

Theorem 6.6.5. *([45, Theorem 2.1]) Let L be an m-Fréchet algebra with ublau, and V an essential Fréchet left L-module. Then for each $v \in V$ and for each neighbourhood U of the origin in V there is an $a \in L$ and a $w \in V$ such that $v = aw$, $w \in \overline{Lv}$, and $w - v \in U$.*

Theorem 6.6.6. *([59, Theorem 8.1]) Let L be a Fréchet m-algebra. Assume L has a right or left ubau. Then L is K-excisive.*

Proof. In view of Lemma (6.5.1), it suffices to show that L satisfies property (56). This follows by applying Theorem 6.6.5 to the essential L-module $L^{\oplus n}$. $\square$

Exercise 6.6.7. Prove that if L and M are as in Exercise (6.6.4), then $L\hat{\otimes}M$ is K-excisive.

Remark 6.6.8. In [12, Lemma 8.1.1] it asserted that if $k \supset \mathbb{Q}$ is a field, and A is a k-algebra, then $\mathrm{Tor}_*^{\tilde{A}_\mathbb{Q}}(\mathbb{Q}, A) = \mathrm{Tor}_*^{\tilde{A}_k}(k, A)$, but the proof uses the identity $\tilde{A}_k \otimes_{\tilde{A}} ? = k \otimes ?$, which is wrong. In *loc. cit.*, the lemma is used in combination with Wodzicki's theorem [59, Theorem 8.1] that a Fréchet algebra L with ublau is H-unital as a $\mathbb{C}$-algebra, to conclude that such L is K-excisive. In Theorem 6.6.6 we gave a different proof of the latter fact.

6.7 Fundamental Theorem and the Toeplitz Ring

Notation. If $G : \mathfrak{Ass} \to \mathfrak{Ab}$ is a functor, and A is a ring, we put

$$NG(A) := \mathrm{coker}(GA \to G(A[t])).$$

Let R be a unital ring. We have a commutative diagram

$$
\begin{array}{ccc}
R & \longrightarrow & R[t] \\
\downarrow & & \downarrow \\
R[t^{-1}] & \longrightarrow & R[t,t^{-1}]
\end{array}
$$

Thus applying the functor K_n we obtain a map

$$K_nR \oplus NK_nR \oplus NK_nR \to K_nR[t,t^{-1}] \tag{59}$$

which sends $NK_nR \oplus NK_nR$ inside $\ker \mathrm{ev}_1$. Thus $K_nR \to K_nR[t,t^{-1}]$ is a split mono, and the intersection of its image with that of $NK_nR \oplus NK_nR$ is 0. On the other hand, the inclusion $\mathscr{T}R \to \Gamma R$ induces a map of exact sequences

$$
\begin{array}{ccccccccc}
0 & \longrightarrow & M_\infty R & \longrightarrow & \mathscr{T}R & \longrightarrow & R[t,t^{-1}] & \longrightarrow & 0 \\
& & \| & & \downarrow & & \downarrow & & \\
0 & \longrightarrow & M_\infty R & \longrightarrow & \Gamma R & \longrightarrow & \Sigma R & \longrightarrow & 0
\end{array}
$$

In particular, we have a homomorphism $R[t,t^{-1}] \to \Sigma R$, and thus a homomorphism

$$\eta : K_nR[t,t^{-1}] \to K_{n-1}R.$$

Note that the maps $R[t] \to \mathscr{T}R$, $t \mapsto \alpha$ and $t \mapsto \alpha^*$, lift the homomorphisms $R[t] \to R[t,t^{-1}]$, $t \mapsto t$ and $t \mapsto t^{-1}$. It follows that $\ker \eta$ contains the image of (59). In [35], Loday introduced a product operation in K-theory of unital rings

$$K_p(R) \otimes K_q(S) \to K_{p+q}(R \otimes S).$$

In particular, multiplying by the class of $t \in K_1(\mathbb{Z}[t,t^{-1}])$ induces a map

$$\cup t : K_{n-1}R \to K_nR[t,t^{-1}]. \tag{60}$$

Loday proves in [35, Theorem 2.3.5] that $\eta \circ (-\cup t)$ is the identity map. Thus the images of (59) and (60) have zero intersection. Moreover, we have the following result, due to Quillen [23], which is known as the fundamental theorem of K-theory.

Theorem 6.7.1. *([23], see also [44]) Let R be a unital ring. The maps (59) and (60) induce an isomorphism*

$$K_nR \oplus NK_nR \oplus NK_nR \oplus K_{n-1}R \overset{\cong}{\to} K_nR[t,t^{-1}] \qquad (n \in \mathbb{Z}).$$

Corollary 6.7.2. *(cf. Theorem 5.3.2)*

$$K_n(R[t,t^{-1}] : \sigma R) = K_{n-1}R \oplus NK_nR \oplus NK_nR \qquad (n \in \mathbb{Z}).$$

Proposition 6.7.3. *(cf. Theorem 5.3.1) Let R be a unital ring, and $n \in \mathbb{Z}$. Then*

$$K_n \mathcal{T} R = K_n R \oplus N K_n R \oplus N K_n R,$$
$$K_n(\mathcal{T} R : \mathcal{T}_0 R) = N K_n R \oplus N K_n R.$$

Proof. Consider the exact sequence

$$0 \to M_\infty R \to \mathcal{T} R \to R[t, t^{-1}] \to 0$$

By Proposition 10.1.2, Exercise 6.4.4 and matrix stability we have a long exact sequence

$$K_n R \to K_n \mathcal{T} R \to K_n R[t, t^{-1}] \to K_{n-1} R \to K_{n-1} \mathcal{T} R \qquad (n \in \mathbb{Z}).$$

By Proposition 5.3.6, the first and the last map are zero. The proposition is immediate from this, from Corollary 6.7.2, and from the discussion above. $\square$

7 Comparison Between Algebraic and Topological K-Theory I

7.1 Stable C^*-Algebras

The following is Higson's homotopy invariance theorem.

Theorem 7.1.1. *[27, Theorem 3.2.2] Let G be a functor from C^*-algebras to abelian groups. Assume that G is split exact and $\mathcal{K}$-stable. Then G is homotopy invariant.*

Lemma 7.1.2. *Let G be a functor from C^*-algebras to abelian groups. Assume that G is M_2-stable. Then the functor $F(A) := G(A \tilde{\otimes} \mathcal{K})$ is $\mathcal{K}$-stable.*

Proof. Let H be an infinite dimensional separable Hilbert space. The canonical isomorphism $\mathbb{C}^2 \otimes_2 H \cong H \oplus H$ induces an isomorphism $\mathcal{K} \tilde{\otimes} \mathcal{K} \to M_2 \mathcal{K}$ which makes the following diagram commute

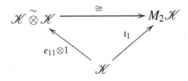

Since G is M_2-stable by hypothesis, it follows that $F(1_A \tilde{\otimes} e_{1,1} \tilde{\otimes} 1_{\mathcal{K}})$ is an isomorphism for all A. $\square$

The following result, due to Suslin and Wodzicki, is (one of the variants of) what is known as Karoubi's conjecture [30].

Theorem 7.1.3. *[47, Theorem 10.9] Let A be a C^*-algebra. Then there is a natural isomorphism $K_n(A \tilde{\otimes} \mathcal{K}) = K_n^{\mathrm{top}}(A \tilde{\otimes} \mathcal{K})$.*

Proof. By definition $K_0 = K_0^{top}$ on all C^*-algebras. By Example 6.6.1 and Theorem 6.6.6, C^*-algebras are K-excisive. In particular K_* is split exact when regarded as a functor of C^*-algebras. By Proposition 6.2.6 (iii), Proposition 2.5.2 (i), and split exactness, K_* is M_∞-stable on C^*-algebras; this implies it is also M_2-stable (Exercise 2.2.3). Thus $K_*(- \widetilde{\otimes} \mathcal{K})$ is $\mathcal{K}$-stable, by Lemma 7.1.2. Hence $K_n(A(0,1] \widetilde{\otimes} \mathcal{K}) = 0$, by split exactness and homotopy invariance (Theorem 7.1.1). It follows that

$$K_{n+1}(A \widetilde{\otimes} \mathcal{K}) = K_n(A(0,1) \widetilde{\otimes} \mathcal{K}) \tag{61}$$

by excision. In particular, for $n \geq 0$,

$$K_n(A \widetilde{\otimes} \mathcal{K}) = K_0(A \widetilde{\otimes} \widetilde{\otimes}_{i=1}^n \mathbb{C}(0,1) \widetilde{\otimes} \mathcal{K}) = K_n^{top}(A \widetilde{\otimes} \mathcal{K}). \tag{62}$$

On the other hand, by Cuntz' theorem 3.2.4, excision applied to the C^*-Toeplitz extension and Lemma 7.1.2, $K_{n+1}(A(0,1) \widetilde{\otimes} \mathcal{K}) = K_n(A \widetilde{\otimes} \mathcal{K} \widetilde{\otimes} \mathcal{K}) = K_n(A \widetilde{\otimes} \mathcal{K})$. Putting this together with (61), we get that $K_*(\mathcal{K} \widetilde{\otimes} A)$ is Bott periodic. It follows that the identity (62) holds for all $n \in \mathbb{Z}$. $\qquad\square$

7.2 Stable Banach Algebras

The following result is a particular case of a theorem of Wodzicki.

Theorem 7.2.1. *([60, Thm. 2], [12, Theorem 8.3.3 and Remark 8.3.4]) Let L be Banach algebra with right or left ubau. Then there is an isomorphism $K_*(L \hat{\otimes} \mathcal{K}) = K_*^{top}(L \hat{\otimes} \mathcal{K})$.*

Proof. Consider the functor $G_L : C^* \to \mathfrak{Ab}, A \mapsto K_*(L \hat{\otimes}(A \widetilde{\otimes} \mathcal{K}))$. By the same argument as in the proof of Theorem 7.1.3, G_L is homotopy invariant. Hence $\mathbb{C} \to \mathbb{C}[0,1]$ induces an isomorphism

$$G_L(\mathbb{C}) = K_*(L \hat{\otimes} \mathcal{K}) \xrightarrow{\cong} G_L(\mathbb{C}[0,1]) = K_*(L \hat{\otimes}(\mathbb{C}[0,1] \widetilde{\otimes} \mathcal{K}))$$
$$= K_*(L \hat{\otimes} \mathcal{K}[0,1]) = K_*((L \hat{\otimes} \mathcal{K})[0,1]).$$

Hence $K_{n+1}(L \hat{\otimes} \mathcal{K}) = K_n(L \hat{\otimes} \mathcal{K}(0,1))$, by Theorem 6.6.6 and Example 6.6.1. Thus $K_n(L \hat{\otimes} \mathcal{K}) = K_n^{top}(L \hat{\otimes} \mathcal{K})$ for $n \geq 0$. Consider the punctured Toeplitz sequence

$$0 \to \mathcal{K} \to \mathscr{T}_0^{top} \to \mathbb{C}(0,1) \to 0$$

By [20, Theorem V.1.5], this sequence admits a continuous linear splitting. Hence it remains exact after applying the functor $L \hat{\otimes} -$. By Theorem 3.2.4, we have

$$K_{-n}(L \hat{\otimes}(\mathcal{K} \widetilde{\otimes} \mathscr{T}_0)) = 0 \qquad (n \geq 0).$$

Thus

$$K_{-n}(L \hat{\otimes} \mathcal{K}) = K_{-n}(L \hat{\otimes}(\mathcal{K} \widetilde{\otimes} \mathcal{K})) = K_0((L \hat{\otimes} \mathcal{K}) \hat{\otimes} \hat{\otimes}_{i=1}^n \mathbb{C}(0,1)) = K_{-n}^{top}(L \hat{\otimes} \mathcal{K}). \quad \square$$

Remark 7.2.2. The theorem above holds more generally for m-Fréchet algebras ([60, Thm. 2], [12, Remark 8.3.4]), with the appropriate definition of topological K-theory (see Sect. 12 below).

Exercise 7.2.3. Let A be a Banach algebra. Consider the map $K_0(A) \to K_{-1}(A(0,1))$ coming from the exact sequence

$$0 \to A(0,1) \to A(0,1] \to A \to 0$$

Put

$$A(0,1)^m = A \hat{\otimes} \left(\hat{\otimes}_{i=1}^m \mathbb{C}(0,1) \right)$$

and define

$$KC_n(A) = \operatorname*{colim}_p K_{-p}(A(0,1)^{n+p}) \qquad (n \in \mathbb{Z})$$

(i) Prove that KC_* satisfies excision, M_∞-stability, continuous homotopy invariance, and nilinvariance
(ii) Prove that $KC_*(A \hat{\otimes} \mathcal{K}) = K_*^{\mathrm{top}}(A)$
(iii) Prove that the composite

$$K_n^{\mathrm{top}}A = K_0(A(0,1)^n) \to KC_n(A) \to KC_n(A \hat{\otimes} \mathcal{K}) = K_n^{\mathrm{top}}(A) \qquad (n \geq 0)$$

is the identity map. In particular $KC_n(A) \to K_n^{\mathrm{top}}(A)$ is surjective for $n \geq 0$.

Remark 7.2.4. J. Rosenberg has conjectured (see [43, Conjecture 23]) that, for $n \leq -1$, the restriction of K_n to commutative C^*-algebras is homotopy invariant. Note that if A is a Banach algebra (commutative or not) such that, $K_{-q}(A(0,1)^p) \to K_{-q}(A(0,1)^p[0,1])$ is an isomorphism for all $p, q \geq 0$, then $KC_n(A) \to K_n^{\mathrm{top}}A$ is an isomorphism for all n. In particular, if Rosenberg's conjecture holds, this will happen for all commutative C^*-algebras A.

8 Topological K-Theory for Locally Convex Algebras

8.1 Diffeotopy KV

We begin by recalling the notion of C^∞-homotopies or diffeotopies (from [15, 16]). Let L be a locally convex algebra. Write $\mathscr{C}^\infty([0,1], L)$ for the algebra of those functions $[0,1] \to L$ which are restrictions of $\mathscr{C}^\infty$-functions $\mathbb{R} \to L$. The algebra $\mathscr{C}^\infty([0,1], L)$ is equipped with a locally convex topology which makes it into a locally convex algebra, and there is a canonical isomorphism

$$\mathscr{C}^\infty([0,1], L) = \mathscr{C}^\infty([0,1], \mathbb{C}) \hat{\otimes} L$$

Two homomorphisms $f_0, f_1 : L \to M$ of locally convex algebras are called *diffeotopic* if there is a homomorphism $H : L \to \mathscr{C}^\infty([0,1], M)$ such that the following diagram commutes

$$\mathscr{C}^\infty([0,1],M)$$

Consider the exact sequences

$$0 \to P^{\mathrm{dif}}L \to \mathscr{C}^\infty([0,1],L) \xrightarrow{\mathrm{ev}_0} L \to 0 \qquad (63)$$

$$0 \to \Omega^{\mathrm{dif}}L \to P^{\mathrm{dif}}L \xrightarrow{\mathrm{ev}_1} L \to 0 \qquad (64)$$

Here $P^{\mathrm{dif}}L$ and $\Omega^{\mathrm{dif}}L$ are the kernels of the evaluation maps. The first of these is split by the natural inclusion $L \to \mathscr{C}^\infty([0,1],L)$, and the second is split the continous linear map sending $l \mapsto (t \mapsto tl)$. We have

$$\Omega^{\mathrm{dif}}L = \Omega^{\mathrm{dif}}\mathbb{C}\hat{\otimes}L, \qquad P^{\mathrm{dif}}L = P^{\mathrm{dif}}\mathbb{C}\hat{\otimes}L.$$

Put

$$\mathrm{GL}(L)_0'' = \mathrm{Im}(\mathrm{GL}P^{\mathrm{dif}}L \to \mathrm{GL}(L))$$

$$KV_1^{\mathrm{dif}}(L) = \mathrm{GL}(L)/\mathrm{GL}(L)_0''.$$

The following is the analogue of Proposition 4.2 for KV_1^{dif} (except for nilinvariance, treated separately in Exercise 8.1.2).

Proposition 8.1.1.
(i) The functor KV_1^{dif} is split exact.
(ii) For each locally convex algebra L, there is a natural surjective map $K_1L \to KV_1^{\mathrm{dif}}L$.
(iii) If

$$0 \to L \to M \to N \to 0 \qquad (65)$$

is an exact sequence such that the map $\mathrm{GL}(M)_0'' \to \mathrm{GL}(N)_0''$ is onto, then the map $K_1N \to K_0L$ of Theorem 2.4.1 factors through $KV_1^{\mathrm{dif}}N$, and the resulting sequence

$$KV_1^{\mathrm{dif}}L \longrightarrow KV_1^{\mathrm{dif}}M \longrightarrow KV_1^{\mathrm{dif}}N$$
$$\downarrow{\partial}$$
$$K_0N \longleftarrow K_0M \longleftarrow K_0L$$

is exact.
(iv) KV_1^{dif} is additive, diffeotopy invariant and M_∞-stable.

Proof. One checks that, mutatis-mutandis, the same argument of the proof of Proposition 4.2 shows this. □

By the same argument as in the algebraic case, we obtain a natural injection

$$KV_1^{\text{dif}}L \hookrightarrow K_0(\Omega^{\text{dif}}L)$$

Higher KV^{dif}-groups are defined by

$$KV_n^{\text{dif}}(L) = KV_1^{\text{dif}}((\Omega^{\text{dif}})^{n-1}L) \qquad (n \geq 2)$$

Exercise 8.1.2.
(i) Show that if L is a locally convex algebra such that $L^n = 0$ and such that $L \to L/L^i$ admits a continuous linear splitting for all $i \leq n-1$, then $KV_1^{\text{dif}}L = 0$.
(ii) Show that if L is as in (i) then the map $KV_1^{\text{dif}}M \to KV_1^{\text{dif}}N$ induced by (65) is an isomorphism.

8.2 Diffeotopy K-Theory

Consider the excision map

$$K_nL \to K_{n-1}(\Omega^{\text{dif}}L) \qquad (n \leq 0)$$

associated to the sequence (64). The *diffeotopy K-theory* of the algebra L is defined by the formula

$$KD_nL = \operatorname*{colim}_{p} K_{-p}((\Omega^{\text{dif}})^{n+p}L) \qquad (n \in \mathbb{Z})$$

It is also possible to express KD in terms of KV^{dif}. First we observe that, since $\Sigma\mathbb{C}$ is a countably dimensional algebra, equipping it with the fine topology makes it into a locally convex algebra [16, Proposition 2.1], and if L is any locally convex algebra then we have

$$\Sigma L = \Sigma\mathbb{C} \otimes_{\mathbb{C}} L = \Sigma\mathbb{C} \hat{\otimes} L.$$

Thus

$$\Omega^{\text{dif}}\Sigma L = \Sigma\Omega^{\text{dif}}L.$$

Taking this into account, and using the same argument as used to prove (44), one obtains

$$KD_nL = \operatorname*{colim}_{r} KV_1^{\text{dif}}(\Sigma^{r+1}(\Omega^{\text{dif}})^{n+r}L) = \operatorname*{colim}_{r} KV_{n+r+1}^{\text{dif}}(\Sigma^{r+1}L).$$

Proposition 8.2.1. *Diffeotopy K-theory has the following properties.*
 (i) It is diffeotopy invariant, nilinvariant and M_∞-stable.
 (ii) It satisfies excision for those exact sequences which admit a continuous linear splitting. That is, if

$$0 \to L \to M \xrightarrow{\pi} N \to 0 \tag{66}$$

is an exact sequence of locally convex algebras and there exists a continuous linear map $s: N \to M$ such that $\pi s = 1_N$, then there is a long exact sequence

$$KD_{n+1}M \to KD_{n+1}N \to KD_nL \to KD_nM \to KD_nN \qquad (n \in \mathbb{Z}).$$

Proof. The proof is essentially the same as that of Theorem (5.1.1). The splitting hypothesis in (ii) guarantees that the functor $L \mapsto \Omega^{\mathrm{dif}}L = \Omega^{\mathrm{dif}}\mathbb{C}\hat{\otimes}L$ and its iterations, send (66) to an exact sequence. $\square$

Comparing KV^{dif} and KD.

The analogue of Proposition 5.2.3 is 8.2.3 below. It is immediate from Lemma 8.2.2, which is the analogue of Lemma 5.2.1; the proof of Lemma 8.2.2 is essentially the same as that of Lemma 5.2.1.

Lemma 8.2.2. *Let L be a locally convex algebra. Assume that for all $n \leq 0$ and all $p \geq 1$, the natural inclusion $\iota_p : L \to L\hat{\otimes}\left(\hat{\otimes}_{i=1}^{p}\mathscr{C}^{\infty}([0,1])\right) = \mathscr{C}^{\infty}([0,1]^{p}, L)$ induces an isomorphism $K_n(L) \xrightarrow{\cong} K_n\mathscr{C}^{\infty}([0,1]^{p}, L)$. Then $KV_1^{\mathrm{dif}}L \to K_0\Omega^{\mathrm{dif}}L$ is an isomorphism, and for every $n \leq 0$ and every $p \geq 0$, $K_n(\mathscr{C}^{\infty}([0,1]^{p}, P^{\mathrm{dif}}A)) = 0$ and $K_n(\Omega^{\mathrm{dif}}A) \to K_n(C^{\infty}([0,1]^{p}, \Omega^{\mathrm{dif}}A))$ is an isomorphism.*

Proposition 8.2.3. *Let L be a locally convex algebra. Assume L satisfies the hypothesis of Lemma 8.2.2. Then*

$$KD_nL = \begin{cases} KV_n^{\mathrm{dif}}L & n \geq 1 \\ K_nL & n \leq 0 \end{cases}$$

8.3 Bott Periodicity

Next we are going to prove a version of Bott periodicity for KD. The proof is analogous to Cuntz' proof of Bott periodicity for K^{top} of C^*-algebras, with the algebra of smooth compact operators and the smooth Toeplitz algebra substituted for the C^*-algebra of compact operators and the Toeplitz C^*-algebra.

8.3.1 Smooth Compact Operators

The algebra $\mathfrak{K}$ of *smooth compact operators* ([37, Sect. 2], [15, Sect. 1.4]) consists of all those $\mathbb{N} \times \mathbb{N}$-matrices $(z_{i,j})$ with complex coefficients such that for all n,

$$\rho_n(z) := \sum_{p,q} p^n q^n |z_{p,q}| < \infty$$

The seminorms ρ_n are submultiplicative, and define a locally convex topology on $\mathfrak{K}$. Since the topology is defined by submultiplicative seminorms, it is an m-algebra. Further because the seminorms above are countably many, it is Fréchet; summing up $\mathfrak{K}$ is an m-Féchet algebra. We have a map

$$e_{11} : \mathbb{C} \to \mathfrak{K}, z \mapsto e_{11}z$$

Whenever we refer to $\mathfrak{K}$-stability below, we shall mean stability with respect to the functor $\mathfrak{K}\hat{\otimes}-$ and the map e_{11}.

8.3.2 Smooth Toeplitz Algebra

The *smooth Toeplitz algebra* [15, Sect. 1.5], is the free m-algebra $\mathscr{T}^{\mathrm{sm}}$ on two generators α, α^* subject to $\alpha\alpha^* = 1$. As in the C^*-algebra case, there is a commutative diagram with exact rows and split exact columns

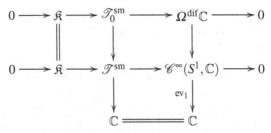

Here $\mathscr{T}_0^{\mathrm{sm}}$ is defined so that the middle column be exact, and we use the exponential map to identify $\Omega^{\mathrm{dif}}\mathbb{C}$ with the kernel of the evaluation map. Moreover the construction of $\mathscr{T}^{\mathrm{sm}}$ given in [15] makes it clear that the rows are exact with a continuous linear splitting, and thus they remain exact after applying $L\hat{\otimes}$, where L is any locally convex algebra.

8.3.3 Bott Periodicity

The following theorem, due to J. Cuntz, appears in [15, Satz 6.4], where it is stated for functors on m-locally convex algebras. The same proof works for functors of all locally convex algebras.

Theorem 8.3.1. *(Cuntz, [15, Satz 6.4]) Let G be a functor from locally convex algebras to abelian groups. Assume that*

- *G is diffeotopy invariant.*
- *G is $\mathfrak{K}$-stable.*
- *G is split exact.*

Then for every locally convex algebra L, we have:

$$G(L\hat{\otimes}\mathscr{T}_0^{\mathrm{sm}}) = 0$$

Theorem 8.3.2. *For every locally convex algebra L, there is a natural isomorphism $KD_*(L\hat{\otimes}\mathfrak{K}) \cong KD_{*+2}(L\hat{\otimes}\mathfrak{K})$.*

Proof. Consider the exact sequence

$$0 \longrightarrow L\hat{\otimes}\mathfrak{K} \longrightarrow L\hat{\otimes}\mathscr{T}_0^{\mathrm{sm}} \longrightarrow \Omega^{\mathrm{dif}}L \longrightarrow 0$$

This sequence is linearly split by construction (see [15, Sect. 1.5]). This splitting property is clearly preserved if we apply the functor $\mathfrak{K}\hat{\otimes}$. Hence by Proposition 8.2.1 (ii), we have a natural map

$$KD_{*+1}(L\hat{\otimes}\mathfrak{K}) = KD_*(\Omega^{\mathrm{dif}}L\hat{\otimes}\mathfrak{K}) \to KD_{*-1}(L\hat{\otimes}\mathfrak{K}\hat{\otimes}\mathfrak{K}) \tag{67}$$

By [15, Lemma 1.4.1], the map $1 \hat{\otimes} e_{11} : \mathfrak{K} \to \mathfrak{K} \hat{\otimes} \mathfrak{K}$ is diffeotopic to an isomorphism. Since KD is diffeotopy invariant, this shows that $KD_*(\mathfrak{K} \hat{\otimes} -)$ is $\mathfrak{K}$-stable. Hence $KD_{*-1}(L \hat{\otimes} \mathfrak{K} \hat{\otimes} \mathfrak{K}) = KD_{*-1}(L \hat{\otimes} \mathfrak{K})$, and by Cuntz' theorem 8.3.1, (67) is an isomorphism.   □

Remark 8.3.3. Cuntz has defined a bivariant topological K-theory for locally convex algebras [16]. This theory associates groups $kk_*^{lc}(L, M)$ to any pair (L, M) of locally convex algebras, and is contravariant in the first variable and covariant in the second. Roughly speaking, $kk_n^{lc}(L, M)$ is defined as a certain colimit of diffeotopy classes of m-fold extensions of L by M ($m \geq n$). There is also an algebraic version of Cuntz' theory, $kk_*(A, B)$, which is defined for all pairs of rings (A, B) [11]. We point out that

$$kk_*^{lc}(\mathbb{C}, M) = KD_*(M \hat{\otimes} \mathfrak{K}). \tag{68}$$

Indeed the proof given in [11, Corollary 8.1.2] that for algebraic kk, $KH_*(A) = kk_*(\mathbb{Z}, A)$ for all rings A, can be adapted to prove (68); one just needs to observe that, for the algebraic suspension, $kk_*^{lc}(L, \Sigma M) = kk_{*-1}^{lc}(L, M)$. Note that, in view of the definition of KD, (68) implies the following "algebraic" formula for kk^{lc}:

$$kk_n^{lc}(\mathbb{C}, L) = \operatorname*{colim}_p K_{-p}((\Omega^{\mathrm{dif}})^{p+q}(L \otimes \mathfrak{K})).$$

9 Comparison Between Algebraic and Topological K-Theory II

9.1 The Diffeotopy Invariance Theorem

Let H be an infinite dimensional separable Hilbert space; write $H \otimes_2 H$ for the completed tensor product of Hilbert spaces. Note any two infinite dimensional separable Hilbert spaces are isomorphic; hence we may regard any operator ideal $\mathscr{J} \lhd \mathscr{B}(H)$ as a functor from Hilbert spaces to $\mathbb{C}$-algebras (see [29, Remark 3.3]). Let $\mathscr{J} \lhd \mathscr{B}$ be an ideal.

- $\mathscr{J}$ is *multiplicative* if $\mathscr{B} \hat{\otimes} \mathscr{B} \to \mathscr{B}(H \otimes_2 H)$ maps $\mathscr{J} \hat{\otimes} \mathscr{J}$ to $\mathscr{J}$.
- $\mathscr{J}$ is *Fréchet* if it is a Fréchet algebra and the inclusion $\mathscr{J} \to \mathscr{B}$ is continuous. A Fréchet ideal is a *Banach* ideal if it is a Banach algebra.

  Write $\omega = (1/n)_n$ for the harmonic sequence.

- $\mathscr{J}$ is *harmonic* if it is a multiplicative Banach ideal such that $\mathscr{J}(\ell^2(\mathbb{N}))$ contains $diag(\omega)$.

Example 9.1.1. Let $p \in \mathbb{R}_{>0}$. Write $\mathscr{L}_p$ for the ideal of those compact operators whose sequence of singular values is p-summable; $\mathscr{L}_p$ is called the p-*Schatten ideal*. It is Banach $\iff p \geq 1$, and is harmonic $\iff p > 1$. There is no interesting locally convex topology on $\mathscr{L}_p$ for $p < 1$.

The following theorem, due to J. Cuntz and A. Thom, is the analogue of Higson's homotopy invariance Theorem 7.1.1 in the locally convex algebra context. The formulation we use here is a consequence of [19, Proposition 5.1.2] and [19, Theorem 4.2.1].

Theorem 9.1.2. *([19]) Let $\mathscr{J}$ be a harmonic operator ideal, and G a functor from locally convex algebras to abelian groups. Assume that*

(i) G is M_2-stable.
(ii) G is split exact.

Then $L \mapsto G(L \hat{\otimes} \mathscr{J})$ is diffeotopy invariant.

We shall need a variant of Theorem 9.1.2 which is valid for all Fréchet ideals $\mathscr{J}$. In order to state it, we introduce some notation. Let $\alpha : L \to M$ be a homomorphism of locally convex algebras. We say that α is an *isomorphism up to square zero* if there exists a continous linear map $\beta : M \hat{\otimes} \dot{M} \to L$ such that the compositions $\beta \circ (\alpha \hat{\otimes} \alpha)$ and $\alpha \circ \beta$ are the multiplication maps of L and M. Note that if α is an isomorphism up to square zero, then its image is a ideal of M, and both its kernel and its cokernel are square-zero algebras.

Definition 9.1.3. *Let G be a functor from locally convex algebras to abelian groups. We call G continously nilinvariant if it sends isomorphisms up to square zero into isomorphisms.*

Example 9.1.4. For any $n \in \mathbb{Z}$, KH_n is a continously nilinvariant functor of locally convex algebras. If $n \leq 0$, the same is true of K_n. In general, if H_* is the restriction to locally convex algebras of any excisive, nilinvariant homology theory of rings, then H_* is continously nilinvariant.

Theorem 9.1.5. *[12, Theorem 6.1.6] Let $\mathscr{J}$ be a Fréchet operator ideal, and G a functor from locally convex algebras to abelian groups. Assume that*

(i) G is M_2-stable.
(ii) G is split exact.
(iii) G is continuously nilinvariant.

Then $L \mapsto G(L \hat{\otimes} \mathscr{J})$ is diffeotopy invariant.

Exercise 9.1.6. Prove:
(i) If L is a locally convex algebra, then $L[t]$ is a locally convex algebra, and there is an isomorphism $L[t] \cong L \hat{\otimes} \mathbb{C}[t]$ where $\mathbb{C}[t]$ is equipped with the fine topology.
(ii) Let G be a diffeotopy invariant functor from locally convex algebras to abelian groups. Prove that G is polynomial homotopy invariant.

The following fact shall be needed below.

Lemma 9.1.7. *Let G be a functor from locally convex algebras to abelian groups, and $\mathscr{J}$ a Fréchet ideal. Assume that G is M_2-stable and that $F(-) := G(-\hat{\otimes}\mathscr{J})$ is diffeotopy invariant. Then F is $\mathfrak{K}$-stable.*

Proof. Let $\iota : \mathbb{C} \to \mathfrak{K}$ be the inclusion; put $\alpha = 1_{\mathscr{J}} \hat{\otimes} \iota$. We have to show that if L is a locally convex algebra, then G maps $1_L \hat{\otimes} \alpha$ to an isomorphism. To do this one constructs a map $\beta : \mathfrak{K} \hat{\otimes} \mathscr{J} \to \mathfrak{K}$, and shows that $G(1_L \hat{\otimes} \beta)$ is inverse to $G(1_L \hat{\otimes} \alpha)$. To define β, proceed as follows. By [12, Proposition 5.1.3], $\mathscr{J} \supset \mathscr{L}^1$, and the tensor product of operators defines a map $\theta : \mathscr{L}^1 \hat{\otimes} \mathscr{J} \to \mathscr{J}$. Write $\phi : \mathfrak{K} \to \mathscr{L}^1$ for the inclusion. Put $\beta = \theta \circ (\phi \hat{\otimes} 1_{\mathscr{J}})$. The argument of the proof of [19, Proposition 6.1.2] now shows that G sends both $1_L \hat{\otimes} \alpha \beta$ and $1_L \hat{\otimes} \beta \alpha$ to identity maps. $\square$

9.2 *KH* of Stable Locally Convex Algebras

Let L be a locally convex algebra. Restriction of functions defines a homomorphism of locally convex algebras $L[t] \to \mathscr{C}^\infty([0,1],L)$, which sends $\Omega L \to \Omega^{\mathrm{dif}}L$. Thus we have a natural map

$$KH_n(L) = \operatorname*{colim}_p K_{-p}(\Omega^{p+n}L) \to \operatorname*{colim}_p K_{-p}((\Omega^{\mathrm{dif}})^{p+n}L) = KD_n(L) \qquad (69)$$

Theorem 9.2.1. *[12, Theorem 6.2.1] Let L be a locally convex algebra, $\mathscr{J}$ a Fréchet ideal, and A a $\mathbb{C}$-algebra. Then*
(i) The functors $KH_n(A \otimes_\mathbb{C} (-\hat{\otimes}\,\mathscr{J}))$ $(n \in \mathbb{Z})$ and $K_m(A \otimes_\mathbb{C} (-\hat{\otimes}\,\mathscr{J}))$ $(m \leq 0)$ are diffeotopy invariant.
(ii) $A \otimes_\mathbb{C} (L\hat{\otimes}\,\mathscr{J})$ is K_n-regular $(n \leq 0)$.
(iii) The map $KH_n(L\hat{\otimes}\,\mathscr{J}) \to KD_n(L\hat{\otimes}\,\mathscr{J})$ of (69) is an isomorphism for all n. Moreover we have

$$KH_n(L\hat{\otimes}\,\mathscr{J}) = KV_n(L\hat{\otimes}\,\mathscr{J}) = KV_n^{\mathrm{dif}}(L\hat{\otimes}J) \qquad (n \geq 1)$$

Proof. Part (i) is immediate from Theorem 5.1.1, Example 9.1.4 and Theorem 9.1.5. It follows from part (i) and Exercise 9.1.6 that $A \otimes_\mathbb{C} (L\hat{\otimes}\,\mathscr{J})$ is K_n-regular for all $n \leq 0$, proving (ii). From part (i) and excision, we get that the two vertical maps in the commutative diagram below are isomorphisms $(n \leq 0)$:

$$
\begin{array}{ccc}
K_n(L\hat{\otimes}\,\mathscr{J}) & \xrightarrow{\ 1\ } & K_n(L\hat{\otimes}\,\mathscr{J}) \\
\downarrow & & \downarrow \\
K_{n-1}(\Omega L\hat{\otimes}\,\mathscr{J}) & \longrightarrow & K_{n-1}(\Omega^{\mathrm{dif}}L\hat{\otimes}\,\mathscr{J})
\end{array}
$$

It follows that the map at the bottom is an isomorphism. This proves the first assertion of (iii). The identity $KH_n(L\hat{\otimes}\,\mathscr{J}) = KV_n(L\hat{\otimes}\,\mathscr{J})$ $(n \geq 1)$ follows from part (i), using Proposition 5.2.3 and Remark 5.2.2. Similarly, part (i) together with Proposition 8.2.3 imply that $KD_n(L\hat{\otimes}\,\mathscr{J}) = KV_n^{\mathrm{dif}}(L\hat{\otimes}\,\mathscr{J})$ $(n \geq 1)$. $\qquad \square$

Corollary 9.2.2.

$$KH_n(A \otimes_\mathbb{C} (L\hat{\otimes}\,\mathscr{J})) = \begin{cases} K_0(A \otimes_\mathbb{C} (L\hat{\otimes}\,\mathscr{J})) & n \text{ even.} \\ K_{-1}(A \otimes_\mathbb{C} (L\hat{\otimes}\,\mathscr{J})) & n \text{ odd.} \end{cases}$$

Proof. Put $B = A \otimes_\mathbb{C} (L\hat{\otimes}\,\mathscr{J})$. By part (ii) of Theorem 9.2.1 above and Remark 5.2.2, or directly by the proof of the theorem, we have that B is K_n-regular for all $n \leq 0$. Thus $KH_n(B) = K_n(B)$ for $n \leq 0$, by Proposition 5.2.3. To finish, we must show that $KH_n(B)$ is 2-periodic. By Lemma 9.1.7, $KH_*(A \otimes_\mathbb{C} (-\hat{\otimes}\,\mathscr{J}))$ is $\mathfrak{K}$-stable. Thus $KH_*(A \otimes_\mathbb{C} (\mathscr{T}_{\mathscr{J}}^{\mathrm{sm}}\hat{\otimes}\,\mathscr{J})) = 0$, by Theorem 8.3.1. Whence $KH_{*+1}(B) = KH_{*-1}(B)$, by excision and diffeotopy invariance. $\qquad \square$

Example 9.2.3. If $\mathscr{J}$ is a Fréchet operator ideal, then by Example 2.5.3, Exercise 2.5.4 and Corollary 9.2.2, we get:

$$KH_n(\mathscr{J}) = \begin{cases} \mathbb{Z} & n \text{ even.} \\ 0 & n \text{ odd.} \end{cases}$$

This formula is valid more generally for "subharmonic" ideals (see [12, Definition 6.5.1] for the definition of this term, and [12, Proposition 7.2.1] for the statement). For example, the Schatten ideals $\mathscr{L}_p$ are subharmonic for all $p > 0$, but are Fréchet only for $p \geq 1$.

10 K-Theory Spectra

In this section we introduce spectra for Quillen's and other K-theories. For a quick introduction to spectra, see [54, 10.9].

10.1 Quillen's K-Theory Spectrum

Let R be a unital ring. Since the loopspace depends only on the connected component of the base point, applying the equivalence of Proposition 6.2.8 to ΣR induces an equivalence

$$\Omega K(\Sigma R) \xrightarrow{\sim} \Omega^2 K(\Sigma^2 R) \tag{70}$$

Moreover, by Sect. 6.3, this map is natural. Put

$$_n\mathbb{K}R := \Omega K(\Sigma^{n+1} R).$$

The equivalence (70) applied to $\Sigma^n R$ yields an equivalence

$$_n\mathbb{K}R \xrightarrow{\sim} \Omega(_{n+1}\mathbb{K}R).$$

The sequence $\mathbb{K}R = \{_n\mathbb{K}R\}$ together with the homotopy equivalences above constitute a spectrum (in the notation of [54, 10.9], Ω-*spectrum* in that of [49, Chap. 8]), the K-theory spectrum; the equivalences are the *bonding maps* of the spectrum. The n-th (stable) homotopy group of $\mathbb{K}R$ is

$$\pi_n\mathbb{K}R = \operatorname*{colim}_p \pi_{n+p}(_n\mathbb{K}R) = K_nR \qquad (n \in \mathbb{Z}).$$

Because its negative homotopy groups are in general nonzero, we say that the spectrum $\mathbb{K}R$ is *nonconnective*. Recall that the homotopy category of spectra $\mathrm{Ho}Spt$ is triangulated, and, in particular, additive. In the first part of the proposition below, we show that $\mathfrak{Ass}_1 \to \mathrm{Ho}Spt$, $R \mapsto \mathbb{K}R$ is an additive functor. Thus we can extend the functor $\mathbb{K}$ to all (not necessarily unital) rings, by

$$\mathbb{K}A := \mathrm{hofiber}(\mathbb{K}(\tilde{A}) \to \mathbb{K}(\mathbb{Z})) \tag{71}$$

Proposition 10.1.1.
(i) The functor $\mathbb{K} : \mathfrak{Ass}_1 \to \mathrm{Ho}Spt$ *is additive.*
(ii) The functor $\mathbb{K} : \mathfrak{Ass} \to \mathrm{Ho}Spt$ *defined in (71) above, is* M_∞-*stable on unital rings.*

Proof. It follows from Proposition 6.2.6 (i) and (iii).   □

If $A \lhd B$ is an ideal, we define the relative K-theory spectrum by

$$\mathbb{K}(B:A) = \text{hofiber}(\mathbb{K}(B) \to \mathbb{K}(B/A)).$$

Proposition 10.1.2. *Every short exact sequence of rings* (16) *with A K-excisive, gives rise to a distinguished triangle*

$$\mathbb{K}A \to \mathbb{K}B \to \mathbb{K}C \to \Omega^{-1}\mathbb{K}A$$

Proof. Immediate from Theorem 6.4.1.   □

10.2 KV-Theory Spaces

Let A be a ring. Consider the simplicial ring

$$\Delta A : [n] \mapsto A \otimes \mathbb{Z}[t_0,\dots,t_n]/\langle 1 - (t_0 + \dots + t_n)\rangle.$$

It is useful to think of elements of $\Delta_n A$ as formal polynomial functions on the algebraic n-simplex $\{(x_0,\dots,x_n) \in \mathbb{Z}^{n+1} : \sum x_i = 1\}$ with values in A. Face and degeneracy maps are given by

$$d_i(f)(t_0,\dots,t_{n-1}) = f(t_0,\dots,t_{i-1},0,t_i,\dots,t_n) \tag{72}$$
$$s_j(f)(t_0,\dots,t_{n+1}) = f(t_0,\dots,t_{i-1},t_i+t_{i+1},\dots,t_{n+1}).$$

Here $f \in \Delta_n A$, $0 \le i \le n$, and $0 \le j \le n-1$.

In the next proposition and below, we shall use the geometric realization of a simplicial space; see [22, Definition I.3.2 (b)] for its definition. We shall also be concerned with simplicial groups; see [54, Chap. 8] for a brief introduction to the latter. The following proposition and the next are taken from D.W. Anderson's paper [1].

Proposition 10.2.1. (*[1, Corollary 1.7]*) *Let A be a ring and $n \ge 1$. Then $KV_nA = \pi_{n-1}\text{GL}\Delta A = \pi_n|B\text{GL}\Delta A|$.*

Proof. The second identity follows from the fact that if G is a group, then $\Omega BG \xrightarrow{\sim} G$ [7] and the fact that, for a simplicial connected space X, one has $\Omega|X| \xrightarrow{\sim} |\Omega X|$. To prove the first identity, proceed by induction on n. Write $\sim$ for the polynomial homotopy relation in GLA and coeq for the coequalizer of two maps. The case $n = 1$ is

$$\pi_0\text{GL}(\Delta A) = \text{coeq}(\text{GL}\Delta_1 A \underset{\text{ev}_1}{\overset{\text{ev}_0}{\rightrightarrows}} \text{GL}A)$$

$$= \text{GL}A/\sim$$
$$= \text{GL}A/\text{GL}(A)_0' = KV_1A.$$

For the inductive step, proceed as follows. Consider the exact sequence of rings

$$0 \to \Omega A \to PA \to A \to 0$$

Using that $GL(-)'_0 = Im(GLP(-) \to GL(-))$ and that $P\Delta = \Delta P$ and $\Omega\Delta = \Delta\Omega$, we obtain exact sequences of simplicial groups

$$1 \longrightarrow GL\Delta\Omega A \longrightarrow GL\Delta PA \longrightarrow GL(\Delta A)'_0 \longrightarrow 1 \qquad (73)$$

$$1 \longrightarrow GL(\Delta A)'_0 \longrightarrow GL\Delta A \longrightarrow KV_1(\Delta A) \longrightarrow 1 \qquad (74)$$

Since KV_1 is homotopy invariant (by Proposition 4.2), we have $\pi_0 KV_1 \Delta A = KV_1 A$ and $\pi_n KV_1 \Delta A = 0$ for $n > 0$. It follows from (74) that

$$\pi_n GL(\Delta A)'_0 = \begin{cases} 0 & n = 0 \\ \pi_n GL(\Delta A) & n \geq 1 \end{cases} \qquad (75)$$

Next, observe that there is a split exact sequence

$$1 \to GL\Delta PA \to GL\Delta A[x] \to GL\Delta A \to 1$$

Here, the surjective map and its splitting are respectively GLd_0 and GLs_0. One checks that the maps

$$h_i : \Delta_n\Delta_1 A \to \Delta_{n+1}A,$$
$$h_i(f)(t_0,\ldots,t_n,x) = f(t_0,\ldots,t_i+t_{i+1},\ldots,t_n,(t_{i+1}+\cdots+t_n)x)$$

$0 \leq i \leq n$ form a simplicial homotopy between the identity and the map $\Delta_n(s_0 d_0)$. Thus GLd_0 is a homotopy equivalence, whence $\pi_* GL\Delta PA = 0$. Putting this together with (75) and using the homotopy exact sequence of (73), we get

$$\pi_n GL\Delta\Omega A = \pi_{n+1}GL\Delta A \qquad (n \geq 0).$$

The inductive step is immediate from this. □

Exercise 10.2.2. Let L be a locally convex algebra. Consider the *geometric n-simplex*

$$\mathbb{A}^n := \{(x_0,\ldots,x_n) \in \mathbb{R}^{n+1} : \sum x_i = 1\} \supset \Delta^n := \{x \in \mathbb{A}^n : x_i \geq 0 \ (0 \leq i \leq n)\}.$$

If L is a locally convex algebra, we write

$$\Delta_n^{dif} L := \mathscr{C}^\infty(\Delta^n, L).$$

Here, $\mathscr{C}^\infty(\Delta^n, -)$ denotes the locally convex vectorspace of all those functions on Δ^n which are restrictions of $\mathscr{C}^\infty$-functions on $\mathbb{A}^n$. The cosimplicial structure on $[n] \mapsto \Delta^n$ induces a simplicial one on $\Delta^{dif}L$. In particular, $\Delta^{dif}L$ is a simplicial locally convex algebra, and $GL(\Delta^{dif}L)$ is a simplicial group.
(i) Prove that $KV_n^{dif}L = \pi_{n-1}GL(\Delta^{dif}L) \ (n \geq 1)$.
(ii) Let A be a Banach algebra. Consider the simplicial Banach algebra $\Delta_*^{top}A = \mathscr{C}(\Delta^*,A)$ and the simplcial group $GL(\Delta^{top}A)$. Prove that $K_n^{top}A = \pi_{n-1}GL(\Delta^{top}A)$ $(n \geq 1)$.

Proposition 10.2.3. (*[1, Corollary 2.3]*) *Let R be a unital ring. Then the map $|BGL\Delta R| \to |K\Delta R|$ is an equivalence.*

Corollary 10.2.4. *If A is a ring and $n \geq 1$, then*

$$KV_n A = \pi_n |K(\Delta\tilde{A} : \Delta A)|$$

Remark 10.2.5. The argument of the proof of Proposition 10.2.3 in [1] applies verbatim to the $\mathscr{C}^\infty$ case, showing that if T is a unital locally convex algebra, then

$$|BGL\Delta^{\mathrm{dif}} T| \xrightarrow{\sim} |K\Delta^{\mathrm{dif}} T|.$$

It follows that if L is any, not necessarily unital locally convex algebra and $\tilde{L}_{\mathbb{C}} = L \oplus \mathbb{C}$ is its unitalization, then

$$KV_n^{\mathrm{dif}} L = \pi_n |K(\Delta^{\mathrm{dif}} \tilde{L}_{\mathbb{C}} : \Delta^{\mathrm{dif}} L)|$$

The analogous formulas for the topological K-theory of Banach algebras are also true and can be derived in the same manner.

10.3 The Homotopy K-Theory Spectrum

Let R be a unital ring. Consider the simplicial spectrum $\mathbb{K}\Delta R$. Put

$$\mathbb{KH}(R) = |\mathbb{K}\Delta R|$$

One checks that $\mathbb{KH} : \mathfrak{Ass}_1 \to \mathrm{Ho}Spt$ is additive. Thus $\mathbb{KH}$ extends to arbitrary rings by

$$\mathbb{KH}(A) = \mathrm{hofiber}(\mathbb{KH}\tilde{A} \to \mathbb{KH}\mathbb{Z}) = |\mathbb{K}(\Delta\tilde{A} : \Delta A)|$$

Remark 10.3.1. If A is any, not necessarily unital ring, one can also consider the spectrum $|\mathbb{K}\Delta A|$; the map

$$\mathbb{K}\Delta A = \mathbb{K}(\widetilde{\Delta A} : \Delta A) \to \mathbb{K}(\Delta\tilde{A} : \Delta A) \tag{76}$$

induces

$$|\mathbb{K}\Delta A| \to |\mathbb{KH}A|. \tag{77}$$

If A happens to be unital, then (76) is an equivalence, whence the same is true of (77). Further, we shall show below that (77) is in fact an equivalence for all $\mathbb{Q}$-algebras A.

Proposition 10.3.2. *Let A be a ring, and $n \in \mathbb{Z}$. Then $KH_n(A) = \pi_n\mathbb{KH}(A)$.*

Proof. It is immediate from the definition of the spectrum $\mathbb{KH}A$ given above that

$$\pi_*\mathbb{KH}(A) = \ker(\pi_*\mathbb{KH}\tilde{A} \to \pi_*\mathbb{KH}\mathbb{Z})$$

Since a similar formula holds for KH_*, it suffices to prove the proposition for unital rings. Let R be a unital ring. By definition, the spectrum $|\mathbb{KH}(R)|$ is the spectrification of the pre-spectrum whose p-th space is $|\Omega K\Delta\Sigma^{p+1}R|$. Thus

$$\pi_n\mathbb{KH}(R) = \operatorname*{colim}_p \pi_{n+p}|\Omega K\Delta\Sigma^{p+1}R| = \operatorname*{colim}_p \pi_{n+p}\Omega|K\Delta\Sigma^{p+1}R|$$

$$= \operatorname*{colim}_p \pi_{n+p+1}|K\Delta\Sigma^{p+1}R| = \operatorname*{colim}_p KV_{n+p}\Sigma^p R = KH_nR. \quad \square$$

Exercise 10.3.3. Let L be a locally convex algebra. Put

$$\mathbb{KD}L = |\mathbb{K}(\Delta^{\mathrm{dif}}\check{L}_{\mathbb{C}} : \Delta^{\mathrm{dif}}L)|.$$

(i) Show that $\pi_n\mathbb{KD}L = KD_nL \ (n \in \mathbb{Z})$
(ii) Construct a natural map

$$\mathbb{K}\Delta^{\mathrm{dif}}L \to \mathbb{KD}L$$

and show it is an equivalence for unital L.

11 Primary and Secondary Chern Characters

In this section, and for the rest of the paper, all rings considered will be $\mathbb{Q}$-algebras.

11.1 Cyclic Homology

The different variants of cyclic homology of an algebra A are related by an exact sequence, Connes' *SBI* sequence

$$HP_{n+1}A \xrightarrow{S} HC_{n-1}A \xrightarrow{B} HN_nA \xrightarrow{I} HP_nA \xrightarrow{S} HC_{n-2}A \qquad (78)$$

Here HC, HN and HP are respectively cyclic, negative cyclic and periodic cyclic homology. The sequence (78) comes from an exact sequence of complexes of $\mathbb{Q}$-vectorspaces. The complex for cyclic homology is Connes' complex $C^\lambda A$, whose definition we shall recall presently; see [34, Sect. 5.1] for the negative cyclic and periodic cyclic complexes. The complex $C^\lambda A$ is a nonnegatively graded chain complex, given in dimension n by the coinvariants

$$C_n^\lambda A := (A^{\otimes n+1})_{\mathbb{Z}/(n+1)\mathbb{Z}} \qquad (79)$$

of the tensor power –taken over $\mathbb{Z}$, or, what is the same, over $\mathbb{Q}$– under the action of $\mathbb{Z}/(n+1)\mathbb{Z}$ defined by the signed cyclic permutation

$$\lambda(a_0 \otimes \cdots \otimes a_n) = (-1)^n a_n \otimes a_0 \otimes \cdots \otimes a_{n-1}.$$

The boundary map $b : C_n^\lambda A \to C_{n-1}^\lambda A$ is induced by

$$b : A^{\otimes n+1} \to A^{\otimes n}, \quad b(a_0 \otimes \cdots \otimes a_n) = \sum_{i=0}^{n-1}(-1)^i a_0 \otimes \cdots \otimes a_i a_{i+1} \otimes \cdots \otimes a_n$$

$$+ (-1)^n a_n a_0 \otimes \cdots \otimes a_{n-1}$$

Example 11.1.1. The map $C_1^\lambda(A) \to C_0^\lambda(A)$ sends the class of $a \otimes b$ to $[a,b] := ab - ba$. Hence

$$HC_0A = A/[A,A].$$

By definition, $HC_nA = 0$ if $n < 0$. Also by definition, HP is periodic of period 2. The following theorem subsumes the main properties of HP.

Theorem 11.1.2.

(i) (Goodwillie, [25]; see also [17]) The functor $HP_* : \mathbb{Q} - \mathfrak{Ass} \to \mathfrak{Ab}$ is homotopy invariant and nilinvariant.

(ii) (Cuntz–Quillen, [18]) HP satisfies excision for $\mathbb{Q}$-algebras; to each exact sequence (16) of $\mathbb{Q}$-algebras, there corresponds a 6-term exact sequence

$$
\begin{array}{ccc}
HP_0A \longrightarrow HP_0B \longrightarrow HP_0C & & (80) \\
\uparrow \qquad\qquad\qquad\qquad \downarrow & & \\
HP_1C \longleftarrow HP_1B \longleftarrow HP_1A & &
\end{array}
$$

Remark 11.1.3. The sequence (78) comes from an exact sequence of complexes, and thus, via the Dold–Kan correspondence, it corresponds to a homotopy fibration of spectra

$$\Omega^{-1}\mathbb{H}\mathbb{C}A \to \mathbb{H}\mathbb{N}A \to \mathbb{H}\mathbb{P}A \qquad (81)$$

Similarly, the excision sequence (80) comes from a cofibration sequence in the category of pro-supercomplexes [13]; applying the Dold–Kan functor and taking homotopy limits yields a homotopy fibration of Bott-periodic spectra

$$\mathbb{H}\mathbb{P}A \to \mathbb{H}\mathbb{P}B \to \mathbb{H}\mathbb{P}C \qquad (82)$$

The sequence (80) is recovered from (82) after taking homotopy groups.

11.2 Primary Chern Character and Infinitesimal K-Theory

The main or primary character is a map going from K-theory to negative cyclic homology

$$c_n : K_nA \to HN_nA \qquad (n \in \mathbb{Z}).$$

(See [34, Chap. 8, Chap. 11] for its definition). This group homomorphism is induced by a map of spectra

$$\mathbb{K}A \to \mathbb{H}\mathbb{N}A$$

Put $\mathbb{K}^{\mathrm{inf}}A := \mathrm{hofiber}(\mathbb{K}A \to \mathbb{H}\mathbb{N}A)$ for its fiber; we call $K_*^{\mathrm{inf}}A$ the *infinitesimal K-theory* of A. Thus, by definition,

$$\mathbb{K}^{\mathrm{inf}}A \to \mathbb{K}A \to \mathbb{H}\mathbb{N}A \qquad (83)$$

is a homotopy fibration. The main properties of K^{inf} are subsumed in the following theorem.

Theorem 11.2.1.

(i) (Goodwillie, [26]) The functor $K_n^{\inf} : \mathbb{Q} - \mathfrak{Ass} \to \mathfrak{Ab}$ is nilinvariant ($n \in \mathbb{Z}$).

(ii) ([10]) $K^{\inf}$ satisfies excision for $\mathbb{Q}$-algebras. Thus to every exact sequence of $\mathbb{Q}$-algebras (16) there corresponds a triangle

$$\mathbb{K}^{\inf}A \to \mathbb{K}^{\inf}B \to \mathbb{K}^{\inf}C \to \Omega^{-1}\mathbb{K}^{\inf}A$$

in $\mathrm{Ho}(Spt)$ and therefore an exact sequence

$$K_{n+1}^{\inf}C \to K_n^{\inf}A \to K_n^{\inf}B \to K_n^{\inf}C \to K_{n-1}^{\inf}A$$

11.3 Secondary Chern Characters

Starting with the fibration sequence (83), one builds up a commutative diagram with homotopy fibration rows and columns

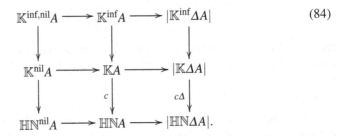

(84)

The middle column is (83); that on the right is (83) applied to ΔA; the horizontal map of homotopy fibrations from middle to right is induced by the inclusion $A \to \Delta A$, and its fiber is the column on the left.

Lemma 11.3.1. *([12, Lemma 2.1.1]) Let A be a simplicial algebra; write $\pi_* A$ for its homotopy groups. Assume $\pi_n A = 0$ for all n. Then $\mathrm{HCA} \xrightarrow{\sim} 0$ and $\mathrm{HNA} \xrightarrow{\sim} \mathrm{HPA}$.*

Proposition 11.3.2. *Let A be a $\mathbb{Q}$-algebra. Then there is a weak equivalence of fibration sequences*

$$
\begin{array}{ccccc}
\mathrm{HN}^{\mathrm{nil}}A & \longrightarrow & \mathrm{HNA} & \longrightarrow & \mathrm{HN}\Delta A \\
\downarrow\wr & & \downarrow\wr & & \downarrow\wr \\
\Omega^{-1}\mathrm{HCA} & \longrightarrow & \mathrm{HNA} & \longrightarrow & \mathrm{HP}A
\end{array}
$$

Proof. By Lemma 11.3.1 and Theorem 11.1.2, we have equivalences

$$\mathrm{HN}\Delta A \xrightarrow{\ \sim\ } \mathrm{HP}\Delta A \xleftarrow{\ \sim\ } \mathrm{HP}A$$

The proposition is immediate from this.   □

Proposition 11.3.3. *If A is a $\mathbb{Q}$-algebra, then the natural map $|\mathbb{K}\Delta A| \to \mathbb{K}\mathrm{HA}$ of (77) above is an equivalence.*

Proof. We already know that the map is an equivalence for unital algebras. Thus since $\mathbb{K}\mathbb{H}$ is excisive, it suffices to show that $\mathbb{K}\varDelta(-)$ is excisive. Using Proposition 11.3.2 and diagram (84), we obtain a homotopy fibration

$$\mathbb{K}^{\text{inf}}\varDelta A \to \mathbb{K}\varDelta A \to \mathbb{H}\mathbb{P}A$$

Note $\mathbb{H}\mathbb{P}$ is excisive by Cuntz–Quillen's theorem 11.1.2 (ii). Moreover, $\mathbb{K}^{\text{inf}}\varDelta(-)$ is also excisive, because K^{inf} is excisive (Theorem 11.2.1 (ii)), and because $\varDelta(-)$ preserves exact sequences and $|-|$ preserves fibration sequences. It follows that $\mathbb{K}\varDelta(-)$ is excisive; this completes the proof. $\qquad\square$

In view of Propositions 11.3.2 and 11.3.3, we may replace diagram (84) by a homotopy equivalent diagram

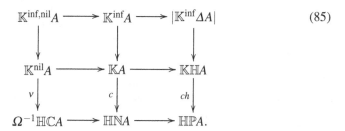

$$(85)$$

The induced maps $v_* : K^{\text{nil}}A \to HC_{*-1}A$ and $ch_* : KH_*A \to HP_*A$ are the *secondary* and the *homotopy* Chern characters. By definition, they fit together with the primary character c_* into a commutative diagram with exact rows

$$\begin{array}{ccccccccc} KH_{n+1}A & \longrightarrow & K_n^{\text{nil}}A & \longrightarrow & K_nA & \longrightarrow & KH_nA & \longrightarrow & \tau K_{n-1}^{\text{nil}}A \\ {\scriptstyle ch_{n+1}}\downarrow & & {\scriptstyle v_n}\downarrow & & {\scriptstyle c_n}\downarrow & & {\scriptstyle ch_n}\downarrow & & {\scriptstyle v_{n-1}}\downarrow \\ HP_{n+1}A & \underset{S}{\longrightarrow} & HC_{n-1}A & \underset{B}{\longrightarrow} & HN_nA & \underset{I}{\longrightarrow} & HP_nA & \underset{S}{\longrightarrow} & HC_{n-2}A. \end{array} \quad (86)$$

Remark 11.3.4. The construction of secondary characters given above goes back to Weibel's paper [56], where a diagram similar to (86), involving Karoubi–Villamayor K-theory KV instead of KH (which had not yet been invented by Weibel), appeared (see also [31] and [32]). For K_0-regular algebras and $n \geq 1$, the latter diagram is equivalent to (86).

Recall that, according to the notation of Sect. 4, an algebra is K_n^{inf}-regular if $K_n^{\text{inf}}A \to K_n^{\text{inf}}\varDelta_pA$ is an isomorphism for all $p \geq 0$. We say that A is K^{inf}-*regular* if it is K_n^{inf}-regular for all n.

Proposition 11.3.5. *Let A be a $\mathbb{Q}$-algebra. If A is K^{inf}-regular, then the secondary character $v_* : K_*^{\text{nil}}A \to HC_{*-1}A$ is an isomorphism.*

Proof. The hypothesis implies that the map $\mathbb{K}^{\text{inf}}A \to \mathbb{K}^{\text{inf}}\varDelta_nA$ is a weak equivalence $(n \geq 0)$. Thus, viewing $\mathbb{K}^{\text{inf}}A$ as a constant simplicial spectrum and taking realizations, we obtain an equivalence $\mathbb{K}^{\text{inf}}A \xrightarrow{\sim} |\mathbb{K}^{\text{inf}}\varDelta A|$. Hence $\mathbb{K}^{\text{inf},\text{nil}}A \xrightarrow{\sim} 0$ and therefore v is an equivalence. $\qquad\square$

Example 11.3.6. The notion of K^{inf}-regularity of $\mathbb{Q}$-algebras was introduced in [12, Sect. 3], where some examples are given and some basic properties are proved; we recall some of them. First of all, for $n \leq -1$, K_n^{inf}-regularity is the same as K_n-regularity. A K_0^{inf}-regular algebra is K_0-regular, but not conversely. If R is unital and K_1^{inf}-regular, then the two sided ideal $\langle [R,R] \rangle$ generated by the additive commutators $[r,s] = rs - sr$ is the whole ring R. In particular, no nonzero unital commutative ring is K_1^{inf}-regular. Both infinite sum and nilpotent algebras are K^{inf}-regular. If (16) is an exact sequence of $\mathbb{Q}$-algebras such that any two of A, B, C are K^{inf}-regular, then so is the third.

We shall see in Theorem 12.2.1 that any stable locally convex algebra is K^{inf}-regular.

11.4 Application to KD

Proposition 11.4.1. *Let L be a locally convex algebra. Then the natural map $\mathbb{K}\Delta^{\mathrm{dif}}L \to \mathbb{K}\mathbb{D}L$ of Exercise 10.3.3 (ii) is an equivalence.*

Proof. By Exercise 10.3.3 (ii), the proposition is true for unital L. Thus it suffices to show that $\mathbb{K}\Delta^{\mathrm{dif}}(-)$ satisfies excision for those exact sequences (65) which admit a continuous linear splitting. Applying the sequence (83) to $\Delta^{\mathrm{dif}}L$ and taking realizations yields a fibration sequence

$$|\mathbb{K}^{\mathrm{inf}}\Delta^{\mathrm{dif}}L| \to |\mathbb{K}\Delta^{\mathrm{dif}}L| \to |\mathbb{H}\mathbb{N}\Delta^{\mathrm{dif}}L|$$

One checks that $\pi_*\Delta^{\mathrm{dif}}L = 0$ (see [12, Lemma 4.1.1]). Hence the map $I : \mathbb{H}\mathbb{N}\Delta^{\mathrm{dif}}L \to \mathbb{H}\mathbb{P}\Delta^{\mathrm{dif}}L$ is an equivalence, by Lemma 11.3.1. Now proceed as in the proof of Proposition 11.3.3, taking into account that $\Delta^{\mathrm{dif}}(-)$ preserves exact sequences with continuous linear splitting. $\qquad\square$

Corollary 11.4.2. *Assume that the map $K_nL \to K_n\Delta_p^{\mathrm{dif}}L$ is an isomorphism for all $n \in \mathbb{Z}$ and all $p \geq 0$. Then $\mathbb{K}L \to \mathbb{K}\mathbb{D}L$ is an equivalence.*

Proof. Analogous to the first part of the proof of Proposition 11.3.5. $\qquad\square$

12 Comparison Between Algebraic and Topological K-Theory III

12.1 Stable Fréchet Algebras

The following is the general version of Theorem 7.2.1, also due to Wodzicki.

Theorem 12.1.1. *([60, Thm. 2], [12, Theorem 8.3.3 and Remark 8.3.4]) Let L be an m-Fréchet algebra with uniformly bounded left or right approximate unit. Then there is a natural isomorphism:*

$$K_n(L\hat{\otimes}\mathscr{K}) \xrightarrow{\sim} KD_n(L\hat{\otimes}\mathscr{K}), \quad \forall n \in \mathbb{Z}.$$

Proof. Write $\mathfrak{C}$ for the full subcategory of those locally convex algebras which are m-Fréchet algebras with left ubau. In view of Corollary 11.4.2, it suffices to show that for all $n \in \mathbb{Z}$ and $p \geq 0$, the map

$$K_n(L \hat{\otimes} \mathcal{K}) \rightarrow K_n(\Delta_p^{\mathrm{dif}} L \hat{\otimes} \mathcal{K}) \tag{87}$$

is an isomorphism for each $L \in \mathfrak{C}$. Note that, since $\Delta_p^{\mathrm{dif}} \mathbb{C}$ is a unital m-Fréchet algebra and its unit is uniformly bounded, the functor $\Delta_p^{\mathrm{dif}}(-) = -\hat{\otimes}\Delta_p^{\mathrm{dif}}\mathbb{C}$ maps $\mathfrak{C}$ into itself. Since $L \rightarrow \Delta_p^{\mathrm{dif}} L$ is a diffeotopy equivalence, this means that to prove (87) is to prove that $K_n(-\hat{\otimes}\mathcal{K}) : \mathfrak{C} \rightarrow \mathfrak{Ab}$ is diffeotopy invariant. Applying the same argument as in the proof of Theorem 7.2.1, we get that the natural map

$$K_*(L \hat{\otimes} \mathcal{K}) \rightarrow K_*((L \hat{\otimes} \mathcal{K})[0,1])$$

is an isomorphism. It follows that $K_*(-\hat{\otimes}\mathcal{K})$ is invariant under continous homotopies, and thus also under diffeotopies. $\qquad \square$

Exercise 12.1.2. Prove that if L is as in Theorem 12.1.1 and $M = L \hat{\otimes} \mathcal{K}$, then $KD_*(M(0,1)) = KD_{*+1}M$.

Exercise 12.1.3. Prove that the map $K_n(L \hat{\otimes} \mathcal{K}) \rightarrow KD_n(L \hat{\otimes} \mathcal{K})$ is an isomorphism for every unital Fréchet algebra L, even if the unit of L is not uniformly bounded. (Hint: use Exercise 6.6.7).

Remark 12.1.4. N.C. Phillips has defined a K^{top} for m-Fréchet algebras [37] which extends that of Banach algebras discussed in Sect. 3 above. We shall see presently that, for L as in Theorem 12.1.1,

$$K_*^{\mathrm{top}}(L \hat{\otimes} \mathcal{K}) = KD_*(L \hat{\otimes} \mathcal{K}) = K_*(L \hat{\otimes} \mathcal{K}).$$

Phillips' theory is Bott periodic and satisfies $K_0^{\mathrm{top}}(M) = K_0(M \hat{\otimes} \mathfrak{K})$ and $K_1^{\mathrm{top}}(M) = K_0((M \hat{\otimes} \mathfrak{K})(0,1))$ for every Fréchet algebra M. On the other hand, for L as in the theorem, we have $KD_0(L \hat{\otimes} \mathcal{K}) = K_0(L \hat{\otimes} \mathcal{K})$ and $KD_1(L \hat{\otimes} \mathcal{K}) = K_0((L \hat{\otimes} \mathcal{K})(0,1))$. But by Lemma 9.1.7, $K_0(M \hat{\otimes} \mathcal{K}) = K_0(M \hat{\otimes} \mathcal{K} \hat{\otimes} \mathfrak{K})$ for every locally convex algebra M. This proves that $KD_n(L \hat{\otimes} \mathcal{K}) = K_n^{\mathrm{top}}(L \hat{\otimes} \mathcal{K})$ for $n = 0, 1$; by Bott periodicity, we get the equality for all n.

12.2 Stable Locally Convex Algebras: The Comparison Sequence

Theorem 12.2.1. *(see [12, Theorem 6.3.1]) Let A be a $\mathbb{C}$-algebra, L be a locally convex algebra, and $\mathcal{J}$ a Fréchet operator ideal. Then*
(i) $A \otimes_{\mathbb{C}} (L \hat{\otimes} \mathcal{J})$ is K^{inf}-regular.
(ii) For each $n \in \mathbb{Z}$, there is a 6-term exact sequence

$$
\begin{array}{ccc}
K_{-1}(A \otimes_{\mathbb{C}} (L \hat{\otimes} \mathcal{J})) \longrightarrow HC_{2n-1}(A \otimes_{\mathbb{C}} (L \hat{\otimes} \mathcal{J})) \longrightarrow K_{2n}(A \otimes_{\mathbb{C}} (L \hat{\otimes} \mathcal{J})) \\
\uparrow \qquad\qquad\qquad\qquad\qquad\qquad\qquad\qquad\qquad\qquad\qquad \downarrow \\
K_{2n-1}(A \otimes_{\mathbb{C}} (L \hat{\otimes} \mathcal{J})) \longleftarrow HC_{2n-2}(A \otimes_{\mathbb{C}} (L \hat{\otimes} \mathcal{J})) \longleftarrow K_0(A \otimes_{\mathbb{C}} (L \hat{\otimes} \mathcal{J})).
\end{array}
\tag{88}
$$

Proof. According to Theorem 11.2.1, K^{inf} is nilinvariant and satisfies excision. Hence, by Theorem 9.1.5, $L \mapsto K_*^{\text{inf}}(A \otimes_{\mathbb{C}} (L \hat{\otimes} \mathscr{J}))$ is diffeotopy invariant, whence it is invariant under polynomial homotopies. This proves (i). Put $B = A \otimes_{\mathbb{C}} (L \hat{\otimes} \mathscr{J})$. By (i) and Proposition 11.3.5, $v_* : K_*^{\text{nil}} B \to HC_{*-1} B$ is an isomorphism. Hence from (86) we get a long exact sequence

$$KH_{m+1}B \longrightarrow HC_{m-1}B \longrightarrow K_m B \longrightarrow KH_m B \xrightarrow{Sch_m} HC_{m-2}B \qquad (89)$$

By Corollary 9.2.2, $KH_{2n}B = K_0 B$ and $KH_{2n-1}B = K_{-1}B$; the sequence of the theorem follows from this, using the sequence (89). $\qquad \square$

Corollary 12.2.2. *For each $n \in \mathbb{Z}$, there is a 6-term exact sequence*

$$KD_1(L \hat{\otimes} \mathscr{J}) \longrightarrow HC_{2n-1}(L \hat{\otimes} \mathscr{J}) \longrightarrow K_{2n}(L \hat{\otimes} \mathscr{J}) \qquad (90)$$

$$K_{2n-1}(L \hat{\otimes} \mathscr{J}) \longleftarrow HC_{2n-2}(L \hat{\otimes} \mathscr{J}) \longleftarrow KD_0(L \hat{\otimes} \mathscr{J}).$$

Proof. By Theorem 9.2.1 (iii), $KD_*(L \hat{\otimes} \mathscr{J}) = KH_*(L \hat{\otimes} \mathscr{J})$. Now use Corollary 9.2.2. $\qquad \square$

Example 12.2.3. We saw in Theorem 12.1.1 that the comparison map $K_*(L \hat{\otimes} \mathscr{K}) \to KD_*(L \hat{\otimes} \mathscr{K})$ is an isomorphism whenever L is an m-Fréchet algebra with left ubau. Thus

$$HC_*(L \hat{\otimes} \mathscr{K}) = 0 \qquad (91)$$

by Corollary 12.2.2. It is also possible to prove (91) directly and deduce Theorem 12.1.1 from the corollary above; see [12, Theorem 8.3.3].

Example 12.2.4. If we set $L = \mathbb{C}$ in Theorem 12.2.1 above, we obtain an exact sequence

$$K_{-1}(A \otimes_{\mathbb{C}} \mathscr{J}) \longrightarrow HC_{2n-1}(A \otimes_{\mathbb{C}} \mathscr{J}) \longrightarrow K_{2n}(A \otimes_{\mathbb{C}} \mathscr{J}) \qquad (92)$$

$$K_{2n-1}(A \otimes_{\mathbb{C}} \mathscr{J}) \longleftarrow HC_{2n-2}(A \otimes_{\mathbb{C}} \mathscr{J}) \longleftarrow K_0(A \otimes_{\mathbb{C}} \mathscr{J}).$$

Further specializing to $A = \mathbb{C}$ and using Example 2.5.3 and Exercise 2.5.4 yields

$$0 \to HC_{2n-1}\mathscr{J} \to K_{2n}\mathscr{J} \to \mathbb{Z} \xrightarrow{\alpha_n} HC_{2n-2}\mathscr{J} \to K_{2n-1}\mathscr{J} \to 0.$$

Here we have written α_n for the composite of $S \circ ch_{2n}$ with the isomorphism $\mathbb{Z} \cong K_0\mathscr{J}$. If for example $\mathscr{J} \subset \mathscr{L}_p$ ($p \geq 1$) then α_n is injective for $n \geq (p+1)/2$, by a result of Connes and Karoubi [10, Sect. 2.9] (see also [12, Proposition 7.2.1]). Setting $p = 1$ we obtain, for each $n \geq 1$, an isomorphism

$$K_{2n}\mathscr{L}_1 = HC_{2n-1}\mathscr{L}_1$$

and an exact sequence

$$0 \to \mathbb{Z} \overset{\alpha_n}{\to} HC_{2n-2}\mathscr{L}_1 \to K_{2n-1}\mathscr{L}_1 \to 0.$$

Note that since $HC_{2n-2}\mathscr{L}_1$ is a $\mathbb{Q}$-vectorspace by definition, the sequence above implies that $K_{2n-1}\mathscr{L}_1$ is isomorphic to the sum of a copy of $\mathbb{Q}/\mathbb{Z}$ plus a $\mathbb{Q}$-vectorspace.

Remark 12.2.5. The exact sequence (92) is valid more generally for subharmonic ideals (see [12, Definition 6.5.1] for the definition of this term, and [12, Theorem 7.1.1] for the statement). In particular, (92) is valid for all the Schatten ideals $\mathscr{L}_p$, $p > 0$. In [12, 7.1.1] there is also a variant of (92) involving relative K-theory and relative cyclic homology; the particular case of the latter when A is K-excisive is due to Wodzicki ([60, Theorem 5]).

Acknowledgements. Work for these notes was partly supported by FSE and by grants PICT03-12330, UBACyT-X294, VA091A05, and MTM00958.

References

1. Anderson, D.: Relationship among K-theories. In: Bass, H. (ed.) Higher K-theories. Lecture Notes in Mathematics, vol. 341, pp. 57–72. Springer, New York (1972)
2. Bass, H.: Algebraic K-theory. W.A. Benjamin, New York (1968)
3. Benson, D.J.: Representations and cohomology. In: Cambridge Studies in Advanced Mathematics, Second edition, vol. 30. Cambridge University Press, Cambridge (1998)
4. Berrick, A.J.: An approach to algebraic K-theory. In: Research Notes in Mathematics, vol. 56. Pitman Books, London (1982)
5. Blackadar, B.: K-theory for operator algebras, Second edition. Cambridge University Press, Cambridge (1998)
6. Bousfield, A., Kan, D.: Homotopy limits, completions and localizations. In: Lecture Notes in Mathematics, vol. 304. Springer, Berlin (1972)
7. Bousfield, A., Friedlander, E.: Homotopy theory of Γ-spaces, spectra, and bisimplicial sets. In: Geometric Applications of Homotopy Theory (Proc. Conf., Evanston, Ill., 1977), II. Lecture Notes in Mathematics, vol. 658, pp. 80–130. Springer, Berlin (1978)
8. Browder, W.: Higher torsion in H-spaces. Trans. Am. Math. Soc. **108**, 353–375 (1963)
9. Connes, A., Karoubi, M.: Caractére mulitplicatif d'un module de Fredholm. K-theory **2**, 431–463 (1988)
10. Cortiñas, G.: The obstruction to excision in K-theory and in cyclic homology. Invent. Math. **454**, 143–173 (2006)
11. Cortiñas, G, Thom, A.: Bivariant algebraic K-theory. J. reine angew. Math. **510**, 71–124 (2007)
12. Cortiñas, G., Thom, A.: Comparison between algebraic and topological K-theory of locally convex algebras. math.KT/067222
13. Cortiñas, G., Valqui, C.: Excision in bivariant periodic cyclic homology: a categorical approach. K-theory **30**, 167–201 (2003)
14. Cuntz, J.: K-theory and C^*-algebras. In: Algebraic K-theory, number theory and analysis. Lecture Notes in Mathematics, vol. 1046, pp. 55–79. Springer, Berlin (1984)

15. Cuntz, J.: Bivariante K-theorie für lokalkonnvexe Algebren und der Chern-Connes-Charakter. Doc. Math. **2**, 139–182 (1997)
16. Cuntz, J.: Bivariant K-theory and the Weyl algebra. K-Theory **35**, 93–137 (2005)
17. Cuntz, J., Quillen, D.: Cyclic homology and nonsingularity. J. Am. Math. Soc. **8**, 373–441 (1995)
18. Cuntz, J., Quillen, D.: Excision in periodic bivariant cyclic cohomology. Invent. Math. **127**, 67–98 (1997)
19. Cuntz, J., Thom, A.: Algebraic K-theory and locally convex algebras. Math. Ann. **334**, 339–371 (2006)
20. Davidson, K. R.: C^*-algebras by example. In: Fields Institute Monographs, vol. 6. American Mathematical Society, Providence (1996)
21. Gabriel, P., Zisman, G.: Calculus of fractions and homotopy theory. In: Ergebnisse der Mathematik und ihrer Grenzgebiete, Band 35. Springer, New York (1967)
22. Gelfand, I., Manin, Y.: Methods of homological algebra. In: Springer Monographs in Mathematics, second edition. Springer, New York (2002)
23. Grayson, D.: Higher algebraic K-theory II. (after Daniel Quillen). Algebraic K-theory (Proc. Conf., Northwestern Univ., Evanston, Ill., 1976). In: Lecture Notes in Mathematics, vol. 551, pp. 217–240. Springer, Berlin (1976)
24. Goerss, P., Jardine, J.: Simplicial homotopy theory. In: Progress in Mathematics, vol. 174. Birkhuser, Basel (1999)
25. Goodwillie, T.: Cyclic homology, derivations, and the free loopspace. Topology **24**, 187–215 (1985)
26. Goodwillie, T.: Relative algebraic K-theory and cyclic homology. Ann. Math. **124**, 347–402 (1986)
27. Higson, N.: Algebraic K-theory of C^*-algebras. Adv. Math. **67**, 1–40 (1988)
28. Hirschorn, P.: Model categories and their localizations. In: Mathematical Surveys and Monographs, vol. 99. American Mathematical Society, Providence (2003)
29. Husemöller, D.: Algebraic K-theory of operator ideals (after Mariusz Wodzicki). In: K-theory, Strasbourg 1992. Astérisque **226**, 193–209 (1994)
30. Karoubi, M.: K-théorie algébrique de certaines algèbres d'opérateurs. In: Algèbres d'opérateurs (Sém., Les Plans-sur-Bex, 1978). Lecture Notes in Mathematics, vol. 725, pp. 254–290. Springer, Berlin (1979)
31. Karoubi, M.: Homologie cyclique et K-théorie. Astérisque, Société Mathématique de France **149** (1987)
32. Karoubi, M.: Sur la K-théorie Multiplicative. Cyclic homology and noncommutative geometry. Fields Inst. Commun. **17**, 59–77 (1997)
33. Karoubi, M., Villamayor, O.: K-théorie algébrique et K-théorie topologique. Math. Scand. **28**, 265–307 (1971)
34. Loday, J.L.: Cyclic homology, 1st edition. Grundlehren der mathematischen Wis-senschaften, vol. 301. Springer, Berlin (1998)
35. Loday, J.L.: K-théorie algébrique et représentations de groupes. Ann. Sci. ENS. 4 Série, tome 9(3), 309–377 (1976)
36. Milnor, J.: Introduction to algebraic K-theory. In: Annals of Mathematics Studies, vol. 72. Princeton University Press, Princeton (1971)
37. Phillips, N.C.: K-theory for Fréchet algebras. Int. J. Math. **2**, 77–129 (1991)
38. Pirashvili, T.I.: Some properties of Karoubi-Villamayor algebraic K-functors (Russian) Soobshch. Akad. Nauk Gruzin. SSR **97**(2), 289–292 (1980)
39. Quillen, D.: Higher algebraic K-theory. I. Algebraic K-theory, I: Higher K-theories (Proc. Conf., Battelle Memorial Inst., Seattle, WA, 1972). In: Lecture Notes in Mathematics, vol. 341, pp. 85–147. Springer, Berlin (1973)

40. Quillen, D.: Higher algebraic K-theory. In: Proceedings of the International Congress of Mathematicians (Vancouver, B.C., 1974), vol. 1, pp. 171–176. Canad. Math. Congress, Montreal, Que. (1975)

41. Rordam, M., Larsen, F., Laustsen, N.J.: An introduction to K-theory for C^*-algebras. In: London Mathematical Society Student Texts, vol. 49. Cambridge University Press, Cambridge (2000)

42. Rosenberg, J.: Algebraic K-theory and its applications. In: Graduate Texts in Mathematics, vol. 147. Springer, New York (1994)

43. Rosenberg, J.: Comparison between algebraic and topological K-theory for Banach algebras and C^*-algebras. In: Friedlander, E.M., Grayson, D.R. (eds.) Handbook of K-Theory. Springer, New York (2005)

44. Schlichting, M.: Higher Algebraic K-Theory II (After Quillen, Thomason and Others). This volume

45. Summers, M.: Factorization in Frchet modules. J. London Math. Soc. **5**, 243–248 (1972)

46. Suslin, A.: Excision in the integral algebraic K-theory. Proc. Steklov Inst. Math. **208**, 255–279 (1995)

47. Suslin, A., Wodzicki, M.: Excision in algebraic K-theory. Ann. Math. **136**, 51–122 (1992)

48. Swan, R.: Excision in algebraic K-theory. J. Pure Appl. Algebra **1**, 221–252 (1971)

49. Switzer, R.: Algebraic topology-homotopy and homology. Grundlehren der mathematischen Wis- senschaften, vol. 212. Springer, Berlin (1975)

50. Vorst, T.: Localization of the K-theory of polynomial extensions. Math. Ann. **244**, 33–54 (1979)

51. Wagoner, J.B.: Delooping classifying spaces in algebraic K-theory. Topology **11**, 349–370 (1972)

52. Wegge-Olsen, N.E.: K-theory and C^*-algebras. Oxford University Press, Oxford (1993)

53. Weibel, C.: The K-book: An introduction to algebraic K-theory. Book-in-progress, available at its author's webpage: http://www.math.rutgers.edu/ weibel/Kbook.html

54. Weibel, C.: An introduction to homological algebra. Cambridge University Press, Cambridge (1994)

55. Weibel, C.: Homotopy Algebraic K-theory. Contemp. Math. **83**, 461–488 (1989)

56. Weibel, C.: Nil K-theory maps to cyclic homology. Trans. Am. Math. Soc. **303**, 541–558 (1987)

57. Weibel, C.A.: Nilpotence and K-theory. J. Algebra **61**(2), 298–307 (1979)

58. Whitehead, G.W.: Elements of Homotopy Theory. Springer, Berlin (1978)

59. Wodzicki, M.: Excision in cyclic homology and in rational algebraic K-theory. Ann. Math. (2) **129**, 591–639 (1989)

60. Wodzicki, M.: Algebraic K-theory and functional analysis. In: First European Congress of Mathematics, Vol. II (Paris, 1992). Progress in Mathematics, vol. 120, pp. 485–496. Birkhäuser, Basel (1994)

61. Zeeman, E.C.: A proof of the comparison theorem for spectral sequences. Proc. Camb. Philos. Soc. **53**, 57–62 (1957)

Higher Algebraic K-Theory
(After Quillen, Thomason and Others)

Marco Schlichting

Mathematics Institute,
Zeeman Building,
University of Warwick,
Coventry CV4 7AL UK,
m.schlichting@warwick.ac.uk

Abstract We present an introduction (with a few proofs) to higher algebraic K-theory of schemes based on the work of Quillen, Waldhausen, Thomason and others. Our emphasis is on the application of triangulated category methods in algebraic K-theory.

1 Introduction

These are the expanded notes for a course taught by the author at the Sedano Winter School on K-theory, January 23–26, 2007, in Sedano, Spain. The aim of the lectures was to give an introduction to higher algebraic K-theory of schemes. I decided to give only a quick overview of Quillen's fundamental results [35, 73], and then to focus on the more modern point of view where structure theorems about derived categories of sheaves are used to compute higher algebraic K-groups.

Besides reflecting my own taste, there are at least two other good reasons for this emphasis. First, there is an ever growing number of results in the literature about the structure of triangulated categories. To name only a few of their authors, we refer the reader to the work of Bondal, Kapranov, Orlov, Kuznetsov, Samarkhin, Keller, Thomason, Rouquier, Neeman, Drinfeld, Toen, van den Bergh, Bridgeland, etc. ... The relevance for K-theory is that virtually all results about derived categories translate into results about higher algebraic K-groups. The link is provided by an abstract Localization Theorem due to Thomason and Waldhausen which – omitting hypothesis – says that a "short exact sequence of triangulated categories gives rise to a long exact sequence of algebraic K-groups". The second reason for this emphasis is that an analog of the Thomason–Waldhausen Localization Theorem also holds for many other (co-) homology theories besides K-theory, among which Hochschild homology, (negative, periodic, ordinary) cyclic homology [49], topological Hochschild (and cyclic) homology [2], triangular Witt groups [6] and higher Grothendieck–Witt groups [77]. All K-theory results that are proved using triangulated category methods therefore have analogs in all these other (co-) homology theories.

P.F. Baum et al., *Topics in Algebraic and Topological K-Theory*,
Lecture Notes in Mathematics 2008, DOI 10.1007/978-3-642-15708-0_4,
© Springer-Verlag Berlin Heidelberg 2011

Here is an overview of the contents of these notes. Sect. 2 is an introduction to Quillen's fundamental article [73]. Here the algebraic K-theory of exact categories is introduced via Quillen's Q-construction. We state some fundamental theorems, and we state/derive results about the G-theory of noetherian schemes and the K-theory of smooth schemes. The proofs in [73] are all elegant and very well-written, so there is no reason to repeat them here. The only additions I have made are a hands-on proof of the fact that Quillen's Q-construction gives the correct K_0-group, and a description of negative K-groups which is absent in Quillen's work.

Section 3 is an introduction to algebraic K-theory from the point of view of triangulated categories. In Sect. 3.1 we introduce the Grothendieck-group K_0 of a triangulated category, give examples and derive some properties which motivate the introduction of higher algebraic K-groups. In Sect. 3.2 we introduce the K-theory space (and the non-connected $\mathbb{K}$-theory spectrum) of a complicial exact category with weak equivalences via Quillen's Q-construction. This avoids the use of the technically heavier $S_\bullet$-construction of Waldhausen [100]. We state in Theorem 3.2.27 the abstract Localization Theorem mentioned above that makes the link between exact sequences of triangulated categories and long exact sequences of algebraic K-groups. In Sect. 3.3 we show that most of Quillen's results in [73] – with the notable exception of *Dévissage* – can be viewed as statements about derived categories, in view of the Localization Theorem. In Sect. 3.4 we give a proof – based on Neeman's theory of compactly generated triangulated categories – of Thomason's Mayer-Vietoris principle for quasi-compact and separated schemes. In Sect. 3.5 we illustrate the use of triangulated categories in the calculation of the K-theory of projective bundles and of blow-ups of schemes along regularly embedded centers. We also refer the reader to results on derived categories of rings and schemes which yield further calculations in K-theory.

Section 4 is a mere collection of statements of mostly recent results in algebraic K-theory the proofs of which go beyond the methods explained in Sects. 2 and 3.

In Appendix A, Sects. 1 and 2 we assemble results from topology and the theory of triangulated categories that are used throughout the text. In Appendix A, Sect. 3, we explain the constructions and elementary properties of the derived functors we will need. Finally, we give in Appendix A, Sect. 4, a proof of the fact that the derived category of complexes of quasi-coherent sheaves (supported on a closed subset with quasi-compact open complement) on a quasi-compact and separated scheme is compactly generated – a fact used in the proof of Thomason's Mayer-Vietoris principle in Sect. 3.4.

2 The K-Theory of Exact Categories

2.1 The Grothendieck Group of an Exact Category

2.1.1 Exact Categories

An *exact category* [73, Sect. 2] is an additive category $\mathscr{E}$ equipped with a family of sequences of morphisms in $\mathscr{E}$, called *conflations* (or admissible exact sequences),

$$X \xrightarrow{i} Y \xrightarrow{p} Z \tag{1}$$

satisfying the properties (a)–(f) below. In a conflation (1), the map i is called *inflation* (or admissible monomorphism) and may be depicted as $\rightarrowtail$, and the map p is called *deflation* (or admissible epimorphism) and may be depicted as $\twoheadrightarrow$.

(a) In a conflation (1), the map i is a kernel of p, and p is a cokernel of i.

(b) Conflations are closed under isomorphisms.

(c) Inflations are closed under compositions, and deflations are closed under compositions.

(d) Any diagram $Z \leftarrow X \overset{i}{\rightarrowtail} Y$ with i an inflation can be completed to a cocartesian square

with j an inflation.

(e) Dually, any diagram $X \rightarrow Z \overset{p}{\leftarrow} Y$ with p a deflation can be completed to a cartesian square

$$
\begin{array}{ccc}
W & \longrightarrow & Y \\
{\scriptstyle q}\downarrow & & \downarrow{\scriptstyle p} \\
X & \longrightarrow & Z
\end{array}
$$

with q a deflation.

(f) The following sequence is a conflation

$$ X \overset{\binom{1}{0}}{\rightarrow} X \oplus Y \overset{(0\,1)}{\rightarrow} Y. $$

Quillen lists another axiom [73, Sect. 2 Exact categories c)] which, however, follows from the axioms listed above [46, Appendix]. For a detailed account of exact categories including the solutions of some of the exercises below, we refer the reader to [17].

An additive functor between exact categories is called *exact* if it sends conflations to conflations.

Let $\mathscr{A}$, $\mathscr{B}$ be exact categories such that $\mathscr{B} \subset \mathscr{A}$ is a full subcategory. We say that $\mathscr{B}$ is a *fully exact subcategory* of $\mathscr{A}$ if $\mathscr{B}$ is closed under extensions in $\mathscr{A}$ (that is, if in a conflation (1) in $\mathscr{A}$, X and Z are isomorphic to objects in $\mathscr{B}$ then Y is isomorphic to an object in $\mathscr{B}$), and if the inclusion $\mathscr{B} \subset \mathscr{A}$ preserves and detects conflations.

2.1.2 Examples

(a) Abelian categories are exact categories when equipped with the family of conflations (1) where $0 \rightarrow X \rightarrow Y \rightarrow Z \rightarrow 0$ is a short exact sequence. Examples of abelian (thus exact) categories are: the category R-Mod of all (left) R-modules where R is a ring; the category R-mod of all finitely generated (left) R-modules where R is a noetherian ring; the category O_X-Mod ($\text{Qcoh}(X)$) of (quasi-coherent) O_X-modules where X is a scheme; the category $\text{Coh}(X)$ of coherent O_X-modules where X is a noetherian scheme.

(b) Let $\mathscr{A}$ be an exact category, and let $\mathscr{B} \subset \mathscr{A}$ be a full additive subcategory closed under extensions in $\mathscr{A}$. Call a sequence (1) in $\mathscr{B}$ a conflation if it is a conflation in $\mathscr{A}$. One checks that $\mathscr{B}$ equipped with this family of conflations is an exact category making $\mathscr{B}$ into a fully exact subcategory of $\mathscr{A}$. In particular, any extension closed subcategory of an abelian category is canonically an exact category.

(c) The category Proj(R) of finitely generated projective left R-modules is extension closed in the category of all R-modules. Similarly, the category Vect(X) of vector bundles (that is, locally free sheaves of finite rank) on a scheme X is extension closed in the category of all O_X-modules. In this way, we consider Proj(R) and Vect(X) as exact categories where a sequence is a conflation if it is a conflation in its ambient abelian category.

(d) An additive category can be made into an exact category by declaring a sequence (1) to be a conflation if it is isomorphic to a sequence of the form 2.1.1 (f). Such exact categories are referred to as *split exact categories*.

(e) Let $\mathscr{E}$ be an exact category. We let Ch$\mathscr{E}$ be the category of chain complexes in $\mathscr{E}$. Objects are sequences (A, d):

$$\cdots \to A^{i-1} \xrightarrow{d^{i-1}} A^i \xrightarrow{d^i} A^{i+1} \to \cdots$$

of morphisms in $\mathscr{E}$ such that $d \circ d = 0$. A morphism $f : (A, d_A) \to (B, d_B)$ is a collection of morphisms $f^i : A^i \to B^i, i \in \mathbb{Z}$, such that $f \circ d_A = d_B \circ f$. A sequence $(A, d_A) \to (B, d_B) \to (C, d_C)$ of chain complexes is a conflation if $A^i \to B^i \to C^i$ is a conflation in $\mathscr{E}$ for all $i \in \mathbb{Z}$. This makes Ch$\mathscr{E}$ into an exact category.

The full subcategory Ch$^b\mathscr{E} \subset$ Ch$\mathscr{E}$ of bounded chain complexes is a fully exact subcategory, where a complex (A, d_A) is bounded if $A^i = 0$ for $i >> 0$ and $i << 0$.

It turns out that the examples in Example 2.1.2 (b), (c) are typical as the following lemma shows. The proof can be found in [94, Appendix A] and [46, Appendix A].

2.1.3 Lemma

Every small exact category can be embedded into an abelian category as a fully exact subcategory.

2.1.4 Exercise

Use the axioms Sect. 2.1.1 (a)–(f) of an exact category or Lemma 2.1.3 above to show the following (and their duals).

(a) A cartesian square as in Sect. 2.1.1 (e) with p a deflation is also cocartesian. Moreover, if $X \to Z$ is an inflation, then $W \to Y$ is also an inflation.

(b) If the composition ab of two morphisms in an exact category is an inflation, and if b has a cokernel, then b is also an inflation. This is Quillen's redundant axiom [73, Sect. 2 Exact categories c)].

(c) Given a composition pq of deflations p, q in $\mathscr{E}$, then there is a conflation ker(q) $\rightarrowtail$ ker(pq) $\twoheadrightarrow$ ker p in $\mathscr{E}$.

2.1.5 Definition of K_0

Let $\mathscr{E}$ be a small exact category. The *Grothendieck group* $K_0(\mathscr{E})$ of $\mathscr{E}$ is the abelian group freely generated by symbols $[X]$ for every object X of $\mathscr{E}$ modulo the relation

$$[Y] = [X] + [Z] \text{ for every conflation } X \rightarrowtail Y \twoheadrightarrow Z. \tag{2}$$

An exact functor $F : \mathscr{A} \to \mathscr{B}$ between exact categories induces a homomorphism of abelian groups $F : K_0(\mathscr{A}) \to K_0(\mathscr{B})$ via $[X] \mapsto [FX]$.

2.1.6 Remark

The conflation $0 \rightarrowtail 0 \twoheadrightarrow 0$ implies that $0 = [0]$ in $K_0(\mathscr{E})$. Let $X \xrightarrow{\cong} Y$ be an isomorphism, then we have a conflation $0 \rightarrowtail X \twoheadrightarrow Y$, and thus $[X] = [Y]$ in $K_0(\mathscr{E})$. So $K_0(\mathscr{E})$ is in fact generated by isomorphism classes of objects in $\mathscr{E}$. The split conflation 2.1.1 (f) implies that $[X \oplus Y] = [X] + [Y]$.

2.1.7 Remark (K_0 for Essentially Small Categories)

By Remark 2.1.6, isomorphic objects give rise to the same class in K_0. It follows that we could have defined $K_0(\mathscr{E})$ as the group generated by isomorphism classes of objects in $\mathscr{E}$ modulo the relation 2.1.5 (2). This definition makes sense for any essentially small (= equivalent to a small) exact category. With this in mind, K_0 is also defined for such categories.

2.1.8 Definition

The groups $K_0(R)$, $K_0(X)$, and $G_0(X)$ are the Grothendieck groups of the essentially small exact categories $\mathrm{Proj}(R)$ of finitely generated projective R-modules where R is any ring, of the category $\mathrm{Vect}(X)$ of vector bundles on a scheme X[1], and of the category $\mathrm{Coh}(X)$ of coherent O_X-modules over a noetherian scheme X.

2.1.9 Examples

For commutative noetherian rings, there are isomorphisms

$$K_0(\mathbb{Z}) \cong \mathbb{Z},$$
$$K_0(R) \cong \mathbb{Z} \qquad \text{where } R \text{ is a local (not necessarily noetherian) ring,}$$
$$K_0(F) \cong \mathbb{Z} \qquad \text{where } F \text{ is a field,}$$
$$K_0(A) \cong \mathbb{Z}^n \qquad \text{where } \dim A = 0 \text{ and } n = \# \mathrm{Spec} A,$$
$$K_0(R) \cong \mathbb{Z} \oplus \mathrm{Pic}(R) \quad \text{where } R \text{ is connected and } \dim R = 1,$$
$$K_0(X) \cong \mathbb{Z} \oplus \mathrm{Pic}(X) \quad \text{where } X \text{ is a connected smooth projective curve.}$$

The group $\mathrm{Pic}(R)$ is the Picard group of a commutative ring R, that is, the group of isomorphism classes of rank 1 projective R-modules with tensor product $\otimes_R$ as group law. The isomorphism in the second to last row is induced by the map $K_0(R) \to \mathbb{Z} \oplus \mathrm{Pic}(R)$ sending a projective module P to its rank $\mathrm{rk} P \in \mathbb{Z}$ and its highest non-vanishing exterior power $\Lambda^{\mathrm{rk} P} P \in \mathrm{Pic}(R)$. Similarly for the last isomorphism.

Proof

The first three follow from the fact that any finitely generated projective module over a commutative local ring or principal ideal domain R is free. So, in all these cases $K_0(R) = \mathbb{Z}$.

---

[1] For this to be the correct K_0-group, one has to make some assumptions about X such as quasi-projective or separated regular noetherian. See Sect. 3.4.

For the fourth isomorphism, in addition we use the decomposition of A into a product $A_1 \times \ldots \times A_n$ of Artinian local rings [3, Theorem 8.7] and the fact $K_0(R \times S) \cong K_0(R) \times K_0(S)$. For the second to last isomorphism, the map $K_0(R) \to \mathbb{Z} \oplus \mathrm{Pic}(R)$ is surjective for any commutative ring R. Injectivity for $\dim R = 1$ follows from Serre's theorem [84, Théorème 1] which implies that a projective module P of rank r over a noetherian ring of Krull dimension d can be written, up to isomorphism, as $Q \oplus R^{r-d}$ for some projective module Q of rank d provided $r \geq d$. For $d = 1$, this means that $P \cong R^{r-1} \oplus \Lambda^r P$.

For the last isomorphism, let $x \in X$ be a closed point with residue field $k(x)$ and $U = X - x$ its open complement. Note that U is affine [41, IV Exercise 1.3]. Anticipating a little, we have a map of short exact sequences

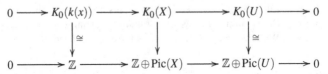

in which the left and right vertical maps are isomorphisms, by the cases proved above. The top row is a special case of Quillen's localization long exact sequence Theorem 2.3.7 (5) using the fact that $K_0 = G_0$ for smooth varieties (Poincaré Duality Theorem 3.3.5), both of which can be proved directly for G_0 and K_0 without the use of the machinery of higher K-theory. The second row is the sum of the exact sequences $0 \to 0 \to \mathbb{Z} \to \mathbb{Z} \to 0$ and $0 \to \mathbb{Z} \to \mathrm{Cl}(X) \to \mathrm{Cl}(U) \to 0$ [41, II Proposition 6.5] in view of the isomorphism $\mathrm{Pic}(X) \cong \mathrm{Cl}(X)$ for smooth varieties [41, II Corollary 6.16]. $\qquad\square$

2.2 Quillen's Q-Construction and Higher K-Theory

In order to define higher K-groups, one constructs a topological space $K(\mathscr{E})$ and defines the K-groups $K_i(\mathscr{E})$ as the homotopy groups $\pi_i K(\mathscr{E})$ of that space. The topological space $K(\mathscr{E})$ is the loop space of the classifying space (Sect. 2.2.2 and Appendix A, Sect. 3) of Quillen's Q-construction. We start with describing the Q-construction.

2.2.1 Quillen's Q-Construction [73, Sect. 2]

Let $\mathscr{E}$ be a small exact category. We define a new category $Q\mathscr{E}$ as follows. The objects of $Q\mathscr{E}$ are the objects of $\mathscr{E}$. A map $X \to Y$ in $Q\mathscr{E}$ is an equivalence class of data $X \xleftarrow{p} W \xrightarrow{i} Y$ where p is a deflation and i an inflation. The datum (W, p, i) is equivalent to the datum (W', p', i') if there is an isomorphism $g : W \to W'$ such that $p = p'g$ and $i = i'g$. The composition of $(W, p, i) : X \to Y$ and $(V, q, j) : Y \to Z$ in $Q\mathscr{E}$ is the map $X \to Z$ represented by the datum $(U, p\bar{q}, j\bar{i})$ where U is the pull-back of q along i as in the diagram

$$
\begin{array}{ccccc}
X & \xleftarrow{\ p\ } & W & \xleftarrow{\ \bar{q}\ } & U \\
& & \downarrow{\scriptstyle i} & & \downarrow{\scriptstyle \bar{i}} \\
& & Y & \xleftarrow{\ q\ } & V & \xrightarrowtail{\ j\ } Z
\end{array}
$$

which exists by 2.1.1 (e). The map $\bar{q}$ (and hence $p\bar{q}$ by 2.1.1 (c)) is a deflation by 2.1.1 (e), and the map $\bar{i}$ (and hence $j\bar{i}$) is an inflation by Exercise 2.1.4 (a). The universal property of cartesian squares implies that composition is well-defined and associative (exercise!). The identity map id_X of an object X of $Q\mathscr{E}$ is represented by the datum $(X, 1, 1)$.

2.2.2 The Classifying Space of a Category

To any small category $\mathscr{C}$, one associates a topological space $B\mathscr{C}$ called the *classifying space of* $\mathscr{C}$. This is a CW-complex constructed as follows (for the precise definition, see Appendix A, Sect. 1.3).

- 0-cells are the objects of $\mathscr{C}$.
- 1-cells are the non-identity morphisms attached to their source and target.
- 2-cells are the 2-simplices (see the figure below) corresponding to pairs (f,g) of composable morphisms such that neither f nor g is an identity morphism.

$(f,g):$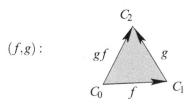

The edges f, g and gf which make up the boundary of the 2-simplex (f,g) are attached to the 1-cells corresponding to f, g, and gf. In case $gf = id_{C_0}$, the whole edge gf is identified with the 0-cell corresponding to C_0.

- 3-cells are the 3-simplices corresponding to triples $C_0 \xrightarrow{f_0} C_1 \xrightarrow{f_1} C_2 \xrightarrow{f_2} C_3$ of composable arrows such that none of the maps f_0, f_1, f_2 is an identity morphism. They are attached in a similar way as in the case of 2-cells, etc.

2.2.3 Exercise

Give the CW-structure of the classifying spaces $B\mathscr{C}$ where the category $\mathscr{C}$ is as follows.

(a) $\mathscr{C}$ is the category with 3 objects A, B, C. The hom sets between two objects of $\mathscr{C}$ contain at most 1 element where the only non-identity maps are $f : A \to B$, $g : B \to C$ and $gf : A \to C$.
(b) $\mathscr{C}$ is the category with 2 objects A, B. The only non-identity maps are $f : A \to B$ and $g : B \to A$. They satisfy $gf = 1_A$ and $fg = 1_B$.

Hint: the category in (a) is the category [2] given in Appendix A, Sect. 1.3. Both categories have contractible classifying space by Lemma A.1.6.

Since we have a category $Q\mathscr{E}$, we have a topological space $BQ\mathscr{E}$. We make the classifying space $BQ\mathscr{E}$ of $Q\mathscr{E}$ into a pointed topological space by choosing a 0-object of $\mathscr{E}$ as base-point. Every object X in $Q\mathscr{E}$ receives an arrow from 0, the map represented by the data $(0,0,0_X)$ where 0_X denotes the zero map $0 \to X$ in $\mathscr{E}$. In particular, the topological space $BQ\mathscr{E}$ is connected, that is, $\pi_0 BQ\mathscr{E} = 0$.

To an object X of $\mathscr{E}$, we associate a loop $l_X = (0,0,0_X)^{-1}(X,0,1)$ based at 0

$$l_X : \qquad 0 \underset{(X,0,1)}{\overset{(0,0,0_X)}{\rightleftarrows}} X$$

in $BQ\mathscr{E}$ by first "walking" along the arrow $(X,0,1)$ and then back along $(0,0,0_X)$ in the opposite direction of the arrow. This loop thus defines an element $[l_X]$ in $\pi_1 BQ\mathscr{E}$.

2.2.4 Proposition

The assignment which sends an object X to the loop l_X induces a well-defined homomorphism of abelian groups $K_0(\mathcal{E}) \to \pi_1 BQ\mathcal{E}$ which is an isomorphism.

Proof

In order to see that the assignment $[X] \mapsto [l_X]$ yields a well defined group homomorphism $K_0(\mathcal{E}) \to \pi_1 BQ\mathcal{E}$, we observe that we could have defined $K_0(\mathcal{E})$ as the free group generated by symbols $[X]$ for each $X \in \mathcal{E}$, modulo the relation $[Y] = [X][Z]$ for any conflation $X \overset{i}{\rightarrowtail} Y \overset{p}{\twoheadrightarrow} Z$. The commutativity is forced by axiom Sect. 2.1.1 (f). So, we have to check that the relation $[l_Y] = [l_X][l_Z]$ holds in $\pi_1 BQ\mathcal{E}$. The loops l_X and l_Z are homotopic to the loops

$$0 \underset{(X,0,1)}{\overset{(0,0,0_X)}{\rightleftarrows}} X \overset{(X,1,i)}{\longrightarrow} Y \qquad \text{and} \qquad 0 \underset{(Z,0,1)}{\overset{(0,0,0_Z)}{\rightleftarrows}} Z \overset{(Y,p,1)}{\longrightarrow} Y \qquad \text{which are}$$

$$0 \underset{(X,0,i)}{\overset{(0,0,0_Y)}{\rightleftarrows}} Y \qquad \text{and} \qquad 0 \underset{(Y,0,1)}{\overset{(X,0,i)}{\rightleftarrows}} Y. \qquad \text{Therefore,}$$

$[l_X][l_Z] = [(0,0,0_Y)^{-1}(X,0,i)][(X,0,i)^{-1}(Y,0,1)] = [(0,0,0_Y)^{-1}(Y,0,1)] = [l_Y]$, and the map $K_0(\mathcal{E}) \to \pi_1 BQ\mathcal{E}$ is well-defined.

Now, we show that the map $K_0(\mathcal{E}) \to \pi_1 BQ\mathcal{E}$ is surjective. By the Cellular Approximation Theorem [104, II Theorem 4.5], every loop in the CW-complex $BQ\mathcal{E}$ is homotopic to a loop with image in the 1-skeleton of $BQ\mathcal{E}$, that is, it is homotopic to a loop which travels along the arrows of $Q\mathcal{E}$. Therefore, every loop in $BQ\mathcal{E}$ is homotopic to a concatenation $a_n^{\pm 1} a_{n-1}^{\pm 1} \cdots a_2^{\pm 1} a_1^{\pm 1}$ of composable paths $a_i^{\pm 1}$ in $BQ\mathcal{E}$ where the a_i's are maps in $Q\mathcal{E}$, and a (resp. a^{-1}) means "walking in the positive (resp. negative) direction of the arrow a". By inserting trivial loops $g \circ g^{-1}$ with $g = (A,0,1) : 0 \to A$, we see that the loop $a_n^{\pm 1} a_{n-1}^{\pm 1} \cdots a_2^{\pm 1} a_1^{\pm 1}$ represents the element

$$[a_n^{\pm 1} g_{n-1} g_{n-1}^{-1} a_{n-1}^{\pm 1} \cdots g_2 g_2^{-1} a_2^{\pm 1} g_1 g_1^{-1} a_1^{\pm 1}]$$
$$= [a_n^{\pm 1} g_{n-1}] \cdot [g_{n-1}^{-1} a_{n-1}^{\pm 1} g_{n-2}] \cdots [g_2^{-1} a_2^{\pm 1} g_1] \cdot [g_1^{-1} a_1^{\pm 1}]$$

in $\pi_1 BQ\mathcal{E}$ where $g_i = (A_i, 0, 1) : 0 \to A_i$ has target A_i which is the endpoint of the path $a_i^{\pm 1}$ and the starting point of $a_{i+1}^{\pm 1}$. Let $(U_i, 0, j_i) : 0 \to X_i$ be the composition $a_i \circ (Y_i, 0, 1)$ in $Q\mathcal{E}$ where Y_i and X_i are the source and target of a_i, respectively. Then we have $[g_{i+1}^{-1} a_i^{\pm 1} g_i] = [(X_i, 0, 1)^{-1}(U_i, 0, j_i)]^{\pm 1}$. This means that the group $\pi_1 BQ\mathcal{E}$ is generated by loops of the form

$$0 \underset{(U,0,j)}{\overset{(X,0,1)}{\rightleftarrows}} X.$$

In the following sequence of homotopies of loops

$$(X,0,1)^{-1}(U,0,j) \sim (X,0,1)^{-1}(U,1,j) \circ (U,0,1)$$
$$\sim (X,0,1)^{-1}(U,1,j) \circ (0,0,0_U) \circ (0,0,0_U)^{-1}(U,0,1)$$
$$\sim (X,0,1)^{-1}(0,0,0_X) \circ (0,0,0_U)^{-1}(U,0,1)$$
$$= l_X^{-1} \circ l_U,$$

the first and third homotopies follow from the identities $(U,0,j) = (U,1,j) \circ (U,0,1)$ and $(0,0,0_X) = (U,1,j) \circ (0,0,0_U)$ in $Q\mathscr{E}$. Since $[l_X^{-1} \circ l_U] = [l_X]^{-1}[l_U]$ is in the image of the map $K_0(\mathscr{E}) \to \pi_1 BQ\mathscr{E}$, we obtain surjectivity.

To show injectivity, we construct a map $\pi_1 BQ\mathscr{E} \to K_0(\mathscr{E})$ so that the composition $K_0(\mathscr{E}) \to K_0(\mathscr{E})$ is the identity. To this end, we introduce a little notation. For a group G, we let $\underline{G}$ be the category with one object $*$ and $\mathrm{Hom}(*,*) = G$. Recall from Appendix A, Sect. 1.5 that $\pi_i B\underline{G} = 0$ for $i \neq 1$ and $\pi_1 B\underline{G} = G$ where the isomorphism $G \to \pi_1 B\underline{G}$ sends an element $g \in G$ to the loop l_g represented by the morphism $g : * \to *$. In order to obtain a map $\pi_1 BQ\mathscr{E} \to K_0(\mathscr{E})$, we construct a functor $F : Q\mathscr{E} \to \underline{K_0(\mathscr{E})}$. The functor sends an object X of $Q\mathscr{E}$ to the object $*$ of $\underline{K_0(\mathscr{E})}$. A map $(W,p,i) : X \to Y$ in $Q\mathscr{E}$ is sent to the map represented by the element $[\ker(p)] \in \overline{K_0(\mathscr{E})}$. Using the notation of 2.2.1, we obtain $F[(V,q,j) \circ (W,p,i)] = F(U,p\bar{q},j\bar{i}) = [\ker(p\bar{q})] = [\ker(\bar{q})] + [\ker(p)] = [\ker(q)] + [\ker(p)] = F(V,q,j) \circ F(W,p,i)$ since, by Exercise 2.1.4 (c), there is a conflation $\ker(\bar{q}) \rightarrowtail \ker(p\bar{q}) \twoheadrightarrow \ker(p)$ and $\ker(\bar{q}) = \ker(q)$ by the universal property of pull-backs. So, F is a functor and it induces a map on fundamental groups of classifying spaces $\pi_1 BQ\mathscr{E} \to \pi_1 \underline{K_0(\mathscr{E})} = K_0(\mathscr{E})$. It is easy to check that the composition $K_0(\mathscr{E}) \to K_0(\mathscr{E})$ is the identity. $\qquad\square$

2.2.5 Definition of $K(\mathscr{E})$

Let $\mathscr{E}$ be a small exact category. The *K-theory space of $\mathscr{E}$* is the pointed topological space

$$K(\mathscr{E}) = \Omega BQ\mathscr{E}$$

with base point the constant loop based at $0 \in Q\mathscr{E}$. The *K-groups of $\mathscr{E}$* are the homotopy groups $K_i(\mathscr{E}) = \pi_i K(\mathscr{E}) = \pi_{i+1} BQ\mathscr{E}$ of the K-theory space of $\mathscr{E}$. An exact functor $\mathscr{E} \to \mathscr{E}'$ induces a functor $Q\mathscr{E} \to Q\mathscr{E}'$ on Q-constructions, and thus, continuous maps $BQ\mathscr{E} \to BQ\mathscr{E}'$ and $K(\mathscr{E}) \to K(\mathscr{E}')$ compatible with composition of exact functors. Therefore, the K-theory space and the K-groups are functorial with respect to exact functors between small exact categories. By Proposition 2.2.4, the group $K_0(\mathscr{E})$ defined in this way coincides with the group defined in Sect. 2.1.5.

2.2.6 Definition of $K(R)$, $K(X)$, $G(X)$

For a ring R and a scheme X, the K-theory spaces $K(R)$ and $K(X)$ are the K-theory spaces associated with the exact categories $\mathrm{Proj}(R)$ of finitely generated projective R-modules and $\mathrm{Vect}(X)$ of vector bundles on X^2. For a noetherian scheme X, the G-theory space $G(X)$ is the K-theory space associated with the abelian category $\mathrm{Coh}(X)$ of coherent O_X-modules.

2.2.7 Remark

There is actually a slight issue with the Definition 2.2.6. The categories $\mathrm{Proj}(R)$ and $\mathrm{Vect}(X)$ have too many objects, so many that they do not form a set; the same is true for $\mathrm{Coh}(X)$. But a topological space is a *set* with a topology. Already the 0-skeletons of $BQ\mathrm{Proj}(R)$ and $BQ\mathrm{Vect}(X)$ – which are the collections of objects of $\mathrm{Proj}(R)$ and $\mathrm{Vect}(X)$ – are too large. To get around this problem, one has to choose "small models" of $\mathrm{Proj}(R)$ and $\mathrm{Vect}(X)$ in order to define the K-theory spaces of R and X. More precisely, one has to choose equivalences of

---

[2] See footnote in Definition 2.1.8.

exact categories $\mathscr{E}_R \simeq \text{Proj}(R)$ and $\mathscr{E}_X \simeq \text{Vect}(X)$ where $\mathscr{E}_R$ and $\mathscr{E}_X$ are small categories, i.e., categories which have a set of objects as opposed to a class of objects. This is possible because $\text{Proj}(R)$ and $\text{Vect}(X)$ only have a *set* of isomorphism classes of objects (Exercise!). Given such a choice of equivalence, one sets $K(R) = K(\mathscr{E}_R)$ and $K(X) = K(\mathscr{E}_X)$. For any other choice of equivalences $\mathscr{E}'_R \simeq \text{Proj}(R)$ and $\mathscr{E}'_X \simeq \text{Vect}(X)$ as above, there are equivalences $\mathscr{E}_R \simeq \mathscr{E}'_R$ and $\mathscr{E}_X \simeq \mathscr{E}'_X$ compatible with the corresponding equivalences with $\text{Proj}(R)$ and $\text{Vect}(X)$ which are unique up to equivalence of functors. It follows from Lemma A.1.6 that these equivalences induce homotopy equivalences of K-theory spaces $K(\mathscr{E}_R) \simeq K(\mathscr{E}'_R)$ and $K(\mathscr{E}_X) \simeq K(\mathscr{E}'_X)$ which are unique up to homotopy. Usually, one avoids these issues by working in a fixed "universe".

In order to reconcile the definition of $K(R)$ given above with the plus-construction given in Cortiñas' lecture [21], we cite the following theorem of Quillen a proof of which can be found in [35].

2.2.8 Theorem ($Q = +$)

There is a natural homotopy equivalence

$$BGL(R)^+ \simeq \Omega_0 BQ \text{Proj}(R),$$

where Ω_0 stands for the connected component of the constant loop in the full loop space. In particular, there are natural isomorphisms for $i \geq 1$

$$\pi_i BGL(R)^+ \cong \pi_{i+1} BQ \text{Proj}(R).$$

Note that the space denoted by $K(R)$ in Cortiñas' lecture [21] is the connected component of 0 of the space we here denote by $K(R)$.

2.2.9 Warning

Some authors define $K(R)$ to be $K_0(R) \times BGL(R)^+$ as functors in R. Strictly speaking, this is wrong: there is no zig–zag of homotopy equivalences between $K_0(R) \times BGL(R)^+$ and $\Omega BQ \text{Proj}(R)$ which is functorial in R.

The problem is not that the usual construction of $BGL(R)^+$ involves choices when attaching 2 and 3-cells. The plus-construction can be made functorial. For instance, Bousfield–Kan's $\mathbb{Z}$-completion $\mathbb{Z}_\infty BGl(R)$ does the job by [10, I 5.5, V 3.3] (compare [10, VII 3.4]). The problem is that one cannot write $K(R)$ functorially as a product of $K_0(R)$ and $BGL(R)^+$. To explain this point, let R be any ring, ΓR be the cone ring of R (see Cortiñas' lecture [21]), and ΣR be the suspension ring of R which is the factor ring $\Gamma R / M_\infty(R)$ of the cone ring by the two-sided ideal $M_\infty R$ of finite matrices. Let $\tilde{R} = \Gamma R \times_{\Sigma R} \Gamma R$. The fibre product square of (unital) rings defining $\tilde{R}$ induces commutative diagrams

$$
\begin{array}{ccc}
K(\tilde{R}) & \longrightarrow & K(\Gamma R) \\
\downarrow & & \downarrow \\
K(\Gamma R) & \longrightarrow & K(\Sigma R)
\end{array}
\qquad \text{and} \qquad
\begin{array}{ccc}
K_0(\tilde{R}) & \longrightarrow & K_0(\Gamma R) \\
\downarrow & & \downarrow \\
K_0(\Gamma R) & \longrightarrow & K_0(\Sigma R).
\end{array}
$$

Using Quillen's Q-construction, or other functorial versions of K-theory, one can show that the left square is homotopy cartesian, and that there is a non-unital ring map $R \to \tilde{R}$ which induces isomorphisms in *K*-theory. If the *K*-theory space $K(R)$ were the product $K_0(R) \times BGL(R)^+$ in a functorial way, the set $K_0(R)$ considered as a topological space with the discrete topology would be a natural retract of $K(R)$. Since the left-hand square in the diagram above is homotopy cartesian, its retract, the right-hand square, would have to be homotopy cartesian as well. This is absurd since $K_0(\Gamma R) = 0$ for all rings R, and $K_0(\tilde{R}) = K_0(R) \neq 0$ for most rings.

2.3 Quillen's Fundamental Theorems

In what follows we will simply cite several fundamental theorems of Quillen. Their proofs in [73] are very readable and highly recommended.

2.3.1 Serre Subcategories and Exact Sequences of Abelian Categories

Let $\mathscr{A}$ be an abelian category. A *Serre subcategory* of $\mathscr{A}$ is a full subcategory $\mathscr{B} \subset \mathscr{A}$ with the property that for a conflation in $\mathscr{A}$

$$M_0 \rightarrowtail M_1 \twoheadrightarrow M_2, \quad \text{we have} \quad M_1 \in \mathscr{B} \iff M_0 \text{ and } M_2 \in \mathscr{B}.$$

It is easy to see that a Serre subcategory $\mathscr{B}$ is itself an abelian category, and that the inclusion $\mathscr{B} \subset \mathscr{A}$ is fully exact. Given a Serre subcategory $\mathscr{B} \subset \mathscr{A}$, one can (up to set theoretical issues which do not exist when $\mathscr{A}$ is small) construct the quotient abelian category $\mathscr{A}/\mathscr{B}$ which has the universal property of a quotient object in the category of exact categories. The quotient abelian category $\mathscr{A}/\mathscr{B}$ is equivalent to the localization $\mathscr{A}[S^{-1}]$ of $\mathscr{A}$ with respect to the class S of morphisms f in $\mathscr{A}$ for which $\ker(f)$ and $\mathrm{coker}(f)$ are isomorphic to objects in $\mathscr{B}$. The class S satisfies a calculus of fractions (Exercise!) and $\mathscr{A}[S^{-1}]$ has a very explicit description; see Appendix A, Sect. 2.6. We will call $\mathscr{B} \to \mathscr{A} \to \mathscr{A}/\mathscr{B}$ an *exact sequence of abelian categories*. More details can be found in [29, 70].

The following two theorems are proved in [73, Sect. 5 Theorem 5] and [73, Sect. 5 Theorem 4].

2.3.2 Theorem (Quillen's Localization Theorem)

Let $\mathscr{A}$ be a small abelian category, and let $\mathscr{B} \subset \mathscr{A}$ be a Serre subcategory. Then the sequence of topological spaces

$$BQ(\mathscr{B}) \to BQ(\mathscr{A}) \to BQ(\mathscr{A}/\mathscr{B})$$

is a homotopy fibration (see Appendix A, Sect. 1.7 for a definition). In particular, there is a long exact sequence of associated K-groups

$$\cdots \to K_{n+1}(\mathscr{A}) \to K_{n+1}(\mathscr{A}/\mathscr{B}) \to K_n(\mathscr{B}) \to K_n(\mathscr{A}) \to K_n(\mathscr{A}/\mathscr{B}) \to \cdots$$

$$\cdots \to K_0(\mathscr{A}) \to K_0(\mathscr{A}/\mathscr{B}) \to 0.$$

2.3.3 Theorem (*Dévissage*)

Let $\mathscr{A}$ be a small abelian category, and $\mathscr{B} \subset \mathscr{A}$ be a full abelian subcategory such that the inclusion $\mathscr{B} \subset \mathscr{A}$ is exact. Assume that every object A of $\mathscr{A}$ has a finite filtration

$$0 = A_0 \subset A_1 \subset ... \subset A_n = A$$

such that the quotients A_i/A_{i-1} are in $\mathscr{B}$. Then the inclusion $\mathscr{B} \subset \mathscr{A}$ induces a homotopy equivalence

$$K(\mathscr{B}) \xrightarrow{\sim} K(\mathscr{A}).$$

In particular, it induces isomorphisms of K-groups $K_i(\mathscr{B}) \cong K_i(\mathscr{A})$.

The following are two applications of Quillen's Localization and *Dévissage* Theorems.

2.3.4 Nilpotent Extensions

Let X be a noetherian scheme and $i : Z \hookrightarrow X$ a closed subscheme corresponding to a nilpotent sheaf of ideals $I \subset O_X$. Assume $I^n = 0$. Then $i_* : \mathrm{Coh}(Z) \to \mathrm{Coh}(X)$ satisfies the hypothesis of the *Dévissage* Theorem because $\mathrm{Coh}(Z)$ can be identified with the subcategory of those coherent sheaves F on X for which $IF = 0$, and every sheaf $F \in \mathrm{Coh}(X)$ has a filtration $0 = I^n F \subset I^{n-1} F \subset ... \subset IF \subset F$ with quotients in $\mathrm{Coh}(Z)$. We conclude that i_* induces a homotopy equivalence $G(Z) \simeq G(X)$. In particular:

2.3.5 Theorem

For a noetherian scheme X, the closed immersion $i : X_{red} \hookrightarrow X$ induces a homotopy equivalence of G-theory spaces

$$i_* : G(X_{red}) \xrightarrow{\sim} G(X).$$

2.3.6 *G*-Theory Localization

Let X be a noetherian scheme, and $j : U \subset X$ be an open subscheme with $i : Z \subset X$ being its closed complement $X - U$. Let $\mathrm{Coh}_Z(X) \subset \mathrm{Coh}(X)$ be the fully exact subcategory of those coherent sheaves F on X which have support in Z, that is, for which $F_{|U} = 0$. Then the sequence

$$\mathrm{Coh}_Z(X) \subset \mathrm{Coh}(X) \xrightarrow{j^*} \mathrm{Coh}(U) \tag{3}$$

is an exact sequence of abelian categories (see Sect. 2.3.8 below). By Theorem 2.3.2, we obtain a homotopy fibration $K\mathrm{Coh}_Z(X) \to K\mathrm{Coh}(X) \to K\mathrm{Coh}(Z)$ of K-theory spaces. For another proof, see Theorem 3.3.2. The inclusion $i_* : \mathrm{Coh}(Z) \subset \mathrm{Coh}_Z(X)$ satisfies *Dévissage* (Exercise!), so we have a homotopy equivalence $K\mathrm{Coh}(Z) \simeq K\mathrm{Coh}_Z(X)$. Put together, we obtain:

2.3.7 Theorem

Let X be a noetherian scheme, and $j : U \subset X$ be an open subscheme with $i : Z \subset X$ being its closed complement $X - U$. Then the following sequence of spaces is a homotopy fibration

$$G(Z) \xrightarrow{i_*} G(X) \xrightarrow{j^*} G(U). \tag{4}$$

In particular, there is an associated long exact sequence of G-theory groups

$$\cdots G_{i+1}(U) \to G_i(Z) \to G_i(X) \to G_i(U) \to G_{i-1}(Z) \cdots \to G_0(U) \to 0 \tag{5}$$

2.3.8 *Proof that (3) is an Exact Sequence of Abelian Categories*

As the "kernel" of the exact functor $\mathrm{Coh}(X) \to \mathrm{Coh}(U)$, the category $\mathrm{Coh}_Z(X)$ is automatically a Serre subcategory of $\mathrm{Coh}(X)$. The composition $\mathrm{Coh}_Z(X) \subset \mathrm{Coh}(X) \to \mathrm{Coh}(U)$ is trivial. Therefore, we obtain an induced functor

$$\mathrm{Coh}(X)/\mathrm{Coh}_Z(X) \to \mathrm{Coh}(U) \tag{6}$$

which we have to show is an equivalence.

The functor (6) is essentially surjective on objects. This is because for any $F \in \mathrm{Coh}(U)$, the O_X-module j_*F is quasi-coherent (X is noetherian). Therefore, j_*F is a filtered colimit $\mathrm{colim}\, G_i$ of its coherent sub-O_X-modules G_i. Every ascending chain of subobjects of a coherent sheaf eventually stops. Therefore, we must have $j^*G_i \cong j^*j_*F = F$ for some i.

The functor (6) is full because for $F, G \in \mathrm{Coh}(X)$, any map $f : j^*F \to j^*G$ in $\mathrm{Coh}(U)$ equals $g_{|U} \circ (t_{|U})^{-1}$ where t and g are maps in a diagram $F \xleftarrow{t} H \xrightarrow{g} G$ of coherent O_X-modules. This diagram can be taken to be the pull-back in $\mathrm{Qcoh}(X)$ of the diagram $F \to j_*j^*F \to j_*j^*G \leftarrow G$ with middle map $j^*(f)$ and outer two maps the unit of adjunction. The object H is coherent as it is a quasi-coherent subsheaf of the coherent sheaf $F \oplus G$. The unit of adjunction $G \to j_*j^*G$ is an isomorphism when restricted to U. Since j^* is an exact functor of abelian categories, the same is true for its pull back $t : F \to H$.

Finally, the functor (6) is faithful by the following argument. The "kernel category" of this functor is trivial, by construction. This implies that the functor (6) is conservative, i.e., detects isomorphisms. Now, let $f : F \to G$ be a map in $\mathrm{Coh}(X)/\mathrm{Coh}_Z(X)$ such that $j^*(f) = 0$. Then $\ker(f) \to F$ and $G \to \mathrm{coker}(f)$ are isomorphisms when restricted to U. Since the functor (6) is conservative, these two maps are already isomorphisms in $\mathrm{Coh}(X)/\mathrm{Coh}_Z(X)$ which means that $f = 0$.

Since a fully faithful and essentially surjective functor is an equivalence, we are done. $\square$

2.3.9 **Theorem** (Homotopy Invariance of G-Theory [73, Proposition 4.1])

Let X and P be noetherian schemes and $f : P \to X$ be a flat map whose fibres are affine spaces (for instance, a geometric vector bundle). Then

$$f^* : G(X) \xrightarrow{\sim} G(P)$$

is a homotopy equivalence. In particular, $G_i(X \times \mathbb{A}^1) \cong G_i(X)$.

2.3.10 *K-Theory of Regular Schemes*

Let X be a regular noetherian and separated scheme. Then the inclusion $\mathrm{Vect}(X) \subset \mathrm{Coh}(X)$ induces a homotopy equivalence $K\,\mathrm{Vect}(X) \simeq K\,\mathrm{Coh}(X)$, that is,

$$K(X) \xrightarrow{\sim} G(X)$$

(see the Poincaré Duality Theorem 3.3.5; classically it also follows from Quillen's Resolution Theorem 2.3.12 below). Thus, Theorems 2.3.7 and 2.3.9 translate into theorems about $K(X)$ when X is regular, noetherian and separated. For instance, Theorem 2.3.9 together with Poincaré Duality implies that the projection $X \times \mathbb{A}^n \to X$ induces isomorphisms

$$K_i(X) \xrightarrow{\cong} K_i(X \times \mathbb{A}^n)$$

whenever X is regular noetherian separated.

Besides the results mentioned above, Quillen proves two fundamental theorems which are also of interest: the Additivity Theorem [73, Sect. 3 Theorem 2 and Corollary 1] and the Resolution Theorem [73, Sect. 4 Theorem 3]. Both are special cases of the Thomason–Waldhausen Localization Theorem 3.2.23 which is stated below. We simply quote Quillen's theorems here, and in Sect. 3.3 we give a proof of them based on the Localization Theorem. However, we have to mention that the Additivity Theorem is used in the proof of the Localization Theorem.

2.3.11 Theorem (Additivity [73, Sect. 3, Corollary 1])

Let $\mathscr{E}$ and $\mathscr{E}'$ be exact categories, and let

$$0 \to F_{-1} \to F_0 \to F_1 \to 0$$

be a sequence of exact functors $F_i : \mathscr{E} \to \mathscr{E}'$ such that $F_{-1}(A) \rightarrowtail F_0(A) \twoheadrightarrow F_1(A)$ is a conflation for all objects A in $\mathscr{E}$. Then the induced maps on K-groups satisfy

$$F_0 = F_{-1} + F_1 : K_i(\mathscr{E}) \to K_i(\mathscr{E}').$$

2.3.12 Theorem (Resolution [73, Sect. 4])

Let $\mathscr{A} \subset \mathscr{B}$ be a fully exact subcategory of an exact category $\mathscr{B}$. Assume that

(a) if $M_{-1} \rightarrowtail M_0 \twoheadrightarrow M_1$ is a conflation in $\mathscr{B}$ with $M_0, M_1 \in \mathscr{A}$, then $M_{-1} \in \mathscr{A}$, and
(b) for every $B \in \mathscr{B}$ there is an exact sequence

$$0 \to A_n \to A_{n-1} \to \cdots A_0 \to B \to 0$$

with $A_i \in \mathscr{A}$.

Then the inclusion $\mathscr{A} \subset \mathscr{B}$ induces a homotopy equivalence of K-theory spaces

$$K(\mathscr{A}) \xrightarrow{\sim} K(\mathscr{B}).$$

2.4 Negative K-Groups

Besides the positive K-groups, one can also define the negative K-groups K_i with $i < 0$. They extend certain K_0 exact sequences to the right (see Cortiñas' lecture [21]). For rings and additive (or split exact) categories they were introduced by Bass [7] and Karoubi [45]. The treatment for exact categories below follows [81].

2.4.1 Idempotent Completion

Let $\mathscr{A}$ be an additive category, and $\mathscr{B} \subset \mathscr{A}$ be a full additive subcategory. We call the inclusion $\mathscr{B} \subset \mathscr{A}$ *cofinal*, or *equivalence up to factors* if every object of $\mathscr{A}$ is a direct factor of an object of $\mathscr{B}$. If $\mathscr{A}$ and $\mathscr{B}$ are exact categories, we require moreover that the inclusion $\mathscr{B} \subset \mathscr{A}$ is fully exact, that is, the inclusion is extension closed, and it preserves and detects conflations. As an example, the category of (finitely generated) free R-modules is cofinal in the category of (finitely generated) projective R-modules.

Given an additive category $\mathscr{A}$, there is a "largest" category $\tilde{\mathscr{A}}$ of $\mathscr{A}$ such that the inclusion $\mathscr{A} \subset \tilde{\mathscr{A}}$ is cofinal. This is the *idempotent completion* $\tilde{\mathscr{A}}$ of $\mathscr{A}$. An additive category is called *idempotent complete* if for every idempotent map $p = p^2 : A \to A$, there is an isomorphism $A \cong X \oplus Y$ under which the map p corresponds to $\left(\begin{smallmatrix} 1 & 0 \\ 0 & 0 \end{smallmatrix}\right) : X \oplus Y \to X \oplus Y$. The objects of the idempotent completion $\tilde{\mathscr{A}}$ of $\mathscr{A}$ are pairs (A, p) with A an object of $\mathscr{A}$ and $p = p^2 : A \to A$ an idempotent endomorphism. Maps $(A, p) \to (B, q)$ in $\tilde{\mathscr{A}}$ are maps $f : A \to B$ in $\mathscr{A}$ such that $fp = f = qf$. Composition is composition in $\mathscr{A}$, and $id_{(A,p)} = p$. Every idempotent $q = q^2 :$ $(A, p) \to (A, p)$ corresponds to $\left(\begin{smallmatrix} 1 & 0 \\ 0 & 0 \end{smallmatrix}\right)$ under the isomorphism $(q, p - q) : (A, q) \oplus (A, p - q) \cong$ (A, p). Therefore, the category $\tilde{\mathscr{A}}$ is indeed idempotent complete. Furthermore, we have a fully faithful embedding $\mathscr{A} \subset \tilde{\mathscr{A}} : A \mapsto (A, 1)$ which is cofinal since the object (A, p) of $\tilde{\mathscr{A}}$ is a direct factor the object $(A, 1)$ of $\mathscr{A}$.

If $\mathscr{E}$ is an exact category, its idempotent completion $\tilde{\mathscr{E}}$ becomes an exact category when we declare a sequence in $\tilde{\mathscr{E}}$ to be a conflation if it is a retract (in the category of conflations) of a conflation of $\mathscr{E}$. For more details, see [94, Appendix A]. Note that the inclusion $\mathscr{E} \subset \tilde{\mathscr{E}}$ is indeed fully exact.

2.4.2 Proposition (Cofinality [36, Theorem 1.1])

Let $\mathscr{A}$ be an exact category and $\mathscr{B} \subset \mathscr{A}$ be a cofinal fully exact subcategory. Then the maps $K_i(\mathscr{B}) \to K_i(\mathscr{A})$ are isomorphisms for $i > 0$ and a monomorphism for $i = 0$. This holds in particular for $K_i(\mathscr{E}) \to K_i(\tilde{\mathscr{E}})$.

2.4.3 Negative K-Theory and the Spectrum $I\!K(\mathscr{E})$

To any exact category $\mathscr{E}$, one can associate a new exact category $S\mathscr{E}$ (see Sect. 2.4.6), called *the suspension of $\mathscr{E}$*, such that there is a natural homotopy equivalence [81]

$$K(\tilde{\mathscr{E}}) \xrightarrow{\sim} \Omega K(S\mathscr{E}). \tag{7}$$

If $\mathscr{E} = \mathrm{Proj}(R)$ one can take $S\mathscr{E} = \mathrm{Proj}(\Sigma R)$ where ΣR is the suspension ring of R; see Cortiñas' lecture [21]) for the definition of ΣR.

One uses the suspension construction to slightly modify the definition of algebraic K-theory in order to incorporate negative K-groups as follows. One sets $I\!K_i(\mathscr{E}) = K_i(\mathscr{E})$ for $i \geq 1$, $I\!K_0(\mathscr{E}) = K_0(\tilde{\mathscr{E}})$ and $I\!K_i(\mathscr{E}) = K_0(\widetilde{S^{-i}\mathscr{E}})$ for $i < 0$. Since $\mathrm{Vect}(X)$, $\mathrm{Coh}(X)$ and $\mathrm{Proj}(R)$ are all idempotent complete, we have the equalities $I\!K_0 \mathrm{Vect}(X) = K_0 \mathrm{Vect}(X) = K_0(X)$, $I\!K_0 \mathrm{Coh}(X) = K_0 \mathrm{Coh}(X) = G_0(X)$ and $I\!K_0 \mathrm{Proj}(R) = K_0 \mathrm{Proj}(R) = K_0(R)$. In these cases, we have not changed the definition of K-theory; we have merely introduced "negative K-groups" $I\!K_i$ for $i < 0$. For this reason, we may write $K_i(X)$ and $K_i(R)$ instead of $I\!K_i(X)$ and $I\!K_i(R)$ for all $i \in \mathbb{Z}$.

In a fancy language, one constructs a spectrum $I\!K(\mathscr{E})$ whose homotopy groups are the groups $I\!K_i(\mathscr{E})$ for $i \in \mathbb{Z}$. The n-th space of this spectrum is $K(S^n\mathscr{E})$, and the structure maps are given by (7). For terminology and basic properties of spectra, we refer the reader to Appendix A, Sect. 1.8.

By Bass' Fundamental Theorem stated in Theorem 3.5.3 below there is a split exact sequence for $i \in \mathbb{Z}$

$$0 \to K_i(R) \to K_i(R[T]) \oplus K_i(R[T^{-1}]) \to K_i(R[T, T^{-1}]) \to K_{i-1}(R) \to 0.$$

One can use this sequence to give a recursive definition of the negative K-groups $K_i(R)$ for $i < 0$, starting with the functor K_0. This was Bass' original definition.

2.4.4 Remark

Although Quillen did not define negative K-groups of exact categories, all K-theory statements in [73] and [35] extend to negative K-theory. The only exceptions are Quillen's Localization Theorem 2.3.2 and the *Dévissage* Theorem 2.3.3. To insure that these two theorems also extend to negative K-theory, we need the abelian categories in question to be noetherian, though it is conjectured that the noetherian hypothesis is unnecessary.

2.4.5 Remark

Not much is known about $I\!K_i(\mathcal{E})$ when $i < 0$, even though we believe their calculations to be easier than those of $K_i(\mathcal{E})$ when $i \geq 0$. However, we do know the following. We have $K_i(R) = 0$ for $i < 0$ when R is a regular noetherian ring [7]. We have $I\!K_{-1}(\mathcal{A}) = 0$ for any abelian category $\mathcal{A}$ [82, Theorem 6], and $I\!K_i(\mathcal{A}) = 0$ for $i < 0$ when $\mathcal{A}$ is a noetherian abelian category [82, Theorem 7]. In particular, $K_{-1}(R) = 0$ for a regular coherent ring R, and $K_i(X) = 0$ for $i < 0$ when X is any regular noetherian and separated scheme. In [19] it is shown that $K_i(X) = 0$ for $i < -d$ when X is a d-dimensional scheme essentially of finite type over a field of characteristic 0, but $K_{-d}(X) = H^d_{cdh}(X, \mathbb{Z})$ can be non-zero [75]. For finite-type schemes over fields of positive characteristic, the same is true provided strong resolution of singularities holds over the base field [30]. It is conjectured that $K_i(\mathbb{Z}G) = 0$ for $i < -1$ when G is a finitely presented group [42]. For results in this direction see [56].

2.4.6 Construction of the Suspension $S\mathcal{E}$

Let $\mathcal{E}$ be an exact category. The countable envelope $\mathcal{F}\mathcal{E}$ of $\mathcal{E}$ is an exact category whose objects are sequences

$$A_0 \rightarrowtail A_1 \rightarrowtail A_2 \rightarrowtail \dots$$

of inflations in $\mathcal{E}$. The morphism set from a sequence A_* to another sequence B_* is

$$\operatorname{Hom}_{\mathcal{F}\mathcal{E}}(A_*, B_*) = \lim_i \operatorname*{colim}_j \operatorname{Hom}_{\mathcal{E}}(A_i, B_j).$$

A sequence in $\mathcal{F}\mathcal{E}$ is a conflation iff it is isomorphic in $\mathcal{F}\mathcal{E}$ to the sequence of maps of sequences $A_* \to B_* \to C_*$ where $A_i \to B_i \to C_i$ is a conflation in $\mathcal{E}$ for $i \in \mathbb{N}$. Colimits of sequences of inflations exist in $\mathcal{F}\mathcal{E}$, and are exact. In particular, $\mathcal{F}\mathcal{E}$ has exact countable direct sums. There is a fully faithful exact functor $\mathcal{E} \to \mathcal{F}\mathcal{E}$ which sends an object $X \in \mathcal{E}$ to the constant sequence $X \xrightarrow{1} X \xrightarrow{1} X \xrightarrow{1} \cdots$. For details of the construction see [46, Appendix B] where $\mathcal{F}\mathcal{E}$ was denoted by $\mathcal{E}^\sim$.

   The suspension $S\mathcal{E}$ of $\mathcal{E}$ is the quotient $\mathcal{F}\mathcal{E}/\mathcal{E}$ of the countable envelope $\mathcal{F}\mathcal{E}$ by the subcategory $\mathcal{E}$. The quotient is taken in the category of small exact categories. The proof of the existence of $\mathcal{F}\mathcal{E}/\mathcal{E}$ and an explicit description is given in [81]. By [81, Theorem 2.1 and Lemma 3.2] the sequence $\mathcal{E} \to \mathcal{F}\mathcal{E} \to S\mathcal{E}$ induces a homotopy fibration $K(\tilde{\mathcal{E}}) \to K(\mathcal{F}\mathcal{E}) \to K(S\mathcal{E})$ of K-theory spaces. Since $\mathcal{F}\mathcal{E}$ has exact countable direct sums, the total space $K(\mathcal{F}\mathcal{E})$ of the fibration is contractible. This yields the homotopy equivalence Sect. 2.4.3 (7).

3 Algebraic K-Theory and Triangulated Categories

3.1 The Grothendieck-Group of a Triangulated Category

Most calculations in the early days of K-theory were based on Quillen's Localization Theorem 2.3.2 for abelian categories together with *Dévissage* Theorem 2.3.3. Unfortunately, not all K-groups are (not even equivalent to) the K-groups of some abelian category, notably $K(X)$ where X is some singular variety. Also, there is no satisfactory generalization of Quillen's Localization Theorem to exact categories which would apply to all situations K-theorists had in mind. This is where triangulated categories come in. They provide a flexible framework that allows us to prove many results which cannot be proved with Quillen's methods alone.

For the rest of this subsection, we will assume that the reader is familiar with Appendix A, Sects. 2.1–2.7.

3.1.1 Definition of $K_0(\mathscr{T})$

Let $\mathscr{T}$ be a small triangulated category. The Grothen-dieck-group $K_0(\mathscr{T})$ of $\mathscr{T}$ is the abelian group freely generated by symbols $[X]$ for every object X of $\mathscr{T}$, modulo the relation $[X] + [Z] = [Y]$ for every distinguished triangle $X \to Y \to Z \to TX$ in $\mathscr{T}$.

3.1.2 Remark

As in Remark 2.1.6, we have $[X] = [Y]$ if there is an isomorphism $f : X \cong Y$ in view of the distinguished triangle $X \xrightarrow{f} Y \to 0 \to TX$. We also have $[X \oplus Y] = [X] + [Y]$ because there is a distinguished triangle $X \to X \oplus Y \to Y \to TX$ which is the direct sum of the distinguished triangles $X \to X \to 0 \to TX$ and $0 \to Y \to Y \to 0$. Moreover, the distinguished triangle $X \to 0 \to TX \to TX$ shows that $[TX] = -[X]$. In particular, every element in $K_0(\mathscr{T})$ can be represented as $[X]$ for some object X in $\mathscr{T}$.

One would like to relate the Grothendieck-group $K_0(\mathscr{E})$ of an exact category $\mathscr{E}$ to the Grothendieck-group of a triangulated category associated with $\mathscr{E}$. This *rôle* is played by the bounded derived category $D^b(\mathscr{E})$ of $\mathscr{E}$.

3.1.3 The Bounded Derived Category of an Exact Category

Let $\mathrm{Ch}^b \mathscr{E}$ be the exact category of bounded chain complexes in $\mathscr{E}$; see Example 2.1.2 (e). Call a bounded chain complex (A, d) in $\mathscr{E}$ *strictly acyclic* if every differential $d^i : A^i \to A^{i+1}$ can be factored as $A^i \twoheadrightarrow Z^{i+1} \rightarrowtail A^{i+1}$ such that the sequence $Z^i \rightarrowtail A^i \twoheadrightarrow Z^{i+1}$ is a conflation in $\mathscr{E}$ for all $i \in \mathbb{Z}$. A bounded chain complex is called *acyclic* if it is homotopy equivalent to a strictly acyclic chain complex. A map $f : (A, d) \to (B, d)$ is called *quasi-isomorphism* if its cone $C(f)$ (see Appendix A, Sect. 2.5 for a definition) is acyclic.

As a category, the *bounded derived category* $D^b(\mathscr{E})$ is the category

$$D^b(\mathscr{E}) = [\mathrm{quis}^{-1}] \mathrm{Ch}^b \mathscr{E}$$

obtained from the category of bounded chain complexes $\mathrm{Ch}^b \mathscr{E}$ by formally inverting the quasi-isomorphisms. A more explicit description of $D^b(\mathscr{E})$ is obtained as follows. Let $\mathscr{K}^b(\mathscr{E})$

be the homotopy category of bounded chain complexes in $\mathscr{E}$. Its objects are bounded chain complexes in $\mathscr{E}$, and maps are chain maps up to chain homotopy. With the same definitions as in Appendix A, Sect. 2.5, the homotopy category $\mathscr{K}^b(\mathscr{E})$ is a triangulated category. Let $\mathscr{K}_{ac}^b(\mathscr{E}) \subset \mathscr{K}^b(\mathscr{E})$ be the full subcategory of acyclic chain complexes. The category $\mathscr{K}_{ac}^b(\mathscr{E})$ is closed under taking cones and shifts T and T^{-1} in $\mathscr{K}^b(\mathscr{E})$. Therefore, it is a full triangulated subcategory of $\mathscr{K}^b(\mathscr{E})$. The bounded derived category of the exact category $\mathscr{E}$ is the Verdier quotient $\mathscr{K}^b(\mathscr{E})/\mathscr{K}_{ac}^b(\mathscr{E})$.

It turns out that distinguished triangles in $D^b(\mathscr{E})$ are precisely those triangles which are isomorphic to the standard triangles constructed as follows. A conflation $X \xrightarrow{i} Y \xrightarrow{p} Z$ of chain complexes in $\mathrm{Ch}^b\mathscr{E}$ yields the standard distinguished triangle

$$ X \xrightarrow{\ i\ } Y \xrightarrow{\ p\ } Z \xrightarrow{q \circ s^{-1}} TX $$

in $D^b(\mathscr{E})$ where s is the quasi-isomorphism $C(i) \to C(i)/C(id_X) \cong Z$, and q is the canonical map $C(f) \to TX$ as in Appendix A, Sect. 2.5. For more details, see [48].

3.1.4 Exercise

Let $\mathscr{E}$ be an exact category. Consider the objects of $\mathscr{E}$ as chain complexes concentrated in degree zero. Show that the map $K_0(\mathscr{E}) \to K_0(D^b\mathscr{E})$ given by $[X] \mapsto [X]$ is an isomorphism. *Hint*: The inverse $K_0(D^b\mathscr{E}) \to K_0(\mathscr{E})$ is given by $[A,d] \mapsto \Sigma_i(-1)^i[A^i]$. The point is to show that this map is well-defined.

3.1.5 Definition

A sequence of triangulated categories $\mathscr{A} \to \mathscr{B} \to \mathscr{C}$ is called *exact* if the composition sends $\mathscr{A}$ to 0, if $\mathscr{A} \to \mathscr{B}$ is fully faithful and coincides (up to equivalence) with the subcategory of those objects in $\mathscr{B}$ which are zero in $\mathscr{C}$, and if the induced functor $\mathscr{B}/\mathscr{A} \to \mathscr{C}$ from the Verdier quotient $\mathscr{B}/\mathscr{A}$ to $\mathscr{C}$ is an equivalence.

3.1.6 Exercise

Let $\mathscr{A} \to \mathscr{B} \to \mathscr{C}$ be an exact sequence of triangulated categories. Then the following sequence of abelian groups is exact

$$ K_0(\mathscr{A}) \to K_0(\mathscr{B}) \to K_0(\mathscr{C}) \to 0. \tag{8} $$

Hint: Show that the map $K_0(\mathscr{C}) \to \mathrm{coker}(K_0(\mathscr{A}) \to K_0(\mathscr{B}))$ given by $[C] \mapsto [B]$ is well-defined where $B \in \mathscr{B}$ is any object whose image in the category $\mathscr{C}$ is isomorphic to the object C.

How can we decide whether a sequence of exact categories induces an exact sequence of bounded derived categories so that we could apply Exercise 3.1.6 and Theorems 3.2.23 and 3.2.27 below? For this, the following facts are often quite useful.

3.1.7 Some Criteria and Facts

Let $\mathscr{A} \to \mathscr{B}$ be an exact functor between exact categories.

(a) If $\mathscr{B}$ is the localization $\Sigma^{-1}\mathscr{A}$ of $\mathscr{A}$ with respect to a set of maps Σ which satisfies a calculus of left (or right) fractions, then $\mathrm{Ch}^b\,\mathscr{B}$ is the localization of $\mathrm{Ch}^b\,\mathscr{A}$ with respect to the set of maps which degree-wise belong to Σ. This set of maps also satisfies a calculus of left (right) fractions, and therefore, $D^b(\mathscr{A}) \to D^b(\mathscr{B})$ is a localization. In particular, $D^b(\mathscr{B})$ is the Verdier quotient of $D^b(\mathscr{A})$ modulo the full triangulated subcategory of objects which are zero in $D^b(\mathscr{B})$.

(b) Suppose that $\mathscr{A}$ is a fully exact subcategory of $\mathscr{B}$. If for any inflation $A \rightarrowtail B$ in $\mathscr{B}$ with $A \in \mathscr{A}$ there is a map $B \to A'$ with $A' \in \mathscr{A}$ such that the composition $A \to A'$ is an inflation in $\mathscr{A}$, then the functor $D^b(\mathscr{A}) \to D^b(\mathscr{B})$ is fully faithful [48, 12.1].

(c) If $\mathscr{A} \to \mathscr{B}$ is a cofinal fully exact inclusion, then $D^b\mathscr{A} \to D^b\mathscr{B}$ is fully faithful and cofinal. If $\mathscr{E}$ is an idempotent complete exact category, then its bounded derived category $D^b\mathscr{E}$ is also idempotent complete [16, Theorem 2.8].

(d) The bounded derived category $D^b\mathscr{E}$ of an exact category $\mathscr{E}$ is generated (as a triangulated category) by the objects of $\mathscr{E}$ (considered as complexes concentrated in degree zero) in the sense that $D^b\mathscr{E}$ is the smallest triangulated subcategory of $D^b\mathscr{E}$ closed under isomorphisms which contains the objects of $\mathscr{E}$ (Exercise!).

We illustrate these facts by giving an example of a sequence of exact categories which induces an exact sequence of bounded derived categories.

3.1.8 Example

Let R be a ring with unit, and let $S \subset R$ be a multiplicative set of central non-zero-divisors in R. Let $\mathscr{H}_S(R) \subset R$-Mod be the full subcategory of those left R-modules which are direct factors of finitely presented S-torsion R-modules of projective dimension at most 1. This category is extension closed in the category of all left R-modules. We consider it as an exact category where a sequence in $\mathscr{H}_S(R)$ is a conflation if it is a conflation of R-modules. Let $\mathscr{P}^1(R) \subset R$-Mod be the full subcategory of those left R-modules M which fit into an exact sequence $0 \to P \to M \to H \to 0$ of R-modules where P is finitely generated projective and $H \in \mathscr{H}_S(R)$. The inclusion $\mathscr{P}^1(R) \subset R$-Mod is closed under extensions and we consider $\mathscr{P}^1(R)$ as a fully exact subcategory of R-Mod. Finally, let $\mathscr{P}'(S^{-1}R) \subset \mathrm{Proj}(S^{-1}R)$ be the full additive subcategory of those finitely generated projective $S^{-1}R$-modules which are localizations of finitely generated projective R-modules.

3.1.9 Lemma

The sequence $\mathscr{H}_S(R) \to \mathscr{P}^1(R) \to \mathscr{P}'(S^{-1}R)$ induces an exact sequence of associated bounded derived categories. Moreover, the inclusion $\mathrm{Proj}(R) \subset \mathscr{P}^1(R)$ induces an equivalence $D^b\,\mathrm{Proj}(R) \overset{\simeq}{\to} D^b\mathscr{P}^1(R)$ of categories. In particular, there is an exact sequence of triangulated categories

$$D^b\mathscr{H}_S(R) \to D^b\,\mathrm{Proj}(R) \to D^b\mathscr{P}'(S^{-1}R). \tag{9}$$

For instance, let R be a Dedekind domain and $S \subset R$ be the set of non-zero elements in R. Then $S^{-1}R = K$ is the field of fractions of R, the category $\mathscr{P}'(S^{-1}R)$ is the category of finite dimensional K-vector spaces, and $\mathscr{H}_S(R)$ is the category of finitely generated torsion R-modules.

Proof of Lemma 3.1.9

Let $\mathscr{P}_0^1(R) \subset \mathscr{P}^1(R)$ be the full subcategory of those R-modules M which fit into an exact sequence $0 \to P \to M \to H \to 0$ with P finitely generated projective, and H an S-torsion R-module of projective dimension at most 1. The inclusion $\mathscr{P}_0^1(R) \subset \mathscr{P}^1(R)$ is fully exact and cofinal. By Sect. 3.1.7 (c), the induced triangle functor $D^b \mathscr{P}_0^1(R) \to D^b \mathscr{P}^1(R)$ is fully faithful and cofinal. By the dual of Sect. 3.1.7 (b), the functor $D^b \operatorname{Proj}(R) \to D^b \mathscr{P}_0^1(R)$ is fully faithful. By Sect. 3.1.7 (d), this functor is also essentially surjective (hence an equivalence) since every object of $\mathscr{P}_0^1(R)$ has projective dimension at most 1. The category $\operatorname{Proj}(R)$ is idempotent complete. By Sect. 3.1.7 (c), the same is true for $D^b \operatorname{Proj}(R)$. It follows that the cofinal inclusions $D^b \operatorname{Proj}(R) \subset D^b \mathscr{P}_0^1(R) \subset D^b \mathscr{P}^1(R)$ are all equivalences.

For any finitely generated projective R-modules P and Q, the natural map $S^{-1} \operatorname{Hom}_R(P, Q) \to \operatorname{Hom}_{S^{-1}R}(S^{-1}P, S^{-1}Q)$ is an isomorphism. Therefore, the category $\mathscr{P}'(S^{-1}R)$ is obtained from $\operatorname{Proj}(R)$ by a calculus of right fractions with respect to the multiplication maps $P \to P : x \mapsto sx$ for $s \in S$ and $P \in \operatorname{Proj}(R)$. By Sect. 3.1.7 (a), the functor $D^b \operatorname{Proj}(R) \cong D^b \mathscr{P}^1(R) \to D^b \mathscr{P}'(S^{-1}R)$ is a localization, that is, $D^b \mathscr{P}'(S^{-1}R)$ is the Verdier quotient $D^b \mathscr{P}^1(R)/\mathscr{K}$ where $\mathscr{K} \subset D^b \mathscr{P}^1(R)$ is the full triangulated subcategory of those objects which are zero in $D^b \mathscr{P}'(S^{-1}R)$. The functor $D^b \mathscr{H}_S(R) \to D^b \mathscr{P}^1(R)$ is fully faithful by Sect. 3.1.7 (b), and it factors through $\mathscr{K}$. We have to show that the full inclusion $D^b \mathscr{H}_S(R) \subset \mathscr{K}$ is an equivalence, that is, we have to show that every object E in $\mathscr{K}$ is isomorphic to an object of $D^b \mathscr{H}_S(R)$. Since $D^b \operatorname{Proj}(R) \cong D^b \mathscr{P}^1(R)$, we can assume that E is a complex of projective R-modules. The acyclic complex $S^{-1}E$ is a bounded complex of projective $S^{-1}R$-modules, and thus it is contractible. The degree-wise split inclusion $i : E \rightarrowtail CE$ of E into its cone CE induces a map of contractible complexes $S^{-1}E \rightarrowtail S^{-1}CE$ which a fortiori is degree-wise split injective. A degree-wise split inclusion of contractible complexes always has a retraction; see Example 3.2.6. Applied to the last map we obtain a retraction $r : S^{-1}CE \to S^{-1}E$ in $\operatorname{Ch}^b \mathscr{P}'(S^{-1}R)$. We can write r as a right fraction $p s_0^{-1}$ with $p : CE \to E$ a chain map and $s_0 : CE \to CE : x \mapsto s_0 x$ the multiplication by s_0 for some $s_0 \in S$. After localization at S, we have $1 = ri = p s_0^{-1} i = s_0^{-1} pi$, and thus, $s_0 = pi$ since the elements of S are central. By the calculus of fractions, there is an $s_1 \in S$ such that $p \circ i \circ s_1 = s_0 s_1$. Since the set $S \subset R$ consists of non-zero-divisors and since E consists of projective modules in each degree, the morphism $s : E \to E$ given by $x \mapsto sx$ with $s \in S$ is injective. Therefore, we obtain a conflation of chain complexes of R-modules $CE \rightarrowtail E \oplus CE / is_1(E) \twoheadrightarrow E / s_0 s_1 E$ where the maps $CE \to E$ and $CE/is_1(E) \to E/s_0 s_1 E$ are induced by p and the other two maps are (up to sign) the natural quotient maps. This shows that in $D^b \mathscr{P}^1(R)$ we have an isomorphism $E \oplus CE / is_1(E) \cong E/s_0 s_1 E$. In particular, the chain complex E is a direct factor of an object of $D^b \mathscr{H}_S(R)$, namely of $E/s_0 s_1 E$. By Sect. 3.1.7 (c), the category $D^b \mathscr{H}_S(R)$ is idempotent complete. Hence, the complex E must be in $D^b \mathscr{H}_S(R)$. $\qquad\square$

Lemma 3.1.9 illustrates a slight inconvenience. The definition of the K-theory of $S^{-1}R$ uses all finitely generated projective $S^{-1}R$-modules and not only those lying in $\mathscr{P}'(S^{-1}R)$. Therefore, one would like to replace $D^b \mathscr{P}'(S^{-1}R)$ with $D^b \operatorname{Proj}(S^{-1}R)$ in Lemma 3.1.9, but these two categories are not equivalent, in general. However, the inclusion $D^b \mathscr{P}'(S^{-1}R) \subset D^b \operatorname{Proj}(S^{-1}R)$ is an equivalence up to factors by Sect. 3.1.7 (c). This observation together with Neeman's Theorem 3.4.5 below motivates the following.

3.1.10 Definition

A sequence of triangulated categories $\mathscr{A} \to \mathscr{B} \to \mathscr{C}$ is *exact up to factors* if the composition is zero, the functor $\mathscr{A} \to \mathscr{B}$ is fully faithful, and the induced functor $\mathscr{B}/\mathscr{A} \to \mathscr{C}$ is an equivalence up to factors. (See Sect. 2.4.1 for the definition of "exact up to factors".)

In this situation, the inclusion $\mathscr{A} \subset \mathscr{A}'$ of $\mathscr{A}$ into the full subcategory $\mathscr{A}'$ of $\mathscr{B}$ whose objects are zero in $\mathscr{C}$ is also an equivalence up to factors [66, Lemma 2.1.33]. Thus, a sequence $\mathscr{A} \to \mathscr{B} \to \mathscr{C}$ of triangulated categories is exact up to factors if, up to equivalences up to factors, $\mathscr{A}$ is the kernel category of $\mathscr{B} \to \mathscr{C}$ and $\mathscr{C}$ is the cokernel category of $\mathscr{A} \to \mathscr{B}$.

3.1.11 Example

Keep the hypothesis and notation of Example 3.1.8. The sequence of triangulated categories

$$D^b \mathscr{H}_S(R) \to D^b \operatorname{Proj}(R) \to D^b \operatorname{Proj}(S^{-1}R)$$

is exact up to factors but not exact, in general.

3.1.12 Idempotent Completion of Triangulated Categories

A triangulated category $\mathscr{A}$ is, in particular, an additive category. So we can speak of its idempotent completion $\tilde{\mathscr{A}}$; see Sect. 2.4.1. It turns out that $\mathscr{A}$ can be equipped with the structure of a triangulated category such that the inclusion $\mathscr{A} \subset \tilde{\mathscr{A}}$ is a triangle functor [16]. A sequence in $\tilde{\mathscr{A}}$ is a distinguished triangle if it is a direct factor of a distinguished triangle in $\mathscr{A}$. Note that the triangulated categories $D^b \operatorname{Vect}(X)$, $D^b \operatorname{Coh}(X)$ and $D^b \operatorname{Proj}(R)$ are all idempotent complete by Sect. 3.1.7 (c).

3.1.13 Exercise

(a) Let $\mathscr{A} \subset \mathscr{B}$ be a cofinal inclusion of triangulated categories. Show that the map $K_0(\mathscr{A}) \to K_0(\mathscr{B})$ is injective [91, Corollary 2.3].
(b) Let $\mathscr{A} \to \mathscr{B} \to \mathscr{C}$ be a sequence of triangulated categories which is exact up to factors. Then the following sequence of abelian groups is exact

$$K_0(\tilde{\mathscr{A}}) \to K_0(\tilde{\mathscr{B}}) \to K_0(\tilde{\mathscr{C}}). \qquad (10)$$

3.1.14 Remark

The statement in Exercise 3.1.13 (a) is part of Thomason's classification of dense subcategories. Call a triangulated subcategory $\mathscr{A} \subset \mathscr{B}$ *dense* if $\mathscr{A}$ is closed under isomorphisms in $\mathscr{B}$ and if the inclusion is cofinal. Thomason's Theorem [91, Theorem 2.1] says that the map which sends a dense subcategory $\mathscr{A} \subset \mathscr{B}$ to the subgroup $K_0(\mathscr{A}) \subset K_0(\mathscr{B})$ is a bijection between the set of dense subcategories of $\mathscr{B}$ and the set of subgroups of $K_0(\mathscr{B})$.

We note that an object of $\mathscr{B}$ of the form $A \oplus A[1]$ is in every dense triangulated subcategory of $\mathscr{B}$ since $[A \oplus A[1]] = [A] - [A] = 0 \in K_0(\mathscr{B})$.

3.2 The Thomason–Waldhausen Localization Theorem

We would like to extend the exact sequence Exercise 3.1.6 (8) to the left, and the exact sequence Exercise 3.1.13 (10) in both directions. However, there is no functor from triangulated categories to spaces (or spectra) which does that and yields Quillen's K-theory of an exact category $\mathscr{E}$ when applied to $D^b \mathscr{E}$ [80, Proposition 2.2]. This is the reason why we need to introduce more structure.

3.2.1 Notation

To abbreviate, we write $\mathrm{Ch}^b(\mathbb{Z})$ for the exact category of bounded chain complexes of finitely generated free $\mathbb{Z}$-modules; see Example 2.1.2 (e). A sequence here is a conflation if it splits in each degree (that is, it is isomorphic in each degree to the sequence Sect. 2.1.1 (f)). There is a symmetric monoidal tensor product

$$\otimes : \mathrm{Ch}^b(\mathbb{Z}) \times \mathrm{Ch}^b(\mathbb{Z}) \to \mathrm{Ch}^b(\mathbb{Z})$$

which extends the usual tensor product of free $\mathbb{Z}$-modules. It is given by the formulas

$$(E \otimes F)^n = \bigoplus_{i+j=n} E^i \otimes F^j, \qquad d(x \otimes y) = (dx) \otimes y + (-1)^{|x|} x \otimes dy. \qquad (11)$$

where $|x|$ denotes the degree of x. The unit of the tensor product is the chain complex $\mathbb{1} = \mathbb{Z} \cdot 1_{\mathbb{Z}}$ which is $\mathbb{Z}$ in degree 0 and 0 elsewhere. There are natural isomorphisms $\alpha :$ $A \otimes (B \otimes C) \cong (A \otimes B) \otimes C$, $\lambda : \mathbb{1} \otimes A \cong A$ and $\rho : A \otimes \mathbb{1} \cong A$ such that certain pentagonal and triangular diagrams commute and such that $\lambda_{\mathbb{1}} = \rho_{\mathbb{1}}$; see [61, VII.1] and Definition 3.2.2 below. Formally, the sextuplet $(\mathrm{Ch}^b(\mathbb{Z}), \otimes, \mathbb{1}, \alpha, \rho, \lambda)$ is a symmetric monoidal category.

Besides the chain complex $\mathbb{1}$, we have two other distinguished objects in $\mathrm{Ch}^b(\mathbb{Z})$. The complex $C = \mathbb{Z} \cdot 1_C \oplus \mathbb{Z} \cdot \eta$ is concentrated in degrees 0 and -1 where it is the free $\mathbb{Z}$-module of rank 1 generated by 1_C and η, respectively. The only non-trivial differential is $d\eta = 1_C$. In fact, C is a commutative differential graded $\mathbb{Z}$-module with unique multiplication such that 1_C is the unit in C. Furthermore, there is the complex $T = \mathbb{Z} \cdot \eta_T$ which is the free $\mathbb{Z}$-module generated by η_T in degree -1 and it is 0 elsewhere. Note that there is a short exact sequences of chain complexes

$$0 \to \mathbb{1} \rightarrowtail C \twoheadrightarrow T \to 0: \quad 1_{\mathbb{Z}} \mapsto 1_C, \quad (1_C, \eta) \mapsto (0, \eta_T).$$

3.2.2 Definition

An exact category $\mathscr{E}$ is called *complicial* if it is equipped with a bi-exact tensor product

$$\otimes : \mathrm{Ch}^b(\mathbb{Z}) \times \mathscr{E} \to \mathscr{E} \qquad (12)$$

which is associative and unital in the sense that there are natural isomorphisms $\alpha : A \otimes (B \otimes X) \cong (A \otimes B) \otimes X$ and $\lambda : \mathbb{1} \otimes X \cong X$ such that the pentagonal diagrams

$$
\begin{array}{ccccc}
A \otimes (B \otimes (C \otimes X)) & \xrightarrow{\ \alpha\ } & (A \otimes B) \otimes (C \otimes X) & \xrightarrow{\ \alpha\ } & ((A \otimes B) \otimes C) \otimes X \\
{\scriptstyle 1 \otimes \alpha} \downarrow & & & & \downarrow {\scriptstyle \alpha \otimes 1} \\
A \otimes ((B \otimes C) \otimes X) & & \xrightarrow{\hspace{3cm} \alpha \hspace{3cm}} & & (A \otimes (B \otimes C)) \otimes X
\end{array}
$$

and triangular diagrams

$$
\begin{array}{ccc}
A \otimes (\mathbb{1} \otimes X) & \xrightarrow{\ \alpha\ } & (A \otimes \mathbb{1}) \otimes X \\
& {\scriptstyle 1 \otimes \lambda} \searrow & \downarrow {\scriptstyle \rho \otimes 1} \\
& & A \otimes X
\end{array}
$$

commute for all $A, B, C \in \mathrm{Ch}^b(\mathbb{Z})$ and $X \in \mathscr{E}$. In other words, a complicial exact category is an exact category $\mathscr{E}$ equipped with a bi-exact action of the symmetric monoidal category $\mathrm{Ch}^b(\mathbb{Z})$ on $\mathscr{E}$; see [35, p. 218] for actions of monoidal categories.

For an object X of $\mathscr{E}$, we write CX and TX instead of $C \otimes X$ and $T \otimes X$. Note that there is a functorial conflation $X \rightarrowtail CX \twoheadrightarrow TX$ which is the tensor product of $\mathbb{1} \rightarrowtail C \twoheadrightarrow T$ with X. For a morphism $f : X \to Y$ in $\mathscr{E}$, we write $C(f)$ for the push-out of f along the inflation $X \rightarrowtail CX$, and we call it the *cone of* f. As a push-out of an inflation, the morphism $Y \to C(f)$ is also an inflation with the same cokernel TX. This yields the conflation in $\mathscr{E}$

$$Y \rightarrowtail C(f) \twoheadrightarrow TX. \tag{13}$$

3.2.3 Example

Let X be a scheme. The usual tensor product of vector bundles $\otimes_{O_X} : \mathrm{Vect}(X) \times \mathrm{Vect}(X) \to \mathrm{Vect}(X)$ extends to a tensor product

$$\otimes : \mathrm{Ch}^b \mathrm{Vect}(X) \times \mathrm{Ch}^b \mathrm{Vect}(X) \longrightarrow \mathrm{Ch}^b \mathrm{Vect}(X)$$

of bounded chain complexes of vector bundles defined by the same formula as in Sect. 3.2.1 (11). The structure map $p : X \to \mathrm{Spec}\,\mathbb{Z}$ associated with the unique ring map $\mathbb{Z} \to \Gamma(X, O_X)$ induces a symmetric monoidal functor $p^* : \mathrm{Proj}(\mathbb{Z}) \to \mathrm{Vect}(X)$ and thus an action

$$\otimes : \mathrm{Ch}^b(\mathbb{Z}) \times \mathrm{Ch}^b \mathrm{Vect}(X) : (M, V) \mapsto p^* M \otimes V$$

which makes $\mathrm{Ch}^b \mathrm{Vect}(X)$ into a complicial exact category.

3.2.4 Example

For any exact category $\mathscr{E}$, the category $\mathrm{Ch}^b \mathscr{E}$ of bounded chain complexes in $\mathscr{E}$ can be made into a complicial exact category as follows. Write $F(\mathbb{Z})$ for the category of finitely generated free $\mathbb{Z}$-modules where each module is equipped with a choice of a basis. So, we have an equivalence $\mathrm{Ch}^b(\mathbb{Z}) \cong \mathrm{Ch}^b F(\mathbb{Z})$. We define an associative and unital tensor product $F(\mathbb{Z}) \times \mathscr{E} \to \mathscr{E}$ by $\mathbb{Z}^n \otimes X = Xe_1 \oplus \ldots \oplus Xe_n$ where Xe_i stands for a copy of X corresponding to the basis element e_i of the based free module $\mathbb{Z}^n = \mathbb{Z}e_1 \oplus \ldots \oplus \mathbb{Z}e_n$. On maps, the tensor product is defined by $(a_{ij}) \otimes f = (a_{ij}f)$. With the usual formulas for the tensor product of chain complexes as in Sect. 3.2.1 (11), this tensor product extends to an associative, unital and bi-exact pairing

$$\otimes : \mathrm{Ch}^b(\mathbb{Z}) \times \mathrm{Ch}^b \mathscr{E} \to \mathrm{Ch}^b \mathscr{E}$$

making the category of bounded chain complexes $\mathrm{Ch}^b \mathscr{E}$ into a complicial exact category.

3.2.5 The Stable Category of a Complicial Exact Category

Let $\mathscr{E}$ be a complicial exact category. Call a conflation $X \rightarrowtail Y \twoheadrightarrow Z$ in $\mathscr{E}$ a *Frobenius conflation* if for every object $U \in \mathscr{E}$ the following holds: Every map $X \to CU$ extends to a map $Y \to CU$, and every map $CU \to Z$ lifts to a map $CU \to Y$. It is shown in Lemma A.2.16 that $\mathscr{E}$ together with the Frobenius conflations is a Frobenius exact category. That is, it is an exact category which has enough injectives and enough projectives, and where injectives and projectives coincide; see Appendix A, Sect. 2.14. The injective-projective objects are precisely the direct factors of objects of the form CU for $U \in \mathscr{E}$. The *stable category* $\underline{\mathscr{E}}$ of the complicial exact category $\mathscr{E}$ is, by definition, the stable category of the Frobenius exact category $\mathscr{E}$; see Appendix A, Sect. 2.14. It has objects the objects of $\mathscr{E}$, and maps are the homotopy classes of maps in $\mathscr{E}$ where two maps $f, g : X \rightarrow Y$ are homotopic if their difference factors through an object of the form CU. As the stable category of a Frobenius exact category, the category $\underline{\mathscr{E}}$ is a triangulated category (Appendix A, Sect. 2.14). Distinguished triangles are those triangles which are isomorphic in $\underline{\mathscr{E}}$ to sequences of the form

$$X \xrightarrow{f} Y \rightarrow C(f) \rightarrow TX \qquad (14)$$

attached to any map $f : X \rightarrow Y$ in $\mathscr{E}$ and extended by the sequence (13).

3.2.6 Example

Continuing the Example 3.2.4, the complicial exact category $\mathrm{Ch}^b \mathscr{E}$ of bounded chain complexes of $\mathscr{E}$ has as associated stable category the homotopy category $\mathscr{K}^b \mathscr{E}$ of Sect. 3.1.3. This is because contractible chain complexes are precisely the injective-projective objects for the Frobenius exact structure of $\mathrm{Ch}^b \mathscr{E}$ (Exercise!). See also Lemma A.2.16.

3.2.7 Definition

An *exact category with weak equivalences* is an exact category $\mathscr{E}$ together with a set $w \subset$ Mor$\mathscr{E}$ of morphisms in $\mathscr{E}$. Morphisms in w are called *weak equivalences*. The set of weak equivalences is required to contain all identity morphisms; to be closed under isomorphisms, retracts, push-outs along inflations, pull-backs along deflations, composition; and to satisfy the "two out of three" property for composition: if two of the three maps among a, b, ab are weak equivalences, then so is the third.

3.2.8 Example

Let $\mathscr{E}$ be an exact category. The exact category $\mathrm{Ch}^b \mathscr{E}$ of bounded chain complexes in $\mathscr{E}$ of Example 2.1.2 (e) together with the set quis of quasi-isomorphisms as defined in Sect. 3.1.3 is an exact category with weak equivalences.

3.2.9 Definition

An exact category with weak equivalences $(\mathscr{E}, w)$ is *complicial* if $\mathscr{E}$ is complicial and if the tensor product Definition 3.2.2 (12) preserves weak equivalences in both variables, that is, if f is a homotopy equivalence in $\mathrm{Ch}^b(\mathbb{Z})$ and g is a weak equivalence in $\mathscr{E}$, then $f \otimes g$ is a weak equivalence in $\mathscr{E}$.

3.2.10 Example

The exact category with weak equivalences $(\mathrm{Ch}^b \mathscr{E}, \mathrm{quis})$ of Example 3.2.8 is complicial with action by $\mathrm{Ch}^b(\mathbb{Z})$ as defined in Sect. 3.2.4.

Before we come to the definition of the K-theory space of a complicial exact category, we introduce some notation. Let $\mathbf{E} = (\mathscr{E}, w)$ be a complicial exact category with weak equivalences. We write $\mathscr{E}^w \subset \mathscr{E}$ for the fully exact subcategory of those objects X in $\mathscr{E}$ for which the map $0 \rightarrow X$ is a weak equivalence.

3.2.11 Exercise

Show that $\mathscr{E}^w$ is closed under retracts in $\mathscr{E}$. More precisely, let X be an object of $\mathscr{E}$. Show that if there are maps $i : X \to A$ and $p : A \to X$ with $pi = 1_X$ and $A \in \mathscr{E}^w$ then $X \in \mathscr{E}^w$. In particular, objects of $\mathscr{E}$ which are isomorphic to objects of $\mathscr{E}^w$ are in $\mathscr{E}^w$.

3.2.12 Definition of $K(\mathscr{E}, w)$, $K(\mathbf{E})$

The *algebraic K-theory space* $K(\mathbf{E}) = K(\mathscr{E}, w)$ of a complicial exact category with weak equivalences $\mathbf{E} = (\mathscr{E}, w)$ is the homotopy fibre of the map of pointed topological spaces $BQ(\mathscr{E}^w) \to BQ(\mathscr{E})$ induced by the inclusion $\mathscr{E}^w \subset \mathscr{E}$ of exact categories. That is,

$$K(\mathbf{E}) = K(\mathscr{E}, w) = F(g) \quad \text{where} \quad g : BQ(\mathscr{E}^w) \to BQ(\mathscr{E})$$

and where $F(g)$ is the homotopy fibre of g as in Appendix A, Sect. 1.7. The *higher algebraic K-groups* $K_i(\mathbf{E})$ of $\mathbf{E}$ are the homotopy groups $\pi_i K(\mathbf{E})$ of the K-theory space of $\mathbf{E}$ for $i \geq 0$.

Exact functors preserving weak equivalences induce maps between algebraic K-theory spaces of complicial exact categories with weak equivalences.

3.2.13 Remark

The K-theory space of an exact category with weak equivalences is usually defined using Waldhausen's $S_\bullet$-construction [100, p. 330, Definition]. A complicial exact category $\mathbf{E} = (\mathscr{E}, w)$ has a "cylinder functor" in the sense of [100, Sect. 1.6] obtained as the tensor product with the usual cylinder in $\mathrm{Ch}^b(\mathbb{Z})$ via the action of $\mathrm{Ch}^b(\mathbb{Z})$ on $\mathscr{E}$. Theorem [100, 1.6.4] together with [100, Sect. 1.9] then show that the K-theory space of any complicial exact category with weak equivalences as defined in Definition 3.2.12 is equivalent to the one in [100].

3.2.14 Theorem [94, Theorem 1.11.7]

Let $\mathscr{E}$ be an exact category. The embedding of $\mathscr{E}$ into $\mathrm{Ch}^b \mathscr{E}$ as degree-zero complexes induces a homotopy equivalence

$$K(\mathscr{E}) \simeq K(\mathrm{Ch}^b \mathscr{E}, \mathrm{quis}).$$

3.2.15 The Triangulated Category $\mathscr{T}(\mathbf{E})$

Let $\mathbf{E} = (\mathscr{E}, w)$ be a complicial exact category with weak equivalences. For objects X of $\mathscr{E}^w$ and A of $\mathrm{Ch}^b(\mathbb{Z})$, the object $A \otimes X$ is in $\mathscr{E}^w$ because the map $0 \to A \otimes X$ is a weak equivalence since it is the tensor product $id_A \otimes (0 \to X)$ of two weak equivalences. It follows that we can consider $\mathscr{E}^w$ as a complicial exact category where the action by $\mathrm{Ch}^b(\mathbb{Z})$ is induced from the action on $\mathscr{E}$. For every object $U \in \mathscr{E}$, the object CU is in $\mathscr{E}^w$ because the map $0 \to CU$ is a weak equivalence as it is the a tensor product $(0 \to C) \otimes 1_U$ of two weak equivalences. More generally, every retract of an object of the form CU is in $\mathscr{E}^w$ by Exercise 3.2.11. Therefore, the two Frobenius categories $\mathscr{E}$ and $\mathscr{E}^w$ have the same injective-projective objects. It follows that the inclusion $\mathscr{E}^w \subset \mathscr{E}$ induces a fully faithful triangle functor of associated stable categories $\underline{\mathscr{E}^w} \subset \underline{\mathscr{E}}$.

3.2.16 Exercise

Show that $\underline{\mathscr{E}}^w$ is closed under retracts in $\underline{\mathscr{E}}$. More precisely, let X be an object of $\underline{\mathscr{E}}$. Show that if there are maps $i : X \to A$ and $p : A \to X$ in $\underline{\mathscr{E}}$ with $pi = 1_X$ and $A \in \underline{\mathscr{E}}^w$ then $X \in \underline{\mathscr{E}}^w$. In particular, objects of $\underline{\mathscr{E}}$ which are isomorphic in $\underline{\mathscr{E}}$ to objects of $\underline{\mathscr{E}}^w$ are already in $\underline{\mathscr{E}}^w$.

3.2.17 Definition

The triangulated category $\mathscr{T}(\mathbf{E})$ associated with a complicial exact category with weak equivalences $\mathbf{E} = (\mathscr{E}, w)$ is the Verdier quotient

$$\mathscr{T}(\mathbf{E}) = \underline{\mathscr{E}}/\underline{\mathscr{E}}^w$$

of the inclusion of triangulated stable categories $\underline{\mathscr{E}}^w \subset \underline{\mathscr{E}}$. By construction, distinguished triangles in $\mathscr{T}(\mathbf{E})$ are those triangles which are isomorphic in $\mathscr{T}(\mathbf{E})$ to triangles of the form Sect. 3.2.5 (14).

One easily checks that the canonical functor $\mathscr{E} \to \mathscr{T}(\mathbf{E}) : X \mapsto X$ induces an isomorphism of categories

$$w^{-1}\mathscr{E} \xrightarrow{\cong} \mathscr{T}(\mathbf{E}).$$

Therefore, as a category, the triangulated category $\mathscr{T}(\mathscr{E}, w)$ of $(\mathscr{E}, w)$ is obtained from $\mathscr{E}$ by formally inverting the weak equivalences.

3.2.18 Exercise

Let $(\mathscr{E}, w)$ be a complicial exact category with weak equivalences. Show that a morphism in $\mathscr{E}$ which is an isomorphism in $\mathscr{T}(\mathscr{E}, w)$ is a weak equivalence. *Hint*: Show that (a) if in a conflation $X \rightarrowtail Y \twoheadrightarrow A$ the object A is in $\underline{\mathscr{E}}^w$ then $X \rightarrowtail Y$ is a weak equivalence, and (b) a map $f : X \to Y$ is an isomorphism in $\mathscr{T}(\mathscr{E}, w)$ iff its cone $C(f)$ is in $\underline{\mathscr{E}}^w$ by Exercise 3.2.16. Conclude using the conflation $X \rightarrowtail CX \oplus Y \twoheadrightarrow C(f)$.

3.2.19 Remark

A conflation $X \overset{i}{\rightarrowtail} Y \overset{p}{\twoheadrightarrow} Z$ in a complicial exact category with weak equivalences (E, w) gives rise to a distinguished triangle

$$X \overset{i}{\longrightarrow} Y \overset{p}{\longrightarrow} Z \longrightarrow TX$$

in $\mathscr{T}(\mathscr{E}, w)$. By definition, this triangle is isomorphic to the standard distinguished triangle Sect. 3.2.5 (14) via the quotient map $C(i) \twoheadrightarrow C(i)/CX \cong Z$ which is weak equivalence in $\mathscr{E}$ and an isomorphism in $\mathscr{T}(\mathscr{E}, w)$.

3.2.20 Example

For an exact category $\mathscr{E}$, the triangulated category $\mathscr{T}(\mathrm{Ch}^b\mathscr{E}, \mathrm{quis})$ associated with the complicial exact category with weak equivalences $(\mathrm{Ch}^b\mathscr{E}, \mathrm{quis})$ of Example 3.2.10 is the usual bounded derived category $D^b\mathscr{E}$ of $\mathscr{E}$ as defined in Sect. 3.1.3.

3.2.21 Example (DG Categories)

Let $\mathscr{C}$ be a dg-category ([50], or see Toen's lecture [92]). There is a canonical embedding $\mathscr{C} \subset \mathscr{C}^{pretr}$ of dg-categories of $\mathscr{C}$ into a "pretriangulated dg-category" $\mathscr{C}^{pretr}$ associated with $\mathscr{C}$ [12, Sect. 4.4]. It is obtained from $\mathscr{C}$ by formally adding iterated shifts and cones of objects of $\mathscr{C}$. The homotopy category $\mathrm{Ho}(\mathscr{C}^{pretr})$ of $\mathscr{C}^{pretr}$ is equivalent to the full triangulated subcategory of the derived category $D(\mathscr{C})$ of dg $\mathscr{C}$-modules which is generated by $\mathscr{C}$. The idempotent completion of $\mathrm{Ho}(\mathscr{C}^{pretr})$ is equivalent to the triangulated category of compact objects in $D(\mathscr{C})$ which is sometimes called the *derived category of perfect $\mathscr{C}$-modules*, and it is also equivalent to the homotopy category of the *triangulated hull* of $\mathscr{C}$ mentioned Toen's lecture [92].

Exercise: Show that $\mathscr{C}^{pretr}$ and the triangulated hull of $\mathscr{C}$ can be made into complicial exact categories with weak equivalences such that the associated triangulated categories are the homotopy categories of $\mathscr{C}^{pretr}$ and of the triangulated hull of $\mathscr{C}$.

3.2.22 Proposition (Presentation of $K_0(\mathbf{E})$)

Let $\mathbf{E} = (\mathscr{E}, w)$ be a complicial exact category with weak equivalences. Then the map $K_0(\mathbf{E}) \to K_0(\mathscr{T}(\mathbf{E})) : [X] \mapsto [X]$ is well-defined and an isomorphism of abelian groups.

Proof:

By Definition 3.2.12 and Proposition 2.2.4, the group $K_0(\mathbf{E})$ is the cokernel of the map $K_0(\mathscr{E}^w) \to K_0(\mathscr{E})$. By Remark 3.2.19, conflations in $\mathscr{E}$ yield distinguished triangles in $\mathscr{T}(\mathbf{E})$. Therefore, the map $K_0(\mathscr{E}) \to K_0(\mathscr{T}(\mathbf{E}))$ given by $[X] \mapsto [X]$ is well-defined. This map clearly sends $K_0(\mathscr{E}^w)$ to zero. It follows that the map in the Proposition is also well-defined.

Now, we show that the inverse map $K_0(\mathscr{T}(\mathbf{E})) \to K_0(\mathbf{E})$ defined by $[X] \mapsto [X]$ is also well-defined. We first observe that the existence of a weak equivalence $f : X \to Y$ implies that $[X] = [Y]$ in $K_0(\mathbf{E})$. This is because there is a conflation $X \rightarrowtail CX \oplus Y \twoheadrightarrow C(f)$ in $\mathscr{E}$ by the definition of the mapping cone $C(f)$. The objects CX and $C(f)$ are in $\mathscr{E}^w$ which implies $[X] = [Y] \in K_0(\mathbf{E})$. More generally, any two objects which are isomorphic in $\mathscr{T}(\mathbf{E})$ give rise to the same element in $K_0(\mathbf{E})$ because they are linked by a zigzag of weak equivalences by Exercise 3.2.18 and the definition of $\mathscr{T}(\mathbf{E})$. Next, we observe that for every object X of $\mathscr{E}$, the existence of the conflation $X \rightarrowtail CX \twoheadrightarrow TX$ in $\mathscr{E}$ with $CX \in \mathscr{E}^w$ shows that $[X] = -[TX]$ in $K_0(\mathbf{E})$. Finally, every distinguished triangle $A \to B \to C \to TA$ in $\mathscr{T}(\mathbf{E})$ is isomorphic in $\mathscr{T}(\mathbf{E})$ to a triangle of the form Sect. 3.2.5 (14) where $Y \rightarrowtail C(f) \twoheadrightarrow TX$ is a conflation in $\mathscr{E}$. Therefore, we have $[A] - [B] + [C] = [X] - [Y] + [C(f)] = [X] - [Y] + [Y] + [TX] = [X] - [Y] + [Y] - [X] = 0$ in $K_0(\mathbf{E})$, and the inverse map is well-defined. $\qquad\square$

Now, we come to the theorem which extends the sequence Exercise 3.1.6 (8) to the left. It is due to Thomason [94, 1.9.8., 1.8.2] based on the work of Waldhausen [100].

3.2.23 Theorem (Thomason–Waldhausen Localization, Connective Version)

Given a sequence $\mathbf{A} \to \mathbf{B} \to \mathbf{C}$ be of complicial exact categories with weak equivalences. Assume that the associated sequence $\mathscr{T}\mathbf{A} \to \mathscr{T}\mathbf{B} \to \mathscr{T}\mathbf{C}$ of triangulated categories is exact. Then the induced sequence of K-theory spaces

$$K(\mathbf{A}) \to K(\mathbf{B}) \to K(\mathbf{C})$$

is a homotopy fibration. In particular, there is a long exact sequence of K-groups

$$\cdots \to K_{i+1}(\mathbf{C}) \to K_i(\mathbf{A}) \to K_i(\mathbf{B}) \to K_i(\mathbf{C}) \to K_{i-1}(\mathbf{A}) \to \cdots$$

ending in $K_0(\mathbf{B}) \to K_0(\mathbf{C}) \to 0$.

The following special case of Theorem 3.2.23 which is important in itself is due to Thomason [94, Theorem 1.9.8].

3.2.24 Theorem (Invariance Under Derived Equivalences)

Let $\mathbf{A} \to \mathbf{B}$ be a functor of complicial exact categories with weak equivalences. Assume that the associated functor of triangulated categories $\mathscr{T}\mathbf{A} \to \mathscr{T}\mathbf{B}$ is an equivalence. Then the induced map $K(\mathbf{A}) \to K(\mathbf{B})$ of K-theory spaces is a homotopy equivalence. In particular, it induces isomorphisms $K_i(\mathbf{A}) \cong K_i(\mathbf{B})$ of K-groups for $i \geq 0$.

3.2.25 Example

Theorem 3.2.23 applied to Example 3.1.8 yields a homotopy fibration

$$K(\mathscr{H}_S(R)) \to K(R) \to K(\mathscr{P}'(S^{-1}R))$$

of K-theory spaces. As mentioned earlier, one would like to replace $\mathscr{P}'(S^{-1}R)$ with $\mathrm{Proj}(S^{-1}R)$ in the homotopy fibration and its associated long exact sequence of homotopy groups. We can do so by the Cofinality Theorem 2.4.2, and we obtain a long exact sequence of K-groups

$$\cdots \to K_{i+1}(S^{-1}R) \to K_i(\mathscr{H}_S(R)) \to K_i(R) \to K_i(S^{-1}R) \to K_i(\mathscr{H}_S(R)) \to \cdots$$

ending in $\cdots \to K_0(R) \to K_0(S^{-1}R)$. The last map $K_0(R) \to K_0(S^{-1}R)$, however, is not surjective, in general. We have already introduced the negative K-groups of an exact category in Sect. 2.4. They do indeed extend this exact sequence to the right. But this is best understood in the framework of complicial exact categories with weak equivalences.

3.2.26 Negative K-Theory of Complicial Exact Categories

To any complicial exact category with weak equivalences $\mathbf{E}$, one can associate a new complicial exact category with weak equivalences $S\mathbf{E}$, called the *suspension* of $\mathbf{E}$, such that there is a natural map

$$K(\mathbf{E}) \to \Omega K(S\mathbf{E}) \tag{15}$$

which is an isomorphism on π_i for $i \geq 1$ and a monomorphism on π_0 [82]; see the construction in Sect. 3.2.33 below. In fact, $K_1(S\mathbf{E}) = K_0((\mathscr{T}\mathbf{E})^{\sim})$ where $(\mathscr{T}\mathbf{E})^{\sim}$ denotes the idempotent completion of $\mathscr{T}\mathbf{E}$. Moreover, the suspension functor sends sequences of complicial exact categories with weak equivalences whose associated sequence of triangulated categories is exact up to factors to sequences with that same property.

One uses the suspension construction to slightly modify the definition of algebraic K-theory in order to incorporate negative K-groups as follows. One sets $\mathbb{K}_i(\mathbf{E}) = K_i(\mathbf{E})$ for $i \geq 1$, $\mathbb{K}_0(\mathbf{E}) = K_0((\mathscr{T}\mathbf{E})^{\sim})$ and $\mathbb{K}_i(\mathbf{E}) = K_0((\mathscr{T}S^{-i}\mathbf{E})^{\sim})$ for $i < 0$. As in the case of exact categories in Sect. 2.4.3, one constructs a spectrum $\mathbb{K}(\mathbf{E})$ whose homotopy groups are the groups $\mathbb{K}_i(\mathbf{E})$ for $i \in \mathbb{Z}$. The n-th space of this spectrum is $K(S^n\mathbf{E})$, and the structure maps are given by (15).

If the category $\mathscr{T}\mathbf{E}$ is idempotent complete, then we may write $K_i(\mathbf{E})$ instead of $\mathbb{K}_i(\mathbf{E})$ for $i \in \mathbb{Z}$. In this case, $\mathbb{K}_i(\mathbf{E}) = K_i(\mathbf{E})$ for all $i \geq 0$. Therefore, we have merely introduced negative K-groups without changing the definition of $\mathbb{K}_0$.

The following theorem extends the exact sequence Exercise 3.1.13 (10) in both directions. For the definition of a homotopy fibration of spectra, see Appendix A, Sect. 1.8.

3.2.27 Theorem (Thomason–Waldhausen Localization, Non-Connective Version)

Let $\mathbf{A} \to \mathbf{B} \to \mathbf{C}$ be a sequence of complicial exact categories with weak equivalences such that the associated sequence of triangulated categories $\mathscr{T}\mathbf{A} \to \mathscr{T}\mathbf{B} \to \mathscr{T}\mathbf{C}$ is exact up to factors. Then the sequence of K-theory spectra

$$\mathbb{K}(\mathbf{A}) \to \mathbb{K}(\mathbf{B}) \to \mathbb{K}(\mathbf{C})$$

is a homotopy fibration. In particular, there is a long exact sequence of K-groups for $i \in \mathbb{Z}$

$$\cdots \to \mathbb{K}_{i+1}(\mathbf{C}) \to \mathbb{K}_i(\mathbf{A}) \to \mathbb{K}_i(\mathbf{B}) \to \mathbb{K}_i(\mathbf{C}) \to \mathbb{K}_{i-1}(\mathbf{A}) \to \cdots$$

3.2.28 Remark

Theorem 3.2.27 is proved in [82, Theorem 9] in view of the fact that for a complicial exact category with weak equivalences $(\mathscr{E}, w)$, the pair $(\mathscr{E}, \mathscr{E}^w)$ is a "Frobenius pair" in the sense of [82, Definition 5] when we equip $\mathscr{E}$ with the Frobenius exact structure.

3.2.29 Theorem (Invariance of $\mathbb{K}$-Theory Under Derived Equivalences)

If a functor $\mathbf{A} \to \mathbf{B}$ of complicial exact categories with weak equivalences induces an equivalence up to factors $\mathscr{T}\mathbf{A} \to \mathscr{T}\mathbf{B}$ of associated triangulated categories, then it induces a homotopy equivalence of $\mathbb{K}$-theory spectra $\mathbb{K}(\mathbf{A}) \xrightarrow{\sim} \mathbb{K}(\mathbf{B})$ and isomorphisms $\mathbb{K}_i(\mathbf{A}) \cong \mathbb{K}_i(\mathbf{B})$ of $\mathbb{K}$-groups for $i \in \mathbb{Z}$.

3.2.30 Agreement

Let $\mathscr{E}$ be an exact category and $(\mathrm{Ch}^b\,\mathscr{E}, \mathrm{quis})$ be the associated complicial exact category of bounded chain complexes with quasi-isomorphisms as weak equivalences. There are natural isomorphisms

$$\mathbb{K}_i(\mathrm{Ch}^b\,\mathscr{E}, \mathrm{quis}) \cong \mathbb{K}_i(\mathscr{E}) \quad \text{for} \quad i \in \mathbb{Z}$$

between the $\mathbb{K}$-groups defined in Sect. 3.2.26 and those defined in Sect. 2.4.3. For $i > 0$, this is Theorem 3.2.14. For $i = 0$, we have $\mathbb{K}_0(\mathscr{E}) = K_0(\tilde{\mathscr{E}}) = K_0(D^b(\tilde{\mathscr{E}})) = K_0(D^b(\mathscr{E})^{\sim}) = \mathbb{K}_0(\mathrm{Ch}^b\,\mathscr{E}, \mathrm{quis})$ by Exercise 3.1.4, Sect. 3.1.7 (c) and Proposition 3.2.22. For $i < 0$, this follows from Theorem 3.2.27, from the case $i = 0$ above, from the fact that the sequence $\mathscr{E} \to \mathscr{F}\mathscr{E} \to S\mathscr{E}$ of Sect. 2.4.6 induces a sequence of associated triangulated categories which is exact up to factors [81, Proposition 2.6, Lemma 3.2] and from the fact that $\mathscr{F}\mathscr{E}$ has exact countable sums which implies $\mathbb{K}_i(\mathscr{F}\mathscr{E}) = 0$ for all $i \in \mathbb{Z}$.

3.2.31 Example

From Example 3.1.11, we obtain a homotopy fibration of K-theory spectra

$$\mathbb{K}(\mathscr{H}_S(R)) \to \mathbb{K}(R) \to \mathbb{K}(S^{-1}R)$$

and an associated long exact sequence of K-groups

$$\cdots \to K_{i+1}(S^{-1}R) \to K_i(\mathscr{H}_S(R)) \to K_i(R) \to K_i(S^{-1}R) \to K_i(\mathscr{H}_S(R)) \to \cdots$$

for $i \in \mathbb{Z}$. Here we wrote K_i instead of $\mathbb{K}_i$ since all exact categories in this example as well as their bounded derived categories are idempotent complete.

3.2.32 $\mathbb{K}$-Theory of DG-Categories

The $\mathbb{K}$-theory $\mathbb{K}(\mathscr{C})$ of a dg-category $\mathscr{C}$ is the $\mathbb{K}$-theory associated with the complicial exact category $\mathscr{C}^{pretr}$ with weak equivalences the homotopy equivalences, that is, those maps which are isomorphisms in $D\mathscr{C}$. By construction, $\mathbb{K}_0(\mathscr{C})$ is K_0 of the triangulated category of compact objects in $D\mathscr{C}$. Instead of $\mathscr{C}^{pretr}$, we could have also used the triangulated hull of $\mathscr{C}$ in the definition of $\mathbb{K}(\mathscr{C})$ because both dg-categories are derived equivalent up to factors.

By the Thomason–Waldhausen Localization Theorem 3.2.27, a sequence of dg categories $\mathscr{A} \to \mathscr{B} \to \mathscr{C}$ whose sequence $D\mathscr{A} \to D\mathscr{B} \to D\mathscr{C}$ of derived categories of dg-modules is exact induces a homotopy fibration of $\mathbb{K}$-theory spectra

$$\mathbb{K}(\mathscr{A}) \to \mathbb{K}(\mathscr{B}) \to \mathbb{K}(\mathscr{C}).$$

This is because the sequence $\mathscr{T}(\mathscr{A}^{pretr}) \to \mathscr{T}(\mathscr{B}^{pretr}) \to \mathscr{T}(\mathscr{C}^{pretr})$ of triangulated categories whose idempotent completion is the sequence of compact objects associated with $D\mathscr{A} \to D\mathscr{B} \to D\mathscr{C}$ is exact up to factors by Neeman's Theorem 3.4.5 (b) below.

3.2.33 Construction of the Suspension $S\mathbf{E}$

Let $\mathbf{E} = (\mathscr{E}, w)$ be a complicial exact category with weak equivalences. Among others, this means that $\mathscr{E}$ is an exact category, and we can construct its countable envelope $\mathscr{F}\mathscr{E}$ as in Sect. 2.4.6. The complicial structure on $\mathscr{E}$ extends to a complicial structure on $\mathscr{F}\mathscr{E}$ by setting

$$A \otimes (E_0 \hookrightarrow E_1 \hookrightarrow E_2 \hookrightarrow \cdots) = (A \otimes E_0 \hookrightarrow A \otimes E_1 \hookrightarrow A \otimes E_2 \hookrightarrow \cdots)$$

for $A \in \mathrm{Ch}^b(\mathbb{Z})$ and $E_* \in \mathscr{F}\mathscr{E}$. Call a map in $\mathscr{F}\mathscr{E}$ a *weak equivalence* if its cone is a direct factor of an object of $\mathscr{F}(\mathscr{E}^w)$. As usual, we write w for the set of weak equivalences in $\mathscr{F}\mathscr{E}$. The pair $\mathscr{F}\mathbf{E} = (\mathscr{F}\mathscr{E}, w)$ defines a complicial exact category with weak equivalences. The fully exact inclusion $\mathscr{E} \to \mathscr{F}\mathscr{E}$ of exact categories of Sect. 2.4.6 defines a functor $\mathbf{E} \to \mathscr{F}\mathbf{E}$ of complicial exact categories with weak equivalences such that the induced functor $\mathscr{T}(\mathbf{E}) \to \mathscr{T}(\mathscr{F}\mathbf{E})$ of associated triangulated categories is fully faithful.

Now, the *suspension* $S\mathbf{E}$ of $\mathbf{E}$ is the complicial exact category with weak equivalences which has as underlying complicial exact category the countable envelope $\mathscr{F}\mathscr{E}$ and as set of weak equivalences those maps in $\mathscr{F}\mathscr{E}$ which are isomorphisms in the Verdier quotient $\mathscr{T}(\mathscr{F}\mathbf{E})/\mathscr{T}(\mathbf{E})$.

3.2.34 Remark (Suspensions of DG-Categories)

If $\mathscr{C}$ is a dg-category, one can also define its suspension as $\Sigma\mathscr{C} = \Sigma \otimes_{\mathbb{Z}} \mathscr{C}$ where $\Sigma = \Sigma\mathbb{Z}$ is the suspension ring of $\mathbb{Z}$ as in Cortiñas' lecture [21]. The resulting spectrum whose n-th space is $K((\Sigma^n\mathscr{C})^{pretr})$ is equivalent to the spectrum $I\!K(\mathscr{C})$ as defined in Sect. 3.2.32. The reason is that the sequence $\mathscr{C} \to \Gamma \otimes \mathscr{C} \to \Sigma \otimes \mathscr{C}$ induces a sequence of pretriangulated dg-categories whose associated sequence of homotopy categories is exact up to factors. This follows from [23] in view of the fact that the sequence of flat dg categories $\mathbb{Z}^{pretr} \to \Gamma^{pretr} \to \Sigma^{pretr}$ induces a sequence of homotopy categories which is exact up to factors. Moreover, $I\!K(\Gamma \otimes \mathscr{C}) \simeq 0$.

Sketch of the proof of Theorem 3.2.27

We first construct the map (15). For that, denote by $\mathscr{E}' \subset \mathscr{F}\mathscr{E}$ the full subcategory of those objects which are zero in $\mathscr{T}SE = \mathscr{T}(\mathscr{F}E)/\mathscr{T}(E)$. The category $\mathscr{E}'$ inherits the structure of a complicial exact category with weak equivalences from $\mathscr{F}E$ which we denote by $\mathbf{E}'$. By construction, the sequence $\mathbf{E}' \to \mathscr{F}\mathbf{E} \to SE$ induces an exact sequence of associated triangulated categories, and hence, a homotopy fibration $K(\mathbf{E}') \to K(\mathscr{F}\mathbf{E}) \to K(SE)$ of K-theory spaces by the Thomason–Waldhausen Localization Theorem 3.2.23. Since $\mathscr{F}\mathscr{E}$ has infinite exact sums which preserve weak equivalences, we have $K(\mathscr{F}\mathbf{E}) \simeq 0$. Therefore, we obtain a homotopy equivalence $K(\mathbf{E}') \xrightarrow{\simeq} \Omega K(SE)$. By construction of $\mathscr{E}'$, the inclusion $\mathscr{E} \subset \mathscr{E}'$ induces a cofinal triangle functor $\mathscr{T}\mathbf{E} \subset \mathscr{T}\mathbf{E}'$; see Appendix A, Sect. 2.7. By Thomason's Cofinality Theorem 3.2.35 below, the map $K(\mathbf{E}) \to K(\mathbf{E}')$ induces isomorphisms on π_i for $i > 0$ and a monomorphism on π_0.

The main step in proving Theorem 3.2.27 consists in showing that for a sequence of complicial exact categories with weak equivalences $\mathbf{A} \to \mathbf{B} \to \mathbf{C}$ where $\mathscr{T}\mathbf{A} \to \mathscr{T}\mathbf{B} \to \mathscr{T}\mathbf{C}$ is exact up to factors, the suspended sequence $SA \to SB \to SC$ induces a sequence $\mathscr{T}SA \to \mathscr{T}SB \to \mathscr{T}SC$ of associated triangulated categories which is also exact up to factors. This is proved in [82, Theorem 3] for complicial exact categories with weak equivalences whose exact structure is the Frobenius exact structure. The proof for general exact structures is *mutatis mutandis* the same.

3.2.35 Theorem (Cofinality [94, 1.10.1, 1.9.8])

Let $\mathbf{A} \to \mathbf{B}$ be a functor of complicial exact categories with weak equivalences such that $\mathscr{T}\mathbf{A} \to \mathscr{T}\mathbf{B}$ is cofinal. Then $K_i(\mathbf{A}) \to K_i(\mathbf{B})$ is an isomorphism for $i \geq 1$ and a monomorphism for $i = 0$.

3.3 Quillen's Fundamental Theorems Revisited

The results of this subsection are due to Quillen [73]. However, we give proofs based on the Thomason–Waldhausen Localization Theorem. This has the advantage that the same results hold for other cohomology theories such as Hochschild homology, (negative, periodic, ordinary) cyclic homology, triangular Witt-groups, hermitian K-theory *etc.* where the analog of the Thomason–Waldhausen Localization Theorem also holds.

3.3.1 G-Theory Localization (Revisited)

It is not true that every exact sequence of abelian categories induces an exact sequence of associated bounded derived categories. For a counter example, see [49, 1.15 Example (c)] where the abelian categories are even noetherian and artinian. However, the exact sequence of abelian categories Sect. 2.3.6 (3) which (together with *Dévissage*) gives rise to the G-theory fibration Theorem 2.3.7 (4) does induce an exact sequence of triangulated categories. So, at least in this case, we can apply the Thomason–Waldhausen Localization Theorem.

3.3.2 Theorem

Let X be a noetherian scheme, $j : U \subset X$ be an open subscheme and $Z = X - U$ be the closed complement. Then the exact sequence of abelian categories Sect. 2.3.6 (3) induces an exact sequence of triangulated categories

$$D^b \mathrm{Coh}_Z(X) \to D^b \mathrm{Coh}(X) \xrightarrow{j^*} D^b \mathrm{Coh}(U).$$

In particular, it induces a homotopy fibration in K-theory

$$K \mathrm{Coh}_Z(X) \to K \mathrm{Coh}(X) \xrightarrow{j^*} K \mathrm{Coh}(U).$$

Proof [49, 1.15 Lemma and Example b)]:

Since the sequence Sect. 2.3.6 (3) is an exact sequence of abelian categories, the functor $\mathrm{Coh}(X) \to \mathrm{Coh}(U)$ is a localization by a calculus of fractions. By Sect. 3.1.7 (a), the functor $D^b \mathrm{Coh}(X) \to D^b \mathrm{Coh}(U)$ is a localization of triangulated categories. Therefore, j^* induces an equivalence $D^b \mathrm{Coh}(X)/D^b_Z \mathrm{Coh}(X) \cong D^b \mathrm{Coh}(U)$ where $D^b_Z \mathrm{Coh}(X) \subset D^b \mathrm{Coh}(X)$ denotes the full triangulated subcategory of those complexes whose cohomology is supported in Z, or equivalently, which are acyclic over U.

The functor $D^b \mathrm{Coh}_Z(X) \to D^b \mathrm{Coh}(X)$ is fully faithful by an application of Sect. 3.1.7 (b). To check the hypothesis of Sect. 3.1.7 (b), let $N \rightarrowtail M$ be an inclusion of coherent O_X-modules with $N \in \mathrm{Coh}_Z(X)$. If $I \subset O_X$ denotes the ideal sheaf of the reduced subscheme $Z_{red} \subset X$ associated with Z, then a coherent O_X-module E has support in Z iff $I^n E = 0$ for some $n \in \mathbb{N}$. By the Artin–Rees Lemma [3, Corollary 10.10] which also works for noetherian schemes with the same proof, there is an integer $c > 0$ such that $N \cap I^n M = I^{n-c}(N \cap I^c M)$ for $n \geq c$. Since N has support in Z, the same is true for $N \cap I^c M$, and we find $N \cap I^n M = I^{n-c}(N \cap I^c M) = 0$ for n large enough. For such an n, the composition $N \subset M \to M/I^n M$ is injective, and we have $M/I^n M \in \mathrm{Coh}_Z(X)$. Hence, the functor $D^b \mathrm{Coh}_Z(X) \to D^b_Z \mathrm{Coh}(X)$ is fully faithful. It is essentially surjective – hence an equivalence – since both categories are generated as triangulated categories by $\mathrm{Coh}_Z(X)$ considered as complexes concentrated in degree zero.

The homotopy fibration of K-theory spaces follows from the Thomason–Wald–hausen Localization Theorem 3.2.23.   □

3.3.3 Remark

The exact sequence of triangulated categories in Theorem 3.3.2 also induces a homotopy fibration of non-connective K-theory spectra $\mathbb{K} \mathrm{Coh}_Z(X) \to \mathbb{K} \mathrm{Coh}(X) \to \mathbb{K} \mathrm{Coh}(U)$ by Theorem 3.2.27. But this does not give us more information since the negative K-groups of noetherian abelian categories such as $\mathrm{Coh}_Z(X)$, $\mathrm{Coh}(X)$ and $\mathrm{Coh}(U)$ are all trivial; see Remark 2.4.5.

3.3.4 Dévissage

The *Dévissage* Theorem 2.3.3 does not follow from the Thomason–Waldhausen Localization Theorem 3.2.27 since *Dévissage* does not hold for Hochschild homology [49, 1.11]. Yet Theorem 3.2.27 holds when K-theory is replaced with Hochschild homology.

Recall that a noetherian scheme X is called *regular* if all its local rings $O_{X,x}$ are regular local rings for $x \in X$. For the definition and basic properties of regular local rings, see [3, 57, 102].

3.3.5 Theorem (Poincaré Duality, [73, Sect. 7.1])

Let X be a regular noetherian separated scheme. Then the fully exact inclusion $\mathrm{Vect}(X) \subset \mathrm{Coh}(X)$ of vector bundles into coherent O_X-modules induces an equivalence of triangulated categories

$$D^b \mathrm{Vect}(X) \cong D^b \mathrm{Coh}(X).$$

In particular, it induces a homotopy equivalence $K(X) \xrightarrow{\sim} G(X)$.

Proof:

We show below that every coherent sheaf F on X admits a surjective map $V \twoheadrightarrow F$ of O_X-modules where V is a vector bundle. This implies that the dual of criterion Sect. 3.1.7 (b) is satisfied, and we see that $D^b \mathrm{Vect}(X) \to D^b \mathrm{Coh}(X)$ is fully faithful. The existence of the surjection also implies that every coherent sheaf F admits a resolution

$$\cdots \to V_i \to V_{i-1} \to \cdots \to V_0 \to F \to 0$$

by vector bundles V_i. By Serre's Theorem [102, Theorem 4.4.16], [57, Theorem 19.2], for every point $x \in X$, the stalk at x of the image E_i of the map $V_i \to V_{i-1}$ is a free $O_{X,x}$-module when $i = \dim O_{X,x}$. Since E_i is coherent, there is an open neighborhood U_x of x over which the sheaf E_i is free and $i = \dim O_{X,x}$. Then E_i is locally free on U_x for all $i \geq \dim O_{X,x}$. Since X is quasi-compact, finitely many of the U_x's suffice to cover X, and we see that E_i is locally free on X for $i \gg 0$. The argument shows that we can truncate the resolution of F at some degree $i \gg 0$, and we obtain a finite resolution of F by vector bundles. Since $D^b \mathrm{Coh}(X)$ is generated by complexes concentrated in degree 0, the last statement implies that $D^b \mathrm{Vect}(X) \to D^b \mathrm{Coh}(X)$ is also essentially surjective, hence an equivalence. By Agreement and Invariance under derived equivalences (Theorems 3.2.24 and 3.2.14), we have $K(X) \simeq G(X)$.

To see the existence of a surjection $V \twoheadrightarrow F$, we can assume that X is connected, hence integral. The local rings $O_{X,x}$ are regular noetherian, hence UFD's. This implies that for any closed $Z \subset X$ of pure codimension 1, there is a line bundle $\mathcal{L}$ and a section $s : O_X \to \mathcal{L}$ such that $Z = X - X_s$ where X_s is the non-vanishing locus $\{x \in X \mid s_x : O_{X,x} \cong \mathcal{L}_x\} \subset X$ of s; see [41, Propositions II 6.11, 6.13]. Since any proper closed subset of X is in such a Z, the open subsets X_s indexed by pairs $(\mathcal{L}, s)$ form a basis for the topology of X where $\mathcal{L}$ runs through the line bundles of X and $s \in \Gamma(X, \mathcal{L})$.

For $a \in F(X_s)$, there is an integer $n \geq 0$ such that $a \otimes s^n \in \Gamma(X_s, F \otimes \mathcal{L}^n)$ extends to a global section of $F \otimes \mathcal{L}^n$; see [41, Lemma 5.14]. This global section defines a map $\mathcal{L}^{-n} \to F$ of O_X-modules such that $a \in F(X_s)$ is in the image of $\mathcal{L}^{-n}(X_s) \to F(X_s)$. It follows that there is a surjection $\bigoplus \mathcal{L}_i \twoheadrightarrow F$ from a sum of line bundles $\mathcal{L}_i$ to F. Since F is coherent and X is quasi-compact, finitely many of the $\mathcal{L}_i$'s are sufficient to yield a surjection. $\square$

3.3.6 Additivity (Revisited)

Let $\mathscr{E}$ be an exact category, and let $E(\mathscr{E})$ denote the exact category of conflations in $\mathscr{E}$. Objects are conflations $A \rightarrowtail B \twoheadrightarrow C$ in $\mathscr{E}$, and maps are commutative diagrams of conflations. A sequence of conflations is called exact if it is exact at the A, B and C-spots. We define exact functors

$$\lambda: \mathscr{E} \to E(\mathscr{E}): A \mapsto (A \overset{1}{\rightarrowtail} A \twoheadrightarrow 0)$$

$$\rho: E(\mathscr{E}) \to \mathscr{E}: (A \rightarrowtail B \twoheadrightarrow C) \mapsto A$$

$$L: E(\mathscr{E}) \to \mathscr{E}: (A \rightarrowtail B \twoheadrightarrow C) \mapsto C$$

$$R: \mathscr{E} \to E(\mathscr{E}): C \mapsto (0 \rightarrowtail C \overset{1}{\twoheadrightarrow} C).$$

Note that λ and L are left adjoint to ρ and R. The unit and counit of adjunctions induce natural isomorphisms $id \overset{\cong}{\to} \rho\lambda$ and $LR \overset{\cong}{\to} id$ and a functorial conflation $\lambda\rho \rightarrowtail id \twoheadrightarrow RL$.

3.3.7 Theorem (Additivity)

The sequence (λ, L) of exact functors induces an exact sequence of triangulated categories

$$D^b\mathscr{E} \to D^b E(\mathscr{E}) \to D^b\mathscr{E}.$$

The associated homotopy fibrations of K-theory spaces and spectra split via (ρ, R), and we obtain homotopy equivalences

$$(\rho, L): K(E(\mathscr{E})) \overset{\sim}{\longrightarrow} K(\mathscr{E}) \times K(\mathscr{E}) \quad and \quad (\rho, L): \mathbb{K}(E(\mathscr{E})) \overset{\sim}{\longrightarrow} \mathbb{K}(\mathscr{E}) \times \mathbb{K}(\mathscr{E})$$

with inverses $\lambda \oplus R$.

Proof:

The functors ρ, λ, L and R are exact. Therefore, they induce triangle functors $D^b\rho$, $D^b\lambda$, D^bL and D^bR on bounded derived categories. Moreover, $D^b\lambda$ and D^bL are left adjoint to $D^b\rho$ and D^bR. The unit and counit of adjunctions $id \overset{\cong}{\to} D^b\rho \circ D^b\lambda$ and $D^bL \circ D^bR \overset{\cong}{\to} id$ are isomorphisms. The natural conflation $\lambda\rho \rightarrowtail id \twoheadrightarrow RL$ induces a functorial distinguished triangle $D^b\lambda \circ D^b\rho \to id \to D^bR \circ D^bL \to D^b\lambda \circ D^b\rho[1]$. By Appendix A, Exercise 2.8, this implies that the sequence of triangulated categories in the theorem is exact. The statements about K-theory follow from the Thomason–Waldhausen Localization Theorems 3.2.23 and 3.2.27. □

Proof of Additivity 2.3.11

The exact sequence of functors $\mathscr{E} \to \mathscr{E}'$ in Theorem 2.3.11 induces an exact functor $F_\bullet: \mathscr{E} \to E(\mathscr{E}')$. Let $M: E(\mathscr{E}') \to \mathscr{E}'$ denote the functor sending a conflation $(A \rightarrowtail B \twoheadrightarrow C)$ to B. By the Additivity Theorem 3.3.7, the composition of the functors

$$E(\mathscr{E}') \overset{(\rho, L)}{\longrightarrow} \mathscr{E}' \times \mathscr{E}' \overset{\lambda \oplus R}{\longrightarrow} E(\mathscr{E}')$$

induces a map on K-theory spaces and spectra which is homotopic to the identity functor. Therefore, the two functors

$$\mathscr{A} \overset{F_\bullet}{\longrightarrow} E(\mathscr{E}') \overset{M}{\longrightarrow} \mathscr{E}' \quad and \quad \mathscr{A} \overset{F_\bullet}{\longrightarrow} E(\mathscr{E}') \overset{(\rho, L)}{\longrightarrow} \mathscr{E}' \times \mathscr{E}' \overset{\lambda \oplus R}{\longrightarrow} E(\mathscr{E}') \overset{M}{\longrightarrow} \mathscr{E}'$$

induce homotopic maps on K-theory spaces and spectra. But these two functors are F_0 and $F_{-1} \oplus F_1$. □

3.3.8 Proposition (Resolution Revisited)

Under the hypothesis of the Resolution Theorem 2.3.12, the inclusion $\mathscr{A} \subset \mathscr{B}$ of exact categories induces an equivalence of triangulated categories

$$D^b(\mathscr{A}) \xrightarrow{\simeq} D^b(\mathscr{B}).$$

In particular, it induces homotopy equivalences of K-theory spaces and spectra

$$K(\mathscr{A}) \xrightarrow{\sim} K(\mathscr{B}) \quad and \quad I\!K(\mathscr{A}) \xrightarrow{\sim} I\!K(\mathscr{B}).$$

Proof:

The hypothesis Theorem 2.3.12 (a) and (b) imply that the dual of criterion Sect. 3.1.7 (b) is satisfied, and the functor $D^b(\mathscr{A}) \to D^b(\mathscr{B})$ is fully faithful. Finally, the hypothesis Theorem 2.3.12 (b) implies that the triangle functor is also essentially surjective; see Sect. 3.1.7 (d).   □

3.4 Thomason's Mayer-Vietoris Principle

Any reasonable cohomology theory for schemes should come with a Mayer–Vietoris long exact sequence for open covers. For K-theory this means that for a scheme $X = U \cup V$ covered by two open subschemes U and V, we should have a long exact sequence of K-groups

$$\cdots \to K_{i+1}(U \cap V) \to K_i(X) \to K_i(U) \oplus K_i(V) \to K_i(U \cap V) \to K_{i-1}(X) \to \cdots$$

for $i \in \mathbb{Z}$. Surprisingly, the existence of such an exact sequence was only proved by Thomason [94] about 20 years after the introduction of higher algebraic K-theory by Quillen. Here, the use of derived categories is essential. For a regular noetherian separated scheme X, the exact sequence also follows from Quillen's Localization Theorem 2.3.7 together with Poincaré Duality Theorem 3.3.5. The purpose of this subsection is to explain the ideas that go into proving Thomason's Mayer-Vietoris exact sequence. Details of some proofs are given in Appendix A, Sects. 3 and 4.

If we defined $I\!K(X)$ naively as the K-theory $I\!K \operatorname{Vect}(X)$ of vector bundles, we would not have such a long exact sequence, in general. For that reason, one has to use perfect complexes instead of vector bundles in the definition of K-theory. For a quasi-projective scheme or a regular noetherian separated scheme, this is the same as vector bundle K-theory; see Proposition 3.4.8. Thomason proves the Mayer–Vietoris exact sequence for general quasi-compact and quasi-separated schemes; see Remark 3.4.13. Here, we will only give definitions and proofs for quasi-compact and separated schemes. This allows us to work with complexes of quasi-coherent sheaves as opposed to complexes of O_X-modules which have quasi-coherent cohomology. This is easier and it is sufficient for most applications.

For the following, the reader is advised to be acquainted with the definitions and statements in Appendix A, Sect. 3. For a quasi-compact and separated scheme X, we denote by $\operatorname{Qcoh}(X)$ the category of quasi-coherent sheaves on X, by $D \operatorname{Qcoh}(X)$ its unbounded derived category (see Appendix A, Sect. 3.1), and for a closed subset $Z \subset X$ we denote by $D_Z \operatorname{Qcoh}(X)$ the full subcategory of $D \operatorname{Qcoh}(X)$ of those complexes which are acyclic when restricted to $X - Z$.

3.4.1 Perfect Complexes

Let X be a quasi-compact and separated scheme. A complex (A, d) of quasi-coherent O_X-modules is called *perfect* if there is a covering $X = \bigcup_{i \in I} U_i$ of X by affine open sub-schemes $U_i \subset X$ such that the restriction of the complex $(A, d)_{|U_i}$ to U_i is quasi-isomorphic to a bounded complex of vector bundles for $i \in I$. The fact that this is independent of the chosen affine cover follows from Appendix A, Sect. 4. Let $Z \subset X$ be a closed subset of X with quasi-compact open complement $X - Z$. We write $\mathrm{Perf}_Z(X) \subset \mathrm{Ch}\,\mathrm{Qcoh}(X)$ for the full subcategory of perfect complexes on X which are acyclic over $X - Z$. The inclusion of categories of complexes is extension closed, and we can consider $\mathrm{Perf}_Z X$ as a fully exact subcategory of the abelian category $\mathrm{Ch}\,\mathrm{Qcoh}(X)$. As in Sect. 3.2.3, the ordinary tensor product of chain complexes makes $(\mathrm{Perf}_Z(X), \mathrm{quis})$ into a complicial exact category with weak equivalences. It is customary to write $D\,\mathrm{Perf}_Z(X)$ for $\mathcal{T}(\mathrm{Perf}_Z(X), \mathrm{quis})$ and $\mathrm{Perf}(X)$ for $\mathrm{Perf}_X(X)$.

3.4.2 Definition

Let X be a quasi-compact and separated scheme, and let $Z \subset X$ be a closed subset of X with quasi-compact open complement $X - Z$. The *K-theory spectrum of X with support in Z* is the $\mathbb{K}$-theory spectrum

$$\mathbb{K}(X \text{ on } Z) = \mathbb{K}(\mathrm{Perf}_Z(X), \mathrm{quis})$$

of the complicial exact category $(\mathrm{Perf}_Z(X), \mathrm{quis})$ as defined in Sect. 3.4.1. In case $Z = X$, we simply write $\mathbb{K}(X)$ instead of $\mathbb{K}(X \text{ on } Z)$. It follows from Proposition 3.4.6 below that the triangulated categories $D\,\mathrm{Perf}_Z(X)$ are idempotent complete. Therefore, we may write $K_i(X \text{ on } Z)$ instead of $\mathbb{K}_i(X \text{ on } Z)$ for the K-groups of X with support in Z and $i \in \mathbb{Z}$.

In order to be able to say anything about the K-theory of perfect complexes, we need to understand, to a certain extend, the structure of the triangulated categories $D\,\mathrm{Perf}_Z(X)$. Lemma 3.4.3 and Proposition 3.4.6 summarize what we will need to know.

3.4.3 Lemma

Let X be a quasi-compact and separated scheme and $Z \subset X$ be a closed subset with quasi-compact open complement $j : U = X - Z \subset X$.

(a) The following sequence of triangulated categories is exact

$$D_Z\,\mathrm{Qcoh}(X) \to D\,\mathrm{Qcoh}(X) \to D\,\mathrm{Qcoh}(U).$$

(b) Let $g : V \subset X$ be a quasi-compact open subscheme such that $Z \subset V$. Then the restriction functor is an equivalence of triangulated categories

$$g^* : D_Z\,\mathrm{Qcoh}(X) \xrightarrow{\simeq} D_Z\,\mathrm{Qcoh}(V).$$

Proof:

For (a), the restriction $j^* : D\,\mathrm{Qcoh}(X) \to D\,\mathrm{Qcoh}(U)$ has a right adjoint $Rj_* : D\,\mathrm{Qcoh}(U) \to D\,\mathrm{Qcoh}(X)$ which for $E \in D\,\mathrm{Qcoh}(U)$ is $Rj_*E = j_*I$ where $E \xrightarrow{\sim} I$ is a $\mathcal{K}$-injective resolution of E. The counit of adjunction $j^*Rj_* \to 1$ is an isomorphism in $D\,\mathrm{Qcoh}(U)$ since for

$E \in D\,\mathrm{Qcoh}(U)$ it is $j^*Rj_*E = j^*j_*I \overset{\cong}{\to} I \xleftarrow{\sim} E$. The claim now follows from general facts about triangulated categories; see Exercise 2.8(a) in Appendix A.

For (b), note that the functor $Rg_* : D\,\mathrm{Qcoh}(V) \to D\,\mathrm{Qcoh}(X)$ sends $D_Z\,\mathrm{Qcoh}(V)$ into $D_Z\,\mathrm{Qcoh}(X)$. This follows from the Base-Change Lemma A.3.7 since for a complex $E \in D_Z\,\mathrm{Qcoh}(V)$, the lemma says $(Rg_*E)|_{X-Z} = R\bar{g}_*(E|_{V-Z}) = 0$ where $\bar{g} : V - Z \subset X - Z$. The unit and counit of adjunction $1 \to Rg_* \circ g^*$ and $g^*Rg_* \to 1$ are isomorphisms in the triangulated category of complexes supported in Z because for such complexes this statement only needs to check when restricted to V where it trivially holds since $Z \subset V$. $\qquad\square$

The reason why Lemma 3.4.3 is so useful lies in the theory of compactly generated triangulated categories and the fact that the categories $D_Z\,\mathrm{Qcoh}(X)$ are indeed compactly generated when X and $X - Z$ are quasi-compact and separated. See 3.4.6 below.

3.4.4 Compactly Generated Triangulated Categories

References are [65] and [64]. Let $\mathscr{A}$ be a triangulated category in which all set indexed direct sums exist. An object A of $\mathscr{A}$ is called *compact* if the canonical map

$$\bigoplus_{i\in I} \mathrm{Hom}(A, E_i) \to \mathrm{Hom}\Big(A, \bigoplus_{i\in I} E_i\Big)$$

is an isomorphism for any set of objects E_i in $\mathscr{A}$ and $i \in I$. Let $\mathscr{A}^c \subset \mathscr{A}$ be the full subcategory of compact objects. It is easy to see that $\mathscr{A}^c$ is an idempotent complete triangulated subcategory of $\mathscr{A}$.

A set S of compact objects is said to *generate* $\mathscr{A}$, or $\mathscr{A}$ is *compactly generated* (by S), if for every object $E \in \mathscr{A}$ we have

$$\mathrm{Hom}(A, E) = 0 \ \ \forall A \in S \implies E = 0.$$

3.4.5 Theorem (Neeman [64])

(a) *Let $\mathscr{A}$ be a compactly generated triangulated category with generating set S of compact objects. Then $\mathscr{A}^c$ is the smallest idempotent complete triangulated subcategory of $\mathscr{A}$ containing S.*

(b) *Let $\mathscr{R}$ be a compactly generated triangulated category, $S_0 \subset \mathscr{R}^c$ be a set of compact objects closed under taking shifts. Let $\mathscr{S} \subset \mathscr{R}$ be the smallest full triangulated subcategory closed under formation of coproducts in $\mathscr{R}$ which contains the set S_0. Then $\mathscr{S}$ and $\mathscr{R}/\mathscr{S}$ are compactly generated triangulated categories with generating sets S_0 and the image of $\mathscr{R}^c$ in $\mathscr{R}/\mathscr{S}$. Moreover, the functor $\mathscr{R}^c/\mathscr{S}^c \to \mathscr{R}/\mathscr{S}$ induces an equivalence between the idempotent completion of $\mathscr{R}^c/\mathscr{S}^c$ and the category of compact objects in $\mathscr{R}/\mathscr{S}$.*

The following proposition will be proved in Appendix A, Sect. 4.

3.4.6 Proposition

Let X be a quasi-compact and separated scheme, and let $Z \subset X$ be a closed subset with quasi-compact open complement $U = X - Z$. Then the triangulated category $D_Z\,\mathrm{Qcoh}(X)$ is compactly generated with category of compact objects the derived category of perfect complexes $D\,\mathrm{Perf}_Z(X)$.

In many interesting cases, the K-theory of perfect complexes is equivalent to the K-theory of vector bundles. This is the case for quasi-projective schemes and for regular noetherian separated schemes both of which are examples of schemes with an ample family of line bundles.

3.4.7 Schemes with an Ample Family of Line Bundles

A quasi-compact scheme X has an *ample family of line bundles* if there is a finite set $L_1, ..., L_n$ of line bundles with global sections $s_i \in \Gamma(X, L_i)$ such that the non-vanishing loci $X_{s_i} = \{x \in X | s_i(x) \neq 0\}$ form an open affine cover of X. See [94, Definition 2.1], [85, II 2.2.4].

Any quasi-compact open (or closed) subscheme of a scheme with an ample family of line bundles has itself an ample family of line bundles, namely the restriction of the ample family to the open (or closed) subscheme. Any scheme which is quasi-projective over an affine scheme has an ample line-bundle. A fortiori it has an ample family of line-bundles. Every separated regular noetherian scheme has an ample family of line bundles. This was shown in the proof of Poincaré Duality Theorem 3.3.5. For more on schemes with an ample family of line-bundles, see [15, 85], [94, 2.1.2] and Appendix A, Sect. 4.2.

3.4.8 Proposition [94, Corollary 3.9]

Let X be a quasi-compact and separated scheme which has an ample family of line bundles. Then the inclusion of bounded complexes of vector bundles into perfect complexes $Ch^b \operatorname{Vect}(X) \subset \operatorname{Perf}(X)$ induces an equivalence of triangulated categories $D^b \operatorname{Vect}(X) \cong D \operatorname{Perf}(X)$. In particular,

$$\mathbb{K} \operatorname{Vect}(X) \simeq \mathbb{K}(X).$$

Proof (see also Appendix A, Proposition 4.7 (a))

Since X has an ample family of line bundles, every quasi-coherent sheaf F on X admits a surjective map $\bigoplus \mathscr{L}_i \to F$ from a direct sum of line bundles to F. The argument is the same as in the last paragraph in the proof of Theorem 3.3.5. This implies that the dual of criterion Sect. 3.1.7 (b) is satisfied, and we have fully faithful functors $D^b \operatorname{Vect}(X) \subset D^b \operatorname{Qcoh}(X) \subset D \operatorname{Qcoh}(X)$. This also implies that the compact objects $\operatorname{Vect}(X)$ generate $D \operatorname{Qcoh}(X)$ as a triangulated category with infinite sums. Since $D^b \operatorname{Vect}(X)$ is idempotent complete [16], the functor $D^b \operatorname{Vect}(X) \to D \operatorname{Perf}(X)$ is an equivalence by Theorem 3.4.5 (a) and Proposition 3.4.6. $\qquad\square$

3.4.9 Theorem (Localization)

Let X be a quasi-compact and separated scheme. Let $U \subset X$ be a quasi-compact open subscheme with closed complement $Z = X - U$. Then there is a homotopy fibration of $\mathbb{K}$-theory spectra

$$\mathbb{K}(X \text{ on } Z) \longrightarrow \mathbb{K}(X) \longrightarrow \mathbb{K}(U).$$

In particular, there is a long exact sequence of K-groups for $i \in \mathbb{Z}$

$$\cdots \to K_{i+1}(U) \to K_i(X \text{ on } Z) \to K_i(X) \to K_i(U) \to K_{i-1}(X \text{ on } Z) \to \cdots$$

Proof:

In view of the Thomason-Waldhausen Localization Theorem 3.2.27, we have to show that the sequence of complicial exact categories with weak equivalences

$$(\mathrm{Perf}_Z(X), \mathrm{quis}) \to (\mathrm{Perf}(X), \mathrm{quis}) \to (\mathrm{Perf}(U), \mathrm{quis})$$

induces a sequence of associated triangulated categories

$$D\,\mathrm{Perf}_Z(X) \to D\,\mathrm{Perf}(X) \to D\,\mathrm{Perf}(U) \tag{16}$$

which is exact up to factors. By Proposition 3.4.6, the sequence (16) is the sequence of categories of compact objects associated with the exact sequence of triangulated categories in Lemma 3.4.3 (a). The claim now follows from Neeman's Theorem 3.4.5 (b).   □

3.4.10 Theorem (Zariski Excision)

Let $j : V \subset X$ be a quasi-compact open subscheme of a quasi-compact and separated scheme X. Let $Z \subset X$ be a closed subset with quasi-compact open complement such that $Z \subset V$. Then restriction of quasi-coherent sheaves induces a homotopy equivalence of $\mathbb{K}$-theory spectra

$$\mathbb{K}(X \text{ on } Z) \xrightarrow{\sim} \mathbb{K}(V \text{ on } Z).$$

In particular, there are isomorphisms of K-groups for all $i \in \mathbb{Z}$

$$K_i(X \text{ on } Z) \xrightarrow{\cong} K_i(V \text{ on } Z).$$

Proof:

By the Invariance Of K-theory Under Derived Equivalences Theorem 3.2.29, it suffices to show that the functor of complicial exact categories with weak equivalences

$$(\mathrm{Perf}_Z(X), \mathrm{quis}) \to (\mathrm{Perf}_Z(V), \mathrm{quis})$$

induces an equivalence of associated triangulated categories. This follows from Lemma 3.4.3 (b) in view of Proposition 3.4.6.   □

3.4.11 Remark

There is a more general excision result where open immersions are replaced with flat maps [94, Theorem 7.1]. It is also a consequence of an equivalence of triangulated categories.

3.4.12 Theorem (Mayer–Vietoris for Open Covers)

Let $X = U \cup V$ be a quasi-compact and separated scheme which is covered by two open quasi-compact subschemes U and V. Then restriction of quasi-coherent sheaves induces a homotopy cartesian square of $\mathbb{K}$-theory spectra

$$\begin{array}{ccc} I\!K(X) & \longrightarrow & I\!K(U) \\ \downarrow & & \downarrow \\ I\!K(V) & \longrightarrow & I\!K(U \cap V). \end{array}$$

In particular, we obtain a long exact sequence of K-groups for $i \in \mathbb{Z}$

$$\cdots \to K_{i+1}(U \cap V) \to K_i(X) \to K_i(U) \oplus K_i(V) \to K_i(U \cap V) \to K_{i-1}(X) \to \cdots$$

Proof:

By the Localization Theorem 3.4.9, the horizontal homotopy fibres of the square are $I\!K(X \text{ on } Z)$ and $I\!K(V \text{ on } Z)$ with $Z = X - U = U - U \cap V \subset V$. The claim follows from Zariski-excision 3.4.10.

$\square$

3.4.13 Remark (Separated Versus Quasi-Separated)

Thomason proves Theorems 3.4.9, 3.4.10 and 3.4.12 for quasi-compact and quasi-separated schemes. A scheme X is quasi-separated if the intersection of any two quasi-compact open subsets of X is quasi-compact. For instance, any scheme whose underlying topological space is noetherian is quasi-separated. Of course, every separated scheme is quasi-separated.

In the generality of quasi-compact and quasi-separated schemes X one has to work with perfect complexes of O_X-*modules* rather than with perfect complexes of *quasi-coherent sheaves*. The reason is that the Base-Change Lemma A.3.7 – which is used at several places in the proofs of Lemma 3.4.3 (b) and Proposition 3.4.6 – does not hold for $D \operatorname{Qcoh}(X)$ when X is quasi-compact and quasi-separated, in general.

Verdier gives a counter example in [85, II Appendice I]. He constructs a quasi-compact and quasi-separated scheme Z (whose underlying topological space is even noetherian), a covering $Z = U \cup V$ of Z by open affine subschemes $j : U = \operatorname{Spec} A \hookrightarrow Z$ and V, and an injective A-module I such that for its associated sheaf $\tilde{I}$ on U, the natural map

$$j_* \tilde{I} = R j_{\operatorname{Qcoh}_*} \tilde{I} \xrightarrow{\not\cong} R j_{Mod_*} \tilde{I} \tag{17}$$

is not a quasi- isomorphism where $j_{\operatorname{Qcoh}_*}$ and j_{Mod_*} are j_* on the category of quasi-coherent modules and O_Z-modules, respectively. If $R j_{\operatorname{Qcoh}_*}$ did satisfy the Base-Change Lemma, then the map (17) would be a quasi- isomorphism on U and V hence a quasi- isomorphism on Z, contradicting (17). Verdier also shows that for this scheme, the forgetful functor

$$D \operatorname{Qcoh}(Z) \to D_{qc}(O_Z\text{-Mod})$$

from the derived category of quasi-coherent modules to the derived category of complexes of O_Z-modules with quasi-coherent cohomology is not fully faithful. In particular, it is not an equivalence contrary to the situation when Z is quasi-compact and separated [14, Corollary 5.5], [4, Proposition 1.3].

3.5 Projective Bundle Theorem and Regular Blow-Ups

We will illustrate the use of triangulated categories in the calculation of higher algebraic K-groups with two more examples: in the proof of the Projective Bundle Formula 3.5.1 and in the (sketch of the) proof of the Blow-up Formula 3.5.4. There is, of course, much more to say about triangulated categories in K-theory. For instance, Example 3.1.11 has been generalized in [67] to stably flat non-commutative Cohn localizations $R \to S^{-1}R$ from which one can derive Waldhausen's calculations in [99]; see [74]. As another example, Swan's calculation of the K-theory of a smooth quadric hypersurface $Q \subset \mathbb{P}^n_k$ [89] can be derived from Kapranov's description of $D\operatorname{Perf}(Q)$ given in [44]. For certain homogeneous spaces, see [44, 52, 76]. After all, any statement about the structure of triangulated categories translates into a statement about higher algebraic K-groups via the Thomason–Waldhausen Localization Theorem 3.2.27.

3.5.1 Theorem (Projective Bundle Theorem [73, Sect. 8 Theorem 2.1])

Let X be a quasi-compact and separated scheme, and let $\mathscr{E} \to X$ be a geometric vector bundle over X of rank $n + 1$. Let $p : \mathbb{P}\mathscr{E} \to X$ be the associated projective bundle with twisting sheaf $O_{\mathscr{E}}(1)$. Then we have an equivalence

$$\prod_{l=0}^{n} O_{\mathscr{E}}(-l) \otimes Lp^* : \ \prod_{l=0}^{n} \mathbb{K}(X) \overset{\sim}{\longrightarrow} \mathbb{K}(\mathbb{P}\mathscr{E}).$$

For the proof we will need the following useful lemma which is a special case of Proposition A.4.7 (a).

3.5.2 Lemma

Let X be a scheme with an ample line-bundle L. Then the category $D^b \operatorname{Vect}(X)$ is generated – as an idempotent complete triangulated category – by the set $L^{\otimes k}$ of line-bundles for $k < 0$.

Proof of the Projective Bundle Theorem 3.5.1

By the Mayer–Vietoris Theorem 3.4.12, the question is local in X. Therefore, we may assume that $X = \operatorname{Spec} A$ is affine and that $p : \mathbb{P}\mathscr{E} \to X$ is the canonical projection $\operatorname{Proj}(A[T_0, ..., T_n]) = \mathbb{P}^n_A \overset{p}{\to} \operatorname{Spec} A$. In this case, X and $\mathbb{P}\mathscr{E}$ have an ample line-bundle A and $O(1)$, and their derived categories of perfect complexes agree with the bounded derived categories of vector bundles by Proposition 3.4.8. Since the twisting sheaf $O(1)$ is ample, we can apply Lemma 3.5.2, and we see that the triangulated category $D \operatorname{Vect}(\mathbb{P}^n)$ is generated – as an idempotent complete triangulated category – by the family of line bundles $\{O_{\mathbb{P}^n}(-l) \mid l \geq 0\}$. Consider the polynomial ring $S = A[T_0, ..., T_n]$ as a graded ring with $\deg T_i = 1$. The sequence $T_0, ..., T_n$ is a regular sequence in S. Therefore, the Koszul complex $\otimes_{i=0}^{n}(S(-1) \overset{T_i}{\to} S)$ induces an exact sequence of graded S-modules

$$0 \to S(-n-1) \to \overset{n+1}{\underset{1}{\bigoplus}} S(-n) \to \overset{\binom{n+1}{2}}{\underset{1}{\bigoplus}} S(-n+1) \to \cdots \to \overset{n+1}{\underset{1}{\bigoplus}} S(-1) \to S \to A \to 0.$$

Taking associated sheaves, we obtain an exact sequence of vector bundles on $\mathbb{P}_A^n$

$$0 \to O(-n-1) \to \overset{n+1}{\underset{1}{\bigoplus}} O(-n) \to \overset{\binom{n+1}{2}}{\underset{1}{\bigoplus}} O(-n+1) \to \cdots \to \overset{n+1}{\underset{1}{\bigoplus}} O(-1) \to O_{\mathbb{P}^n} \to 0.$$

This shows that $D^b \operatorname{Vect}(\mathbb{P}_A^n)$ is generated as an idempotent complete triangulated category by $O(-n), \ldots, O(-1), O_{\mathbb{P}^n}$. For $i \leq j$, let $D^b_{[i,j]} \subset D^b \operatorname{Vect}(\mathbb{P}^n)$ be the full idempotent complete triangulated subcategory generated by $O(l)$ where $i \leq l \leq j$. We have a filtration

$$0 \subset D^b_{[0,0]} \subset D^b_{[-1,0]} \subset \ldots \subset D^b_{[-n,0]} = D^b \operatorname{Vect}(\mathbb{P}^n).$$

The unit of adjunction $F \to Rp_* Lp^* F$ is a quasi-isomorphism for $F = A$ because $A \to H^0(Rp_* Lp^* A) = H^0(Rp_* O_{\mathbb{P}^n}) = H^0(\mathbb{P}^n, O_{\mathbb{P}^n})$ is an isomorphism and $H^i(\mathbb{P}^n, O_{\mathbb{P}^n}) = 0$ for $i \neq 0$ [38, Proposition III 2.1.12]. Since $D^b \operatorname{Proj}(A)$ is generated as an idempotent complete triangulated category by A, we see that the unit of adjunction $F \to Rp_* Lp^* F$ is a quasi-isomorphism for all $F \in D^b \operatorname{Proj}(A)$. This implies that $Lp^* = p^* : D^b \operatorname{Proj}(A) \to D^b \operatorname{Vect}(\mathbb{P}^n)$ is fully faithful and, hence, an equivalence onto its image $D^b_{[0,0]}$. Since $O(l)$ is an invertible sheaf, we obtain equivalences $O(-l) \otimes Lp^* : D^b \operatorname{Proj}(A) \to D^b_{[-l,-l]}$.

By the calculation of the cohomology of the projective space $\mathbb{P}_A^n$ (loc.cit.), we have $H^*(\mathbb{P}_A^n, O(-k)) = 0$ for $k = 1, \ldots, n$. Therefore, the homomorphism sets in $D^b \operatorname{Vect}(\mathbb{P}_A^n)$ satisfy

$$\operatorname{Hom}(O(-j)[r], O(-l)[s]) = H^{s-r}(\mathbb{P}_A^n, O(-l+j)) = 0$$

for $0 \leq j < l \leq n$. This implies that the composition

$$D^b_{[-l,-l]} \subset D^b_{[-l,0]} \to D^b_{[-l,0]}/D^b_{[-l+1,0]}$$

is an equivalence; see Exercise A.2.8 (b).

To finish the proof, we simply translate the statements about triangulated categories above into statements about K-theory. For $i \leq j$, let $\operatorname{Ch}^b_{[i,j]} \subset \operatorname{Ch}^b \operatorname{Vect}(\mathbb{P}^n)$ be the full subcategory of those chain complexes which lie in $D^b_{[i,j]}$. Write w for the set of maps in $\operatorname{Ch}^b_{[-l,0]}$ which are isomorphisms in the quotient triangulated category $D^b_{[-l,0]}/D^b_{[-l+1,0]}$. By construction, the sequence

$$(\operatorname{Ch}^b_{[-l+1,0]}, \operatorname{quis}) \to (\operatorname{Ch}^b_{[-l,0]}, \operatorname{quis}) \to (\operatorname{Ch}^b_{[-l,0]}, w) \tag{18}$$

induces an exact sequence of associated triangulated categories, and by Theorem 3.2.27, it induces a homotopy fibration in K-theory for $l = 1, \ldots, n$. We have seen that the composition

$$O(-l) \otimes p^* : (\operatorname{Ch}^b \operatorname{Proj}(A), \operatorname{quis}) \to (\operatorname{Ch}^b_{[-l,0]}, \operatorname{quis}) \to (\operatorname{Ch}^b_{[-l,0]}, w)$$

induces an equivalence of associated triangulated categories. By Theorem 3.2.29, the composition induces an equivalence in K-theory. It follows that the K-theory fibration associated with (18) splits, and we obtain a homotopy equivalence

$$(O(-l) \otimes p^*, 1) : \mathbb{K}(A) \times \mathbb{K}(\operatorname{Ch}^b_{[-l+1,0]}, \operatorname{quis}) \xrightarrow{\sim} \mathbb{K}(\operatorname{Ch}^b_{[-l,0]}, \operatorname{quis})$$

for $l = 1, \ldots, n$. Since $\operatorname{Ch}^b_{[-n,0]} = \operatorname{Ch}^b \operatorname{Vect}(\mathbb{P}_A^n)$, this implies the theorem. $\qquad\square$

3.5.3 Theorem (Bass' Fundamental Theorem)

Let X be a quasi-compact and separated scheme. Then there is a split exact sequence for all $n \in \mathbb{Z}$

$$0 \to \mathbb{K}_n(X) \to \mathbb{K}_n(X[T]) \oplus \mathbb{K}_n(X[T^{-1}]) \to \mathbb{K}_n(X[T, T^{-1}]) \to \mathbb{K}_{n-1}(X) \to 0.$$

Proof

The projective line $\mathbb{P}^1_X$ over X has a standard open covering given by $X[T]$ and $X[T^{-1}]$ with intersection $X[T, T^{-1}]$. Thomason's Mayer-Vietoris Theorem 3.4.12 applied to this covering yields a long exact sequence

$$\to \mathbb{K}_n(\mathbb{P}^1_X) \xrightarrow{\beta} \mathbb{K}_n(X[T]) \oplus \mathbb{K}_n(X[T^{-1}]) \to \mathbb{K}_n(X[T, T^{-1}]) \to \mathbb{K}_{n-1}(\mathbb{P}^1_X) \to$$

By the Projective Bundle Theorem 3.5.1, the group $\mathbb{K}_n(\mathbb{P}^1_X)$ is $\mathbb{K}_n(X) \oplus \mathbb{K}_n(X)$ with basis $[O_{\mathbb{P}^1}]$ and $[O_{\mathbb{P}^1}(-1)]$. Making a base-change, we can write $\mathbb{K}_n(\mathbb{P}^1_X)$ as $\mathbb{K}_n(X) \oplus \mathbb{K}_n(X)$ with basis $[O_{\mathbb{P}^1}]$ and $[O_{\mathbb{P}^1}] - [O_{\mathbb{P}^1}(-1)]$. Since on $X[T]$ and on $X[T^{-1}]$ the two line-bundles $O_{\mathbb{P}^1}$ and $O_{\mathbb{P}^1}(-1)$ are isomorphic, the map β in the long Mayer–Vietoris exact sequence above is trivial on the direct summand $K(X)$ corresponding to the base element $[O_{\mathbb{P}^1}] - [O_{\mathbb{P}^1}(-1)]$. The map β is split injective on the other summand $K(X)$ corresponding to the base element $[O_{\mathbb{P}^1}]$. Therefore, the long Mayer–Vietoris exact sequence breaks up into shorter exact sequences. These are the exact sequences in the theorem. The splitting of the map $\mathbb{K}_n(X[T, T^{-1}]) \to \mathbb{K}_{n-1}(X)$ is given by the cup product with the element $[T] \in K_1(\mathbb{Z}[T, T^{-1}])$. $\qquad\square$

The following theorem is due to Thomason [90]. For the (sketch of the) proof given below, we follow [19, Sect. 1].

3.5.4 Theorem (Blow-Up Formula)

Let $i : Y \subset X$ be a regular embedding of pure codimension d with X quasi-compact and separated. Let $p : X' \to X$ be the blow-up of X along Y and $j : Y' \subset X'$ the exceptional divisor. Write $q : Y' \to Y$ for the induced map. Then the square of $\mathbb{K}$-theory spectra

$$
\begin{array}{ccc}
\mathbb{K}(X) & \xrightarrow{\ Li^*\ } & \mathbb{K}(Y) \\
{\scriptstyle Lp^*}\big\downarrow & & \big\downarrow{\scriptstyle Lq^*} \\
\mathbb{K}(X') & \xrightarrow[\ Lj^*\ ]{} & \mathbb{K}(Y')
\end{array}
$$

is homotopy cartesian. Moreover, there is a homotopy equivalence

$$\mathbb{K}(X') \simeq \mathbb{K}(X) \times \prod_{1}^{d-1} \mathbb{K}(Y).$$

Proof (sketch)

To simplify the argument, we note that the question as to whether the square of K-theory spectra is homotopy cartesian is local in X by Thomason's Mayer-Vietoris Theorem 3.4.12. Therefore, we can assume that $X = \operatorname{Spec} A$ and $Y = \operatorname{Spec} A/I$ are affine and that $I \subset A$ is an ideal generated by a regular sequence $x_1, ..., x_d$ of length d. In this case, all schemes X, X', Y, Y' have an ample line-bundle. By Proposition 3.4.8, the K-theory of perfect complexes on X, X', Y, Y' agrees with the vector bundle K-theory on those schemes. Furthermore, the map $Y' \to Y$ is the canonical projection $\mathbb{P}_Y^{d-1} \to Y$.

Let $S = \bigoplus_{i \geq 0} I^n$. Then we have $X' = \operatorname{Proj} S$ and $Y' = \operatorname{Proj} S/IS$. The exact sequence $0 \to IS = S(1) \to S \to S/IS \to 0$ of graded S-modules induces an exact sequence of sheaves $0 \to O_{X'}(1) \to O_{X'} \to j_*O_{Y'} \to 0$ on X' and an associated distinguished triangle $O_{X'}(1) \to O_{X'} \to Rj_*O_{Y'} \to O_{X'}(1)[1]$ in $D^b \operatorname{Vect}(X')$. Restricted to Y', this triangle becomes the following distinguished triangle: $O_{Y'}(1) \to O_{Y'} \to j^*Rj_*O_{Y'} \to O_{Y'}(1)[1]$. Since $O_{Y'}(1) \to O_{Y'}$ is the zero map (as $Y' = \mathbb{P}_Y^{d-1}$), we have an isomorphism $j^*Rj_*O_{Y'} \cong O_{Y'} \oplus O_{Y'}(1)[1]$ in $D^b \operatorname{Vect}(Y')$. This shows that Lj^* respects the filtration of triangulated subcategories

$$
\begin{array}{ccccccc}
D_{X'}^0 & \subset & D_{X'}^1 & \subset & \cdots & \subset & D_{X'}^{d-1} & = & D^b \operatorname{Vect}(X') \\
\downarrow Lj^* & & \downarrow Lj^* & & . & & \downarrow Lj^* & & \\
D_{Y'}^0 & \subset & D_{Y'}^1 & \subset & \cdots & \subset & D_{Y'}^{d-1} & = & D^b \operatorname{Vect}(Y')
\end{array}
$$

defined by setting $D_{X'}^l$ and $D_{Y'}^l$ to be the full idempotent complete triangulated subcategories of $D^b \operatorname{Vect}(X')$ and $D^b \operatorname{Vect}(Y')$ generated by $O_{X'}$ and the complexes $O_{X'}(-k) \otimes Rj_*O_{Y'}$ for $k = 1, ..., l$ in the first case, and by $O_{Y'}(-k)$ for $k = 0, ..., l$ in the second case. The fact that $D_{Y'}^{d-1} = D^b \operatorname{Vect}(Y')$ was shown in the proof of Theorem 3.5.1. A similar argument – using the ampleness of $O_{X'}(1)$ and the Koszul complex associated with the regular sequence $x_1, ..., x_d \in S$ – shows that we have $D_{X'}^{d-1} = D^b \operatorname{Vect}(X')$. One checks that, on associated graded pieces, Lj^* induces equivalences of triangulated categories for $l = 1, ..., d-1$

$$ Lj^* : D_{X'}^l/D_{X'}^{l-1} \xrightarrow{\simeq} D_{Y'}^l/D_{Y'}^{l-1}. \tag{19} $$

A cohomology calculation shows that the units of adjunction $E \to Rp_*Lp^*E$ and $F \to Rq_*Lq^*F$ are isomorphisms for $E = O_X$ and $F = O_Y$. Since X and Y are affine, the triangulated categories $D^b \operatorname{Vect}(X)$ and $D^b \operatorname{Vect}(Y)$ are generated (up to idempotent completion) by O_X and O_Y. Therefore, the units of adjunction $E \to Rp_*Lp^*E$ and $F \to Rq_*Lq^*F$ are isomorphisms for all $E \in D^b \operatorname{Vect}(X)$ and $F \in D^b \operatorname{Vect}(Y)$. It follows that Lp^* and Lq^* are fully faithful and induce equivalences onto their images

$$ Lp^* : D^b \operatorname{Vect}(X) \xrightarrow{\simeq} D_{X'}^0, \quad \text{and} \quad Lq^* : D^b \operatorname{Vect}(Y) \xrightarrow{\simeq} D_{Y'}^0. \tag{20} $$

This finishes the triangulated category background.

In order to prove the K-theory statement, define categories $\operatorname{Ch}_{X'}^l \subset \operatorname{Ch}^b \operatorname{Vect}(X')$ and $\operatorname{Ch}_{Y'}^l \subset \operatorname{Ch}^b \operatorname{Vect}(Y')$ as the fully exact complicial subcategories of those complexes which lie in $D_{X'}^l$ and $D_{Y'}^l$, respectively. Then $\mathscr{T}(\operatorname{Ch}_{X'}^l, \operatorname{quis}) = D_{X'}^l$ and $\mathscr{T}(\operatorname{Ch}_{Y'}^l, \operatorname{quis}) = D_{Y'}^l$. The functor j^* respects the filtration of exact categories with weak equivalences

$$(\text{Ch}^0_{X'}, \text{quis}) \quad \subset \quad (\text{Ch}^1_{X'}, \text{quis}) \quad \subset \quad \cdots \quad \subset \quad (\text{Ch}^{d-1}_{X'}, \text{quis}) \qquad (21)$$

$$\downarrow j^* \qquad\qquad\qquad \downarrow j^* \qquad\qquad\qquad\qquad\qquad \downarrow j^*$$

$$(\text{Ch}^0_{Y'}, \text{quis}) \quad \subset \quad (\text{Ch}^1_{Y'}, \text{quis}) \quad \subset \quad \cdots \quad \subset \quad (\text{Ch}^{d-1}_{Y'}, \text{quis}).$$

If we denote by quis^l the set of maps in Ch^l which are isomorphisms in D^l/D^{l-1}, then $\mathscr{T}(\text{Ch}^l, \text{quis}^l) = D^l/D^{l-1}$. By the Theorem on Invariance Of K-theory Under Derived Equivalences 3.2.29, the equivalence (19) yields an equivalence of K-theory spectra $j^* : \mathbb{K}(\text{Ch}^l_{X'}, \text{quis}^l) \xrightarrow{\simeq} \mathbb{K}(\text{Ch}^l_{Y'}, \text{quis}^l)$ for $l = 1, ..., d-1$. The sequence $(\text{Ch}^{l-1}, \text{quis}) \to (\text{Ch}^l, \text{quis}) \to (\text{Ch}^l, \text{quis}^l)$ induces a homotopy fibration of K-theory spectra by the Thomason–Waldhausen Localization Theorem 3.2.27. Therefore, all individual squares in (21) induce homotopy cartesian squares of K-theory spectra. As a composition of homotopy cartesian squares, the outer square also induces a homotopy cartesian squares of K-theory spectra. By (20), the outer square of (21) yields the K-theory square in the theorem.

The formula for $\mathbb{K}(X')$ in terms of $\mathbb{K}(X)$ and $\mathbb{K}(Y)$ follows from the fact that $Lp^* : \mathbb{K}(X) \to \mathbb{K}(X')$ is split injective with retraction given by Rp_* and the fact that the cofibre of Lp^* is the cofibre of $Lq^* : \mathbb{K}(Y) \to \mathbb{K}(Y')$ which is given by the Projective Bundle Theorem 3.5.1. □

4 Beyond Triangulated Categories

4.1 Statement of Results

Of course, not all results in algebraic K-theory can be obtained using triangulated category methods. In this subsection we simply state some of these results. For more overviews on a variety of topics in K-theory, we refer the reader to the K-theory handbook [24].

4.1.1 Brown–Gersten–Quillen Spectral Sequence [73]

Let X be a noetherian scheme, and write $X^p \subset X$ for the set of points of codimension p in X. There is a filtration $0 \subset ... \subset \text{Coh}^2(X) \subset \text{Coh}^1(X) \subset \text{Coh}^0(X) = \text{Coh}(X)$ of $\text{Coh}(X)$ by the Serre abelian subcategories $\text{Coh}^i(X) \subset \text{Coh}(X)$ of those coherent sheaves whose support has codimension $\geq i$. This filtration together with Quillen's Localization and *Dévissage* Theorems leads to the Brown–Gersten–Quillen (BGQ) spectral sequence

$$E_1^{p,q} = \bigoplus_{x \in X^p} K_{-p-q}(k(x)) \Rightarrow G_{-p-q}(X).$$

If X is regular and of finite type over a field, inspection of the differential d_1 yields an isomorphism

$$E_2^{p,-p} \cong \text{CH}^p(X)$$

where $\text{CH}^p(X)$ is the Chow-group of codimension p cycles modulo rational equivalence as defined in [28].

4.1.2 Gersten's Conjecture and Bloch's Formula

The Brown–Gersten–Quillen spectral sequence yields a complex

$$0 \to G_n(X) \to \bigoplus_{x \in X^0} K_n(k(x)) \xrightarrow{d_1} \bigoplus_{x \in X^1} K_{n-1}(k(x)) \xrightarrow{d_1} \cdots$$

The Gersten conjecture says that this complex is exact for $X = \operatorname{Spec} R$ where R is a regular local noetherian ring. The conjecture is proved in case R (is regular local noetherian and) contains a field [69] building on the geometric case proved in [73]. For other examples of rings satisfying the Gersten conjecture, see [86]. For K-theory with finite coefficients, Gersten's conjecture holds for the local rings of a smooth variety over a discrete valuation ring [32].

As a corollary, Quillen [73] obtains for a regular scheme X of finite type over a field a calculation of the E_2-term of the BGQ-spectral sequence as $E_2^{p,q} \cong H_{Zar}^p(X, \mathscr{K}_{-q,X})$, and he obtains Bloch's formula

$$\mathrm{CH}^p(X) \cong H_{Zar}^p(X, \mathscr{K}_{p,X})$$

where $\mathscr{K}_{p,X}$ denotes the Zariski sheaf associated with the presheaf $U \mapsto K_p(U)$.

4.1.3 Computation of $K(\mathbb{F}_q)$

Quillen computed the K-groups of finite fields in [72]. They are given by the formulas $K_0(\mathbb{F}_q) \cong \mathbb{Z}$, $K_{2n}(\mathbb{F}_q) = 0$ for $n > 0$ and $K_{2n-1}(\mathbb{F}_q) \cong \mathbb{Z}/(q^n - 1)\mathbb{Z}$ for $n > 0$.

4.1.4 The Motivic Spectral Sequence

Let X be a smooth scheme over a perfect field. Then there is a spectral sequence [27], [55]

$$E_2^{p,q} = H_{mot}^{p-q}(X, \mathbb{Z}(-q)) \Rightarrow K_{-p-q}(X)$$

where $H_{mot}^p(X, \mathbb{Z}(q))$ denotes the motivic cohomology of X as defined in [62, 98]. It is proved in *loc.cit.* that this group is isomorphic to Bloch's higher Chow group $\mathrm{CH}^q(X, 2q - p)$ as defined in [13]. Rationally, the spectral sequence collapses and yields an isomorphism [13, 53]

$$K_n(X)_{\mathbb{Q}} \cong \bigoplus_i \mathrm{CH}^i(X, n)_{\mathbb{Q}}.$$

4.1.5 Milnor K-Theory and the Bloch–Kato Conjecture

Let F be a commutative field. The Milnor K-theory $K_*^M(F)$ of F is the graded ring generated in degree 1 by symbols $\{a\}$ for $a \in F^\times$ a unit in F, modulo the relations $\{ab\} = \{a\} + \{b\}$ and $\{c\} \cdot \{1 - c\} = 0$ for $c \neq 1$. One easily computes $K_0^M(F) = \mathbb{Z}$ and $K_1^M(F) = F^\times$. Since $K_1(F) = F^\times$, since Quillen's K-groups define a graded ring $K_*(F)$ which is commutative in the graded sense, and since the Steinberg relation $\{c\} \cdot \{1 - c\} = 0$ holds in $K_2(F)$, we obtain a morphism $K_*^M(F) \to K_*(F)$ of graded rings extending the isomorphisms in degrees 0 and 1 above. Matsumoto's Theorem says that this map is also an isomorphism in degree 2, that is, the map $K_2^M(F) \to K_2(F)$ is an isomorphism; see [59].

In a similarly way, the ring structure on motivic cohomology yields a map

$$K_n^M(F) \to H_{mot}^n(F, \mathbb{Z}(n))$$

from Milnor K-theory to motivic cohomology. This map is an isomorphism for all $n \in \mathbb{N}$ and any field F by results of Nesterenko–Suslin [68] and Totaro [93].

Let $m = p^\nu$ be a prime power with p different from the characteristic of F, and let F_s be a separable closure of F. Then we have an exact sequence of Galois modules $1 \to \mu_m \to F_s^\times \xrightarrow{m} F_s^\times \to 1$ where μ_m denotes the group of m-th roots of unity. The first boundary map in the associated long exact sequence of étale cohomology groups induces a map $F^\times \to H^1_{et}(F, \mu_m)$. Using the multiplicative structure of étale cohomology, this map extends to a map of graded rings $K^M_*(F) \to H^*_{et}(F, \mu_m^{\otimes *})$ which induces the "norm residue homomorphism"

$$K^M_n(F)/m \to H^n_{et}(F, \mu_m^{\otimes n}).$$

The Bloch–Kato conjecture [11] for the prime p says that this map is an isomorphism for all n. The conjecture for $m = 2^\nu$ was proved by Voevodsky [96], and proofs for $m = p^\nu$ odd have been announced by Rost and Voevodsky.

As a consequence of the Bloch–Kato conjecture, Suslin and Voevodsky show in [88] (see also [33]) that the natural map from motivic cohomology with finite coefficients to étale cohomology is an isomorphism in a certain range:

$$H^i_{mot}(X, \mathbb{Z}/m(j)) \xrightarrow{\cong} H^i_{et}(X, \mu_m^{\otimes j}) \quad \text{for} \quad i \leq j \text{ and } m = p^\nu \tag{22}$$

where X is a smooth scheme over a field F of characteristic $\neq p$. For $i = j+1$, this map is still injective. If char $F = p$, Geisser and Levine show in [32] that

$$H^i_{mot}(F, \mathbb{Z}/p^\nu(j)) = 0 \quad \text{for} \quad i \neq j \quad \text{and} \quad K^M_n(F)/p^\nu \cong K_n(F, \mathbb{Z}/p^\nu).$$

4.1.6 Quillen–Lichtenbaum

The Bloch–Kato conjecture implies the Quillen–Lich–tenbaum conjecture. Let X be a smooth quasi-projective scheme over the complex numbers $\mathbb{C}$. A comparison of the motivic spectral sequence 4.1.4 with the Atiyah–Hirzebruch spectral sequence converging to complex topological K-theory using the isomorphisms (22) and Grothendieck's isomorphism between étale cohomology with finite coefficients and singular cohomology implies an isomorphism

$$K^{alg}_n(X, \mathbb{Z}/m) \xrightarrow{\cong} K^{top}_n(X_{\mathbb{C}}, \mathbb{Z}/m) \quad \text{for} \quad n \geq \dim X - 1$$

between the algebraic K-theory with finite coefficients of X and the topological complex K-theory of the analytic topological space $X_{\mathbb{C}}$ of complex points associated with X; see for instance [71, Theorem 4.1]. For schemes over fields F other than the complex numbers, there is an analogous isomorphism where topological K-theory is replaced with étale K-theory and $\dim X$ with $cd_m X$ provided char $F \nmid m$; see for instance [54, Corollary 13.3].

4.1.7 Computation of $K(\mathbb{Z})$

Modulo the Bloch–Kato conjecture for odd primes (which is announced as proven by Rost and Voevodsky) and the Vandiver conjecture, the K-groups of $\mathbb{Z}$ for $n \geq 2$ are given as follows [51, 60, 103]

$n \bmod 8$	1	2	3	4	5	6	7	0
$K_n(\mathbb{Z})$	$\mathbb{Z} \oplus \mathbb{Z}/2$	$\mathbb{Z}/2c_k$	$\mathbb{Z}/2w_{2k}$	0	$\mathbb{Z}$	$\mathbb{Z}/c_k$	$\mathbb{Z}/w_{2k}$	0

where k is the integer part of $1 + \frac{n}{4}$, and the numbers c_k and w_{2k} are the numerator and denominator of $\frac{B_k}{4k}$ with B_k the k-th Bernoulli number. The B_k's are the coefficients of the power series

$$\frac{t}{e^t - 1} = 1 - \frac{t}{2} + \sum_{k=1}^{\infty} (-1)^{k+1} B_k \frac{t^{2k}}{(2k)!}$$

The Vandiver is still wide open, though it seems to be hard to come by a counter example; see [101, Remark on p. 159] for a discussion of the probability for finding such a counter example. The Vandiver conjecture is only used in the calculation of $K_{2m}(\mathbb{Z})$. It is in fact equivalent to $K_{4m}(\mathbb{Z}) = 0$ for all $m > 0$. In contrast, the calculation of $K_{2m+1}(\mathbb{Z})$ is independent of the Vandiver conjecture but it does use the Bloch–Kato conjecture.

4.1.8 Cdh Descent [19]

The following is due to Häsemeyer [39]. Let k be a field of characteristic 0, and write Sch_k for the category of separated schemes of finite type over k. Let F be a contravariant functor from Sch_k to the category of spectra (or chain complexes of abelian groups). Let $Y \to X \leftarrow X'$ be maps of schemes in Sch_k and $Y' = Y \times_X X'$ be the fibre product. Consider the following square of spectra (or chain complexes)

$$\begin{array}{ccc} F(X) & \longrightarrow & F(Y) \\ \downarrow & & \downarrow \\ F(X') & \longrightarrow & F(Y') \end{array} \qquad (23)$$

obtained by functoriality of F. Suppose that F satisfies the following.

(a) *Nisnevich Descent.* Let $f : X' \to X$ be an étale map and $Y \to X$ be an open immersion. Assume that f induces an isomorphism $f : (X' - Y')_{red} \cong (X - Y)_{red}$. Then the square (23) is homotopy cartesian.
(b) *Invariance under nilpotent extensions.* The map $X_{red} \to X$ induces an equivalence $F(X) \simeq F(X_{red})$.
(c) *Excision for ideals.* Let $f : R \to S$ be a map of commutative rings, $I \subset R$ be an ideal such that $f : I \to f(I)$ is an isomorphism and $f(I)$ is an ideal in S. Consider $X = \operatorname{Spec} R$, $Y = \operatorname{Spec} R/I$, $X' = \operatorname{Spec} S$, $Y' = \operatorname{Spec} S/f(I)$ and the induced maps between them. Then (23) is homotopy cartesian.
(d) *Excision for blow-ups along regularly embedded centers.* Let $Y \subset X$ be a regular embedding of pure codimension. A closed immersion is regular of pure codimension d if, locally, its ideal sheaf is generated by a regular sequence of length d. Let X' be the blow-up of X along Y and $Y' \subset X'$ be the exceptional divisor. Then (23) is homotopy cartesian.

If a functor F satisfies (a)–(d), then the square (23) is homotopy cartesian for any *abstract blow-up square* in Sch_k. A fibre square of schemes as above is called abstract blow-up if $Y \subset X$ is a closed immersion, $X' \to X$ is proper and $X' - Y' \to X - Y$ is an isomorphism.

A functor F is said to satisfy *cdh-descent* if it satisfies Nisnevich descent (see (a) above) and if it sends abstract blow-up squares to homotopy cartesian squares. Thus, a functor for which (a)–(d) hold satisfies cdh-descent for separated schemes of finite type over a field of characteristic 0.

Example (Infinitesimal K-theory [19])

By Remark 3.4.11, Theorems 3.2.27 and 3.5.4, $I\!K$-theory satisfies (a) and (d). But neither (b) nor (c) hold for $I\!K$-theory. The same is true for cyclic homology and its variants since (a) and (d) are formal consequences of the Localization Theorem 3.2.27. Therefore, the homotopy fibre K^{inf} of the Chern character $I\!K \to HN$ from $I\!K$-theory to negative cyclic homology satisfies (a) and (d). By a theorem of Goodwillie [34], K^{inf} satisfies (b), and by a theorem of Cortiñas [22], K^{inf} satisfies (c). Therefore, infinitesimal K-theory K^{inf} satisfies cdh-descent in characteristic 0.

This was used in [19] to prove that $K_i(X) = 0$ for $i < -d$ when X is a d-dimensional scheme essentially of finite type over a field of characteristic 0. Moreover, we have $K_{-d}(X) = H^d_{cdh}(X, \mathbb{Z})$.

Examples

Cdh-descent in characteristic 0 also holds for homotopy K-theory KH [39], periodic cyclic homology HP [19] and stabilized Witt groups [78].

4.1.9 Homotopy Invariance and Vorst's Conjecture

Recall from Sect. 2.3.10 that algebraic K-theory is homotopy invariant for regular rings. More precisely, if R is a commutative regular noetherian ring, then the inclusion of constant polynomials $R \to R[T_1, ..., T_n]$ induces for all $i \in \mathbb{Z}$ an isomorphism on K-groups

$$K_i(R) \xrightarrow{\cong} K_i(R[T_1, ..., T_n]). \tag{24}$$

In fact, the converse – a (special case of a) conjecture of Vorst [97]– is true in the following sense [20]. Let R be (a localization of) a ring of finite type over a field of characteristic zero. If the map (24) is an isomorphism for all $n \in \mathbb{N}$ and all $i \in \mathbb{Z}$ (in fact $i = 1 + \dim R$ suffices), then R is a regular ring.

A Appendix

A.1 Background from Topology

In this appendix we recall the definition of a simplicial set and of a classifying space of a category. Details can be found for instance in [25, 31, 58, 102]. We also recall in Sect. A.1.7, the definition of a homotopy fibration and in Sect. A.1.8, the definition of a spectrum.

A.1.1 Simplicial Sets

Let Δ be the category whose objects are the ordered sets $[n] = \{0, 1, 2, ..., n\}$ for $n \geq 0$. A morphism in this category is an order preserving map of sets. Composition in Δ is composition of maps of sets. For $i = 0, ..., n$ the unique order preserving injective maps $d_i : [n-1] \to [n]$ which leave out i are called *face maps*. For $j = 0, ..., n-1$ the unique order preserving surjective maps $s_j : [n] \to [n-1]$ for which the pre-image of $j \in [n-1]$ contains two elements are called *degeneracy maps*. Every map in Δ is a composition of face and degeneracy maps. Thus, Δ is generated by face and degeneracy maps modulo some relations which the reader can find in the references cited above.

A *simplicial set* is a functor $X : \Delta^{op} \to$ Sets where Sets stands for the category of sets. Thus, for every integer $n \geq 0$ we are given a set X_n, and for every order preserving map $\theta : [n] \to [m]$ we are given a map of sets $\theta^* : X_m \to X_n$ such that $(\theta \circ \sigma)^* = (\sigma)^* \circ (\theta)^*$. Since Δ is generated by face and degeneracy maps, it suffices to specify θ^* for face and degeneracy maps and to check the relations alluded to above. A map of simplicial sets $X \to Y$ is a natural transformation of functors.

A *cosimplicial space* is a functor $\Delta \to$ Top where Top stands for the category of compactly generated Hausdorff topological spaces. A Hausdorff topological space is compactly generated if a subset is closed iff its intersection with every compact subset is closed in that compact subset. Every compact Hausdorff space and every CW-complex is compactly generated. For details, see [61, VIII.8], [104, I.4]. The standard cosimplicial space is the functor $\Delta_* : \Delta \to$ Top where

$$\Delta_n = \{(t_0, ..., t_n) \in \mathbb{R}^n \mid t_i \geq 0, \; t_0 + \cdots t_n = 1\} \subset \mathbb{R}^n$$

is equipped with the subspace topology coming from $\mathbb{R}^n$.

An order preserving map $\theta : [n] \to [m]$ defines a continuous map

$$\theta_* : \Delta_n \to \Delta_m : (s_0, ..., s_n) \mapsto (t_0, ..., t_m) \quad \text{with} \quad t_i = \sum_{\theta(j)=i} s_j$$

such that $(\theta \circ \sigma)_* = \theta_* \circ \sigma_*$. The space Δ_n is homeomorphic to the usual n-dimensional ball with boundary $\partial \Delta_n = \bigcup_{0 \leq i \leq n} (d_i)_* \Delta_{n-1} \subset \Delta_n$ homeomorphic to the $n-1$-dimensional sphere.

The *topological realization* of a simplicial set X is the quotient topological space

$$|X| = \bigsqcup_{j \geq 0} X_j \times \Delta_j / \sim$$

where the equivalence relation $\sim$ is generated by $(\theta^* x, t) = (x, \theta_* t)$ for $x \in X_j$, $t \in \Delta_i$ and $\theta : [i] \to [j]$. A simplex $x \in X_n$ is called *non-degenerate* if $x \notin s_j^* X_{n-1}$ for all $j = 0, ... n - 1$. Write $X_n^{nd} \subset X_n$ for the set of non-degenerate n-simplices. Let $|X|_n \subset |X|$ be the image of $\bigsqcup_{n \geq j \geq 0} X_j \times \Delta_j$ in $|X|$. Note that $|X|_0 = X_0$. One checks that the square

$$
\begin{array}{ccc}
X_n^{nd} \times \partial \Delta_n & \hookrightarrow & X_n^{nd} \times \Delta_n \\
\downarrow & & \downarrow \\
|X|_{n-1} & \hookrightarrow & |X|_n
\end{array}
$$

is cocartesian. Therefore, the space $|X|_n$ is obtained from $|X|_{n-1}$ by attaching exactly one n-cell Δ_n along its boundary $\partial \Delta_n$ for each non-degenerate n-simplex in X. In particular, $|X| = \bigcup_{n \geq 0} |X|_n$ has the structure of a CW-complex.

If X and Y are simplicial sets, the product simplicial set $X \times Y$ has n-simplices $X_n \times Y_n$ with structure maps given by $\theta^*(x, y) = (\theta^* x, \theta^* y)$. A proof of the following proposition can be found in [25, Proposition 4.3.15].

A.1.2 Proposition

For simplicial sets X and Y the projection maps $X \times Y \to X$ and $X \times Y \to Y$ induce a map of topological spaces $|X \times Y| \to |X| \times |Y|$ which is a homeomorphism provided the cartesian product $|X| \times |Y|$ is taken in the category of compactly generated Hausdorff topological spaces.

A.1.3 The Classifying Space of a Category

Consider the ordered set $[n]$ as a category whose objects are the integers $0, 1, ..., n$. There is a unique map $i \to j$ if $i \le j$. Then a functor $[n] \to [m]$ is nothing else than an order preserving map. Thus, we can consider Δ as the category whose objects are the categories $[n]$ for $n \ge 0$, and where the morphisms in Δ are the functors $[n] \to [m]$.

Let $\mathscr{C}$ be a small category. Its nerve is the simplicial set $N_*\mathscr{C}$ whose n-simplices $N_n\mathscr{C}$ are the functors $[n] \to \mathscr{C}$. A functor $\theta : [n] \to [m]$ defines a map $N_m\mathscr{C} \to N_n\mathscr{C}$ given by $F \mapsto F \circ \theta$. We have $(\theta \circ \sigma)^* = (\sigma)^* \circ (\theta)^*$ and $N_*\mathscr{C}$ is indeed a simplicial set. An n-simplex in $N_*\mathscr{C}$, that is, a functor $[n] \to \mathscr{C}$, is nothing else than a string of composable arrows

$$C_0 \xrightarrow{f_0} C_1 \xrightarrow{f_1} \cdots \xrightarrow{f_{n-1}} C_n \qquad (25)$$

in $\mathscr{C}$. The face map d_i^* deletes the object C_i and, if $i \ne 0, n$, it composes the maps f_{i-1} and f_i. The degeneracy map s_i doubles C_i and it inserts the identity map 1_{C_i}. In particular, the n-simplex (25) is non-degenerate iff none of the maps f_i is the identity map for $i = 0, ..., n-1$.

The *classifying space* $B\mathscr{C}$ of a small category $\mathscr{C}$ is the topological realization

$$B\mathscr{C} = |N_*\mathscr{C}|$$

of the nerve simplicial set $N_*\mathscr{C}$ of $\mathscr{C}$. Any functor $\mathscr{C} \to \mathscr{C}'$ induces maps $N_*\mathscr{C} \to N_*\mathscr{C}'$ and $B\mathscr{C} \to B\mathscr{C}'$ on associated nerves and classifying spaces.

The classifying space construction commutes with products. This is because a functor $[n] \to \mathscr{C} \times \mathscr{C}'$ is the same as a pair of functors $[n] \to \mathscr{C}, [n] \to \mathscr{C}'$. Therefore, we have $N_*(\mathscr{C} \times \mathscr{C}') = N_*\mathscr{C} \times N_*\mathscr{C}'$ and $B(\mathscr{C} \times \mathscr{C}') = B\mathscr{C} \times B\mathscr{C}'$ by Proposition A.1.2.

A.1.4 Example $B[1]$

The nerve of the category $[1]$ has two non-degenerate 0-simplices, namely the objects 0 and 1. It has exactly one non-degenerate 1-simplex, namely the map $0 \to 1$. All other simplices are degenerate. Thus, the classifying space $B[1]$ of $[1]$ is obtained from the two point set $\{0, 1\}$ by attaching a 1-cell Δ_1 along its boundary $\partial\Delta_1$. The attachment is such that the two points of $\partial\Delta_1$ are identified with the two points $\{0, 1\}$. We see that $B[1]$ is homeomorphic to the usual interval $\Delta_1 \cong [0, 1]$.

A.1.5 Example BG

For a group G, we let $\underline{G}$ be the category with one object $*$ and where $\mathrm{Hom}(*, *) = G$. Then $\pi_i B\underline{G} = 0$ for $i \ne 1$ and $\pi_1 B\underline{G} = G$ where the isomorphism $G \to \pi_1 B\underline{G}$ sends an element $g \in G$ to the loop l_g represented by the morphism $g : * \to *$. For details, see for instance [102, Exercise 8.2.4, Example 8.3.3].

A.1.6 Lemma

A natural transformation $\eta : F_0 \to F_1$ between functors $F_0, F_1 : \mathscr{C} \to \mathscr{C}'$ induces a homotopy $BF_0 \simeq BF_1$ between the associated maps on classifying spaces $BF_0, BF_1 : B\mathscr{C} \to B\mathscr{C}'$. In particular, an equivalence of categories $\mathscr{C} \to \mathscr{C}'$ induces a homotopy equivalence $B\mathscr{C} \to B\mathscr{C}'$.

Proof:

A natural transformation $\eta : F_0 \to F_1$ defines a functor $H : [1] \times \mathscr{C} \to \mathscr{C}'$ which sends the object (i, X) to $F_i(X)$ where $i = 0, 1$ and $X \in \mathscr{C}$. There are two types of morphisms in $[1] \times \mathscr{C}$, namely (id_i, f) and $(0 \to 1, f)$ where $i = 0, 1$ and $f : X \to Y$ is a map in $\mathscr{C}$. They are sent to $F_i(f)$ for $i = 0, 1$ and to $\eta_Y F_0(f) = F_1(f) \eta_X$, respectively. It is easy to check that H is indeed a functor. Now, H induces a map $[0, 1] \times B\mathscr{C} = B[1] \times B\mathscr{C} = B([1] \times \mathscr{C}) \to B\mathscr{C}'$ on classifying spaces whose restrictions to $\{0\} \times B\mathscr{C}$ and $\{1\} \times B\mathscr{C}$ are BF_0 and BF_1. Thus, BF_0 and BF_1 are homotopic maps.

If $F : \mathscr{C} \to \mathscr{C}'$ is an equivalence of categories, then there are a functor $G : \mathscr{C}' \to \mathscr{C}$ and natural isomorphisms $FG \cong 1$ and $1 \cong GF$. Thus, the map $BG : B\mathscr{C}' \to B\mathscr{C}$ is a homotopy inverse of BF. $\qquad\square$

A.1.7 Homotopy Fibres and Homotopy Fibrations

Let $g : Y \to Z$ be a map of pointed topological spaces. The homotopy fibre $F(g)$ of g is the pointed topological space

$$F(g) = \{(\gamma, y) | \ \gamma : [0, 1] \to Z \ \text{s.t.} \ \gamma(0) = *, \gamma(1) = g(y)\} \subset Z^{[0,1]} \times Y$$

with base-point the pair $(*, *)$ where the first $*$ is the constant path $t \mapsto *$ for $t \in [0, 1]$. There is a continuous map of pointed spaces $F(g) \to Y$ given by $(\gamma, y) \mapsto y$ which fits into a natural long exact sequence of homotopy groups [104, Corollary IV.8.9]

$$\cdots \to \pi_{i+1} Z \to \pi_i F(g) \to \pi_i Y \to \pi_i Z \to \pi_{i-1} F(g) \to \cdots \qquad (26)$$

ending in $\pi_0 Y \to \pi_0 Z$. For more details, see [104, Chap. I.7].

A sequence of pointed spaces $X \xrightarrow{f} Y \xrightarrow{g} Z$ such that the composition is the constant map to the base-point of Z is called *homotopy fibration* if the natural map $X \to F(g)$ given by $x \mapsto (*, f(x))$ is a homotopy equivalence. In this case, there is a long exact sequence of homotopy groups as in (26) with X in place of $F(g)$.

A.1.8 Spectra and Homotopy Cartesian Squares of Spectra

A *spectrum* is a sequence $E_0, E_1, E_2, \ldots$ of pointed topological spaces together with pointed maps $\sigma_i : E_i \to \Omega E_{i+1}$ called *bonding maps* or *structure maps*. The spectrum (E, σ) is called Ω-*spectrum* if the bonding maps σ_i are homotopy equivalences for all $i \in \mathbb{N}$. For $i \in \mathbb{Z}$, the homotopy group $\pi_i E$ of the spectrum (E, σ) is the colimit

$$\pi_i E = \text{colim}(\pi_{i+l} \Omega^{k-l} E_k \xrightarrow{\sigma} \pi_{i+l} \Omega^{k-l+1} E_{k+1} \xrightarrow{\sigma} \pi_{i+l} \Omega^{k-l+2} E_{k+2} \to \cdots).$$

This colimit is independent of k and l as long as $i + l \geq 0$ and $k \geq l$. Thus, it also makes sense for $i < 0$. If $(\mathscr{E}, \sigma)$ is an Ω-spectrum, then $\pi_i E = \pi_i E_0$ for $i \geq 0$ and $\pi_i E = \pi_0 E_{-i}$ for $i < 0$.

A map of spectra $f : (E, \sigma) \to (E', \sigma')$ is a sequence of pointed maps $f_i : E_i \to E_i'$ such that $\sigma_i' f_i = (\Omega f_{i+1}) \sigma_i$. The map of spectra is called *equivalence of spectra* if it induces an isomorphism on all homotopy groups π_i for $i \in \mathbb{Z}$. The homotopy fibre $F(f)$ of a map of spectra $f : (E, \sigma) \to (E', \sigma')$ is the sequence of pointed topological spaces $F(f_0), F(f_1), F(f_2), \ldots$ together with bonding maps $F(f_i) \to \Omega F(f_{i+1}) = F(\Omega f_{i+1})$ between the homotopy fibres of f_i and Ωf_{i+1} given by the maps σ_i and σ_i'. Taking a colimit over the exact sequences (26) yields the exact sequence of abelian groups for $i \in \mathbb{Z}$

$$\cdots \to \pi_{i+1} E' \to \pi_i F(f) \to \pi_i E \to \pi_i E' \to \pi_{i-1} F(f) \to \cdots \qquad (27)$$

A sequence of spectra $E'' \to E \xrightarrow{f} E'$ is a *homotopy fibration* if the composition $E'' \to E'$ is (homotopic to) the zero spectrum (the spectrum with all spaces a point), and the induced map $E'' \to F(f)$ is an equivalence of spectra. In this case, we can replace $F(f)$ by E'' in the long exact sequence (27). A commutative square of spectra

$$
\begin{array}{ccc}
E_{00} & \xrightarrow{f_0} & E_{01} \\
{\scriptstyle g_0}\downarrow & & \downarrow{\scriptstyle g_1} \\
E_{10} & \xrightarrow{f_1} & E_{11}
\end{array}
$$

is called *homotopy cartesian* if the induced map $F(f_0) \to F(f_1)$ on horizontal homotopy fibres (or equivalently, the map $F(g_0) \to F(g_1)$ on vertical homotopy fibres) is an equivalence of spectra. From the exact sequence (27) and the equivalence $F(f_0) \xrightarrow{\sim} F(f_1)$, we obtain a long exact sequence of homotopy groups of spectra for $i \in \mathbb{Z}$

$$
\cdots \to \pi_{i+1}(E_{11}) \to \pi_i(E_{00}) \to \pi_i(E_{01}) \oplus \pi_i(E_{10}) \to \pi_i(E_{11}) \to \pi_{i-1}(E_{00}) \to \cdots
$$

For more on spectra, see [1, III], [9, 43, 79].

A.2 Background on Triangulated Categories

Our main references here are [48, 66, 95].

A.2.1 Definition

A *triangulated category* is an additive category $\mathscr{A}$ together with an auto-equivalence[3] $T : \mathscr{A} \to \mathscr{A}$ and a class of sequences

$$
X \xrightarrow{u} Y \xrightarrow{v} Z \xrightarrow{w} TX \tag{28}
$$

of maps in $\mathscr{A}$ called *distinguished triangles*. They are to satisfy the axioms TR1 – TR4 below.

TR1. Every sequence of the form (28) which is isomorphic to a distinguished triangle is a distinguished triangle. For every object A of $\mathscr{A}$, the sequence $A \xrightarrow{1} A \to 0 \to TA$ is a distinguished triangle. Every map $u : X \to Y$ in $\mathscr{A}$ is part of a distinguished triangle (28).

TR2. A sequence (28) is distinguished if and only if $Y \xrightarrow{v} Z \xrightarrow{w} TX \xrightarrow{-Tu} TY$ is a distinguished triangle.

TR3. For any two distinguished triangles $X \xrightarrow{u} Y \xrightarrow{v} Z \xrightarrow{w} TX$ and $X' \xrightarrow{u'} Y' \xrightarrow{v'} Z' \xrightarrow{w'} TX'$ and for any pair of maps $f : X \to X'$ and $g : Y \to Y'$ such that $gu = u'f$ there is a map $h : Z \to Z'$ such that $hv = v'g$ and $(Tf)w = w'h$.

TR4. Octahedron axiom, see [48, 95] and in Sect. A.2.2 below.

In a distinguished triangle (28) the object Z is determined by the map u up to (non-canonical) isomorphism. We call Z "the" *cone* of u.

---

[3] We may sometimes write $A[1]$ instead of TA especially when A is a complex.

A.2.2 Good Maps of Triangles and the Octahedron Axiom

A useful reformulation of the octahedron axiom TR4 (which we haven't stated...) is as follows [66, Definition 1.3.13 and Remark 1.4.7]. Call a map of distinguished triangles

$$
\begin{array}{ccccccc}
A_0 & \xrightarrow{a_0} & A_1 & \xrightarrow{a_1} & A_2 & \xrightarrow{a_2} & TA_0 \\
\downarrow{f_0} & & \downarrow{f_1} & & \downarrow{f_2} & & \downarrow{Ta_0} \\
B_0 & \xrightarrow{b_0} & B_1 & \xrightarrow{b_1} & B_2 & \xrightarrow{b_2} & TB_0
\end{array}
\tag{29}
$$

good if the mapping cone (in the sense of complexes)

$$
B_0 \oplus A_1 \xrightarrow{\begin{pmatrix} b_0 & f_1 \\ 0 & -a_1 \end{pmatrix}} B_1 \oplus A_2 \xrightarrow{\begin{pmatrix} b_1 & f_2 \\ 0 & -a_2 \end{pmatrix}} B_2 \oplus TA_0 \xrightarrow{\begin{pmatrix} b_2 & Tf_0 \\ 0 & -Ta_0 \end{pmatrix}} TB_0 \oplus TA_1
\tag{30}
$$

is a distinguished triangle. The reformulation of the octahedron axiom [63, Theorem 1.8] says that in a triangulated category every commutative diagram

$$
\begin{array}{ccccccc}
A_0 & \xrightarrow{a_0} & A_1 & \xrightarrow{a_1} & A_2 & \xrightarrow{a_2} & TA_0 \\
\downarrow{f_0} & & \downarrow{f_1} & & & & \downarrow{Ta_0} \\
B_0 & \xrightarrow{b_0} & B_1 & \xrightarrow{b_1} & B_2 & \xrightarrow{b_2} & TB_0
\end{array}
$$

in which the rows are distinguished triangles can be completed into a good morphism of distinguished triangles.

We will need the following special case below. If in a good map of distinguished triangles as in (29) the map f_2 is an isomorphism then the triangle

$$
A_0 \xrightarrow{\begin{pmatrix} -f_0 \\ a_0 \end{pmatrix}} B_0 \oplus A_1 \xrightarrow{(b_0\ f_1)} B_1 \xrightarrow{a_2 f_2^{-1} b_1} TA_0
$$

is distinguished. This is because, in case f_2 is an isomorphism, this triangle is a direct factor of the distinguished triangle obtained by rotating via TR2 the distinguished triangle (30). Therefore, it is a distinguished triangle itself [16, Lemma 1.6].

A.2.3 Definition

Let $\mathscr{R}$ and $\mathscr{S}$ be triangulated categories. A *triangle functor* [48, Sect. 8] from $\mathscr{R}$ to $\mathscr{S}$ is a pair (F, φ) where $F : \mathscr{R} \to \mathscr{S}$ is an additive functor and $\varphi : FT \xrightarrow{\cong} TF$ is a natural isomorphism such that for any distinguished triangle (28) in $\mathscr{R}$, the triangle $FX \to FY \to FZ \to TFX$ given by the maps $(Fu, Fv, \varphi_X Fw)$ is distinguished in $\mathscr{S}$. Triangle functors can be composed in the obvious way.

If a triangle functor has an adjoint, then the adjoint can be made into a triangle functor in a canonical way [66, Lemma 5.3.6], [47, 6.7]. In particular, if a triangulated category has infinite sums, then an arbitrary direct sum of distinguished triangles is a distinguished triangle.

A.2.4 Exercise

Let $F : \mathscr{S} \to \mathscr{T}$ be a triangle functor. If the functor is conservative (that is, a map f in $\mathscr{S}$ is an isomorphism iff $F(f)$ is) and full, then F is fully faithful.

A.2.5 Example (The Homotopy Category of an Additive Category)

Let $\mathscr{A}$ be an additive category. We denote by $\mathscr{K}(\mathscr{A})$ the homotopy category of chain complexes in $\mathscr{A}$. Its objects are the chain complexes in $\mathscr{A}$. Maps in $\mathscr{K}(\mathscr{A})$ are chain maps up to chain homotopy. The category $\mathscr{K}(\mathscr{A})$ is a triangulated category where a sequence is a distinguished triangle if it is isomorphic in $\mathscr{K}(\mathscr{A})$ to a cofibre sequence

$$X \xrightarrow{f} Y \xrightarrow{j} C(f) \xrightarrow{q} TX.$$

Here, $C(f)$ is the *mapping cone* of the chain map $f : X \to Y$ which is $C(f)^i = Y^i \oplus X^{i+1}$ in degree i and has differential $d^i = \begin{pmatrix} d_Y & f \\ 0 & -d_X \end{pmatrix}$. The object TX is the *shift* of X which is $(TX)^i = X^{i+1}$ in degree i and has differential $d^i = -d_X^{i+1}$. The maps $j : Y \to C(f)$ and $q : C(f) \to TX$ are the canonical inclusions and projections in each degree.

A.2.6 Calculus of Fractions

Let $\mathscr{C}$ be a category and $w \subset \mathrm{Mor}\,\mathscr{C}$ be a class of morphisms in $\mathscr{C}$. The *localization of $\mathscr{C}$ with respect to w* is the category obtained from $\mathscr{C}$ by formally inverting the morphisms in w. This is a category $\mathscr{C}[w^{-1}]$ together with a functor $\mathscr{C} \to \mathscr{C}[w^{-1}]$ which satisfies the following universal property. For any functor $\mathscr{C} \to \mathscr{D}$ which sends maps in w to isomorphisms, there is a unique functor $\mathscr{C}[w^{-1}] \to \mathscr{D}$ such that the composition $\mathscr{C} \to \mathscr{C}[w^{-1}] \to \mathscr{D}$ is the given functor $\mathscr{C} \to \mathscr{D}$. In general, the category $\mathscr{C}[w^{-1}]$ may or may not exist. It always exists if $\mathscr{C}$ is a small category.

  If the class w satisfies a "calculus of right (or left) fractions", there is an explicit description of $\mathscr{C}[w^{-1}]$ as we shall explain now. A class w of morphisms in a category $\mathscr{C}$ is said to satisfy a *calculus of right fractions* if (a) – (c) below hold.

(a) The class w is closed under composition. The identity morphism 1_X is in w for every object X of $\mathscr{C}$.
(b) For all pairs of maps $u : X \to Y$ and $s : Z \to Y$ such that $s \in w$, there are maps $v : W \to Z$ and $t : W \to X$ such that $t \in w$ and $sv = ut$.
(c) For any three maps $f, g : X \to Y$ and $s : Y \to Z$ such that $s \in w$ and $sf = sg$, there is a map $t : W \to X$ such that $t \in w$ and $ft = gt$.

If the class w satisfies the dual of (a) – (c) then it is said to satisfy a *calculus of left fractions*. If w satisfies both, a calculus of left and right fractions, then w is said to satisfy a *calculus of fractions*.

  If a class w of maps in a category $\mathscr{C}$ satisfies a calculus of right fractions, then the localized category $\mathscr{C}[w^{-1}]$ has the following description. Objects are the same as in $\mathscr{C}$. A map $X \to Y$ in $\mathscr{C}[w^{-1}]$ is an equivalence class of data $X \xleftarrow{s} M \xrightarrow{f} Y$ written as a *right fraction* fs^{-1} where f and s are maps in $\mathscr{C}$ such that $s \in w$. The datum fs^{-1} is equivalent to the datum $X \xleftarrow{t} N \xrightarrow{g} Y$ iff there are map $\bar{s} : P \to N$ and $\bar{t} : P \to M$ such that $\bar{s}$ (or $\bar{t}$) is in w and such that $s\bar{t} = \bar{s}t$ and $f\bar{t} = g\bar{s}$. The composition $(fs^{-1})(gt^{-1})$ is defined as follows. By (b) above, there are maps h and r in $\mathscr{C}$ such that $r \in w$ and $sh = gr$. Then $(fs^{-1})(gt^{-1}) = (fh)(tr)^{-1}$. In this description it is not clear whether $\mathrm{Hom}_{\mathscr{C}[w^{-1}]}(X, Y)$ is actually a set. However, it is a set if $\mathscr{C}$ is a small category. But in general, this issue has to be dealt with separately.

A.2.7 Verdier Quotient

Let $\mathscr{A}$ be a triangulated category and $\mathscr{B} \subset \mathscr{A}$ be a full triangulated subcategory. The class w of maps whose cones are isomorphic to objects in $\mathscr{B}$ satisfies a calculus of fractions. The *Verdier quotient* $\mathscr{A}/\mathscr{B}$ is, by definition, the localized category $\mathscr{A}[w^{-1}]$. It is a triangulated category where a sequence is a distinguished triangle if it is isomorphic to the image of a distinguished triangle of $\mathscr{A}$ under the localization functor $\mathscr{A} \to \mathscr{A}[w^{-1}]$; see [95], [66, Sect. 2]. If $\mathscr{B}' \subset \mathscr{A}$ denotes the full subcategory of those objects which are zero in the Verdier quotient $\mathscr{A}/\mathscr{B}$, then we have $\mathscr{B} \subset \mathscr{B}'$, the category $\mathscr{B}'$ is a triangulated category and every object of $\mathscr{B}'$ is a direct factor of an object of $\mathscr{B}$ [66, 2.1.33].

A.2.8 Exercise

The following exercises are variations on a theme called "Bousfield localization"; see [66, Sect. 9].

(a) Let $L : \mathscr{S} \to \mathscr{T}$ be a triangle functor which has a right adjoint R such that the counit of adjunction $LR \to id$ is an isomorphism. Let $\lambda : \mathscr{R} \subset \mathscr{S}$ be the full subcategory of $\mathscr{S}$ of those objects which are zero in $\mathscr{T}$. Then the sequence $\mathscr{R} \to \mathscr{S} \to \mathscr{T}$ is an exact sequence of triangulated categories (in the sense of Definition 3.1.5). Furthermore, the inclusion $\lambda : \mathscr{R} \subset \mathscr{S}$ has a right adjoint $\rho : \mathscr{S} \to \mathscr{R}$, and the counit and unit of adjunction fit into a functorial distinguished triangle in $\mathscr{S}$

$$\lambda\rho \to 1 \to RL \to \lambda\rho[1].$$

(b) Let $\mathscr{T}$ be a triangulated category, and let $\mathscr{T}_0, \mathscr{T}_1 \subset \mathscr{T}$ be full triangulated subcategories. Assume that $\mathrm{Hom}(A_0, A_1) = 0$ for all objects $A_0 \in \mathscr{T}_0$ and $A_1 \in \mathscr{T}_1$. If $\mathscr{T}$ is generated as a triangulated category by the union of $\mathscr{T}_0$ and $\mathscr{T}_1$, then the composition $\mathscr{T}_1 \subset \mathscr{T} \to \mathscr{T}/\mathscr{T}_0$ is an equivalence. Moreover, an inverse induces a left adjoint $\mathscr{T} \to \mathscr{T}/T_0 \cong \mathscr{T}_1$ to the inclusion $\mathscr{T}_1 \subset \mathscr{T}$.

(c) Let $\mathscr{A} \xrightarrow{\lambda} \mathscr{B} \xrightarrow{L} \mathscr{C}$ be a sequence of triangle functors. Assume that λ and L have right adjoints ρ and R such that the unit $1 \to \rho\lambda$ and counit $LR \to 1$ are isomorphisms. Assume furthermore that for every object B of $\mathscr{B}$ the unit and counit of adjunction extend to a distinguished triangle in $\mathscr{B}$

$$\lambda\rho(B) \to B \to RL(B) \to \lambda\rho(B)[1].$$

Then the sequence of triangulated categories (λ, L) is exact.

A.2.9 Example (The Derived Category of an Abelian Category)

Let $\mathscr{A}$ be an abelian category. Its *unbounded derived category* $D(\mathscr{A})$ is obtained from the category $\mathrm{Ch}\,\mathscr{A}$ of chain complexes in $\mathscr{A}$ by formally inverting the quasi-isomorphisms. Recall that a chain map $f : A \to B$ is a quasi-isomorphism if it induces isomorphisms $H^i(f) : H^iA \to H^iB$ in cohomology for all $i \in \mathbb{Z}$ where for a chain complex (C, d) we have $H^iC = \ker d^i / \mathrm{im}\, d^{i-1}$. Since homotopy equivalences are quasi-isomorphisms, the category $D(\mathscr{A})$ is also obtained from the homotopy category $\mathscr{K}(\mathscr{A})$ by formally inverting the quasi-isomorphisms. Let $\mathscr{K}_{ac}(\mathscr{A}) \subset \mathscr{K}(\mathscr{A})$ be the full subcategory of acyclic chain complexes. This is the category of those chain complexes C for which $H^iC = 0$ for all $i \in \mathbb{Z}$. The inclusion

$\mathcal{K}_{ac}(\mathcal{A}) \subset \mathcal{K}(\mathcal{A})$ is closed under taking cones. Furthermore, a chain complex A is acyclic iff TA is. Therefore, $\mathcal{K}_{ac}(\mathcal{A})$ is a full triangulated subcategory of $\mathcal{K}(\mathcal{A})$. Since a map is a quasi-isomorphism iff its cone is acyclic, we see that the category $D(\mathcal{A})$ is the Verdier quotient $\mathcal{K}(\mathcal{A})/\mathcal{K}_{ac}(\mathcal{A})$. In particular, the category $D(\mathcal{A})$ is a triangulated category (provided it exists, that is, provided it has small homomorphism sets).

There are versions $D^b\mathcal{A}, D^+\mathcal{A}, D^-\mathcal{A}$ of $D\mathcal{A}$ which are obtained from the category of bounded, bounded below, bounded above chain complexes in $\mathcal{A}$ by formally inverting the quasi-isomorphisms. Again, they are the Verdier quotients $\mathcal{K}^{b+-}(\mathcal{A})/\mathcal{K}_{ac}^{b+-}(\mathcal{A})$ of the corresponding homotopy categories by the homotopy category of acyclic chain complexes.

A.2.10 Exercise

Let $\mathcal{A}$ be an abelian category. Show that the obvious triangle functors $D^b\mathcal{A}, D^+\mathcal{A}, D^-\mathcal{A} \rightarrow D\mathcal{A}$ are fully faithful. Hint: Use the existence of the truncation functors $\tau^{\geq n} : D\mathcal{A} \rightarrow D^+\mathcal{A}$ and $\tau^{\leq n} : D\mathcal{A} \rightarrow D^-\mathcal{A}$ which for a complex E are the quotient complex $\tau^{\geq n}E = \cdots 0 \rightarrow \text{coker}\, d^{n-1} \rightarrow E^{n+1} \rightarrow \cdots$ and the subcomplex $\tau^{\leq n}E = \cdots \rightarrow E^{n-1} \rightarrow \ker(d^n) \rightarrow 0 \rightarrow \cdots$ of E; see [8, Exemple 1.3.2].

A.2.11 The Derived Category of a Grothendieck Abelian Category

Recall that a *Grothendieck abelian category* is an abelian category $\mathcal{A}$ in which all set-indexed colimits exist, where filtered colimits are exact and which has a generator. An object U is a generator of $\mathcal{A}$ if for every object X of $\mathcal{A}$ there is a surjection $\oplus_I U \rightarrow X$ with I some index set. A set of objects is called *set of generators* if their direct sum is a generator. The unbounded derived category $D\mathcal{A}$ of a Grothendieck abelian category has small hom sets [102, Remark 10.4.5], [5, 26].

For a Grothendieck abelian category $\mathcal{A}$, the derived category $D\mathcal{A}$ has the following explicit description. Following [87], a complex $I \in \text{Ch}\,\mathcal{A}$ is called $\mathcal{K}$-*injective* if for every map $f : X \rightarrow I$ and every quasi-isomorphism $s : X \rightarrow Y$ there is a unique map (up to homotopy) $g : Y \rightarrow I$ such that $gs = f$ in $\mathcal{K}(\mathcal{A})$. This is equivalent to the requirement that $\text{Hom}_{\mathcal{K}\mathcal{A}}(A, I) = 0$ for all acyclic chain complexes A. For instance, a bounded below chain complex of injective objects in $\mathcal{A}$ is $\mathcal{K}$-injective. But $\mathcal{K}$-injective chain complexes do not need to consist of injective objects (for instance, every contractible chain complex is $\mathcal{K}$-injective), nor does an unbounded chain complex of injective objects need to be $\mathcal{K}$-injective.

In a Grothendieck abelian category, every chain complex has a $\mathcal{K}$-injective resolution [5, 26]. This means that for every chain complex X in $\mathcal{A}$ there is a quasi-isomorphism $X \rightarrow I$ where I is a $\mathcal{K}$-injective complex. Let $\mathcal{K}_{inj}(\mathcal{A}) \subset \mathcal{K}(\mathcal{A})$ be the full subcategory of all $\mathcal{K}$-injective chain complexes. This is a triangulated subcategory. By definition, a quasi-isomorphism $I \xrightarrow{\sim} X$ from a $\mathcal{K}$-injective complex I to an arbitrary complex X always has a retraction up to homotopy. Therefore, the composition of triangle functors $\mathcal{K}_{inj}(\mathcal{A}) \subset \mathcal{K}(\mathcal{A}) \rightarrow D(\mathcal{A})$ is fully faithful. This composition is also essentially surjective because every chain complex in $\mathcal{A}$ has a $\mathcal{K}$-injective resolution. Therefore, the triangle functor $\mathcal{K}_{inj}(\mathcal{A}) \rightarrow D(\mathcal{A})$ is an equivalence.

A.2.12 Right Derived Functors

Let $F : \mathcal{A} \rightarrow \mathcal{B}$ be an additive functor between abelian categories. The functor induces a triangle functor $\mathcal{K}\mathcal{A} \rightarrow \mathcal{K}\mathcal{B}$ between the homotopy categories of unbounded chain complexes in $\mathcal{A}$ and $\mathcal{B}$. Denote by $L_{\mathcal{A}}$ and $L_{\mathcal{B}}$ the localization triangle functors $\mathcal{K}\mathcal{A} \rightarrow D\mathcal{A}$

and $\mathscr{K}\mathscr{B} \to D\mathscr{B}$. Furthermore, denote by $F : \mathscr{K}\mathscr{A} \to D\mathscr{B}$ the composition of $\mathscr{K}\mathscr{A} \to \mathscr{K}\mathscr{B}$ with the localization functor $L_{\mathscr{B}}$. The *right-derived functor of F* is a pair (RF, λ) as in the diagram

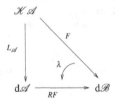

where $RF : D\mathscr{A} \to D\mathscr{B}$ is a triangle functor and $\lambda : F \to RF \circ L_{\mathscr{A}}$ is a natural transformation of triangle functors which has the following universal property. For any pair (G, γ) where $G : D\mathscr{A} \to D\mathscr{B}$ is a triangle functor and $\gamma : F \to G \circ L_{\mathscr{A}}$ is a natural transformation of triangle functors, there is a unique natural transformation of triangle functors $\eta : RF \to G$ such that $\gamma = \eta \circ \lambda$. Of course, the pair (RF, λ) is uniquely determined by the universal property up to isomorphisms of natural transformations of triangle functors.

If $F : \mathscr{A} \to \mathscr{B}$ is any additive functor between Grothendieck abelian categories, then the right derived functor (RF, λ) of F always exists. For $E \in D\mathscr{A}$, it is given by $RF(E) = F(I)$ where $E \to I$ is a $\mathscr{K}$-injective resolution of E. The natural transformation λ at E is the image $FE \to FI$ under F of the resolution map $E \to I$. More generally, one has the following.

A.2.13 Exercise

Let $F : \mathscr{K}\mathscr{A} \to D\mathscr{B}$ be a triangle functor. Assume that there is a triangle endofunctor $G : \mathscr{K}\mathscr{A} \to \mathscr{K}\mathscr{A}$ such that $FG : \mathscr{K}\mathscr{A} \to D\mathscr{B}$ sends quasi-isomorphisms to isomorphisms. Assume furthermore that there is a natural quasi-isomorphism $\lambda : id \xrightarrow{\sim} G$ such that the two natural transformations $G\lambda$ and λ_G of functors $G \to GG$ satisfy $FG\lambda = F\lambda_G$. Then the pair $(FG, F\lambda)$ represents the right derived functor of F.

In the remainder of the subsection, we collect some basic facts about Frobenius exact categories and their triangulated stable categories. They constitute the framework for the complicial exact categories considered in the text.

A.2.14 Frobenius Exact Categories

An object P in an exact category $\mathscr{E}$ is called *projective* if for every deflation $q : Y \twoheadrightarrow Z$ and every map $f : P \to Z$ there is a map $g : P \to Y$ such that $f = qg$. An exact category $\mathscr{E}$ has *enough projectives* if for every object E of $\mathscr{E}$ there is a deflation $P \twoheadrightarrow E$ with P projective. Dually, an object I in $\mathscr{E}$ is called *injective* if for every inflation $j : X \rightarrowtail Y$ and every map $f : X \to I$ there is a map $g : Y \to I$ such that $f = gj$. An exact category $\mathscr{E}$ has *enough injectives* if for every object E of $\mathscr{E}$ there is an inflation $E \rightarrowtail I$ with I injective.

An exact category $\mathscr{E}$ is called *Frobenius exact category* if it has enough injectives and enough projectives, and an object is injective iff it is projective. Call two maps $f, g : X \to Y$ in a Frobenius exact category $\mathscr{E}$ *homotopic* if their difference factors through a projective-injective object. Homotopy is an equivalence relation. The *stable category $\underline{\mathscr{E}}$* of a Frobenius exact category $\mathscr{E}$ is the category whose objects are the objects of $\mathscr{E}$ and whose maps are the homotopy classes of maps in $\mathscr{E}$. The stable category of a Frobenius exact category is a triangulated category as follows. To define the shift $T : \underline{\mathscr{E}} \to \underline{\mathscr{E}}$, we choose for every object X of $\mathscr{E}$ an inflation $X \rightarrowtail I(X)$ into an injective object, and we set $TX = I(X)/X$. Distinguished triangles in the stable category $\underline{\mathscr{E}}$ are those triangles which are isomorphic in $\underline{\mathscr{E}}$ to sequences of the form

$$X \xrightarrow{f} Y \to I(X) \sqcup_X Y \to I(X)/X$$

where $f : X \to Y$ is any map in $\mathscr{E}$. For more details, we refer the reader to [48] and [40, Sect. 9].

A.2.15 Complicial Exact Categories as Frobenius Categories

Recall from Sect. 3.2.1 the bounded complex of free $\mathbb{Z}$-modules $C = \mathbb{Z} \cdot 1_C \oplus \mathbb{Z} \cdot \eta$ where 1_C and η have degrees 0 and -1, respectively. There is a degree-wise split inclusion of chain complexes $i : \mathbb{1} = \mathbb{Z} \rightarrowtail C$ defined by $1 \mapsto 1_C$. Similarly, denote by $P \in \mathrm{Ch}^b(\mathbb{Z})$ the complex $P = Hom(C, \mathbb{1})$ which is concentrated in degrees 0 and 1 where it is a free $\mathbb{Z}$-module of rank 1. There is a degree-wise split surjection $p : P \to \mathbb{1} = \mathbb{Z}$ defined by $f \mapsto f(1_C)$.

Let $\mathscr{E}$ be a complicial exact category. This means that $\mathscr{E}$ comes equipped with an action by the category $\mathrm{Ch}^b(\mathbb{Z})$ of bounded complexes of free $\mathbb{Z}$-modules of finite rank; see Definition 3.2.2. We have natural inflations $i_E = i \otimes 1_E : E \rightarrowtail CE$ and natural deflations $p_E = p \otimes 1_E : PE \twoheadrightarrow E$ for every object E of $\mathscr{E}$. Call an inflation $j : X \rightarrowtail Y$ in $\mathscr{E}$ Frobenius inflation if for every object $U \in \mathscr{E}$ and every map $f : X \to CU$ there is a map $g : Y \to CU$ such that $f = gj$. Similarly, call a deflation $q : Y \twoheadrightarrow Z$ in $\mathscr{E}$ Frobenius deflation if for every object $U \in \mathscr{E}$ and every map $f : CU \to Z$ there is a map $g : CU \to Y$ such that $f = qg$.

A.2.16 Lemma

Let $\mathscr{E}$ be a complicial exact category.

(a) *For every object E of $\mathscr{E}$, the natural inflation $i_X : X \rightarrowtail CX$ is a Frobenius inflation, and the natural deflation $p_X : PX \twoheadrightarrow X$ is a Frobenius deflation.*

(b) *Frobenius inflations (deflations) are closed under composition.*

(c) *Frobenius inflations (deflations) are preserved under push-outs (pull-backs)*

(d) *Split injections (surjections) are Frobenius inflations (deflations).*

(e) *For a conflation $X \rightarrowtail Y \twoheadrightarrow Z$ in $\mathscr{E}$, the map $X \rightarrowtail Y$ is a Frobenius inflation iff the map $Y \twoheadrightarrow Z$ is a Frobenius deflation.*

(f) *The category $\mathscr{E}$ equipped with the Frobenius conflations as defined in Sect. 3.2.5 is a Frobenius exact category. In this exact structure, an object is injective (projective) iff it is a direct factor of an object of the form CU with $U \in \mathscr{E}$.*

Proof:

For (a), we note that C is a commutative dg $\mathbb{Z}$-algebra with unique multiplication $\mu : C \otimes C \to C$ and unit map $i : \mathbb{1} \to C : 1 \mapsto 1_C$. Let $f : X \to CU$ be a map in $\mathscr{E}$. We define the map $f' : CX \to CU$ as the composition $CX \xrightarrow{1 \otimes f} CCU \xrightarrow{\mu \otimes 1} CU$. Then we have $f' i_X = (\mu \otimes 1_U) \circ (1_C \otimes f) \circ (i \otimes 1_X) = (\mu \otimes 1_U) \circ (i \otimes 1_C \otimes 1_U) \circ f = f$ since the composition $C \xrightarrow{i \otimes 1} C \otimes C \xrightarrow{\mu} C$ is the identity. This shows that i_X is a Frobenius inflation. The proof that $PX \to X$ is a Frobenius deflation is similar using the fact that $P = Hom(C, \mathbb{1})$ is a co-algebra. Sections (b), (c) and (d) are clear.

For (e), we note that the map $CX \to TX$ is a Frobenius deflation. This is because this map is isomorphic to $PTX \to TX$ via the isomorphism $C \to Hom(C,T) = PT$ which is adjoint to $C \otimes C \xrightarrow{\mu} C \to T$. Let $X \rightarrowtail Y$ be a Frobenius inflation. By definition, there is a map $Y \to CX$ such that the composition $X \to Y \to CX$ is the canonical Frobenius inflation $i_X : X \rightarrowtail CX$. Passing to quotients, we see that $Y \twoheadrightarrow Z$ is a pull-back of $CX \twoheadrightarrow TX$. Since the latter is a Frobenius deflation, we can apply (c), and we see that $Y \twoheadrightarrow Z$ is a Frobenius deflation as well. The other implication in (e) is dual. For (f), we note that (a) – (e) imply that $\mathscr{E}$ together with the Frobenius conflations is an exact category. By definition, objects of the form CU are injective and projective for the Frobenius exact structure, hence any of its direct factors is injective and projective. For an object I of $\mathscr{E}$ which is injective in the Frobenius exact structure, the Frobenius inflation $I \rightarrowtail CI$ has a retraction since, by the definition of injective objects, the map $1 : I \to I$ to the injective I extends to CI. Therefore, the injective object I is a direct factor of CI. Similarly for projective objects. $\square$

A.3 The Derived Category of Quasi-Coherent Sheaves

A.3.1 Separated Schemes and Their Quasi-Coherent Sheaves

Let X be a quasi-compact and separated scheme. In the category $\mathrm{Qcoh}(X)$ of quasi-coherent O_X-modules, all small colimits exist and filtered colimits are exact (as they can be calculated locally on quasi-compact open subsets). Every quasi-coherent O_X-module is a filtered colimit of its quasi-coherent submodules of finite type [37, 9.4.9]. Therefore, the set of quasi-coherent O_X-modules of finite type forms a set of generators for $\mathrm{Qcoh}(X)$. Hence, the category $\mathrm{Qcoh}(X)$ is a Grothendieck abelian category. In particular, its derived category $D\,\mathrm{Qcoh}(X)$ exists, and it has an explicit description as in Example A.2.9.

A.3.2 Examples of Hom-Sets in $D\,\mathrm{Qcoh}(X)$

For a complex E of quasi-coherent O_X-modules, the set of homomorphisms $\mathrm{Hom}(O_X, E)$ in the triangulated category $D\,\mathrm{Qcoh}(X)$ is given by the formula

$$\mathrm{Hom}(O_X, E) = H^0(Rg_*E)$$

where $g : X \to \mathrm{Spec}\,\mathbb{Z}$ is the structure map of X. We can see this by replacing E with a $\mathscr{K}$-injective resolution $E \xrightarrow{\sim} I$. Then both sides are $H^0(I(X))$.

More generally, for a vector bundle A on X, the homomorphism set $\mathrm{Hom}(A, E)$ in $D\,\mathrm{Qcoh}(X)$ can be calculated as above using the equality

$$\mathrm{Hom}(A, E) = \mathrm{Hom}(O_X, E \otimes A^\vee)$$

where $A^\vee$ is the dual sheaf $Hom(A, O_X)$ of A. Again, we can see this by choosing a K-injective resolution $E \xrightarrow{\sim} I$ of E and noting that $E \otimes A^\vee \xrightarrow{\sim} I \otimes A^\vee$ is a $\mathscr{K}$-injective resolution of $E \otimes A^\vee$ when A (and thus $A^\vee$) is a vector bundle.

A.3.3 The Čech Resolution

Let X be a quasi-compact scheme, and let $\mathcal{U} = \{U_0, ..., U_n\}$ be a finite cover of X by quasi-compact open subsets $U_i \subset X$. For a $k+1$ tuple $\underline{i} = (i_0, ..., i_k)$ such that $0 \leq i_0, ..., i_k \leq n$, write $j_{\underline{i}} : U_{\underline{i}} = U_{i_0} \cap ... \cap U_{i_k} \subset X$ for the open immersion of the intersection of the corresponding U_i's. Let F be a quasi-coherent O_X module. We consider the sheafified Čech complex $\check{C}(\mathcal{U}, F)$ associated with the cover $\mathcal{U}$ of X. In degree k it is the quasi-coherent O_X-module

$$\check{C}(\mathcal{U}, F)_k = \bigoplus_{\underline{i}} j_{\underline{i}*} j_{\underline{i}}^* F$$

where the indexing set is taken over all $k+1$-tuples $\underline{i} = (i_0, ..., i_k)$ such that $0 \leq i_0 < \cdots < i_k \leq n$. The differential $d_k : \check{C}(\mathcal{U}, F)_k \to \check{C}(\mathcal{U}, F)_{k+1}$ for the component $\underline{i} = (i_0, ..., i_{k+1})$ is given by the formula

$$(d_k(x))_{\underline{i}} = \sum_{l=0}^{k+1} (-1)^l \, j_{\underline{i}*} j_{\underline{i}}^* \, x_{(i_0, ..., \hat{i}_l, ..., i_{k+1})}.$$

Note that the complex $\check{C}(\mathcal{U}, F)$ is concentrated in degrees $0, ..., n$.

The units of adjunction $F \to j_{i*} j_i^* F$ define a map $F \to \check{C}(\mathcal{U}, F)_0 = \bigoplus_{i=0}^n j_{i*} j_i^* F$ into the degree zero part of the Čech complex such that $d_0(F) = 0$. Therefore, we obtain a map of complexes of quasi-coherent O_X-modules $F \to \check{C}(\mathcal{U}, F)$. This map is a quasi-isomorphism for any quasi-coherent O_X-module F as can be checked by restricting the map to the open subsets U_i of the cover $\mathcal{U}$ of X.

More generally, if F is a complex, then $\check{C}(\mathcal{U}, F)$ is a bicomplex, and we can consider its total complex $\mathrm{Tot}\check{C}(\mathcal{U}, F)$ which, by a slight abuse of notation, we will still denote by $\check{C}(\mathcal{U}, F)$. The map $F \to \check{C}(\mathcal{U}, F)$ is a map of bicomplexes. Taking total complexes, we obtain a natural quasi-isomorphism of complexes of quasi-coherent O_X-modules

$$\lambda_F : F \xrightarrow{\sim} \check{C}(\mathcal{U}, F). \tag{31}$$

which is called the *Čech resolution of F* associated with the open cover $\mathcal{U}$.

A.3.4 Exercise

Let X be a scheme and $\mathcal{U}$ be a finite open cover of X. Write $\check{C}$ for the functor $F \mapsto \check{C}(\mathcal{U}, F)$ from complexes of quasi-coherent sheaves to itself defined in Sect. A.3.3 above. Show that for any complex F of quasi-coherent sheaves on X, the following two maps $\check{C}(F) \to \check{C}(\check{C}(F))$ are chain homotopic:

$$\check{C}(\lambda_F) \sim \lambda_{\check{C}(F)}.$$

A.3.5 Explicit Description of Rg_*

Let $g : X \to Y$ be a map of quasi-compact schemes such that there is a finite cover $\mathcal{U} = \{U_0, ..., U_n\}$ of X with the property that the restrictions $g_{\underline{i}} : U_{\underline{i}} \to Y$ of g to all finite intersections $U_{\underline{i}} = U_{i_0} \cap ... \cap U_{i_k}$ of the U_i's are affine maps where $\underline{i} = (i_0, ..., i_k)$ and $i_0, ..., i_k \in \{0, ..., n\}$. If X and Y are quasi-compact and separated such a cover always exists. In this case, any cover $\mathcal{U} = \{U_0, ..., U_n\}$ of X by affine open subschemes $U_i \subset X$ such that each U_i maps into an open affine subscheme of Y has this property.

Using the cover $\mathcal{U}$ instead of $\mathcal{K}$-injective resolutions, one can construct the right-derived functor $Rg_* : D\operatorname{Qcoh}(X) \to D\operatorname{Qcoh}(Y)$ of $g_* : \operatorname{Qcoh}(X) \to \operatorname{Qcoh}(Y)$ as follows. By assumption, for every $k+1$-tuple $\underline{i} = (i_0,...,i_k)$, the restriction $g_{\underline{i}} = g \circ j_{\underline{i}} : U_{\underline{i}} \to Y$ of g to $U_{\underline{i}}$ is an affine map. Therefore, the functor

$$g_*\check{C}(\mathcal{U}) : \operatorname{Qcoh}(X) \to \operatorname{Ch}\operatorname{Qcoh}(Y) : F \mapsto g_*\check{C}(\mathcal{U},F)$$

is exact. Taking total complexes, this functor extends to a functor on all complexes

$$g_*\check{C}(\mathcal{U}) : \operatorname{Ch}\operatorname{Qcoh}(X) \to \operatorname{Ch}\operatorname{Qcoh}(Y) : F \mapsto g_*\check{C}(\mathcal{U},F)$$

which preserves quasi-isomorphisms as it is exact and sends acyclics to acyclics. This functor is equipped with a natural quasi-isomorphism given by the Čech resolution

$$\lambda_F : F \xrightarrow{\sim} \check{C}(\mathcal{U},F).$$

By Exercises A.2.13 and A.3.4, the right derived functor Rg_* of g_* is represented by the pair $(g_*\check{C}(\mathcal{U}), g_*\lambda)$.

A.3.6 Lemma

Let $g : X \to Y$ be a map of quasi-compact and separated schemes. Then for any set E_i, $i \in I$, of complexes of quasi-coherent O_X-modules, the following natural map of complexes of O_Y-modules is a quasi-isomorphism

$$\bigoplus_I Rg_*(E_i) \xrightarrow{\sim} Rg_*\left(\bigoplus_I E_i\right).$$

Proof:

This follows from the explicit construction of Rg_* given in Sect. A.3.5, for which the map in the lemma is already an isomorphism in $\operatorname{Ch}\operatorname{Qcoh}(Y)$.   □

A.3.7 Lemma (Base-Change for Open Immersions)

Let $X = U \cup V$ be a quasi-compact separated scheme which is covered by two quasi-compact open subschemes U and V. Denote by $j : U \hookrightarrow X$, $j_V : U \cap V \hookrightarrow V$, $i : V \hookrightarrow X$, $i_U : U \cap V \hookrightarrow U$ the corresponding open immersions. Then for every complex E of quasi-coherent O_U-modules, the natural map

$$i^* \circ Rj_*E \xrightarrow{\cong} Rj_{V*} \circ i_U^* E$$

of complexes of quasi-coherent O_V-modules is an isomorphism in $D\operatorname{Qcoh}(V)$.

Proof

We first make the following remark. For a quasi-coherent O_U-module M, we have the canonical map $i^* j_* M \to j_{V*} i_U^* M$ which is adjoint to the map $j_V^* i^* j_* M = i_U^* j^* j_* M \to i_U^* M$ obtained by applying i_U^* to the counit map $j^* j_* M \to M$. Calculating sections over open subsets, we see that the map $i^* j_* M \to j_{V*} i_U^* M$ is an isomorphism for every quasi-coherent O_U-module M.

For the proof of the lemma, recall that U is quasi-compact. Therefore we can find a finite open affine cover $\mathscr{U} = \{U_0, ..., U_n\}$ of U. For this cover, all inclusions $U_{i_0} \cap ... \cap U_{i_k} \subset X$ are affine maps because X is separated. By Sect. A.3.5 and the remark above, for a complex E of quasi-coherent O_U-modules, we have an isomorphism

$$i^* R j_* E = i^* j_* \check{C}(\mathscr{U}, E) \xrightarrow{\cong} j_{V*} i_U^* \check{C}(\mathscr{U}, E) = j_{V*} \check{C}(\mathscr{U} \cap V, i_U^* E)$$

where $\mathscr{U} \cap V$ is the cover $\{U_0 \cap V, ..., U_n \cap V\}$ of V. Pull-backs of affine maps are affine maps. Hence, all inclusions $(U_{i_0} \cap V) \cap ... \cap (U_{i_k} \cap V) \subset V$ are affine maps. By Sect. A.3.5, the functor $j_{V*} \check{C}(\mathscr{U} \cap V)$ represents $R j_{V*}$. So, the isomorphism above represents an isomorphism of functors $i^* \circ R j_* \xrightarrow{\cong} R j_{V*} \circ i_U^*$.   □

A.3.8 Lemma

Let $X = V_1 \cup V_2$ be a quasi-compact and separated scheme which is covered by two quasi-compact open subschemes $V_1, V_2 \subset X$. Denote by $j_i : V_i \hookrightarrow X$ and $j_{12} : V_1 \cap V_2 \hookrightarrow X$ the corresponding open immersions. Then for $E \in D \operatorname{Qcoh}(X)$, there is a distinguished triangle in $D \operatorname{Qcoh}(X)$

$$E \longrightarrow R j_{1*}(j_1^* E) \oplus R j_{2*}(j_2^* E) \longrightarrow R j_{12*}(j_{12}^* E) \longrightarrow E[1].$$

Proof

Consider the commutative square in $D \operatorname{Qcoh}(X)$

$$
\begin{array}{ccc}
E & \longrightarrow & R j_{1*}(j_1^* E) \\
\downarrow & & \downarrow \\
R j_{2*}(j_2^* E) & \longrightarrow & R j_{12*}(j_{12}^* E)
\end{array}
$$

in which the maps are induced by the unit of adjunction maps $1 \to R j_* \circ j^*$ and the base-change isomorphism in Lemma A.3.7. By Sect. A.2.2, we can complete this square to the right into a good map of distinguished triangles. The map on horizontal cones is an isomorphism when restricted to U_1 (since both cones are zero, by Base-Change A.3.7) and when restricted to U_2 (by the Five Lemma and Base-Change A.3.7). Therefore, the map on horizontal cones is an isomorphism in $D \operatorname{Qcoh}(X)$. Finally, the sequence $E \to R j_{1*}(j_1^* E) \oplus R j_{2*}(j_2^* E) \to R j_{12*}(j_{12}^* E)$ can be completed to a distinguished triangle by the last paragraph in Sect. A.2.2.   □

A.4 Proof of Compact Generation of D_Z Qcoh(X)

In this appendix, we prove Proposition 3.4.6, first for schemes with an ample family of line bundles and then, by a formal induction argument, for general quasi-compact and separated schemes. To summarize, in Lemma A.4.10 we show that D_Z Qcoh(X) is compactly generated and in Lemmas A.4.8 and A.4.9 we show that the compact objects in D_Z Qcoh(X) are precisely those complexes which are isomorphic (in the derived category) to bounded complexes of vector bundles when restricted to the open subsets of an affine open cover of X. Part of the exposition is taken from [83]. When $Z = X$, the reader may also find proofs in [65, Corollary 2.3 and Proposition 2.5] and [18, Theorems 3.1.1 and 3.1.3].

We first recall the usual technique of extending a section of a quasi-coherent sheaf from an open subset cut out by a divisor to the scheme itself. For a proof, see [37, Théorème 9.3.1], [41, Lemma II.5.14].

A.4.1 Lemma

Let X be a quasi-compact and separated scheme, $s \in \Gamma(X, L)$ be a global section of a line bundle L on X, and $X_s = \{x \in X \mid s(x) \neq 0 \in L_x/m_x L_x\}$ be the non-vanishing locus of s. Let F be a quasi-coherent sheaf on X. Then the following hold.

(a) *For every $f \in \Gamma(X_s, F)$, there is an $n \in \mathbb{N}$ such that $f \otimes s^n$ extends to a global section of $F \otimes L^{\otimes n}$.*

(b) *For every $f \in \Gamma(X, F)$ such that $f_{|X_s} = 0$, there is an $n \in \mathbb{N}$ such that $f \otimes s^n = 0$.*

A.4.2 Schemes with an Ample Family of Line Bundles

A scheme X has an *ample family of line bundles* if there is a finite set $L_1, ..., L_n$ of line bundles on X and if there are global sections $s_i \in \Gamma(X, L_i)$ such that the non-vanishing loci $X_{s_i} = \{x \in X \mid s_i(x) \neq 0 \in L_x/m_x L_x\}$ form an open affine cover of X; see [94, Definition 2.1], [85, II 2.2.4]. Note that such an X is necessarily quasi-compact.

Recall that if $f \in \Gamma(X, L)$ is a global section of a line bundle L on a scheme X, then the open inclusion $X_f \subset X$ is an affine map (as can be seen by choosing an open affine cover of X trivializing the line bundle L). As a special case, the open subscheme X_f is affine whenever X is affine. Thus, for the affine cover $X = \bigcup X_{s_i}$ associated with an ample family of line-bundles as above, all finite intersections of the X_{s_i}'s are affine.

Let X be a scheme which has an ample family of line bundles. Then there is a set $\{L_i \mid i \in I\}$ of line bundles on X together with global sections $s_i \in \Gamma(X, L_i)$ such that the set $\{X_{s_i} \mid i \in I\}$ of non-vanishing loci forms an open affine basis for the topology of X [94, 2.1.1 (b)]. If X is affine, this follows from the definition of the Zariski topology. For a general X (with an ample family of line bundles), the sections which give rise to a basis of topology on an open affine X_s can be extended (up to a power of s) to global sections, by Lemma A.4.1. Therefore, every open subset of a basis for X_s is also the non-vanishing locus of a global section of some line bundle on X.

Let X be a scheme which has an ample family of line bundles $L_1, ..., L_n$. Then for every quasi-coherent sheaf F on X, there is a surjective map $M \twoheadrightarrow F$ of quasi-coherent sheaves where M is a (possibly infinite) direct sum of line bundles of the form L_i^k for $i = 1, ..., n$ and $k < 0$. This follows from the definition of an ample family of line bundles and Lemma A.4.1.

A.4.3 Truncated Koszul Complexes

Let X be a quasi-compact and separated scheme, and let L_i, $i = 1, ..., l$ be a finite set of line bundles together with global sections $s_i \in \Gamma(X, L_i)$. Let $U = \bigcup_{i=1}^{l} X_{s_i}$ be the union of the non-vanishing loci X_{s_i} of the s_i's, and $j : U \subset X$ be the corresponding open immersion. The global sections s_i define maps $s_i : O_X \to L_i$ of line-bundles whose O_X-duals are denoted by s_i^{-1} : $L_i^{-1} \to O_X$. We consider the maps s_i^{-1} as (cohomologically graded) chain-complexes with O_X placed in degree 0. For an l-tuple $n = (n_1, ..., n_l)$ of negative integers, the Koszul complex

$$\bigotimes_{i=1}^{l} (L_i^{n_i} \xrightarrow{s^{n_i}} O_X) \tag{32}$$

is acyclic over U. This is because the map $s^{n_i} = (s_i^{-1})^{\otimes |n_i|} : L_i^{n_i} \to O_X$ is an isomorphism when restricted to X_{s_i}, hence the Koszul complex (32) is acyclic (even contractible) over each X_{s_i}. Let $K(s^n)$ denote the bounded complex which is obtained from the Koszul complex (32) by deleting the degree zero part O_X and placing the remaining non-zero part in degrees $-l+1, ..., 0$. The last differential d^{-1} of the Koszul complex defines a map

$$K(s^n) = \left[\bigotimes_{i=1}^{l} (L_i^{n_i} \xrightarrow{s^{n_i}} O_X) \right]^{\leq -1} [-1] \xrightarrow{\varepsilon} O_X$$

of complexes of vector bundles. This map of complexes is a quasi-isomorphism over U, since its cone, the Koszul complex, is acyclic over U. For a complex M of quasi-coherent O_X-modules, we write ε_M for the tensor product map $\varepsilon_M = 1_M \otimes \varepsilon : M \otimes K(s^n) \to M \otimes O_X \cong M$.

The following proposition is a generalization of Lemma A.4.1. It is implicit in the proof of [94, Proposition 5.4.2]. We omit the proof (which is not very difficult, but not very enlightening either). Details can be found in *loc.cit.* and in [83, Lemma 9.6].

A.4.4 Proposition

Let X be a quasi-compact and quasi-separated scheme, and $L_1, ... L_n$ be a finite set of line bundles together with global sections $s_i \in \Gamma(X, L_i)$ for $i = 1, ..., n$. Let $U = \bigcup_{i=1}^{n} X_{s_i}$ be the union of the non-vanishing loci X_{s_i} of the s_i's, and $j : U \subset X$ be the corresponding open immersion. Let M be a complex of quasi-coherent O_X-modules and let A be a bounded complex of vector bundles on X. Then the following hold.

(a) *For every map $f : j^*A \to j^*M$ of complexes of O_U-modules between the restrictions of A and M to U, there is an l-tuple of negative integers $n = (n_1, ..., n_l)$ and a map $\tilde{f} : A \otimes K(s^n) \to M$ of complexes of O_X-modules such that $f \circ j^*(\varepsilon_A) = j^*(\tilde{f})$.*
(b) *For every map $f : A \to M$ of complexes of O_X-modules such that $j^*(f) = 0$, there is an l-tuple of negative integers $n = (n_1, ..., n_l)$ such that $f \circ \varepsilon_A = 0$.*

A.4.5 Lemma

Let X be a quasi-compact scheme, and $s \in \Gamma(X, L)$ be a global section of a line-bundle L such that X_s is affine. Let $N \to E$ be a map of complexes of quasi-coherent O_X-modules such that its restriction to X_s is a quasi-isomorphism. If E is a bounded complex of vector bundles on X, then there is an integer $k > 0$ and a map of complexes $E \otimes L^{-k} \to N$ whose restriction to X_s is a quasi-isomorphism.

Proof:

Write $j : X_s \subset X$ for the open inclusion. Since X_s is affine, say $X_s = \operatorname{Spec} A$, we have an equivalence of categories between quasi-coherent O_{X_s}-modules and A-modules under which the map $j^*N \to j^*E$ becomes a quasi-isomorphism of complexes of A-modules with j^*E a bounded complex of projectives. Any quasi-isomorphism of complexes of A-modules with target a bounded complex of projectives has a retraction up to homotopy. Therefore, the choice of a homotopy right inverse $f : j^*E \to j^*N$ yields a quasi-isomorphisms. By Lemma A.4.1, there is a map of complexes $\tilde{f} : E \otimes L^k \to N$ such that $j^*\tilde{f} = f \cdot s^k$ for some $k < 0$. In particular, $\tilde{f}$ is a quasi-isomorphism when restricted to X_s. $\qquad\square$

Lemma A.4.5 has the following generalization.

A.4.6 Lemma

Let X be a quasi-compact and separated scheme. Let $U = \bigcup_{i=1}^n X_{s_i}$ be the union of affine non-vanishing loci X_{s_i} associated with global sections $s_i \in \Gamma(X, L_i)$ of line bundles L_i on X where $i = 1, \ldots, n$. Denote by $j : U \subset X$ the open immersion. Let $b : M \to B$ be a map of complexes of quasi-coherent O_X-modules such that its restriction j^*b to U is a quasi-isomorphism. If B is a bounded complex of vector bundles on X, then there is a map of complexes $a : A \to M$ such that its restriction to U is a quasi-isomorphism and A is a bounded complex of vector bundles on X.

Proof:

We prove the lemma by induction on n. For $n = 1$, this is Lemma A.4.5. Let $U_0 = \bigcup_{i=1}^{n-1} X_{s_i} \subset X$. By the induction hypothesis, there is map $a_0 : A_0 \to M$ from a bounded complex of vector bundles A_0 such that a_0 is a quasi-isomorphism when restricted to U_0. The induced map $b_0 : C(a_0) \to C(ba_0)$ on mapping cones is a quasi-isomorphism when restricted to U (hence when restricted to X_{s_n}) since b is. Moreover, both cones are acyclic on U_0. Note that we have a distinguished triangle $A_0 \to M \to C(a_0) \to A_0[1]$ in $\mathcal{K} \operatorname{Qcoh}(X)$. Since X_{s_n} is affine and $C(ba_0)$ a bounded complex of vector bundles, Lemma A.4.5 implies the existence of a map $a_1 : A_1 \to C(a_0)$ of complexes with $A_1 = C(ba_0) \otimes L_n^k$ a bounded complex of vector bundles on X which is acyclic when restricted to U_0, and a_1 is a quasi-isomorphism when restricted to X_{s_n}. It follows that a_1 is a quasi-isomorphism when restricted to U. Let A be a complex such that $A \to A_1 \to A_0[1]$ extends to a distinguished triangle in $\mathcal{K} \operatorname{Qcoh}(X)$ where the last map is $A_1 \to C(a_0) \to A_0[1]$. We can choose A to be a bounded complex of vector bundles because A_0 and A_1 are also of this form. Let $a : A \to M$ be a map such that $(a, a_1, 1_{A_0[1]})$ is a map of triangles. By the Five-lemma, the map $a : A \to M$ is a quasi-isomorphism when restricted to U. $\square$

The following proposition is a more precise version of Proposition 3.4.6 in case X has an ample line bundle. For a closed subscheme $Z \subset X$ of a scheme X, denote by $D_Z^b \operatorname{Vect}(X) \subset D^b \operatorname{Vect}(X)$ the full triangulated subcategory of those complexes of vector bundles which are acyclic over $X - Z$. Recall from Sect. 3.4.4 the definition of a "compact object" and of a "compactly generated triangulated category".

A.4.7 Proposition

Let X be a quasi-compact and separated scheme which has an ample family of line bundles $L_1, \ldots, L_n$. Let $Z \subset X$ be a closed subset with quasi-compact open complement $j : U = X - Z \subset X$. Then the following hold.

(a) $D\operatorname{Qcoh}(X)$ *is a compactly generated triangulated category with generating set of compact objects the set*

$$\mathscr{L} = \{\, L_i^{k_i}[l] \mid i = 1, \ldots, n, \ k_i < 0, \ k_i, l \in \mathbb{Z} \,\}.$$

The inclusion $\operatorname{Vect}(X) \subset \operatorname{Qcoh}(X)$ *yields a triangle functor* $D^b\operatorname{Vect}(X) \subset D\operatorname{Qcoh}(X)$ *which is fully faithful and induces an equivalence of* $D^b\operatorname{Vect}(X)$ *with the triangulated subcategory of compact objects in* $D\operatorname{Qcoh}(X)$. *In particular,* $D^b\operatorname{Vect}(X)$ *is generated – as an idempotent complete triangulated category – by the set of line bundles* $L_i^{k_i}$ *where* $i = 1, \ldots, n$, *where* $k_i < 0$ *and* $k_i \in \mathbb{Z}$.

(b) *The following sequence of triangulated categories is exact up to factors*

$$D_Z^b\operatorname{Vect}(X) \to D^b\operatorname{Vect}(X) \to D^b\operatorname{Vect}(U).$$

(c) *The triangulated category* $D_Z\operatorname{Qcoh}(X)$ *is compactly generated, and the inclusion* $\operatorname{Vect}(X) \subset \operatorname{Qcoh}(X)$ *of vector bundles into quasi-coherent sheaves yields a fully faithful triangle functor* $D_Z^b\operatorname{Vect}(X) \subset D_Z\operatorname{Qcoh}(X)$ *which induces an equivalence of* $D_Z^b\operatorname{Vect}(X)$ *with the triangulated subcategory of compact objects in* $D_Z\operatorname{Qcoh}(X)$.

Proof:

For (a), we first note that a vector bundle A on X is compact in $D\operatorname{Qcoh}(X)$. This is because the functor $E \mapsto \operatorname{Hom}(A, E)$ is, in the notation of Sect. A.3.2, the same as the functor $E \mapsto H^0(Rg_*(E \otimes A^\vee))$. The latter functor commutes with infinite direct sums since its component functors $E \mapsto E \otimes A^\vee$, Rg_* and $H^0 : D(\mathbb{Z}\text{-Mod}) \to \mathbb{Z}\text{-Mod}$ have this property. Secondly, recall that the compact objects form a triangulated subcategory. Therefore, every complex of vector bundles is compact in $D\operatorname{Qcoh}(X)$. Next, we will check that the set $\mathscr{L}$ which consists of compact objects generates $D\operatorname{Qcoh}(X)$. For that, let E be a complex such that every map $L \to E$ is zero in $D\operatorname{Qcoh}(X)$ when $L \in \mathscr{L}$. We have to show that $E = 0$ in $D\operatorname{Qcoh}(X)$, that is, that $H^*E = 0$. Since $\mathscr{L}$ is closed under shifts, it suffices to show that the cohomology sheaf $H^0E = \ker(d^0)/\operatorname{im}(d^{-1})$ is zero where d^i is the i-th differential of E. By ampleness of the family $L_1, \ldots, L_n$, we can choose a surjection $M \twoheadrightarrow \ker(d^0)$ of quasi-coherent O_X-modules with M a (possibly infinite) direct sum of line bundles of the form $L_i^{k_i}$ where $i = 1, \ldots, n$ and $k_i < 0$. Composing the inclusion of complexes $\ker(d^0) \to E$ with this surjection defines a map of complexes $M \twoheadrightarrow \ker(d^0) \to E$ which induces a surjective map $M = H^0M \twoheadrightarrow \ker(d^0) \twoheadrightarrow H^0E$ of cohomology sheaves. Since every map $L_i^{k_i} \to E$ is zero in $D\operatorname{Qcoh}(X)$, the induced surjective map $M \twoheadrightarrow H^0E$ is the zero map, hence $H^0E = 0$. Altogether, the arguments above show that $D\operatorname{Qcoh}(X)$ is compactly generated by the set $\mathscr{L}$. Finally, the triangle functors $D^b\operatorname{Vect}(X) \subset D^b\operatorname{Qcoh}(X) \subset D\operatorname{Qcoh}(X)$ are fully faithful. The first by the dual of Sect. 3.1.7 (b) and the second by Sect. A.2.10. The remaining statements in (a) follow directly from Neeman's Theorem 3.4.5 (a) and from Sect. 3.1.7 (c).

For part (b), denote by U-quis the set of maps of complexes of vector bundles on X which are quasi-isomorphisms when restricted to $U = X - Z$. By construction, the following sequence of triangulated categories is exact

$$D_Z^b\operatorname{Vect}(X) \to D^b\operatorname{Vect}(X) \to \mathscr{T}(\operatorname{Ch}^b\operatorname{Vect}(X), U\text{-quis}),$$

and the triangle functor $\mathscr{T}(\operatorname{Ch}^b\operatorname{Vect}(X), U\text{-quis}) \to D^b\operatorname{Vect}(U)$ is conservative. Using Proposition A.4.4 and Lemma A.4.6 we see that the last triangle functor is full. Any conservative and full triangle functor is fully faithful, by Sect. A.2.4. Hence, the last triangle functor is fully faithful. The restriction to U of an ample family of line bundles on X is an ample family of line bundles on U. Therefore, part (a) shows that the triangle functor is also cofinal. It follows that the sequence in part (b) of the Proposition is exact up to factors.

For part (c), we already know that the functor $D_Z^b \operatorname{Vect}(X) \to D_Z \operatorname{Qcoh}(X)$ is fully faithful since both categories are full subcategories of $D \operatorname{Qcoh}(X)$. By part (a), every object in $D_Z^b \operatorname{Vect}(X)$ is compact in $D \operatorname{Qcoh}(X)$. Since the inclusion $D_Z \operatorname{Qcoh}(X) \subset D \operatorname{Qcoh}(X)$ commutes with infinite sums, the objects of $D_Z^b \operatorname{Vect}(X)$ are also compact in $D_Z \operatorname{Qcoh}(X)$. Let $\mathscr{S} \subset D_Z \operatorname{Qcoh}(X)$ be the smallest full triangulated subcategory closed under arbitrary coproducts in $D_Z \operatorname{Qcoh}(X)$ which contains $D_Z^b \operatorname{Vect}(X)$. Then $\mathscr{S}$ is compactly generated with category of compact objects $D_Z^b \operatorname{Vect}(X)$. By Neeman's Theorem 3.4.5 (b), the triangulated category $D \operatorname{Qcoh}(X)/\mathscr{S}$ is compactly generated. It has as category of compact objects the idempotent completion of $D^b \operatorname{Vect}(X)/D_Z^b \operatorname{Vect}(X)$. By part (b), this category is $D^b \operatorname{Vect}(U)$. The functor $D \operatorname{Qcoh}(X)/\mathscr{S} \to D \operatorname{Qcoh}(U)$ preserves coproducts and compact objects, and it induces an equivalence of categories of compact objects. Any triangle functor between compactly generated triangulated categories which commutes with coproducts and which induces an equivalence on compact objects is an equivalence. Therefore, the triangle functor $D \operatorname{Qcoh}(X)/\mathscr{S} \to D \operatorname{Qcoh}(U)$ is an equivalence. It follows that $\mathscr{S} = D_Z \operatorname{Qcoh}(X)$.   □

For the remainder of the subsection, write $D_X(A, F)$ for maps in $D \operatorname{Qcoh}(X)$ from A to F, and similarly for U, V and $U \cap V$ in place of X.

A.4.8 Lemma

Let $X = U \cup V$ be a quasi-compact and separated scheme covered by two quasi-compact open subschemes U, V. Then a complex $A \in D \operatorname{Qcoh}(X)$ is compact iff $A_{|U} \in D \operatorname{Qcoh}(U)$ and $A_{|V} \in D \operatorname{Qcoh}(V)$ are compact.

Proof:

Write $j : U \hookrightarrow X$ for the open immersion. Let $A \in D \operatorname{Qcoh}(X)$ be a compact object, and let $F_i \in D \operatorname{Qcoh}(U)$ be a set of complexes on U where $i \in I$. In the sequence of equations

$$
\begin{aligned}
D_U(j^*A, \oplus_I F_i) &= D_X(A, Rj_* \oplus_I F_i) \\
&= D_X(A, \oplus_I Rj_* F_i) \\
&= \oplus_I D_X(A, Rj_* F_i) \\
&= \oplus_I D_U(j^*A, F_i),
\end{aligned}
$$

the first and last are justified by adjointness of j^* and Rj_*, the second by Sect. A.3.6, and the third by compactness of A. This shows that $A_{|U}$ is compact. The same argument also shows that $A_{|V}$ is compact.

For the other direction, assume that $A_{|U}$ and $A_{|V}$ are compact. Then $A_{|U \cap V}$ is also compact, by the argument above. Let $F_i \in D \operatorname{Qcoh}(X)$ be a set of complexes of quasi-coherent sheaves on X where $i \in I$. We have

$$D_X(A, Rj_* j^* \oplus_I F_i) = D_U(j^*A, j^* \oplus_I F_i) = D_U(j^*A, \oplus_I j^* F_i), \qquad (*)$$

by adjointness of j^* and Rj_* and the fact that j^* commutes with infinite sums (as it is a left adjoint). Similarly, for V and $U \cap V$ in place of U. For every $i \in I$, Lemma A.3.8 provides us with a distinguished triangle

$$F_i \longrightarrow Rj_{1*}(j_1^* F_i) \oplus Rj_{2*}(j_2^* F_i) \longrightarrow Rj_{12*}(j_{12}^* F_i) \longrightarrow F_i[1] \qquad (**)$$

where $j_1 : U \subset X$, $j_2 : V \subset X$ and $j_{12} : U \cap V \subset X$ are the corresponding open immersions. Taking direct sum, we obtain a distinguished triangle

$$\oplus_I F_i \to \oplus_I Rj_{1*}(j_1^* F_i) \oplus \oplus_I Rj_{2*}(j_2^* F_i) \to \oplus_I Rj_{12*}(j_{12}^* F_i) \to \oplus_I F_i[1]$$

which receives a canonical map from $(**)$. Using Lemma A.3.6, we have a canonical isomorphism $\oplus_I Rg_*(g^* F_i) \overset{\cong}{\to} Rg_*(g^* \oplus_I F_i)$ for $g = j_1, j_2, j_{12}$, and the last distinguished triangle becomes

$$\oplus_I F_i \to Rj_{1*}(j_1^* \oplus_I F_i) \oplus Rj_{2*}(j_2^* \oplus_I F_i) \to Rj_{12*}(j_{12}^* \oplus_I F_i) \to \oplus_I F_i[1].$$

Applying the functor $D_X(A, \ )$ to the last triangle, the triangles $(**)$ and the natural map from $(**)$ to the last triangle, we obtain a map of long exact sequences of abelian groups. In view of the identification $(*)$ above, this is the commutative diagram

$$
\begin{array}{ccccccc}
\cdots \to & \oplus_I D_X(A, F_i) & \to & \oplus_I D_U(A, F_i) \oplus \oplus_I D_V(A, F_i) & \to & \oplus_I D_{U \cap V}(A, F_i) & \to \cdots \\
& \downarrow & & \downarrow \cong & & \downarrow \cong & \\
\cdots \to & D_X(A, \oplus_I F_i) & \to & D_U(A, \oplus_I F_i) \oplus D_V(A, \oplus_I F_i) & \to & D_{U \cap V}(A, \oplus_I F_i) & \to \cdots
\end{array}
$$

where we wrote $D_U(A, F_i)$ in place of $D_U(A_{|U}, F_{i|U})$, similarily for V and $U \cap V$. All but every third vertical map in the diagram is an isomorphism, by compactness of $A_{|U}, A_{|V}$ and $A_{|U \cap V}$. By the Five Lemma, the remaining vertical maps are also isomorphisms. Hence, A is compact. $\qquad \square$

Let $j : U \subset X$ be an open immersion of quasi-compact and separated schemes with closed complement $Z = X - U$. Recall that $j^* : D \operatorname{Qcoh}(X) \to D \operatorname{Qcoh}(U)$ has a right adjoint Rj_* such that the counit of adjunction $j^* Rj_* \to 1$ is an isomorphism. By Exercise A.2.8 (a), the inclusion $J : D_Z \operatorname{Qcoh}(X) \subset D \operatorname{Qcoh}(X)$ has a right adjoint which we denote by $R : D \operatorname{Qcoh}(X) \to D_Z \operatorname{Qcoh}(X)$. It is part of a functorial distinguished triangle

$$JR(E) \to E \to Rj_* j^* E \to JR(E)[1]. \qquad (33)$$

A.4.9 Lemma

Let X be a quasi-compact and separated scheme, $Z \subset X$ a closed subset with quasi-compact open complement $j : U = X - Z \subset X$. Then an object $A \in D_Z \operatorname{Qcoh}(X)$ is compact in $D_Z \operatorname{Qcoh}(X)$ iff it is compact in $D \operatorname{Qcoh}(X)$.

Proof:

Let A be an object of $D_Z \operatorname{Qcoh}(X)$. If A is compact in $D \operatorname{Qcoh}(X)$ then A is also compact in $D_Z \operatorname{Qcoh}(X)$ because the inclusion $D_Z \operatorname{Qcoh}(X) \subset D \operatorname{Qcoh}(X)$ commutes with infinite coproducts.

For an object $B \in D_Z \mathrm{Qcoh}(X)$, we have

$$D_X(B, Rj_* j^* E) = D_U(j^* B, j^* E) = 0.$$

Therefore, the long exact sequence of hom-sets associated with the distinguished triangle (33) yields an isomorphism $D_X(B, JRE) \cong D_X(B, E)$. Since j^* and Rj_* commute with infinite coproducts, the distinguished triangle (33) shows that $IR : D\,\mathrm{Qcoh}(X) \to D\,\mathrm{Qcoh}(X)$ also commutes with infinite coproducts. Let A be a compact object of $D_Z \mathrm{Qcoh}(X)$, and let F_i be a set of objects in $D\,\mathrm{Qcoh}(X)$ where $i \in I$. Then

$$D_X(JA, \oplus_I F_i) = D_X(JA, JR \oplus_I F_i) = D_X(JA, \oplus_I JRF_i) = D_X(JA, J \oplus_I RF_i)$$
$$= D_Z \mathrm{Qcoh}(X)(A, \oplus_I RF_i) = \oplus_I D_Z \mathrm{Qcoh}(X)(A, RF_i)$$
$$= \oplus_I D_X(JA, F_i).$$

Thus, A is also compact in $D\,\mathrm{Qcoh}(X)$. □

A.4.10 Lemma

Let X be a quasi-compact and separated scheme, $Z \subset X$ a closed subset with quasi-compact open complement $X - Z \subset X$. Then the triangulated category $D_Z \mathrm{Qcoh}(X)$ is compactly generated.

Proof:

The lemma is true when X has an ample family of line bundles, by Proposition A.4.7. In particular, it is true for affine schemes and their quasi-compact open subschemes. The proof for general quasi-compact and separated X is by induction on the number of elements in a finite cover of X by open subschemes which have an ample family of line bundles. We only need to prove the induction step. Assume $X = U \cup V$ is covered by two open subschemes U and V such that the lemma holds for U, V and $U \cap V$ in place of X. Denote by i, $\bar{i}$, j, and $\bar{j}$ the open immersions $V \hookrightarrow X$, $U \cap V \hookrightarrow U$, $U \hookrightarrow X$, and $U \cap V \hookrightarrow V$, respectively.

Consider the diagram of triangulated categories

$$
\begin{array}{ccccc}
D_{Z-U}\mathrm{Qcoh}(X) & \xrightarrow{\ J\ } & D_Z\mathrm{Qcoh}(X) & \xrightarrow{\ j^*\ } & D_{Z\cap U}\mathrm{Qcoh}(U) \\
\downarrow{\scriptstyle\simeq} & & \downarrow{\scriptstyle i^*} & & \downarrow{\scriptstyle \bar{i}^*} \\
D_{Z\cap V - U\cap V}\mathrm{Qcoh}(V) & \longrightarrow & D_{Z\cap V}\mathrm{Qcoh}(V) & \xrightarrow{\ \bar{j}^*\ } & D_{Z\cap U\cap V}\mathrm{Qcoh}(U\cap V)
\end{array}
$$

in which the rows are exact, by (the argument in the proof of) Lemma 3.4.3 (a), and the left vertical map is an equivalence, by Lemma 3.4.3 (b).

Let A be a compact object of $D_{Z\cap U}\mathrm{Qcoh}(U)$. We will show that $E = A \oplus A[1]$ is (up to isomorphism) the image j^*C of a compact object C of $D_Z\mathrm{Qcoh}(X)$. By induction hypothesis, the lower row in the diagram is an exact sequence of compactly generated triangulated in which the functors preserve infinite coproducts and compact objects (Lemmas A.4.8 and A.4.9). By Lemmas A.4.8 and A.4.9, $\bar{i}^*A$ is compact. By Neeman's Theorem 3.4.5 (b) and Remark 3.1.14, there is a compact object B of $D_{Z\cap V}\mathrm{Qcoh}(V)$ and an isomorphism $g : \bar{j}^*B \xrightarrow{\cong} \bar{i}^*E$. Define the object C of $D_Z\mathrm{Qcoh}(X)$ to be the third object in the distinguished triangle in $D_Z\mathrm{Qcoh}(X)$

$$C \longrightarrow Ri_*B \oplus Rj_*E \longrightarrow Ri\bar{j}_*(\bar{i}^*E) \longrightarrow C[1]$$

in which the middle map on the summands Ri_*B and Rj_*E are given by the maps g : $(i\bar{j})^*Ri_*B = \bar{j}^*B \to \bar{j}^*E$ and id : $(i\bar{j})^*Rj_*E = (j\bar{i})^*Rj_*E = \bar{i}^*E \to \bar{i}^*E$, in view of the adjunction between $R(i\bar{j})_*$ and $(i\bar{j})^*$. By the Base-Change Lemma A.3.7, we have $j^*C \cong E$ and $i^*C \cong B$. By Lemmas A.4.8 and A.4.9, C is compact. Summarizing, every compact object of $D_{Z\cap U}\mathrm{Qcoh}(U)$ is a direct factor of the image of a compact object of $D_Z\mathrm{Qcoh}(X)$.

To finish the proof that $D_Z\mathrm{Qcoh}(X)$ is compactly generated, let E be an object of $D_Z\mathrm{Qcoh}(X)$ such that every map from a compact object of $D_Z\mathrm{Qcoh}(X)$ to E is trivial. We have to show that $E = 0$. Since compact objects of $D_{Z-U}\mathrm{Qcoh}(X)$ are also compact objects of $D_Z\mathrm{Qcoh}(X)$ (Lemma A.4.9), all maps from compact objects of $D_{Z-U}\mathrm{Qcoh}(X)$ to E vanish. The category $D_{Z-U}\mathrm{Qcoh}(X)$ is compactly generated. This is because it is equivalent to $D_{Z\cap V\cup(V-U)}\mathrm{Qcoh}(V)$ which is compactly generated, by induction hypothesis. Therefore, all maps from all objects of $D_{Z-U}\mathrm{Qcoh}(X)$ to E are trivial. For the right adjoint R of J, we therefore have $R(E) = 0$. The distinguished triangle (33) then shows that the unit of adjunction $E \to Rj_*j^*E$ is an isomorphism. I claim that $j^*E = 0$. For that, it suffices to show that $D_U(A, j^*E) = 0$ for all compact $A \in D_{Z\cap U}\mathrm{Qcoh}(U)$, since $D_{Z\cap U}\mathrm{Qcoh}(U)$ is compactly generated, by induction hypothesis. All such compact A's are direct factors of objects of the form j^*C with $C \in D_Z\mathrm{Qcoh}(X)$ compact. Therefore, it suffices to show that $D_U(j^*C, j^*E) = 0$ for all compact $C \in D_Z\mathrm{Qcoh}(X)$. But $D_U(j^*C, j^*E) = D_X(C, Rj_*j^*E) = D_X(C, E) = 0$. So $j^*E = 0$. In view of the isomorphism $E \cong Rj_*j^*E$, we have $E = 0$.   □

Acknowledgements The author acknowledges support from NSF and MPIM-Bonn.

References

1. Adams, J.F.: Stable homotopy and generalised homology. In: Chicago Lectures in Mathematics. University of Chicago Press, Chicago (1974)
2. Mandell, M.A., Blumberg, A.J.: Localization theorems in topological Hochschild homology and topological cyclic homology. `arXiv:0802.3938v2` (2008)
3. Atiyah, M.F., Macdonald, I.G.: Introduction to commutative algebra. Addison-Wesley Publishing Co., Reading (1969)
4. Alonso Tarrío, L., Jeremías López, A., Lipman, J.: Local homology and cohomology on schemes. Ann. Sci. École Norm. Sup. (4), **30**(1), 1–39 (1997)
5. Alonso Tarrío, L., Jeremías López, A., Souto Salorio, M.J.: Localization in categories of complexes and unbounded resolutions. Can. J. Math. **52**(2), 225–247 (2000)
6. Balmer, P.: Triangular Witt groups. I. The 12-term localization exact sequence. K-Theory **19**(4), 311–363 (2000)
7. Bass, H.: Algebraic K-theory. W.A. Benjamin, New York (1968)
8. Beĭlinson, A.A., Bernstein, J., Deligne, P.: Faisceaux pervers. In: Analysis and topology on singular spaces, I (Luminy, 1981). Astérisque, vol. 100, pp. 5–171. Soc. Math. France, Paris (1982)
9. Bousfield, A.K., Friedlander, E.M.: Homotopy theory of Γ-spaces, spectra, and bisimplicial sets. In: Geometric applications of homotopy theory (Proc. Conf., Evanston, Ill., 1977), II. Lecture Notes in Mathematics, vol. 658, pp. 80–130. Springer, Berlin (1978)
10. Bousfield, A.K., Kan, D.M.: Homotopy limits, completions and localizations. In: Lecture Notes in Mathematics, vol. 304. Springer, Berlin (1972)

11. Bloch, S., Kato, K.: p-adic étale cohomology. Inst. Hautes Études Sci. Publ. Math. (63), 107–152 (1986)

12. Bondal, A.I., Larsen, M., Lunts, V.A.: Grothendieck ring of pretriangulated categories. Int. Math. Res. Not. 29, 1461–1495 (2004)

13. Bloch, S.: Algebraic cycles and higher K-theory. Adv. Math. 61(3), 267–304 (1986)

14. Bökstedt, M., Neeman, A.: Homotopy limits in triangulated categories. Compos. Math. 86(2), 209–234 (1993)

15. Borelli, M.: Divisorial varieties. Pac. J. Math. 13, 375–388 (1963)

16. Balmer, P., Schlichting, M.: Idempotent completion of triangulated categories. J. Algebra 236(2), 819–834 (2001)

17. Buehler, T.: Exact categories. Expositiones Mathematicae, 28(1), 1–69 (2010)

18. Bondal, A., van den Bergh, M.: Generators and representability of functors in commutative and noncommutative geometry. Mosc. Math. J. 3(1), 1–36, 258 (2003)

19. Cortiñas, G., Haesemeyer, C., Schlichting, M., Weibel, C.: Cyclic homology, cdh-cohomology and negative K-theory. Ann. Math. (2) 167(2), 549–573 (2008)

20. Cortiñas, G., Haesemeyer, C., Weibel, C.: K-regularity, cdh-fibrant Hochschild homology, and a conjecture of Vorst. J. Am. Math. Soc. 21(2), 547–561 (2008)

21. Cortiñas, G.: Algebraic v. topological K-theory: a friendly match. In this volume

22. Cortiñas, G.: The obstruction to excision in K-theory and in cyclic homology. Invent. Math. 164(1), 143–173 (2006)

23. Drinfeld, V.: DG quotients of DG categories. J. Algebra 272(2), 643–691 (2004)

24. Friedlander, E.M., Grayson, D.R. (eds.): Handbook of K-theory, vol. 1,2. Springer, Berlin (2005)

25. Fritsch, R., Piccinini, R.A.: Cellular structures in topology. In: Cambridge Studies in Advanced Mathematics, vol. 19. Cambridge University Press, Cambridge (1990)

26. Franke, J.: On the Brown representability theorem for triangulated categories. Topology 40(4), 667–680 (2001)

27. Friedlander, E.M., Suslin, A.: The spectral sequence relating algebraic K-theory to motivic cohomology. Ann. Sci. École Norm. Sup. (4) 35(6), 773–875 (2002)

28. Fulton, W.: Intersection theory. In: Ergebnisse der Mathematik und ihrer Grenzgebiete. 3. Folge. A Series of Modern Surveys in Mathematics [Results in Mathematics and Related Areas. 3rd Series. A Series of Modern Surveys in Mathematics], second edn., vol. 2. Springer, Berlin (1998)

29. Gabriel, P.: Des catégories abéliennes. Bull. Soc. Math. France 90, 323–448 (1962)

30. Geisser, T., Hesselholt, L.: On the vanishing of negative k-groups. arXiv:0811.0652 (2008)

31. Goerss, P.G., Jardine, J.F.: Simplicial homotopy theory. In: Progress in Mathematics, vol. 174. Birkhäuser, Basel (1999)

32. Geisser, T., Levine, M.: The K-theory of fields in characteristic p. Invent. Math. 139(3), 459–493 (2000)

33. Geisser, T., Levine, M.: The Bloch-Kato conjecture and a theorem of Suslin-Voevodsky. J. Reine Angew. Math. 530, 55–103 (2001)

34. Goodwillie, T.G.: Relative algebraic K-theory and cyclic homology. Ann. Math. (2) 124(2), 347–402 (1986)

35. Grayson, D.: Higher algebraic K-theory. II (after Daniel Quillen). In: Algebraic K-theory (Proc. Conf., Northwestern Univ., Evanston, Ill., 1976). Lecture Notes in Mathematics, vol. 551, pp. 217–240. Springer, Berlin (1976)

36. Grayson, D.R.: Localization for flat modules in algebraic K-theory. J. Algebra 61(2), 463–496 (1979)

37. Grothendieck, A.: Éléments de géométrie algébrique. I. Le langage des schémas. Inst. Hautes Études Sci. Publ. Math. **4**, 228 (1960)
38. Grothendieck, A.: Éléments de géométrie algébrique. III. Étude cohomologique des faisceaux cohérents. I. Inst. Hautes Études Sci. Publ. Math. **11**, 167 (1961)
39. Haesemeyer, C.: Descent properties of homotopy *K*-theory. Duke Math. J. **125**(3), 589–620 (2004)
40. Happel, D.: On the derived category of a finite-dimensional algebra. Comment. Math. Helv. **62**(3), 339–389 (1987)
41. Hartshorne, R.: Algebraic geometry. In: Graduate Texts in Mathematics, vol. 52. Springer, New York (1977)
42. Hsiang, W.C.: Geometric applications of algebraic *K*-theory. In: Proceedings of the International Congress of Mathematicians, (Warsaw, 1983), vol. 1,2, pp. 99–118. PWN, Warsaw (1984)
43. Hovey, M., Shipley, B., Smith, J.: Symmetric spectra. J. Am. Math. Soc. **13**(1), 149–208 (2000)
44. Kapranov, M.M.: On the derived categories of coherent sheaves on some homogeneous spaces. Invent. Math. **92**(3), 479–508 (1988)
45. Karoubi, M.: Foncteurs dérivés et *K*-théorie. Catégories filtrées. C. R. Acad. Sci. Paris Sér. A-B **267**, A328–A331 (1968)
46. Keller, B.: Chain complexes and stable categories. Manuscripta Math. **67**(4), 379–417 (1990)
47. Keller, B.: Derived categories and universal problems. Commun. Algebra **19**(3), 699–747 (1991)
48. Keller, B.: Derived categories and their uses. In: Handbook of algebra, vol. 1, pp. 671–701. North-Holland, Amsterdam (1996)
49. Keller, B.: On the cyclic homology of exact categories. J. Pure Appl. Algebra **136**(1), 1–56 (1999)
50. Keller, B.: On differential graded categories. In: International Congress of Mathematicians, vol. II, pp. 151–190. European Mathematical Society, Zürich (2006)
51. Kurihara, M.: Some remarks on conjectures about cyclotomic fields and *K*-groups of **Z**. Compos. Math. **81**(2), 223–236 (1992)
52. Kuznetsov, A.: Derived categories of quadric fibrations and intersections of quadrics. Adv. Math. **218**(5), 1340–1369 (2008)
53. Levine, M.: Bloch's higher Chow groups revisited. In: *K*-theory (Strasbourg, 1992). Astérisque **226**, 10, 235–320 (1994)
54. Levine, M.: *K*-theory and motivic cohomology of schemes. *K*-theory archive preprint 336 (1999)
55. Levine, M.: The homotopy coniveau tower. J. Topol. **1**(1), 217–267 (2008)
56. Lück, W., Reich, H.: The Baum-Connes and the Farrell-Jones conjectures in *K*- and *L*-theory. In: Handbook of *K*-theory, vol. 1,2, pp. 703–842. Springer, Berlin (2005)
57. Matsumura, H.: Commutative ring theory. In: Cambridge Studies in Advanced Mathematics, second edn., vol. 8. Cambridge University Press, Cambridge (1989) Translated from the Japanese by M. Reid.
58. Peter May, J.: Simplicial objects in algebraic topology. In: Van Nostrand Mathematical Studies, vol. 11. D. Van Nostrand Co., Inc., Princeton (1967)
59. Milnor, J.: Introduction to algebraic *K*-theory. In: Annals of Mathematics Studies, vol. 72. Princeton University Press, Princeton (1971)
60. Mitchell, S.A.: Hypercohomology spectra and Thomason's descent theorem. In: Algebraic *K*-theory (Toronto, ON, 1996). Fields Institute Communications, vol. 16 , pp. 221–277. American Mathematical Society, Providence, RI (1997)

61. Mac Lane, S.: Categories for the working mathematician. In: Graduate Texts in Mathematics, second edn., vol. 5. Springer, New York (1998)
62. Mazza, C., Voevodsky, V., Weibel, C.: Lecture notes on motivic cohomology. Clay Mathematics Monographs, vol. 2. American Mathematical Society, Providence, RI (2006)
63. Neeman, A.: Some new axioms for triangulated categories. J. Algebra 139(1), 221–255 (1991)
64. Neeman, A.: The connection between the K-theory localization theorem of Thomason, Trobaugh and Yao and the smashing subcategories of Bousfield and Ravenel. Ann. Sci. École Norm. Sup. (4) 25(5), 547–566 (1992)
65. Neeman, A.: The Grothendieck duality theorem via Bousfield's techniques and Brown representability. J. Am. Math. Soc. 9(1), 205–236 (1996)
66. Neeman, A.: Triangulated categories. In: Annals of Mathematics Studies, vol. 148. Princeton University Press, Princeton, NJ (2001)
67. Neeman, A., Ranicki, A.: Noncommutative localisation in algebraic K-theory. I. Geom. Topol. 8, 1385–1425 (electronic) (2004)
68. Nesterenko, Y.P., Suslin, A.A.: Homology of the general linear group over a local ring, and Milnor's K-theory. Izv. Akad. Nauk SSSR Ser. Mat. 53(1), 121–146 (1989)
69. Panin, I.A.: The equicharacteristic case of the Gersten conjecture. Tr. Mat. Inst. Steklova, 241(Teor. Chisel, Algebra i Algebr. Geom.) pp. 169–178 (2003)
70. Popescu, N.: Abelian categories with applications to rings and modules. London Mathematical Society Monographs, vol. 3. Academic Press, London (1973)
71. Pedrini, C., Weibel, C.: The higher K-theory of complex varieties. K-Theory. 21(4), 367–385 (2000) Special issues dedicated to Daniel Quillen on the occasion of his sixtieth birthday, Part V
72. Quillen, D.: On the cohomology and K-theory of the general linear groups over a finite field. Ann. Math. (2) 96, 552–586 (1972)
73. Quillen, D.: Higher algebraic K-theory. I. In: Algebraic K-theory, I: Higher K-theories (Proc. Conf., Battelle Memorial Inst., Seattle, Wash., 1972). Lecture Notes in Math., vol. 341, pp. 85–147. Springer, Berlin (1973)
74. Ranicki, A.: Noncommutative localization in topology. In: Non-commutative localization in algebra and topology. London Mathematics Society Lecture Note Series, vol. 330, pp. 81–102. Cambridge University Press, Cambridge (2006)
75. Reid, L.: N-dimensional rings with an isolated singular point having nonzero K_{-N}. K-Theory. 1(2), 197–205 (1987)
76. Samokhin, A.: Some remarks on the derived categories of coherent sheaves on homogeneous spaces. J. Lond. Math. Soc. (2) 76(1), 122–134 (2007)
77. Schlichting, M.: Hermitian K-theory, derived equivalences and Karoubi's fundamental theorem. In preparation
78. Schlichting, M.: Witt groups of singular varieties. In preparation.
79. Schwede, S.: Book project about symmetric spectra. www.math.uni-bonn.de/people/schwede
80. Schlichting, M.: A note on K-theory and triangulated categories. Invent. Math. 150(1), 111–116 (2002)
81. Schlichting, M.: Delooping the K-theory of exact categories. Topology 43(5), 1089–1103 (2004)
82. Schlichting, M.: Negative K-theory of derived categories. Math. Z. 253(1), 97–134 (2006)
83. Schlichting, M.: The Mayer-Vietoris principle for Grothendieck-Witt groups of schemes. Invent. Math. 179(2), 349–433 (2010)

84. Serre, J.-P.: Modules projectifs et espaces fibrés à fibre vectorielle. In: Séminaire, P., Dubreil, M.-L., Dubreil-Jacotin et C. Pisot, 1957/58, Fasc. 2, Exposé 23, p. 18. Secrétariat mathématique, Paris (1958)

85. Théorie des intersections et théorème de Riemann-Roch. Séminaire de Géométrie Algébrique du Bois-Marie 1966–1967 (SGA 6). Dirigé par P. Berthelot, A. Grothendieck et L. Illusie. Avec la collaboration de Ferrand, D., Jouanolou, J.P., Jussila, O., Kleiman, S., Raynaud, M. et Serre, J.P. Lecture Notes in Mathematics, vol. 225. Springer, Berlin (1971)

86. Sherman, C.: Group representations and algebraic K-theory. In: Algebraic K-theory, Part I (Oberwolfach, 1980). Lecture Notes in Mathematics, vol. 966, pp. 208–243. Springer, Berlin (1982)

87. Spaltenstein, N.: Resolutions of unbounded complexes. Compos. Math. **65**(2), 121–154 (1988)

88. Suslin, A., Voevodsky, V.: Bloch-Kato conjecture and motivic cohomology with finite coefficients. In: The arithmetic and geometry of algebraic cycles (Banff, AB, 1998). NATO Science Series C: Mathematical and Physical Sciences, vol. 548, pp. 117–189. Kluwer Academic, Dordrecht (2000)

89. Swan, R.G.: K-theory of quadric hypersurfaces. Ann. Math. (2), **122**(1), 113–153 (1985)

90. Thomason, R.W.: Les K-groupes d'un schéma éclaté et une formule d'intersection excédentaire. Invent. Math. **112**(1):195–215 (1993)

91. Thomason, R.W.: The classification of triangulated subcategories. Compos. Math. **105**(1), 1–27 (1997)

92. Toen, B.: Lectures on DG-categories. In this volume

93. Totaro, B.: Milnor K-theory is the simplest part of algebraic K-theory. K-Theory **6**(2), 177–189 (1992)

94. Thomason, R.W., Trobaugh, T.: Higher algebraic K-theory of schemes and of derived categories. In: The Grothendieck Festschrift, Vol. III. Progress in Mathematics, vol. 88, pp. 247–435. Birkhäuser, Boston, MA (1990)

95. Verdier, J.-L.: Des catégories dérivées des catégories abéliennes. Astérisque **239**, pp. xii+253 (1996) With a preface by Illusie, L., edited and with a note by Maltsiniotis, G

96. Voevodsky, V.: Motivic cohomology with $\mathbf{Z}/2$-coefficients. Publ. Math. Inst. Hautes Études Sci. **(98)**, 59–104 (2003)

97. Vorst, T.: Localization of the K-theory of polynomial extensions. Math. Ann. **244**(1), 33–53 (1979) With an appendix by Wilberd van der Kallen

98. Voevodsky, V., Suslin, A., Friedlander, E.M.: In: Cycles, transfers, and motivic homology theories. Annals of Mathematics Studies, vol. 143. Princeton University Press, Princeton, NJ (2000)

99. Waldhausen, F.: Algebraic K-theory of generalized free products. I, II. Ann. Math. (2) **108**(1), 135–204 (1978)

100. Waldhausen, F.: Algebraic K-theory of spaces. In: Algebraic and geometric topology (New Brunswick, N.J., 1983). Lecture Notes in Mathematics, vol. 1126, pp. 318–419. Springer, Berlin (1985)

101. Washington, L.C.: Introduction to cyclotomic fields. Graduate Texts in Mathematics, vol. 83. Springer, New York (1982)

102. Weibel, C.A.: An introduction to homological algebra. Cambridge Studies in Advanced Mathematics, vol. 38. Cambridge University Press, Cambridge (1994)

103. Weibel, C.: Algebraic K-theory of rings of integers in local and global fields. In: Handbook of K-theory, vol. 1,2, pp. 139–190. Springer, Berlin (2005)

104. Whitehead, G.W.: In: Elements of homotopy theory. Graduate Texts in Mathematics, vol. 61. Springer, New York (1978)

Lectures on DG-Categories

Toën Bertrand

Université Montpellier 2, Case Courrier 051,
Place Eugène Bataillon, 34095 MONTPELLIER Cedex France,
btoen@math.univ-montp2.fr

1 Introduction

The purpose of these four lectures is to provide an introduction to the theory of dg-categories.

There are several possible points of view to present the subject, and my choice has been to emphasised its relations with the localization problem (in the sense of category theory). In the same way that the notion of complexes can be introduced for the need of derived functors, dg-categories will be introduced here for the need of a *derived version* of the localization construction. The purpose of the first lecture is precisely to recall the notion of the localization of a category and to try to explain its bad behaviour throught several examples. In the second part of the first lecture I will introduce the notion of dg-categories and quasi-equivalences, and explain how they can be used in order to state a refined version of the notion of localization. The existence and properties of this new localization will be studied in the next lectures.

The second lecture is concerned with reminders about model category theory, and its applications to the study of dg-categories. The first part is a very brief overview of the basic notions and results of the theory, and the second part presents the specific model categories appearing in the context of dg-categories.

Lecture three goes into the heart of the subject and is concerned with the study of the homotopy category of dg-categories. The key result is a description of the set of morphisms in this homotopy category as the set of isomorphism classes of certain objects in a derived category of bi-modules. This result possesses several important consequences, such as the existence of localizations and of derived internal Homs for dg-categories. The very last part of this third lecture presents the notion of triangulated dg-categories, which is a refined (and better) version of the usual notion of triangulated categories.

The last lecture contains a few applications of the general theory explaining how the problems with localization mentioned in the first lecture are solved when working with dg-categories. We start to show that triangulated dg-categories have functorial cones, unlike the case of triangulated categories. We also show that many invariants (such as K-theory, Hochschild homology, ...) are invariant of dg-categories, though

P.F. Baum et al., *Topics in Algebraic and Topological K-Theory*,
Lecture Notes in Mathematics 2008, DOI 10.1007/978-3-642-15708-0_5,
© Springer-Verlag Berlin Heidelberg 2011

it is know that they are not invariant of triangulated categories. We also give a gluing statement, providing a way to glue objects in dg-categories in a situation where it is not possible to glue objects in derived categories. To finish I will present the notion of saturated dg-categories and explain how they can be used in order to define a *secondary K-theory*.

2 Lecture 1: DG-Categories and Localization

The purpose of this first lecture is to explain one motivation for working with dg-categories concerned with the localization construction in category theory (in the sense of Gabriel–Zisman, see below). I will start by presenting some very concrete problems often encountered when using the localization construction. In a second part I will introduce the homotopy category of dg-categories, and propose it as a setting in order to define a better behaved localization construction. This homotopy category of dg-categories will be further studied in the next lectures.

2.1 The Gabriel–Zisman Localization

Let C be a category and S be a subset of the set of morphisms in C^1. A *localization of C with respect to S* is the data of a category $S^{-1}C$ and a functor

$$l : C \longrightarrow S^{-1}C$$

satisfying the following property: for any category D the functor induced by composition with l

$$l^* : \underline{Hom}(S^{-1}C, D) \longrightarrow \underline{Hom}(C, D)$$

is fully faithful and its essential image consists of all functors $f : C \longrightarrow D$ such that $f(s)$ is an isomorphism in D for any $s \in S$ (here $\underline{Hom}(A, B)$ denotes the category of functors from a category A to another category B).

Using the definition it is not difficult to show that if a localization exists then it is unique, up to an equivalence of categories, which is itself unique up to a unique isomorphism. It can also be proved that a localization always exists. One possible proof of the existence of localizations is as follows. Let I be the category with two objects 0 and 1 and a unique morphism $u : 0 \to 1$. In the same way, let $\bar{I}$ be the category with two objects 0 and 1 and with a unique isomorphism $\bar{u} : 0 \to 1$. There exists a natural functor $I \longrightarrow \bar{I}$ sending 0 to 0, 1 to 1 and u to $\bar{u}$. Let now C be a category and S be a set of morphisms in C. For any $s \in S$, with source $x \in C$ and target $y \in C$, we define a functor $i_s : I \longrightarrow C$ sending 0 to x, 1 to y and u to s. We get this way a diagram of categories and functors

---

[1] In these lectures I will not take into account set theory problems, and will do as if all categories were *small*. I warn the ready that, at some point, we will have to consider *non-small* categories, and thus that these set theory problems should be solved somehow. On possible solution is for instance by fixing various Grothendieck universes (see [2, Exp. 1]).

$$\coprod_{s \in S} I \longrightarrow \coprod_s \overline{I}.$$

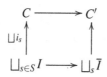

We consider this as a diagram in the category of categories (objects are categories and morphisms are functors), and we form the push-out

$$\coprod_{s \in S} I \longrightarrow \coprod_s \overline{I}$$

It is not hard to show that for any category D the category of functors $\underline{Hom}(C',D)$ is *isomorphic* to the full sub-category of $\underline{Hom}(C,D)$ consisting of all functors sending elements of S to isomorphisms in D. In particular, the induced functor $C \longrightarrow C'$ is a localization in the sense we defined above.

The only non-obvious point with this argument is the fact that the category of categories possesses push-outs and even all kind of limits and colimits. One possible way to see this is by noticing that the category of small categories is monadic over the category of (oriented) graphs, and to use a general result of existence of colimits in monadic categories (see e.g. [8, II-Prop. 7.4]).

In general localizations are extremely difficult to describe in a useful manner, and the existence of localizations does not say much in practice (though it is sometimes useful to know that they exist). The push-out constructions mentioned above can be explicited to give a description of the localization C'. Explicitly, C' has the same objects as C itself. Morphisms between two objects x and y in C' are represented by strings of arrows in C

$$x \longrightarrow x_1 \longleftarrow x_2 \longrightarrow x_3 \longleftarrow \cdots \longleftarrow x_n \longrightarrow y,$$

for which all the arrows going backwards are assumed to be in S. To get the right set of morphisms in C' we need to say when two such strings define the same morphism (see [9, Sect. I.1.1] for details). This description for the localization is rather concrete, however it is most often useless in practice.

The following short list of examples show that localized categories are often encountered and provide interesting categories in general.

Examples:

(a) If all morphisms in S are isomorphisms then the identity functor $C \to C$ is a localization.

(b) If S consists of all morphisms in C, then $S^{-1}C$ is the groupoid completion of C. When C has a unique object with a monoid M of endomorphisms, then $S^{-1}C$ has unique object with the group M^+ as automorphisms (M^+ is the group completion of the monoid M).

(c) Let R be a ring and $C(R)$ be the category of (unbounded) complexes over R. Its objects are families of R-modules $\{E^n\}_{n\in\mathbb{Z}}$ together with maps $d^n : E^n \to E^{n+1}$ such that $d^{n+1}d^n = 0$. Morphisms are simply families of morphisms commuting with the d's. Finally, for $E \in C(R)$, we can define its n-th cohomology by $H^n(E) := Ker(d^n)/Im(d^{n-1})$, which is an R-module. The construction $E \mapsto H^n(E)$ provides a functor H^n from $C(R)$ to R-modules.

A morphism $f : E \longrightarrow F$ in $C(R)$ is called a *quasi-isomorphism* if for all $i \in \mathbb{Z}$ the induced map

$$H^i(f) : H^i(E) \longrightarrow H^i(F)$$

is an isomorphism. We let S be the set of quasi-isomorphisms in $C(R)$. Then $S^{-1}C(R)$ is the *derived category* of R and is denoted by $D(R)$. Understanding the hidden structures of derived categories is one of the main objectives of dg-category theory.

Any R-module M can be considered as a complex concentrated in degree 0, and thus as an object in $D(R)$. More generally, if $n \in \mathbb{Z}$, we can consider the object $M[n]$ which is the complex concentrated in degree $-n$ and with values M. It can be shown that for two R-modules M and N there exists a natural isomorphism

$$Hom_{D(R)}(M,N[n]) \simeq Ext^n(M,N).$$

(d) Let Cat be the category of categories: its objects are categories and its morphisms are functors. We let S be the set of categorical equivalences. The localization category $S^{-1}Cat$ is called the *homotopy category of categories*. It can be shown quite easily that $S^{-1}Cat$ is equivalent to the category whose objetcs are categories and whose morphismes are isomorphism classes of functors (see Exercise 2.1.2).

(e) Let Top be the category of topological spaces and continuous maps. A morphism $f : X \longrightarrow Y$ is called a *weak equivalence* if it induces isomorphisms on all homotopy groups (with respect to all base points). If S denotes the set of weak equivalences then $S^{-1}Top$ is called the *homotopy category of spaces*. It can be shown that $S^{-1}Top$ is equivalent to the category whose objects are CW-complexes and whose morphisms are homotopy classes of continuous maps.

One comment before going on. Let us denote by $Ho(Cat)$ the category $S^{-1}Cat$ considered in example (4) above. Let C be a category and S be a set of morphisms in C. We define a functor

$$F : Ho(Cat) \longrightarrow Set$$

sending a category D to the set of all isomorphism classes of functors $C \longrightarrow D$ sending S to isomorphisms. The functor F is therefore a sub-functor of the functor h^C corepresented by C. Another way to consider localization is by stating that the functor F is corepresentable by an object $S^{-1}C \in Ho(Cat)$. This last point of view is a bit less precise as the original notion of localizations, as the object $S^{-1}C$ satisfies a universal property only on the level of isomorphism classes of functors and not on the level of categories of functors themselves. However, this point of view is often useful and enough in practice.

Exercise 2.1.1 *Let C and D be two categories and S (resp. T) be a set of morphisms in C (resp. in D) containing the identities.*

(a) Prove that the natural functor

$$C \times D \longrightarrow (S^{-1}C) \times (T^{-1}D)$$

is a localization of $C \times D$ with respect to the set $S \times T$. In other words localizations commutes with finite products.

(b) We assume that there exist two functors

$$f : C \longrightarrow D \qquad C \longleftarrow D : g$$

with $f(S) \subset T$ and $g(T) \subset S$. We also assume that there exists two natural transformations $h : fg \Rightarrow id$ and $k : gf \Rightarrow id$ such that for any $x \in C$ (resp. $y \in D$) the morphism $k(y) : g(f(x)) \rightarrow x$ (resp. $h(y) : f(g(y)) \rightarrow y$) is in S (resp. in T). Prove that the induced functors

$$f : S^{-1}C \longrightarrow T^{-1}D \qquad S^{-1}C \longleftarrow T^{-1}D : g$$

are equivalences inverse to each other.

(c) If S consists of all morphisms in C and if C has a final or initial object then $C \longrightarrow *$ is a localization of C with respect to S.

Exercise 2.1.2 *Let Cat be the category of categories and functors, and let [Cat] be the category whose objects are categories and whose morphisms are isomorphism classes of functors (i.e. $Hom_{[Cat]}(C,D)$ is the set of isomorphism classes of objects in $\underline{Hom}(C,D)$). Show that the natural projection*

$$Cat \longrightarrow [Cat]$$

is a localization of Cat along the subset of equivalences of categories (prove directly that it has the correct universal property).

2.2 Bad Behavior of the Gabriel–Zisman Localization

In these lectures we will be mainly interested in localized categories of the type $D(R)$ for some ring R (or some more general object, see lecture 2). I will therefore explain the bad behaviour of the localization using examples of derived categories. However, this bad behaviour is a general fact and also applies to other examples of localized categories.

Though the localization construction is useful to construct interesting new categories, the resulting localized categories are in general badly behaved. Often, the category to be localized has some nice properties, such as the existence of limits and colimits or being abelian, but these properties are lost after localization. Here is a sample of problems often encountered in practice.

(a) The derived category $D(R)$ lacks the standard categorical constructions of limits and colimits. There exists a non-zero morphism $e : \mathbb{Z}/2 \longrightarrow \mathbb{Z}/2[1]$ in $D(\mathbb{Z})$, corresponding to the non-zero element in $Ext^1(\mathbb{Z}/2, \mathbb{Z}/2)$ (recall that $Ext^i(M, N) \simeq [M, N[i]]$, where $N[i]$ is the complex whose only non-zero part is N in degree $-i$, and $[-, -]$ denotes the morphisms in $D(R)$). Suppose that the morphism e has a kernel, i.e. that a fiber product

$$
\begin{array}{ccc}
X & \longrightarrow & \mathbb{Z}/2 \\
\downarrow & & \downarrow{\scriptstyle e} \\
0 & \longrightarrow & \mathbb{Z}/2[1]
\end{array}
$$

exists in $D(\mathbb{Z})$. Then, for any integer i, we have a short exact sequence

$$
0 \longrightarrow [\mathbb{Z}, X[i]] \longrightarrow [\mathbb{Z}, \mathbb{Z}/2[i]] \longrightarrow [\mathbb{Z}, \mathbb{Z}/2[i+1]],
$$

or in other words

$$
0 \longrightarrow H^i(X) \longrightarrow H^i(\mathbb{Z}/2) \longrightarrow H^{i+1}(\mathbb{Z}/2).
$$

This implies that $X \longrightarrow \mathbb{Z}/2$ is a quasi-isomorphism, and thus an isomorphism in $D(\mathbb{Z})$. In particular $e = 0$, which is a contradiction.

A consequence of this is that $D(R)$ is not an abelian category, though the category of complexes itself $C(R)$ is abelian.

(b) The fact that $D(R)$ has no limits and colimits might not be a problem by itself, as it is possible to think of interesting categories which do not have limits and colimits (e.g. any non-trivial groupoid has no final object). However, the case of $D(R)$ is very frustrating as it seems that $D(R)$ is very close to having limits and colimits. For instance it is possible to show that $D(R)$ admits *homotopy limits and homotopy colimits* in the following sense. For a category I, let $C(R)^I$ be the category of functors from I to $C(R)$. A morphism $f : F \longrightarrow G$ (i.e. a natural transformation between two functors $F, G : I \longrightarrow C(R)$) is called a *levelwise quasi-isomorphism* if for any $i \in I$ the induced morphism $f(i) : F(i) \longrightarrow G(i)$ is a quasi-isomorphism in $C(R)$. We denote by $D(R, I)$ the category $C(R)^I$ localized along levelwise quasi-isomorphisms. The constant diagram functor $C(R) \longrightarrow C(R)^I$ is compatible with localizations on both sides and provides a functor

$$
c : D(R) \longrightarrow D(R, I).
$$

It can then been shown that the functor c has a left and a right adjoint denoted by

$$
Hocolim_I : D(R, I) \longrightarrow D(R) \qquad D(R) \longleftarrow D(R, I) : Holim_I,
$$

called the *homotopy colimit* and the *homotopy limit* functor. Homotopy limits and colimits are very good replacement of the notions of limits and colimits, as they are the best possible approximation of the colimit and limit functors that are compatible with the notion of quasi-isomorphisms. However, this is quite unsatisfactory as the category $D(R,I)$ depends on more than the category $D(R)$ alone (note that $D(R,I)$ is not equivalent to $D(R)^I$), and in general it is impossible to recontruct $D(R,I)$ from $D(R)$.

(c) To the ring R are associated several invariants such as its K-theory spectrum, its Hochschild (resp. cyclic) homology .... It is tempting to think that these invariants can be directly defined on the level of derived categories, but this is not the case (see [24]). However, it has been noticed that these invariants only depend on R up to some notion of equivalence that is much weaker than the notion of isomorphism. For instance, any functor $D(R) \longrightarrow D(R')$ which is induced by a complex of (R,R')-bi-modules induces a map on K-theory, Hochschild homology and cyclic homology. However, it is not clear that every functor $D(R) \longrightarrow D(R')$ comes from a complex of (R,R')-bi-modules (there are counter examples when R and R' are dg-algebras, see [7, Remarks 2.5 and 6.8]). Definitely, the derived category of complexes of (R,R')-bi-modules is not equivalent to the category of functors $D(R) \longrightarrow D(R')$. This is again an unsatisfactory situation and it is then quite difficult (if not impossible) to understand the true nature of these invariants (i.e. of which mathematical structures are they truly invariants?).

(d) Another important problem with the categories $D(R)$ is their non local nature. To explain this let $\mathbb{P}^1$ be the projective line (e.g. over $\mathbb{Z}$). As a scheme $\mathbb{P}^1$ is the push-out

$$\begin{array}{ccc} Spec\,\mathbb{Z}[X,X^{-1}] & \longrightarrow & Spec\,\mathbb{Z}[T] \\ \downarrow & & \downarrow \\ Spec\,\mathbb{Z}[U] & \longrightarrow & \mathbb{P}^1, \end{array}$$

where T is sent to X and U is sent to X^{-1}. According to the push-out square, the category of quasi-coherent sheaves on $\mathbb{P}^1$ can be described as the (2-categorical) pull-back

$$\begin{array}{ccc} QCoh(\mathbb{P}^1) & \longrightarrow & Mod(\mathbb{Z}[T]) \\ \downarrow & & \downarrow \\ Mod(\mathbb{Z}[U]) & \longrightarrow & Mod(\mathbb{Z}[X,X^{-1}]). \end{array}$$

In other words, a quasi-coherent module on $\mathbb{P}^1$ is the same thing as a triple (M,N,u), where M (resp. N) is a $\mathbb{Z}[T]$-module (resp. $\mathbb{Z}[U]$-module), and u is an isomorphism

$$u : M \otimes_{\mathbb{Z}[T]} \mathbb{Z}[X,X^{-1}] \simeq N \otimes_{\mathbb{Z}[U]} \mathbb{Z}[X,X^{-1}]$$

of $\mathbb{Z}[X,X^{-1}]$-modules. This property is extremely useful in order to reduce problems of quasi-coherent sheaves on schemes to problems of modules over rings. Unfortunately, this property is lost when passing to the derived categories. The square

$$D_{qcoh}(\mathbb{P}^1) \longrightarrow D(\mathbb{Z}[T])$$
$$\downarrow \qquad\qquad\qquad \downarrow$$
$$D(\mathbb{Z}[U]) \longrightarrow D(\mathbb{Z}[X,X^{-1}]),$$

is not cartesian (in the 2-categorical sense) anymore (e.g. there exist non zero morphisms $\mathscr{O}_X \longrightarrow \mathscr{O}_X(-2)[1]$ that go to zero as a morphism in $D(\mathbb{Z}[U]) \times_{D(\mathbb{Z}[X,X^{-1}])} D(\mathbb{Z}[T]))$. The derived categories of the affine pieces of $\mathbb{P}^1$ do not determine the derived category of quasi-coherent sheaves on $\mathbb{P}^1$.

The list of problems above suggests the existence of a some sort of categorical structure lying in between the category of complexes $C(R)$ and its derived category $D(R)$, which is rather close to $D(R)$ (i.e. in which the quasi-isomorphisms are inverted in some sense), but for which (1)–(4) above are no longer a problem. There exist several possible approaches, and my purpose is to present one of them using dg-categories.

Exercise 2.2.1 *Let $I = B\mathbb{N}$ be the category with a unique object $*$ and with the monoid $\mathbb{N}$ of natural numbers as endomorphism of this object. There is a bijection between the set of functors from I to a category C and the set of pairs (x,h), where x in an object in C and h is an endomorphism of x.*

Let R be a commutative ring.

(a) Show that there is a natural equivalence of categories

$$D(R,I) \simeq D(R[X]),$$

where $D(R,I)$ is the derived category of I-diagram of complexes of R-modules as described in example (2) above. Deduce from this that $D(R,I)$ is never an abelian category (unless $R = 0$).

(b) Prove that $D(R)$ is abelian when R is a field (show that $D(R)$ is equivalent to the category of $\mathbb{Z}$-graded R-vector spaces).

(c) Deduce that $D(R,I)$ and $D(R)^I$ can not be equivalent in general.

(d) Let now I be the category with two objects 0 and 1 and a unique morphism from 1 to 0. Using a similar approach as above show that $D(R,I)$ and $D(R)^I$ are not equivalent in general.

2.3 DG-Categories and DG-Functors

We now fix a base commutative ring k. Unless specified, all the modules and tensor products will be over k.

2.3.1 DG-Categories

We start by recalling that a *dg-category* T (over k) consists of the following data.

- A set of objects $Ob(T)$, also sometimes denoted by T itself
- For any pair of objects $(x,y) \in Ob(T)^2$ a complex $T(x,y) \in C(k)$

- For any triple $(x,y,z) \in Ob(T)^3$ a composition morphism $\mu_{x,y,z} : T(x,y) \otimes T(y,z) \longrightarrow T(x,z)$.
- For any object $x \in Ob(T)$, a morphism $e_x : k \longrightarrow T(x,x)$

These data are required to satisfy the following associativity and unit conditions.

(a) (Associativity) For any four objects (x,y,z,t) in T, the following diagram

$$
\begin{array}{ccc}
T(x,y) \otimes T(y,z) \otimes T(z,t) & \xrightarrow{\ id \otimes \mu_{y,z,t}\ } & T(x,y) \otimes T(y,t) \\
\downarrow{\scriptstyle \mu_{x,y,z} \otimes id} & & \downarrow{\scriptstyle \mu_{x,y,t}} \\
T(x,z) \otimes T(z,t) & \xrightarrow[\ \mu_{x,z,t}\ ]{} & T(x,t)
\end{array}
$$

commutes.

(b) (Unit) For any $(x,y) \in Ob(T)^2$ the two morphisms

$$T(x,y) \simeq k \otimes T(x,y) \xrightarrow{e_x \otimes id} T(x,x) \otimes T(x,y) \xrightarrow{\mu_{x,x,y}} T(x,y)$$

$$T(x,y) \simeq T(x,y) \otimes k \xrightarrow{id \otimes e_y} T(x,y) \otimes T(y,y) \xrightarrow{\mu_{x,y,y}} T(x,y)$$

are equal to the identities.

In a more explicit way, a dg-category T can also be described as follows. It has a set of objects $Ob(T)$. For any two objects x and y, and any $n \in \mathbb{Z}$ it has a k-module $T(x,y)^n$, though as morphisms of degree n from x to y. For three objects x, y and z, and any integers n and m there is a composition map

$$T(x,y)^n \times T(y,z)^m \longrightarrow T(x,z)^{n+m}$$

which is bilinear and associative. For any object x, there is an element $e_x \in T(x,x)^0$, which is a unit for the composition. For any two objects x and y there is a differential $d : T(x,y)^n \longrightarrow T(x,y)^{n+1}$, such that $d^2 = 0$. And finally, we have the graded Leibnitz rule

$$d(f \circ g) = d(f) \circ g + (-1)^m f \circ d(g),$$

for f and g two composable morphisms, with f of degree m. Note that this implies that $d(e_x) = 0$, and thus that e_x is always a zero cycle in the complex $T(x,x)$.

On a more conceptual side, a dg-category is a $C(k)$-enriched category in the sense of [17], where $C(k)$ is the symmetric monoidal category of complexes of k-modules. All the basic notions of dg-categories presented in these notes can be expressed in terms of enriched category theory, but we will not use this point of view. We, however, encourage the reader to consult [17] and to (re)consider the definitions of dg-categories, dg-functors, tensor product of dg-categories ... in the light of enriched category theory.

Examples:

(a) A very simple example is the opposite dg-category T^{op} of a dg-category T. The set of objects of T^{op} is the same as the one of T, and we set

$$T^{op}(x,y) := T(y,x)$$

together with the obvious composition maps

$$T(y,x) \otimes T(z,y) \simeq T(z,y) \otimes T(y,x) \longrightarrow T(z,x),$$

where the first isomorphism is the symmetry isomorphism of the monoidal structure on the category of complexes (see [5, Sect. X.4.1] for the signs rule).

(b) A fundamental example of dg-category over k is the one given by considering the category of complexes over k itself. Indeed, we define a dg-category $\underline{C(k)}$ by setting its set of objects to be the set of complexes of k-modules. For two complexes E and F, we define $\underline{C(k)}(E,F)$ to be the complex $\underline{Hom}^*(E,F)$ of morphisms from E to F. Recall, that for any $n \in \mathbb{Z}$ the k-module of elements of degree n in $\underline{Hom}^*(E,F)$ is given by

$$\underline{Hom}^n(E,F) := \prod_{i \in \mathbb{Z}} Hom(E^i, F^{i+n}).$$

The differential

$$d : \underline{Hom}^n(E,F) \longrightarrow \underline{Hom}^{n+1}(E,F)$$

sends a family $\{f^i\}_{i \in \mathbb{Z}}$ to the family $\{d \circ f^i - (-1)^n f^{i+1} \circ d\}_{i \in \mathbb{Z}}$. Note that the zero cycles in $\underline{Hom}^*(E,F)$ are precisely the morphisms of complexes from E to F. The composition of morphisms induces composition morphisms

$$\underline{Hom}^n(E,F) \times \underline{Hom}^m(E,F) \longrightarrow \underline{Hom}^{n+m}(E,F).$$

It is easy to check that these data defines a dg-category $\underline{C(k)}$.

(c) There is slight generalization of the previous example for the category $C(R)$ of complexes of (left) R-modules, where R is any associative and unital k-algebra. Indeed, for two complexes of R-modules E and F, there is a complex $\underline{Hom}^*(E,F)$ defined as in the previous example. The only difference is that now $\underline{Hom}^*(E,F)$ is only a complex of k-modules and not of R-modules in general (except when R is commutative). These complexes define a dg-category $\underline{C(R)}$ whose objects are complexes of R-modules.

(d) A far reaching generalization of the two previous examples is the case of complexes of objects in any k-linear Grothendieck category (i.e. an abelian co-complete category with a small generator and for which filtered colimits are exact, or equivalently a localization of a modules category, see [10]). Indeed, for such a category $\mathscr{A}$ and two complexes E and F of objects in $\mathscr{A}$, we define a complex of k-modules $\underline{Hom}^*(E,F)$ as above

$$\underline{Hom}^n(E,F) := \prod_{i \in \mathbb{Z}} Hom(E^i, F^{i+n}),$$

with the differential given by the same formula as in example (2). The composition of morphisms induce morphisms

$$\underline{Hom}^n(E,F) \times \underline{Hom}^m(F,G) \longrightarrow \underline{Hom}^{n+m}(E,G).$$

It is easy to check that these data define a dg-category whose objects are complexes in $\mathscr{A}$. It will be denoted by $\underline{C(\mathscr{A})}$.

(e) From a dg-category T, we can construct a category $Z^0(T)$ of 0-cycles as follows. It has the same objects as T, and for two such objects x and y the set of morphisms between x and y in $Z^0(T)$ is defined to be the set of 0-cycles in $T(x,y)$ (i.e. degree zero morphisms $f \in T(x,y)^0$ such that $d(f) = 0$. The Leibniz rule implies that the composition of two 0-cycles is again a 0-cycle, and thus we have induced composition maps

$$Z^0(T(x,y)) \times Z^0(T(y,z)) \longrightarrow Z^0(T(x,z)).$$

These composition maps define the category $Z^0(T)$. The category $Z^0(T)$ is often named the *underlying category of T*. We observe that $Z^0(T)$ is more precisely a k-linear category (i.e. that *Homs* sets are endowed with k-module structures such that the composition maps are bilinear).

For instance, let $\mathscr{A}$ be a Grothendieck category and $\underline{C(\mathscr{A})}$ its associated dg-category of complexes as defined in example (4) above. The underlying category of $\underline{C(\mathscr{A})}$ is then isomorphic to the usual category $C(\mathscr{A})$ of complexes and morphisms of complexes in $\mathscr{A}$.

(f) Conversely, if C is a k-linear category we view C as a dg-category in a rather obvious way. The set of objects is the same of the one of C, and the complex of morphisms from x to y is simply the complex $C(x,y)[0]$, which is $C(x,y)$ in degree 0 and 0 elsewhere. In the sequel, every k-linear category will be considered as a dg-category in this obvious way. Note that, this way the category of k-linear categories and k-linear functors form a full sub-category of dg-categories and dg-functors (see Sect. 1.3.2 below).

(g) A dg-category T with a unique object is essentially the same thing as a dg-algebra. Indeed, if x is the unique object the composition law on $T(x,x)$ induces a unital and associative dg-algebra structure on $T(x,x)$. Conversely, if B is a unital and associative dg-algebra we can construct a dg-category T with a unique object x and with $T(x,x) := B$. The multiplication in B is then used to define the composition on $T(x,x)$.

(h) Here is now a non-trivial example of a dg-category arising from geometry. In this example $k = \mathbb{R}$. Let X be a differential manifold (say $\mathscr{C}^\infty$). Recall that a flat vector bundle on X consists the data of a smooth (real) vector bundle V on X together with a connexion

$$\nabla : A^0(X,V) \longrightarrow A^1(X,V),$$

(where $A^n(X,V)$ is the space of smooth n-forms on X with coefficients in V) such that $\nabla^2 = 0$. For two such flat bundles (V,∇_V) and (W,∇_W) we define a complex $A^*_{DR}(V,W)$ by

$$A^*_{DR}(V,W)^n := A^n(X, \underline{Hom}(V,W)),$$

where $\underline{Hom}(V,W)$ is the vector bundle of morphisms from V to W. The differential

$$\mathrm{d} : A^n_{DR}(V,W) \longrightarrow A^{n+1}_{DR}(V,W)$$

is defined by sending $\omega \otimes f$ to $\mathrm{d}(\omega) \otimes f + (-1)^n \omega \wedge \nabla(f)$. Here, $\nabla(f)$ is the 1-form with coefficients in $\underline{Hom}(V,W)$ defined by

$$\nabla(f) := \nabla_W \circ f - (f \otimes id) \circ \nabla_V.$$

The fact that $\nabla^2_V = \nabla^2_W = 0$ implies that $A^*_{DR}(V,W)$ is a complex. Moreover, we define a composition

$$A^n_{DR}(U,V) \times A^m_{DR}(V,W) \longrightarrow A^{n+m}(U,W)$$

for three flat bundles U, V and W by

$$(\omega \otimes f).(\omega' \otimes g) := (\omega \wedge \omega') \otimes (f \circ g).$$

It is easy to check that these data defines a dg-category $T_{DR}(X)$ (over $\mathbb{R}$) whose objects are flat bundles on X, and whose complex of morphisms from (V, ∇_V) to (W, ∇_W) are the complexes $A^*_{DR}(V,W)$.

By construction the underlying category of $T_{DR}(X)$ is the category of flat bundles and flat maps. By the famous Riemann–Hilbert correspondence (see [6] for the analog statement in the complex analytic case) this category is thus equivalent to the category of finite dimensional linear representations of the fundamental group of X, or equivalently of finite dimensional local systems (i.e. of locally constant sheaves of finite dimensional $\mathbb{C}$-vector spaces). Moreover, for two flat bundles (V, ∇_V) and (W, ∇_W), corresponding to two local systems L_1 and L_2, the cohomology group $H^i(T_{DR}(X)(V,W)) = H^i(A^*_{DR}(V,W))$ is isomorphic to the Ext group $Ext^i(L_1, L_2)$, computed in the category of abelian sheaves over X. Therefore, we see that even when X is simply connected the dg-category $T_{DR}(X)$ contains interesting informations about the cohomology of X (even though the underlying category of $T_{DR}(X)$ is simply the category of finite dimensional vector spaces).

(i) The previous example has the following complex analog. Now we let $k = \mathbb{C}$, and X be a complex manifold. We define a dg-category $T_{Dol}(X)$ in the following way. The objects of $T_{Dol}(X)$ are the holomorphic complex vector bundles on X. For two such holomorphic bundles V and W we let

$$T_{Dol}(X)(V,W) := A^*_{Dol}(V,W),$$

where $A^*_{Dol}(V,W)$ is the Dolbeault complex with coefficients in the vector bundle of morphisms from V to W. Explicitly,

$$A^q_{Dol}(V,W) := A^{0,q}(X, \underline{Hom}(V,W))$$

is the space of $(0,q)$-forms on X with coefficients in the holomorphic bundle $\underline{Hom}(V,W)$ of morphisms from V to W. The differential

$$A^q_{Dol}(V,W) \longrightarrow A^{q+1}_{Dol}(V,W)$$

is the operator $\overline{\partial}$, sending $\omega \otimes f$ to

$$\overline{\partial}(\omega \otimes f) := \overline{\partial}(\omega) \otimes f + (-1)^q \omega \wedge \overline{\partial}(f),$$

where $\overline{\partial}(f)$ is defined by

$$\overline{\partial}(f) = \overline{\partial}_W \circ f - (f \otimes id) \circ \overline{\partial}_V,$$

with

$$\overline{\partial}_V : A^0(X,V) \longrightarrow A^{0,1}(X,V) \qquad \overline{\partial}_W : A^0(X,W) \longrightarrow A^{0,1}(X,W)$$

being the operators induced by the holomorphic structures on V and W (see [11, Chap 0 Sect. 5]). As in the previous example we can define a composition

$$A^*_{Dol}(U,V) \times A^*_{Dol}(V,W) \longrightarrow A^*_{Dol}(U,W)$$

for three holomorphic bundles U, V and W on X. These data defines a dg-category $T_{Dol}(X)$ (over $\mathbb{C}$).

By construction, the underlying category of $T_{Dol}(X)$ has objects the holomorphic vector bundles, and the morphisms in this category are the $\mathscr{C}^\infty$-morphisms of complex vector bundles $f : V \longrightarrow W$ satisfying $\overline{\partial}(f) = 0$, or equivalently the holomorphic morphisms. Moroever, for two holomorphic vector bundles V and W the cohomology group $H^i(T_{Dol}(X))$ is isomorphic to $Ext^i_{\mathscr{O}_X}(\mathscr{V},\mathscr{W})$, the i-th ext-group between the associated sheaves of holomorphic sections (or equivalently the ext-group in the category of holomorphic coherent sheaves). For instance, if $\mathbf{1}$ is the trivial vector bundle of rank 1 and V is any holomorphic vector bundle, we have

$$H^i(T_{Dol}(\mathbf{1},V)) \simeq H^i_{Dol}(X,V),$$

the i-th Dolbeault cohomology group of V.

The dg-category $T_{Dol}(X)$ is important as it provides a rather explicit model for the derived category of coherent sheaves on X. Indeed, the homotopy category $H^0(T_{Dol}(X))$ (see definition 2.3.1) is equivalent to the full sub-category of $D^b_{coh}(X)$, the bounded coherent derived category of X, whose objects are holomorphic vector bundles. Also, for two such holomorphic vector bundles V and W and all i we have

$$Hom_{D^b_{coh}(X)}(V,W[i]) \simeq H^i(T_{Dol}(X)(V,W)) \simeq Ext^i_{\mathscr{O}_X}(\mathscr{V},\mathscr{W}).$$

(j) Here is one last example of a dg-category in the topological context. We construct a dg-category $dgTop$, whose set of objects is the set of all topological spaces. For two topological spaces X and Y, we define a complex of morphisms $dgTop(X,Y)$ in the following way. We first consider $Hom^\Delta(X,Y)$, the simplicial set (see [13] for the notion of simplicial sets) of continuous maps between X and Y: by definition the set of n-simplices in $Hom^\Delta(X,Y)$ is the set of continuous maps $X \times \Delta^n \longrightarrow Y$, where $\Delta^n := \{x \in [0,1]^{n+1}/\sum x_i = 1\}$ is the standard simplex of dimension n in $\mathbb{R}^{n+1}$. The face and degeneracy operators of $Hom^\Delta(X,Y)$ are defined using the face embeddings ($0 \le i \le n$)

$$d_i : \Delta^n \longrightarrow \Delta^{n+1}$$
$$x \longmapsto (x_0,\ldots,x_{i-1},0,x_i,\ldots x_n),$$

and the natural projections ($0 \le i \le n$)

$$s_i : \Delta^{n+1} \longrightarrow \Delta^n$$
$$x \longmapsto (x_0,\ldots,x_i+x_{i+1},x_{i+2},\ldots,x_{n+1}).$$

Now, for any two topological spaces X and Y we set

$$dgTop(X,Y) := C_*(Hom^\Delta(X,Y)),$$

the homology chain complex of $Hom^\Delta(X,Y)$ with coefficients in k. Explicitly, $C_n(Hom^\Delta(X,Y))$ is the free k-module generated by continuous maps $f : X \times \Delta^n \longrightarrow Y$. The differential of such a map is given by the formula

$$d(f) := \sum_{0 \le i \le n} (-1)^i d_i(f),$$

where $d_i(f)$ is the map $X \times \Delta^{n-1} \longrightarrow Y$ obtained by composition

$$X \times \Delta^{n-1} \xrightarrow{id \times d_i} X \times \Delta^n \xrightarrow{f} Y.$$

For three topological spaces X, Y and Z, there exists a composition morphism at the level of simplicial sets of continuous maps

$$Hom^\Delta(X,Y) \times Hom^\Delta(Y,Z) \longrightarrow Hom^\Delta(X,Z).$$

This induces a morphism on the level of chain complexes

$$C_*(Hom^\Delta(X,Y) \times Hom^\Delta(Y,Z)) \longrightarrow C_*(Hom^\Delta(X,Z)).$$

Composing this morphism with the famous Eilenberg–MacLane map (see [20, Sect. 29])

$$C_*(Hom^\Delta(X,Y)) \otimes C_*(Hom^\Delta(Y,Z)) \longrightarrow C_*(Hom^\Delta(X,Y) \times Hom^\Delta(Y,Z))$$

defines a composition

$$dgTop(X,Y) \otimes dgTop(Y,Z) \longrightarrow dgTop(X,Z).$$

The fact that the Eilenberg–MacLane morphisms are associative and unital (they are moreover commutative, see [20, Sect. 29]) implies that this defines a dg-category $dgTop$.

2.3.2 DG-Functors

For two dg-categories T and T', a *morphism of dg-categories* (or simply a *dg-functor*) $f : T \longrightarrow T'$ consists of the following data.

- A map of sets $f : Ob(T) \longrightarrow Ob(T')$.
- For any pair of objects $(x,y) \in Ob(T)^2$, a morphism in $C(k)$

$$f_{x,y} : T(x,y) \longrightarrow T'(f(x), f(y)).$$

These data are required to satisfy the following associativity and unit conditions.

(a) For any $(x,y,z) \in Ob(T)^3$ the following diagram

$$
\begin{array}{ccc}
T(x,y) \otimes T(y,z) & \xrightarrow{\;\mu_{x,y,z}\;} & T(x,z) \\
\downarrow{\scriptstyle f_{x,y} \otimes f_{y,z}} & & \downarrow{\scriptstyle f_{x,z}} \\
T'(f(x),f(y)) \otimes T'(f(y),f(z)) & \xrightarrow{\;\mu'_{f(x),f(y),f(z)}\;} & T'(f(x),f(z))
\end{array}
$$

commutes.

(b) For any $x \in Ob(T)$, the following diagram

$$
\begin{array}{ccc}
k & \xrightarrow{\;e_x\;} & T(x,x) \\
& {\scriptstyle e'_{f(x)}} \searrow & \downarrow{\scriptstyle f_{x,x}} \\
& & T'(f(x),f(x))
\end{array}
$$

commutes.

Examples:

(a) Let T be any dg-category and $x \in T$ be an object. We define a dg-functor

$$f = h^x : T \longrightarrow \underline{C(k)}$$

in the following way (recall that $\underline{C(k)}$ is the dg-category of complexes over k). The map on the set of objects sends an object $y \in T$ to the complex $T(x,y)$. For two objects y and z in T we define a morphism

$$f_{y,z} : T(y,z) \longrightarrow \underline{C(k)}(f(y), f(x)) = \underline{Hom}^*(T(x,y), T(x,z)),$$

which by definition is the adjoint to the composition morphism

$$m_{x,y,z} : T(x,y) \otimes T(y,z) \longrightarrow T(x,z).$$

The associativity and unit condition on composition of morphisms in T imply that this defines a morphism of dg-categories

$$h^x : T \longrightarrow \underline{C(k)}.$$

Dually, we can also define a morphism of dg-categories

$$h_x : T^{op} \longrightarrow \underline{C(k)}$$

by sending y to $T(y,x)$.

(b) For any dg-category T there exists a dg-functor

$$T \otimes T^{op} \longrightarrow \underline{C(k)},$$

sending a pair of objects (x,y) to the complex $T(y,x)$. Here, $T \otimes T^{op}$ denotes the dg-category whose set of objects is $Ob(T) \times Ob(T')$, and whose complex of morphisms are given by

$$(T \otimes T^{op})((x,y),(x',y')) := T(x,x') \otimes T(y',y).$$

We refer to Exercise 2.3.4 and Sect. 3.2 for more details about the tensor product of dg-categories.

(c) Let R and S be two associative and unital k-algebras, and $f : R \longrightarrow S$ be a k-morphism. The morphism f induces two functors

$$f^* : C(R) \longrightarrow C(S) \qquad C(R) \longleftarrow C(S) : f_*,$$

adjoint to each others. The functor f_* sends a complex of S-modules to the corresponding complex of R-modules obtained by restricting the scalars from S to R by the morphism f. Its left adjoint f^* sends a complex of R-modules E to the complex $S \otimes_R E$. It is not difficult to show that the functors f_* and f^* are compatible with the complex of morphisms $\underline{Hom}^*$ and thus define morphisms of dg-categories

$$f^* : \underline{C(R)} \longrightarrow \underline{C(S)} \qquad \underline{C(R)} \longleftarrow \underline{C(S)} : f_*.$$

More generally, if $f : \mathscr{A} \longrightarrow \mathscr{B}$ is any k-linear functor between Grothendieck categories, there is an induced morphism of dg-categories

$$f : \underline{C(\mathscr{A})} \longrightarrow \underline{C(\mathscr{B})}.$$

(d) Let $f : X \longrightarrow Y$ be a $\mathscr{C}^\infty$-morphism between two differential manifolds. Then, the pull-back for flat bundles and differential forms defines a morphism of dg-categories constructed in our example 8

$$f^* : T_{DR}(Y) \longrightarrow T_{DR}(X).$$

In the same way, if now f is a holomorphic morphism between two complex varieties, then there is a dg-functor

$$f^* : T_{Dol}(Y) \longrightarrow T_{Dol}(X)$$

ontained by pulling-back the holomorphic vector bundles and differential forms.

DG-functors can be composed in an obvious manner, and dg-categories together with dg-functors form a category denoted by $dg - cat_k$ (or $dg - cat$ if the base ring k is clear).

For a dg-category T, we define a category $H^0(T)$ in the following way. The set of objects of $H^0(T)$ is the same as the set of objects of T. For two objects x and y the set of morphisms in $H^0(T)$ is defined by

$$H^0(T)(x,y) := H^0(T(x,y)).$$

Finally, the composition of morphisms in $H^0(T)$ is defined using the natural morphisms

$$H^0(T(x,y)) \otimes H^0(T(y,z)) \longrightarrow H^0(T(x,y) \otimes T(y,z))$$

composed with the morphism

$$H^0(\mu_{x,y,z}) : H^0(T(x,y) \otimes T(y,z)) \longrightarrow H^0(T(x,z)).$$

Definition 2.3.1 *The category $H^0(T)$ is called the* homotopy category of T.

Examples:

(a) If C is a k-linear category considered as a dg-category as explained in our example 6 above, then $H^0(C)$ is naturally isomorphic to C itself.
(b) We have $H^0(T^{op}) = H^0(T)^{op}$ for any dg-category T.
(c) For a k-algebra R, the homotopy category $H^0(C(R))$ is usually denoted by $K(R)$, and is called the *homotopy category of complexes of R-modules*. More generally, if $\mathscr{A}$ is a Grothendieck category, $H^0(C(\mathscr{A}))$ is denoted by $K(\mathscr{A})$, and is called the homotopy category of complexes in $\mathscr{A}$.
(d) If X is a differentiable manifold, then $H^0(T_{DR}(X))$ coincides with $Z^0(T_{DR}(X))$ and is isomorphic to the category of flat bundles and flat maps between them. As we already mentioned, this last category is equivalent by the Riemann–Hilbert correspondence to the category of local systems on X.
     When X is a complex manifold, we also have that $H^0(T_{Dol}(X))$ coincides with $Z^0(T_{Dol}(X))$ and is isomorphic to the category of holomorphic vector bundles and holomorphic maps between them.
(e) The category $H^0(dgTop)$ is the category whose objects are topological spaces and whose set of morphisms between X and Y is the free k-module over the set of homotopy classes of maps from X to Y.

One of the most important notions in dg-category theory is the notion of quasi-equivalences, a mixture in between quasi-isomorphisms and categorical equivalences.

Definition 2.3.2 *Let $f : T \longrightarrow T'$ be a dg-functor between dg-categories*

(a) The morphism f is quasi-fully faithful *if for any two objects x and y in T the morphism $f_{x,y} : T(x,y) \longrightarrow T'(f(x), f(y))$ is a quasi-isomorphism of complexes.*
(b) The morphism f is quasi-essentially surjective *if the induced functor $H^0(f) : H^0(T) \longrightarrow H^0(T')$ is essentially surjective.*
(c) The morphism f is a quasi-equivalence *if it is quasi-fully faithful and quasi-essentially surjective.*

We will be mainly interested in dg-categories up to quasi-equivalences. We therefore introduce the following category.

Definition 2.3.3 *The* homotopy category of dg-categories *is the category* $dg - cat$ *localized along quasi-equivalences. It is denoted by* $Ho(dg - cat)$. *Morphisms in* $Ho(dg - cat)$ *between two dg-categories* T *and* T' *will often be denoted by*

$$[T, T'] := Hom_{Ho(dg-cat)}(T, T').$$

Note that the construction $T \mapsto H^0(T)$ provides a functor $H^0(-) : dg - cat \longrightarrow Cat$, which descends as a functor on homotoy categories

$$H^0(-) : Ho(dg - cat) \longrightarrow Ho(Cat).$$

Remark 1. In the last section we have seen that the localization construction is not well behaved, but in the definition above we consider $Ho(dg - cat)$ which is obtained by localization. Therefore, the category $Ho(dg - cat)$ will not be well behaved itself. In order to get the most powerful approach the category $dg - cat$ should have been itself localized in a more refined maner (e.g. as a higher category, see [29, Sect. 2]). We will not need such an evolved approach, and the category $Ho(dg - cat)$ will be enough for most of our purpose.

Examples:

(a) Let $f : T \longrightarrow T'$ be a quasi-fully faithful dg-functor. We let T_0' be the full (i.e. with the same complexes of morphisms as T') sub-dg-category of T' consisting of all objects $x \in T'$ such that x is isomorphic in $H^0(T')$ to an object in the image of the induced functor $H^0(f) : H^0(T) \longrightarrow H^0(T')$. Then the induced dg-functor $T \longrightarrow T_0'$ is a quasi-equivalence.

(b) Let $f : R \longrightarrow S$ be a morphism of k-algebras. If the morphism of dg-categories

$$f^* : \underline{C(R)} \longrightarrow \underline{C(S)}$$

is quasi-fully faithful then the morphism f is an isomorphism. Indeed, if f^* is quasi-fully faithful we have that

$$\underline{Hom}^*(R, R) \longrightarrow \underline{Hom}^*(S, S)$$

is a quasi-isomorphism. Evaluating this morphism of complexes at H^0 we find that the induced morphism

$$R \simeq H^0(\underline{Hom}^*(R, R)) \longrightarrow H^0(\underline{Hom}^*(S, S)) \simeq S$$

is an isomorphism. This last morphism being f itself, we see that f is an isomorphism.

(c) Suppose that T is a dg-category such that for all objects x and y we have $H^i(T(x, y)) = 0$ for all $i \neq 0$. We are then going to show that T and $H^0(T)$ are isomorphic in $Ho(dg - cat)$. We first define a dg-category $T_{\leq 0}$ in the following way. The dg-category $T_{\leq 0}$ possesses the same set of objects as T itself. For two objects x and y we let

$$T_{\leq 0}(x,y)^n := T(x,y) \ if \ n < 0 \qquad T_{\leq 0}(x,y)^n := 0 \ if \ n > 0$$

and

$$T_{\leq 0}(x,y)^0 := Z^0(T(x,y)) = Ker(d : T(x,y)^0 \to T(x,y)^1).$$

The differential on $T_{\leq 0}(x,y)$ is simply induced by the one on $T(x,y)$. It is not hard to see that the composition morphisms of T induces composition morphisms

$$T_{\leq 0}(x,y)^n \times T_{\leq 0}(y,z)^m \longrightarrow T_{\leq 0}(x,z)^{n+m}$$

which makes these data into a dg-category $T_{\leq 0}$ (this is because the composition of two 0-cocycles is itself a 0-cocycle). Moreover, there is a natural dg-functor

$$T_{\leq 0} \longrightarrow T$$

which is the identity on the set of objects and the natural inclusions of complexes

$$T_{\leq 0}(x,y) \subset T(x,y)$$

on the level of morphisms. Now, we consider the natural dg-functor (here, as always, the k-linear category $H^0(T)$ is considered as a dg-category in the obvious way)

$$T_{\leq 0} \longrightarrow H^0(T)$$

which is the identity on the set of objects and the natural projection

$$T_{\leq 0}(x,y) \longrightarrow H^0(T(x,y)) = H^0(T_{\leq 0}(x,y)) = T_{\leq 0}(x,y)^0 / Im(T(x,y)^{-1} \to T_{\leq 0}(x,y)^0)$$

on the level of morphisms. We thus have a diagram of dg-categories and dg-functors

$$H^0(T) \longleftarrow T_{\leq 0} \longrightarrow T,$$

which by assumptions on T are all quasi-equivalences. This implies that T and $H^0(T)$ becomes isomorphic as objects in $Ho(dg-cat)$.

(d) Suppose that $f : X \longrightarrow Y$ is a $\mathscr{C}^\infty$ morphism between differentiable manifolds, such that there exists another $\mathscr{C}^\infty$ morphism $g : Y \longrightarrow X$ and two $\mathscr{C}^\infty$ morphisms

$$h : X \times \mathbb{R} \longrightarrow X \qquad k : Y \times \mathbb{R} \longrightarrow Y$$

with

$$h_{X \times \{0\}} = gf, \ h_{X \times \{1\}} = id \qquad k_{Y \times \{0\}} = fg, \ k_{Y \times \{1\}} = id.$$

Then the dg-functor

$$f^* : T_{DR}(Y) \longrightarrow T_{DR}(X)$$

is a quasi-equivalence. Indeed, we know that $H^0(T_{DR}(X))$ is equivalent to the category of linear representations of the fundamental group of X. Therefore, as the morphism f is in particular a homotopy equivalence it induces an isomorphisms on the level of the fundamental groups, and thus the induced functor

$$f^* : H^0(T_{DR}(Y)) \longrightarrow H^0(T_{DR}(X))$$

is an equivalence of categories. The fact that the dg-functor f^* is also quasi-fully faithful follows from the homotopy invariance of de Rham cohomology, and more precisely from the fact that the projection $p : X \times \mathbb{R} \longrightarrow X$ induces a quasi-equivalence of dg-categories

$$p^* : T_{DR}(X) \longrightarrow T_{DR}(X \times \mathbb{R}).$$

We will not give more details in these notes.

As particular case of the above statement we see that the projection $\mathbb{R}^n \longrightarrow *$ induces a quasi-equivalence

$$T_{DR}(*) \longrightarrow T_{DR}(\mathbb{R}^n).$$

As $T_{DR}(*)$ is itself isomorphic to the category of finite dimensional real vector spaces, we see that $T_{DR}(\mathbb{R}^n)$ is quasi-equivalent to the category of finite dimensional vector spaces.

(e) Let now X be a connected complex manifold and $p : X \longrightarrow *$ be the natural projection. Then the induced dg-functor

$$p^* : T_{Dol}(*) \longrightarrow T_{Dol}(X)$$

is quasi-fully faithful if and only if $H^i(X, \mathcal{O}_X) = 0$ for all $i \neq 0$ (here $\mathcal{O}_X$ is the sheaf of holomorphic functions on X). Indeed, all the vector bundles are trivial on $*$. Moreover, for $\mathbf{1}^r$ and $\mathbf{1}^s$ two trivial vector bundles of rank r and s on $*$ we have

$$T_{Dol}(X)(p^*(\mathbf{1}^r), p^*(\mathbf{1}^s)) \simeq T_{Dol}(X)(\mathbf{1}, \mathbf{1})^{rs},$$

where 1 also denotes the trivial holomorphic bundle on X. Therefore, p^* is quasi-fully faithful if and only if $H^i(T_{Dol}(X))(\mathbf{1}, \mathbf{1}) = 0$ for all $i \neq 0$. As we have

$$H^i(T_{Dol}(X)(\mathbf{1}, \mathbf{1})) = H^i_{Dol}(X, \mathbf{1}) = H^i(X, \mathcal{O}_X)$$

this implies the statement. As an example, we see that

$$T_{Dol}(*) \longrightarrow T_{Dol}(\mathbb{P}^n)$$

is quasi-fully faithful (here $\mathbb{P}^n$ denotes the complex projective space), but

$$T_{Dol}(*) \longrightarrow T_{Dol}(E)$$

is not for any complex elliptic curve E.

More generally, if $f : X \longrightarrow Y$ is any proper holomorphic morphism between complex manifolds, then the dg-functor

$$f^* : T_{Dol}(Y) \longrightarrow T_{Dol}(X)$$

is quasi-fully faithful if and only if we have

$$\mathbb{R}^i f_*(\mathscr{O}_X) = 0 \; \forall \, i > 0,$$

where $\mathbb{R}^i f_*(\mathscr{O}_X)$ denotes the higher direct images of the coherent sheaf $\mathscr{O}_X$ of holomorphic functions on X (see e.g. [11]). We will not prove this statement in these notes. As a consequence we see that f^* is quasi-fully faithful if it is a blow-up along a smooth complex sub-manifold of Y, or if it is a bundle in complex projective spaces.

(f) For more quasi-equivalences between dg-categories in the context of non-abelian Hodge theory see [26].

Exercise 2.3.4 (a) Let T and T' be two dg-categories. Show how to define a dg-category $T \otimes T'$ whose set of objects is the product of the sets of objects of T and T', and for any two pairs (x,y) and (x',y')

$$(T \otimes T')((x,y),(x',y')) := T(x,y) \otimes T'(x',y').$$

(b) Show that the construction $(T,T') \mapsto T \otimes T'$ defines a symmetric monoidal structure on the category $dg - cat$.

(c) Show that the symmetric monoidal structure $\otimes$ on $dg - cat$ is closed (i.e. that for any two dg-categories T and T' there exists a dg-category $\underline{Hom}(T,T')$ together with functorial isomorphisms

$$Hom(T'',\underline{Hom}(T,T')) \simeq Hom(T'' \otimes T,T').$$

Exercise 2.3.5 Let $k \to k'$ be a morphism of commutative rings, and $dg - cat_k$ (resp. $dg - cat_{k'}$) the categories of dg-categories over k (resp. over k').

(a) Show that there exists a forgetful functor

$$dg - cat_{k'} \longrightarrow dg - cat_k$$

which consists of seing complexes over k' as complexes over k using the morphism $k \to k'$.

(b) Show that this forgetful functor admits a left adjoint

$$- \otimes_k k' : dg - cat_k \longrightarrow dg - cat_{k'}.$$

(c) Let $1_{k'}$ be the dg-category over k with a single object and with k' as k-algebra of endomorphisms of this object. Show that for any dg-category T over k, there exists a natural isomorphism of dg-categories over k

$$T \otimes_k k' \simeq T \otimes 1_{k'},$$

where the tensor product on the right is the one of dg-categories over k as defined in exercise 2.3.4, and the left hand side is considered as an object in $dg - cat_k$ throught the forgetful functor.

(d) *Show that the forgetful functor*

$$dg - cat_{k'} \longrightarrow dg - cat_k$$

also possesses a right adjoint

$$(-)^{k'/k} : dg - cat_k \longrightarrow dg - cat_{k'}$$

(*show that for any $T \in dg - cat_k$ the dg-category $\underline{Hom}(1_{k'}, T)$ can be naturally endowed with a structure of dg-category over k'*).

Exercise 2.3.6 *Let T be a dg-category and $u \in Z^0(T(x, y))$ a morphism in its underlying category. Show that the following four conditions are equivalent.*

(a) *The image of u in $H^0(T(x, y))$ is an isomorphism in $H^0(T)$.*
(b) *There exists $v \in Z^0(T(y, x))$ and two elements $h \in T(x, x)^{-1}$, $k \in T(y, y)^{-1}$ such that*

$$d(h) = vu - e_x \qquad d(k) = uv - e_y.$$

(c) *For any object $z \in T$, the composition with u*

$$u \circ - : T(z, x) \longrightarrow T(z, y)$$

is a quasi-isomorphism of complexes.
(d) *For any object $z \in T$, the composition with u*

$$- \circ u : T(y, z) \longrightarrow T(x, z)$$

is a quasi-isomorphism of complexes.

Exercise 2.3.7 *We denote by B the commutative k-dg-algebra whose underlying graded k-algebra is a (graded commutative) polynomial algebra in two variables $k[X, Y]$, with X in degree 0, Y in degree -1 and $d(Y) = X^2$. We consider B as a dg-category with a unique object.*

(a) *Show that there exists a natural quasi-equivalence*

$$p : B \longrightarrow k[X]/(X^2) =: k[\varepsilon],$$

where $k[\varepsilon]$ is the commutative algebra of dual numbers, considered as a dg-category with a unique object.
(b) *Show that p does not admit a section in $dg - cat$. Deduce from this that unlike the case of categories, there are quasi-equivalences $T \longrightarrow T'$ in $dg - cat$ such that the inverse of f in $Ho(dg - cat)$ can not be represented by a dg-functor $T' \longrightarrow T$ in $dg - cat$ (i.e. quasi-inverses do not exist in general).*

Exercise 2.3.8 *Show that two k-linear categories are equivalent (as k-linear categories) if and only if they are isomorphic in $Ho(dg - cat)$.*

2.4 Localizations as a dg-Category

For a k-algebra R, the derived category $D(R)$ is defined as a localization of the category $C(R)$, and thus has a universal property in $Ho(Cat)$. The purpose of this series of lectures is to show that $C(R)$ can also be localized as a dg-category $C(R)$ in order to get an object $L(R)$ satisfying a universal property in $Ho(dg - cat)$. The two objects $L(R)$ and $D(R)$ will be related by the formula

$$H^0(L(R)) \simeq D(R),$$

and we will see that the extra information encoded in $L(R)$ is enough in order to solve all the problems mentioned in Sect. 1.2.

Let T be any dg-category, S be a subset of morphisms in the category $H^0(T)$, and let us define a subfunctor $F_{T,S}$ of the functor $[T, -]$, corepresented by $T \in Ho(dg - cat)$. We define

$$F_{T,S} : Ho(dg - cat) \longrightarrow Ho(Cat)$$

by sending a dg-category T' to the subset of morphisms $[T, T']$ consisting of all morphism f whose induced functor $H^0(f) : H^0(T) \longrightarrow H^0(T')$ sends morphisms of S to isomorphisms in $H^0(T')$. Note that the functor $H^0(f)$ is only determined as a morphism in $Ho(Cat)$, or in other words up to isomorphism. However, the property that $H^0(f)$ sends elements of S to isomorphisms is preserved under isomorphisms of functors, and thus only depends on the class of $H^0(f)$ as a morphism in $Ho(Cat)$.

Definition 2.4.1 *For T and S as above, a* localization *of T along S is a dg-category $L_S T$ corepresenting the functor $F_{T,S}$.*

To state the previous definition in more concrete terms, a localization is the data of a dg-category $L_S T$ and a dg-functor $l : T \longrightarrow L_S T$, such that for any dg-category T' the induced map

$$l^* : [L_S T, T'] \longrightarrow [T, T']$$

is injective and identifies the left hand side with the subset $F_{T,S}(T') \subset [T, T']$.

An important first question is the existence of localization as above. We will see that like localizations of categories they always exist. This, of course, requires to know how to compute the set $[T, T']$ of morphisms in $Ho(dg - cat)$. As the category $Ho(dg - cat)$ is itself defined by localization this is not an easy problem. We will give a solution to this question in the next lectures, based on an approach using model category theory.

3 Lecture 2: Model Categories and dg-Categories

The purpose of this second lecture is to study in more details the category $Ho(dg - cat)$. Localizations of categories are very difficult to describe in general. The purpose of model category theory is precisely to provide a general tool to describe localized categories. By definition, a model category is a category together with three classes of morphisms, fibrations, cofibrations and (weak) equivalences satisfying some axioms mimicking the topological notions of Serre fibrations, cofibrations and weak homotopy equivalences. When M is a model category, with W as equivalences, then the localized category $W^{-1}M$ possesses a very nice description in terms of homotopy classes of morphisms between objects belonging to a certain class of nicer objects called fibrant and cofibrant. A typical example is when $M = Top$ is the category of topological spaces and W is the class of weak equivalences (see example 5 of Sect. 2.1). Then all objects are fibrant, but the cofibrant objects are the retracts of CW-complexes. It is well known that the category $W^{-1}Top$ is equivalent to the category of CW-complexes and homotopy classes of continuous maps between them.

In this lecture, I will start by some brief reminders on model categories. I will then explain how model category structures appear in the context of dg-categories by describing the model category of dg-categories (due to G. Tabuada, [27]) and the model category of dg-modules. We will also see how model categories can be used in order to construct interesting dg-categories. In the next lecture these model categories will be used in order to understand maps in $Ho(dg-cat)$, and to prove the existence of several important constructions such as localization and internal Homs.

3.1 Reminders on Model Categories

In this section we use the conventions of [13] for the notion of model category. We also refer the reader to this book for the proofs of the statements we will mention.

We let M be a category with arbitrary limits and colimits. Recall that a (closed) model category structure on M is the data of three classes of morphisms in M, the fibrations Fib, the cofibrations Cof and the equivalences W, satisfying the following axioms (see [13]).

(a) If $X \xrightarrow{f} Y \xrightarrow{g} Z$ are morphisms in M, then f, g and gf are all in W if and only if two of them are in W.
(b) The fibrations, cofibrations and equivalences are all stable by retracts.
(c) Let

$$
\begin{array}{ccc}
A & \xrightarrow{f} & X \\
i \downarrow & & \downarrow p \\
B & \xrightarrow{g} & Y
\end{array}
$$

be a commutative square in M with $i \in Cof$ and $p \in Fib$. If either i or p is also in W then there is a morphism $h : B \longrightarrow X$ such that $ph = g$ and $hi = f$.
(d) Any morphism $f : X \longrightarrow Y$ can be factorized in two ways as $f = pi$ and $f = qj$, with $p \in Fib$, $i \in Cof \cap W$, $q \in Fib \cap W$ and $j \in Cof$. Moreover, the existence of these factorizations are required to be functorial in f.

The morphisms in $Cof \cap W$ are usually called *trivial cofibrations* and the morphisms in $Fib \cap W$ *trivial fibrations*. Objects x such that $\emptyset \longrightarrow x$ is a cofibration are called *cofibrant*. Dually, objects y such that $y \longrightarrow *$ is a fibration are called *fibrant*. The factorization axiom (4) implies that for any object x there is a diagram

$$Qx \xrightarrow{\ i\ } x \xrightarrow{\ p\ } Rx,$$

where i is a trivial fibration, p is a trivial cofibration, Qx is a cofibrant object and Rx is a fibrant object. Moreover, the functorial character of the factorization states that the above diagram can be, and will always be, chosen to be functorial in x.

Exercise 3.1.1 *Let M be a model category and $i : A \longrightarrow B$ a morphism. We assume that for every commutative square*

$$
\begin{array}{ccc}
A & \xrightarrow{\ f\ } & X \\
{\scriptstyle i}\big\downarrow & & \big\downarrow{\scriptstyle p} \\
B & \xrightarrow[\ g\ ]{} & Y,
\end{array}
$$

with p a fibration (resp. a trivial fibration) there is a morphism $h : B \longrightarrow X$ such that $ph = g$ and $hi = f$. Then i is a trivial cofibration (resp. a cofibration). (Hint: factor i using axiom (4) and use the stability of Cof and W by retracts). As a consequence, the class Fib is determined by W and Cof, and similarly the class Cof is determined by W and Fib.

By definition, the homotopy category of a model category M is the localized category

$$Ho(M) := W^{-1}M.$$

A model category structure is a rather simple notion, but in practice it is never easy to check that three given classes Fib, Cof and W satisfy the four axioms above. This can be explained by the fact that the existence of a model category structure on M has a very important consequence on the localized category $W^{-1}M$. For this, we introduce the notion of homotopy between morphisms in M in the following way. Two morphisms $f, g : X \longrightarrow Y$ are called *homotopic* if there is a commutative diagram in M

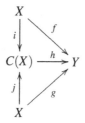

satisfying the following two properties:

(a) There exists a morphism $p : C(X) \longrightarrow X$, which belongs to $Fib \cap W$, such that $pi = pj = id$.
(b) The induced morphism

$$i \bigsqcup j : X \bigsqcup X \longrightarrow C(X)$$

is a cofibration.

When X is cofibrant and Y is fibrant in M (i.e. $\emptyset \longrightarrow X$ is a cofibration and $Y \longrightarrow *$ is a fibration), it can be shown that being homotopic as defined above is an equivalence relation on the set of morphisms from X to Y. This equivalence relation is shown to be compatible with composition, which implies the existence of a category $M^{cf} / \sim$, whose objects are cofibrant and fibrant objects and morphisms are homotopy classes of morphisms.

It is easy to see that if two morphisms f and g are homotopic in M then they are equal in $W^{-1}M$. Indeed, in the diagram above defining the notion of being homotopic, the image of p in $Ho(M)$ is an isomorphism. Therefore, so are the images of i and j. Moreover, the inverses of the images of i and j in $Ho(M)$ are equal (because equal to the image of p), which implies that i and j have the same image in $Ho(M)$. This implies that the image of f and of g are also equal. From this, we deduce that the localization functor

$$M \longrightarrow Ho(M)$$

restricted to the sub-category of cofibrant and fibrant objects M^{cf} induces a well defined functor

$$M^{cf} / \sim \longrightarrow Ho(M).$$

The main statement of model category theory is that this last functor is an equivalence of categories.

Our first main example of a model category will be $C(k)$, the category of complexes over some base commutative ring k. The fibrations are taken to be the degree-wise surjective morphisms, and the equivalences are taken to be the quasi-isomorphisms. This determines the class of cofibrations as the morphisms having the correct lifting property. It is an important theorem that this defines a model category structure on $C(k)$ (see [13]). The homotopy category of this model category is by definition $D(k)$ the derived category of k. Therefore, maps in $D(k)$ can be described as homotopy classes of morphisms between fibrant and cofibrant complexes. As the cofibrant objects in $C(k)$ are essentially the complexes of projective modules (see [13] or Exercise 3.1.2 below) and that every object is fibrant, this gives back essentially the usual way of describing maps in derived categories.

Exercise 3.1.2 (a) *Prove that if E is a cofibrant object in $C(k)$ then for any $n \in \mathbb{Z}$ the k-module E^n is projective.*

(b) Prove that if E is a complex which is bounded above (i.e. there is an n_0 such that $E^n = 0$ for all $n \geq n_0$), and such that E^n is projective for all n, then E is cofibrant.

(c) Contemplate the example in [13, Rem. 2.3.7] of a complex of projective modules which is not a cofibrant object in $C(k)$.

Here are few more examples of model categories.

Examples:

(a) The category *Top* of topological spaces is a model category whose equivalences are the weak equivalences (i.e. continuous maps inducing isomorphisms on all homotopy groups) and whose fibrations are the Serre fibrations (see [13, Def. 2.3.4]). All objects are fibrant for this model category, and the typical cofibrant objects are the CW-complexes. Its homotopy category $Ho(Top)$ is also equivalent to the category of CW-complexes and homotopy classes of continuous maps between them.

(b) For any model category M and any (small) category I we consider M^I the category of I-diagrams in M (i.e. of functors from I to M). We define a morphism $f : F \longrightarrow G$ in M^I to be a fibration (resp. an equivalence) if for all $i \in I$ the induced morphism $f_i : F(i) \longrightarrow G(i)$ is a fibration (resp. an equivalence) in M. When M satisfies a technical extra condition, precisely when M is *cofibrantly generated* (see [13, Sect. 2.1]), then these notions define a model category structure on M^I. The construction $M \mapsto M^I$ is very useful as it allows to construct new model categories from old ones.

(c) Let *Cat* be the category of categories. We define a morphism in *Cat* to be an equivalence if it is a categorical equivalence, and a cofibration if it is injective on the set of objects. This defines a model category structure on *Cat* (see [15]).

(d) Let $\mathscr{A}$ be any Grothendieck category and $M = C(\mathscr{A})$ be its category of complexes. Then it can be shown that there exists a model category structure on M whose equivalences are the quasi-isomorphisms and the cofibrations are the monomorphisms (see [14]).

Exercise 3.1.3 *Let M be a model category and $Mor(M)$ be the category of morphisms in M (objects are morphisms and morphisms are commutative squares in M). We define a morphism $(f, g) : u \longrightarrow v$*

$$
\begin{array}{ccc}
A & \xrightarrow{\ f\ } & A' \\
{\scriptstyle u}\downarrow & & \downarrow{\scriptstyle v} \\
B & \xrightarrow{\ g\ } & B',
\end{array}
$$

to be an equivalence (resp. a fibration) if both f and g are equivalences (resp. fibrations) in M. Show that this defines a model category structure on $Mor(M)$. Show *moreover that a morphism (f,g) a cofibration if and only if f is cofibration and the induced morphism*

$$B \bigsqcup_A A' \longrightarrow B'$$

are cofibrations in M.

Before going back to dg-catgeories we will need a more structured notion of a model category structure, the notion of a $C(k)$-*model category structure*. Suppose that M is a model category. A $C(k)$-model category structure on M is the data of a functor

$$- \otimes - : C(k) \times M \longrightarrow M$$

satisfying the following two conditions.

(a) The functor $\otimes$ above defines a closed $C(k)$-module structure on M (see [13, Sect. 4]). In other words, we are given functorial isomorphisms in M

$$E \otimes (E' \otimes X) \simeq (E \otimes E') \otimes X \qquad k \otimes X \simeq X$$

for any $E, E' \in C(k)$ and $X \in M$ (satisfying the usual associativity and unit conditions, see [13, Sect. 4]). We are also given for two objects X and Y in M a complex $\underline{Hom}(X,Y) \in C(k)$, together with functorial isomorphisms of complexes

$$Hom(E, \underline{Hom}(X,Y)) \simeq Hom(E \otimes X, Y)$$

for $E \in C(k)$, and $X, Y \in M$.

(b) For any cofibration $i : E \longrightarrow E'$ in $C(k)$, and any cofibration $j : A \longrightarrow B$ in M, the induced morphism

$$E \otimes B \bigsqcup_{E \otimes A} E' \otimes A \longrightarrow E' \otimes B$$

is a cofibration in M, which is moreover an equivalence if i or j is so.

Condition (1) above is a purely categorical stucture, and simply asserts the existence of an enrichment of M into $C(k)$ in a rather strong sense. The second condition is a compatibility condition between this enrichement and the model structures on $C(k)$ and M (which is the non trivial part to check in practice).

Examples:

(a) The category $C(k)$ can be considered as enriched over itself by using the tensor product of complexes $- \otimes - : C(k) \times C(k) \longrightarrow C(k)$. For this tensoring it is a $C(k)$-model category (this is another way to state that $C(k)$ is a monoidal model category in the sense of [13, Def. 4.2.6]).
(b) Let X be a topological space. We let $Sh(X,k)$ be the category of sheaves of k-modules and $C(Sh(X,k))$ be the category of complexes in $Sh(X,k)$. As $Sh(X,k)$ is a Grothendieck category, the category $C(Sh(X,k))$ can be endowed with a model category structure for which the equivalences are the

quasi-isomorphisms and the cofibrations are the monomorphisms of complexes. The category $Sh(X,k)$ has a natural tensoring over the category of k-modules, and this structure extends to a tensoring of $C(Sh(X,k))$ over the category $C(k)$. Explicitely, if $\mathscr{F}$ is any sheaf of complexes of k-modules over X and $E \in C(k)$, we let $E \otimes \mathscr{F}$ to be the sheaf associated with the presheaf $U \mapsto E \otimes \mathscr{F}(U) \in C(k)$. It can be shown that this tensoring makes $C(Sh(X,k))$ into a $C(k)$-model category.

One main consequence for a model category M to be a $C(k)$-model category is that its homotopy category $Ho(M)$ comes equipped with a natural enrichment over $D(k) = Ho(C(k))$. Explicitely, for two objects x and y in M we set

$$\mathbb{R}\underline{Hom}(x,y) := \underline{Hom}(Qx,Ry),$$

where Qx is a cofibrant replacement of x and Ry is a fibrant replacement of y. The object $\mathbb{R}\underline{Hom}(x,y) \in D(k)$ can be seen to define an enrichment of $Ho(M)$ into $D(k)$ (see [13, Thm. 4.3.4] for details). A direct consequence of this is the important formula

$$H^0(\mathbb{R}\underline{Hom}(x,y)) \simeq Hom_{Ho(M)}(x,y).$$

Therefore, we see that if x and y are cofibrant and fibrant, then set of morphisms between x and y in $Ho(M)$ can be identified with $H^0(\mathbb{R}\underline{Hom}(x,y))$.

Exercise 3.1.4 *Let $f : M \longrightarrow N$ be a functor between two model categories.*

(a) *Show that if f preserves cofibrations and trivial cofibrations then it also preserves equivalences between cofibrant objects.*
(b) *Assume that f preserves cofibrations and trivial cofibrations and that it does admit a right adjoint $g : N \longrightarrow M$. Show that g preserves fibrations and trivial fibrations.*
(c) *Under the same conditions as in (2), define*

$$\mathbb{L}f : Ho(M) \longrightarrow Ho(N)$$

by sending an object x to $f(Qx)$ where Qx is a cofibrant replacement of x. In the same way, define

$$\mathbb{R}g : Ho(M) \longrightarrow Ho(N)$$

by sending an object y to $g(Ry)$ where Ry is a fibrant replacement of y. Show that $\mathbb{L}f$ and $\mathbb{R}g$ are adjoint functors.

3.2 Model Categories and dg-Categories

We start by the model category of dg-categories itself. The equivalences for this model structure are the quasi-equivalences. The fibrations are defined to be the morphisms $f : T \longrightarrow T'$ satisfying the following two properties. The cofibrations are then defined to be the morphisms with the correct lifting property.

(a) For any two objects x and y in T, the induced morphism

$$f_{x,y} : T(x,y) \longrightarrow T'((f(x),f(y))$$

is a fibration in $C(k)$ (i.e. is surjective).

(b) For any isomorphism $u' : x' \to y'$ in $H^0(T')$, and any $y \in H^0(T)$ such that $f(y) = y'$, there is an isomorphism $u : x \to y$ in $H^0(T)$ such that $H^0(f)(u) = u'$.

Theorem 3.2.1 *(see [27]) The above definitions define a model category structure on dg − cat.*

This is a key statement in the homotopy theory of dg-categories, and many results in the sequel will depend in an essential way from the existence of this model structure. We will not try to describe its proof in these notes, this would lead us too far.

The theorem 3.2.1 is of course very useful, even though it is not very easy to find cofibrant dg-categories and also to describe the homotopy equivalence relation in general. However, we will see in the next lecture that this theorem implies another statement which provide a very useful way to described maps in $Ho(dg - cat)$. It is this last description that will be used in order to check that localizations in the sense of dg-categories (see definition 2.4.1) always exist.

Exercise 3.2.2 *(a) Let **1** be the dg-category with a unique object and k as endomorphism of this object (this is also the unit for the monoidal structure on dg − cat). Show that **1** is a cofibrant object.*

(b) Let Δ_k^1 be the k-linear category with two objects 0 and 1 and with (all k's are here placed in degree 0)

$$\Delta_k^1(0,0) = k \qquad \Delta_k^1(0,1) = k \qquad \Delta_k^1(1,1) = k \qquad \Delta_k^1(1,0) = 0$$

and obvious compositions (Δ_k^1 is the k-linearization of the category with two objects and a unique non trivial morphism between them). Show that Δ_k^1 is a cofibrant object.

(c) Use exercice 2.3.7 in order to show that $k[\varepsilon]$ is not a cofibrant dg-category (when considered as a dg-category with a unique object).

(d) Let T be the dg-category with four objects x, x', y and y' and with the following non trivial complex of morphisms (here we denote by $k < x >$ the rank 1 free k-module with basis x)

$$T(x,x')^0 = k < f > \quad T(x,y)^0 = k < u > \quad T(x',y')^0 = k < u' > \quad T(y,y')^0 = k < g >$$

$$T(x,y')^0 = k < u'f > \oplus k < gu > \quad T(x,y')^{-1} = k < h > \quad T(x,y')^i = 0 \; for \; i \neq 0, -1$$

such that $d(h) = u'f - gu$. In other words, T is freely generated by four morphisms of degree 0, u, u', f and g, one morphism of degre −1, h, and has a unique relation $d(h) = u'f - gu$. Show that there exists a trivial fibration

$$T \longrightarrow \Delta_k^1 \otimes \Delta_k^1.$$

Show moreover that this trivial fibration possesses no section, and conclude that $\Delta_k^1 \otimes \Delta_k^1$ is not a cofibrant dg-category.

Let now T be a dg-category. A T-dg-module is the data of a dg-functor $F : T \longrightarrow C(k)$. In other words a T-dg-module F consists of the data of complexes $F_x \in C(k)$ for each object x of T, together with morphisms

$$F_x \otimes T(x,y) \longrightarrow F_y$$

for each objects x and y, satisfying the usual associativity and unit conditions. A morphism of T-dg-module consists of a natural transformation between dg-functors (i.e. families of morphisms $F_x \longrightarrow F'_x$ commuting with the maps $F_x \otimes T(x,y) \longrightarrow F_y$ and $F'_x \otimes T(x,y) \longrightarrow F'_y$).

We let $T - Mod$ be the category of T-dg-modules. We define a model category structure on $T - Mod$ by defining equivalences (resp. fibrations) to be the morphisms $f : F \longrightarrow F'$ such that for all $x \in T$ the induced morphism $f_x : F_x \longrightarrow F'_x$ is an equivalence (resp. a fibration) in $C(k)$. It is known that this defines a model category structure (see [28]). This model category is in a natural way a $C(k)$-model category, for which the $C(k)$-enrichment is defined by the formula $(E \otimes F)_x := E \otimes F_x$.

Definition 3.2.3 *The* derived category *of a dg-category T is*

$$D(T) := Ho(T - Mod).$$

The previous definition generalizes the derived categories of rings. Indeed, if R is a k-algebra it can also be considered as a dg-category, sometimes denoted by BR, with a unique object and R as endomorphism of this object (considered as a complex of k-modules concentrated in degree 0). Then $D(BR) \simeq D(R)$. Indeed, a BR-dg-module is simply a complex of R-modules.

Exercise 3.2.4 *Let T be a dg-category.*

(a) *Let $x \in T$ be an object in T and $\underline{h}_x : T^{op} \longrightarrow C(k)$ the T-dg-module represented by x (the one sending y to $T(y,x)$). Prove that $\underline{h}_x$ is cofibrant and fibrant as an object in $T^{op} - Mod$.*

(b) *Prove that $x \mapsto \underline{h}_x$ defines a functor*

$$H^0(T) \longrightarrow D(T^{op}).$$

(c) *Show that for any $F \in D(T^{op})$ there is a functorial bijection*

$$Hom_{D(T^{op})}(\underline{h}_x, F) \simeq H^0(F_x).$$

(d) *Show that the above functor $H^0(T) \longrightarrow D(T^{op})$ is fully faithful.*

Any morphism of dg-categories $f : T \longrightarrow T'$ induces an adjunction on the corresponding model categories of dg-modules

$$f_! : T - Mod \longrightarrow T' - Mod \qquad T - Mod \longleftarrow T' - Mod : f^*,$$

for which the functor f^* is defined by composition with f, and $f_!$ is its left adjoint. This adjunction is a *Quillen adjunction*, i.e. f^* preserves fibrations and trivial fibrations, and therefore can be derived into an adjunction on the level of homotopy categories (see exercice 3.1.4 and [13, Lem. 1.3.10])

$$\mathbb{L}f_! : D(T) \longrightarrow D(T') \qquad D(T) \longleftarrow D(T') : f^* = \mathbb{R}f^*.$$

It can be proved that when f is a quasi-equivalence then f^* and $\mathbb{L}f_!$ are equivalences of categories inverse to each others (see [28, Prop. 3.2]).

Exercise 3.2.5 *Let $f : T \longrightarrow T'$ be a dg-functor. Prove that for any $x \in T$ we have*

$$\mathbb{L}f_!(\underline{h}^x) \simeq \underline{h}^{f(x)}$$

in $D(T')$ (recall that $\underline{h}^x$ is the T-dg-module corepresented by x, sending y to $T(x,y)$).

For a $C(k)$-model category M we can also define a notion of *T-dg-modules with coefficients in M* as being dg-functors $T \longrightarrow M$ (where M is considered as a dg-category using its $C(k)$-enrichment). This category is denoted by M^T (so that $T - Mod = C(k)^T$). When M satisfies some mild assumptions (e.g. being cofibrantly generated, see [13, Sect. 2.1]) we can endow M^T with a model category structure similar to $T - Mod$, for which equivalences and fibrations are defined levelwise in M. The existence of model categories as M^T will be used in the sequel to describe morphisms in $Ho - (dg - cat)$.

Exercise 3.2.6 *Let T and T' be two dg-categories. Prove that there is an equivalence of categories*

$$M^{(T \otimes T')} \simeq (M^T)^{T'}.$$

Show moreover that this equivalence of categories is compatible with the two model category structures on both sides.

We finish this second lecture by describing a way to construct many examples of dg-categories using model categories. For this, let M be a $C(k)$-enriched model category. Using the $C(k)$-enrichment M can also be considered as a dg-category whose set of objects is the same as the set of objects of M and whose complexes of morphisms are $\underline{Hom}(x,y)$. This dg-category will sometimes be denoted by $\underline{M}$, but it turns out not to be the right dg-category associated to the $C(k)$-model category M (at least it is not the one we will be interested in the sequel). Instead, we let $Int(M)$ be the full sub-dg-category of $\underline{M}$ consisting of fibrant and cofibrant objects in M. From the general theory of model categories it can be easily seen that the category $H^0(Int(M))$ is naturally isomorphic to the category of fibrant and cofibrant objects in M and homotopy classes of morphisms between them. In particular there exists a natural equivalence of categories

$$H^0(Int(M)) \simeq Ho(M).$$

The dg-category $Int(M)$ is therefore a dg-enhancement of the homotopy category $Ho(M)$. Of course, not every dg-category is of form $Int(M)$. However, we will see that any dg-catgeory can be, up to a quasi-equivalence, fully embedded into some dg-category of the form $Int(M)$. This explains the importance of $C(k)$-model categories in the study of dg-categories.

Remark 2. The construction $M \mapsto Int(M)$ is an ad-hoc construction, and does not seem very intrinsic (e.g. as it is defined it depends on the choice of fibrations and cofibrations in M, and not only on equivalences). However, we will see in the next lecture that $Int(M)$ can also be characterized by as the localization of the dg-category $\underline{M}$ along the equivalences in M, showing that it only depends on the dg-category $\underline{M}$ and the subset W (and not of the classes Fib and Cof)

Let T be a dg-category. We can consider the $C(k)$-enriched Yoneda embedding

$$\underline{h}_- : T \longrightarrow T^{op} - Mod,$$

which is a dg-functor when $T^{op} - Mod$ is considered as a dg-category using its natural $C(k)$-enrichment. It turns out that for any $x \in T$, the T^{op}-dg-module $\underline{h}_x$ is cofibrant (see exercice 3.2.4) and fibrant (avery T^{op}-dg-module if fibrant by definition). We therefore get a natural dg-functor

$$\underline{h} : T \longrightarrow Int(T^{op} - Mod).$$

It is easy to check that $\underline{h}$ is quasi-fully faithful (it even induces isomorphisms on complexes of morphisms).

Definition 3.2.7 *For a dg-category T the morphism*

$$\underline{h} : T \longrightarrow Int(T^{op} - Mod)$$

is called the Yoneda embedding *of the dg-category T.*

4 Lecture 3: Structure of the Homotopy Category of dg-Categories

In this lecture we will truly start to go into the heart of the subject and describe the category $Ho(dg - cat)$. I will start by a theorem describing the set of maps between two objects in $Ho(dg - cat)$. This fundamental result has two important consequences: the existence of localizations of dg-categories, and the existence of dg-categories of morphisms between two dg-categories, both characterized by universal properties in $Ho(dg - cat)$. At the end of this lecture, I will introduce the notion of *Morita equivalences and triangulated dg-categories*, and present a refine version of the category $Ho(dg - cat)$, better suited for many purposes.

4.1 Maps in the Homotopy Category of dg-Categories

We start by computing the set of maps in $Ho(dg - cat)$ from a dg-category T to a dg-category of the form $Int(M)$. As any dg-category can be fully embedded into some $Int(M)$ this will be enough to compute maps in $Ho(dg - cat)$ between any two objects.

Let M be a $C(k)$-model category. We assume that M satisfies the following three technical conditions (they will always be satisfied for the applications we have in mind).

(a) M is cofibrantly generated, and the domain and codomain of the generating cofibrations are cofibrant objects in M.
(b) For any cofibrant object X in M, and any quasi-isomorphism $E \longrightarrow E'$ in $C(k)$, the induced morphism $E \otimes X \longrightarrow E' \otimes X$ is an equivalence.
(c) Infinite sums preserve weak equivalences in M.

Exercise 4.1.1 *Let R be a k-algebra considered as a dg-category. Show that the $C(k)$-model category $R - Mod = C(R)$ does not satisfy condition (2) above if R is not flat over k.*

Condition (1) this is a very mild condition, as almost all model categories encountered in real life are cofibrantly generated. Condition (2) is more serious, as it states that cofibrant objects of M are flat in some sense, which is not always the case. For example, to be sure that the model category $T - Mod$ satisfies (2) we need to impose the condition that all the complexes $T(x,y)$ are flat (e.g. cofibrant in $C(k)$). Conditions (3) is also rather mild and is often satisfied for model categories of algebraic nature. The following proposition is the main result concerning the description of the set of maps in $Ho(dg - cat)$, and almost all the further results are consequences of it. Note that it is wrong if condition (2) above is not satisfied.

Proposition 4.1.2 *Let T be any dg-category and M be a $C(k)$-model category satisfying conditions (1), (2) and (3) above. Then, there exists a natural bijection*

$$[T, Int(M)] \simeq Iso(Ho(M^T))$$

between the set of morphisms from T to $Int(M)$ in $Ho(dg - cat)$ and the set of isomorphism classes of objects in $Ho(M^T)$.

Ideas of proof (see [28] for details): Let $Q(T) \longrightarrow T$ be a cofibrant model for T. The pull-back functor on dg-modules with coefficients in M induces a functor

$$Ho(M^T) \longrightarrow Ho(M^{Q(T)}).$$

Condition (2) on M insures that this is an equivalence of categories, as shown by the following lemma.

Lemma 4.1.3 *Let* $f : T' \longrightarrow T$ *be a quasi-equivalence between dg-categories and* M *be a* $C(k)$*-model category satisfying conditions* (1)*,* (2) *and* (3) *as above. Then the Quillen adjunction*

$$f_! : Ho(M^{T'}) \longrightarrow Ho(M^T) \qquad Ho(M^{T'}) \longleftarrow Ho(M^T) : f^*$$

is a Quillen equivalence.

Idea of a proof of the lemma: We need to show that the two natural transformations

$$\mathbb{L}f_!f^* \Rightarrow id \qquad id \Rightarrow f^*\mathbb{L}f_!$$

are isomorphism. For this, we first check that this is the case when evaluated at a certain kind of objects. Let $x \in T$ and $X \in M$ be a cofibrant object. We consider the object $\underline{h}^x \otimes X \in Ho(M^T)$, sending $y \in T$ to $T(x,y) \otimes X \in M$. Let $x' \in T'$ be an object such that $f(x')$ and x are isomorphic in $H^0(T')$. Because of our condition (2) on M it is not hard to show that $\underline{h}^x \otimes X$ and $\underline{h}^{f(x')} \otimes X$ are isomorphic in $Ho(M^T)$. Therefore, we have

$$f^*(\underline{h}^x \otimes X) \simeq f^*(\underline{h}^{f(x')} \otimes X).$$

Moreover, $f^*(\underline{h}^{f(x')} \otimes X) \in Ho(M^{T'})$ sends an object $y' \in T'$ to $T(f(x'), f(y')) \otimes X$. Because f is quasi-fully faithful (and because of our assumption (2) on M) we see that $f^*(\underline{h}^{f(x')} \otimes X)$ is isomorphic in $Ho(M^{T'})$ to $\underline{h}^{x'} \otimes X$ which sends y' to $T'(x', y') \otimes X$. Finally, it is not hard to see that $\underline{h}^{x'} \otimes X$ is a cofibrant object and that

$$\mathbb{L}f_!(\underline{h}^{x'} \otimes X) \simeq f_!(\underline{h}^{x'} \otimes X) \simeq \underline{h}^{f(x')} \otimes X.$$

Thus, we have

$$\mathbb{L}f_!f^*(\underline{h}^x \otimes X) \simeq \mathbb{L}f_!(\underline{h}^{x'} \otimes X) \simeq \underline{h}^{f(x')} \otimes X \simeq \underline{h}^x \otimes X,$$

or in other words the adjunction morphism

$$\mathbb{L}f_!f^*(\underline{h}^x \otimes X) \longrightarrow \underline{h}^x \otimes X$$

is an isomorphism. In the same way, we can see that for any $x' \in T'$ the adjunction morphism

$$\underline{h}^{x'} \otimes X \longrightarrow f^*\mathbb{L}f_!(\underline{h}^{x'} \otimes X)$$

is an isomorphism.

To conclude the proof of the lemma we use that the objects $\underline{h}^x \otimes X$ generate the category $Ho(M^T)$ be homotopy colimits and that f^* and $\mathbb{L}f_!$ both commute with homotopy colimits. To see that $\underline{h}^x \otimes X$ generates $Ho(M^T)$ by homotopy colimits we use the condition (1), and the small object argument, which shows that any object is equivalent to a transfinite composition of push-outs along morphisms $\underline{h}^x \otimes X \longrightarrow \underline{h}^x \otimes Y$ for $X \to Y$ a cofibration between cofibrant objects in M. The fact that $\mathbb{L}f_!$ preserves homotopy colimits is formal and follows from the general fact that the left derived functor of a left Quillen functor always preserves homotopy colimits. Finally, the fact that f^* preserves homotopy colimits uses the condition (3) (which up to now has not been used). Indeed, we need to show that f^* preserves infinite homotopy sums and homotopy push-outs. As infinite sums are also infinite homotopy sums in M^T (because of conditions (3)), the fact that f^* preserves infinite homotopy sums follows from the fact that the functor f^* commutes with infinite sums. To show that f^* commutes with homotopy push-outs we use that M^T is a $C(k)$-model category, and thus a stable model category in the sense of [13, Sect. 7]. This implies that homotopy push-outs squares are exactly the homotopy pull-backs squares. As f^* is right Quillen it preserves homotopy pull-backs squares, and thus homotopy push-outs.

Therefore, we deduce from what we have seen that the adjunction morphism

$$\mathbb{L}f_! f^*(E) \longrightarrow E$$

is an isomorphism for any $E \in Ho(M^T)$. In the same we way we see that for any $E' \in Ho(M^{T'})$ the adjunction morphism

$$E' \longrightarrow f^* \mathbb{L}f_!(E')$$

is an isomorphism. This finishes the proof of the lemma.   □

The above lemma imply that we can assume that T is a cofibrant dg-catgeory. As all objects in $dg-cat$ are fibrant $[T, Int(M)]$ is then the quotient of the set of morphisms in $dg-cat$ by the homotopy relations. In particular, the natural map $[T, Int(M)] \longrightarrow Iso(Ho(M^T))$ is surjective (this uses that a cofibrant and fibrant object in M^T factors as $T \to Int(M) \to M$, i.e. is levelwise fibrant and cofibrant). To prove injectivity, we start with two morphisms $u, v : T \longrightarrow Int(M)$ in $dg-cat$, and we assume that the corresponding objects F_u and F_v in M^T are equivalent. Using that any equivalences can be factorized as a composition of trivial cofibrations and trivial fibrations, we easily reduce the problem to the case where there exists a trivial fibration $F_u \longrightarrow F_v$ (the case of cofibration is somehow dual). This morphism can be considered as a dg-functor $T \longrightarrow Int(Mor(M))$, where $Mor(M)$ is the model category of morphisms in M (note that fibrant objects in $Mor(M)$ are fibrations between fibrant objects in M). Moreover, this dg-functor factors throught $T' \subset Int(Mor(M))$, the full sub-dg-category corresponding to equivalences in M. We therefore have a commutative diagram in $dg-cat$

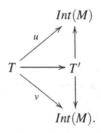

The two morphisms $T' \longrightarrow Int(M)$ are easily seen to be quasi-equivalences, and to possess a common section $Int(M) \longrightarrow T'$ sending an object of M to the its identity morphism. Projecting this diagram in $Ho(dg-cat)$, we see that $[u] = [v]$ in $Ho(dg-cat)$.   □

We will now deduce from proposition 4.1.2 a description of the set of maps $[T,T']$ between two objects in $Ho(dg-cat)$. For this we use the $C(k)$-enriched Yoneda embedding

$$\underline{h}: T' \longrightarrow Int((T')^{op} - Mod),$$

sending an object $x \in T'$ to the $(T')^{op}$-dg-module defined by

$$\underline{h}_x: (T')^{op} \longrightarrow C(k)$$
$$y \longmapsto T'(y,x).$$

The dg-module $\underline{h}$ is easily seen to be cofibrant and fibrant in $(T')^{op} - Mod$, and thus we have $\underline{h}_x \in Int((T')^{op} - Mod)$ as required. The enriched version of the Yoneda lemma implies that $\underline{h}$ is a quasi-fully faithful dg-functor. More precisely, we can show that the induced morphism of complexes

$$T'(x,y) \longrightarrow \underline{Hom}(\underline{h}_x, \underline{h}_y) = Int((T')^{op} - Mod)((\underline{h}_x, \underline{h}_y)$$

is an isomorphims of complexes.

Using the description of maps in $Ho(dg-cat)$ as being homotopy classes of morphisms between cofibrant objects, we see that the morphism $\underline{h}$ induces a injective map

$$[T,T'] \hookrightarrow [T, Int((T')^{op} - Mod)]$$

whose image consists of morphisms $T \longrightarrow Int((T')^{op} - Mod)$ factorizing in $Ho(dg-cat)$ throught the quasi-essential image of $\underline{h}$. We easily get this way the following corollary (see Sect. 3.2 and exercice 2.3.4 for the definition of the tensor product of two dg-categories).

Corollary 1. *Let T and T' be two dg-categories, one of them having cofibrant complexes of morphisms. Then, there exists a natural bijection between $[T,T']$ and the subset of $Iso(Ho(T \otimes (T')^{op} - Mod))$ consisting of $T \otimes (T')^{op}$-dg-modules F such that for any $x \in T$, there is $y \in T'$ such that $F_{x,-}$ and $\underline{h}_y$ are isomorphic in $Ho((T')^{op} - Mod)$.*

Exercise 4.1.4 *Let T be a dg-category.*

(a) Show that $[\mathbf{1},T]$ is in bijection with the set of isomorphism classes of objects in the category $H^0(T)$ (recall that $\mathbf{1}$ is the unit dg-category, with a unique object and k as algebra of endomorphisms).

(b) Show that $[\Delta_k^1, T]$ is in bijection with the set of isomorphism classes of morphisms in the category $H^0(T)$ (recall that Δ_k^1 is the dg-category with two object and freely generated by a unique non trivial morphism).

Exercise 4.1.5 *Let C and D be two k-linear categories, also considered as dg-categories over k. Show that there exists a natural bijection between $[C,D]$ and the set of isomorphism classes of k-linear functors from C to D. Deduce from this that there exists a fully faithful functor*

$$Ho(k-cat) \longrightarrow Ho(dg-cat_k),$$

from the homotopy category of k-linear categories (k-linear categories and isomorphism classes of k-linear functors) and the homotopy category of dg-categories.

Exercise 4.1.6 *Let R be an associative and unital k-algebra, which is also considered as dg-category with a unique object and R as endomorphisms of this object. Show that there is a natural bijection between $[R, Int(C(k))]$ and the set of isomorphism classes of the derived category $D(R)$.*

4.2 Existence of Internal Homs

For two dg-catgeories T and T' we can construct their tensor product $T \otimes T'$ in the following way. The set of objects of $T \otimes T'$ is the product $Ob(T) \times Ob(T')$. For $(x, y) \in Ob(T)^2$ and $(x', y') \in Ob(T')^2$ we set

$$(T \otimes T')((x, x'), (y, y')) := T(x, y) \otimes T(x', y'),$$

with the obvious compositions and units. When k is not a field the functor $\otimes$ does not preserves quasi-equivalences. However, it can be derived by the following formula

$$T \otimes^{\mathbb{L}} T' := Q(T) \otimes Q(T'),$$

where Q is a cofibrant replacement functor on $dg - cat$. This defines a symmetric monoidal structure

$$- \otimes^{\mathbb{L}} - : Ho(dg - cat) \times Ho(dg - cat) \longrightarrow Ho(dg - cat).$$

Proposition 4.2.1 *The monoidal structure $- \otimes^{\mathbb{L}} -$ is closed. In other words, for two dg-categories T and T' there is third dg-category $\mathbb{R}\underline{Hom}(T, T') \in Ho(dg - cat)$, such that for any third dg-category U there exists a bijection*

$$[U, \mathbb{R}\underline{Hom}(T, T')] \simeq [U \otimes^{\mathbb{L}} T, T'],$$

functorial in $U \in Ho(dg - cat)$.

Idea of proof: As for the corollary 1 we can reduce the problem of showing that $\mathbb{R}\underline{Hom}(T, Int(M))$ exists for a $C(k)$-model category M satisfying the same conditions as in proposition 4.1.2. Under the same hypothesis than corollary 1 it can be checked (using proposition 4.1.2) that $\mathbb{R}\underline{Hom}(T, Int(M))$ exists and is given by $Int(M^T)$. □

For two dg-categories T and T', one of them having cofibrant complexes of morphisms it is possible to show that $\mathbb{R}\underline{Hom}(T, T')$ is given by the full sub-dg-category of $Int(T \otimes (T')^{op} - Mod)$ consisting of dg-modules satifying the condition of corollary 1.

Finally, note that when $M = C(k)$ we have

$$\mathbb{R}\underline{Hom}(T, Int(C(k))) \simeq Int(T - Mod).$$

In particular, we find a natural equivalence of categories

$$D(T) \simeq H^0(\mathbb{R}\underline{Hom}(T, Int(C(k)))),$$

which is an important formula.

In the sequel we will use the following notations for a dg-category T

$$L(T) := Int(T - Mod) \simeq \mathbb{R}\underline{Hom}(T, Int(C(k)))$$

$$\widehat{T} := Int(T^{op} - Mod) \simeq \mathbb{R}\underline{Hom}(T^{op}, Int(C(k))).$$

Note that we have natural equivalences

$$H^0(L(T)) \simeq D(T) \qquad H^0(\widehat{T}) \simeq D(T^{op}).$$

Therefore, $L(T)$ and $\widehat{T}$ are dg-enhancement of the derived categories $D(T)$ and $D(T^{op})$. Note also that the Yoneda embedding of definition 3.2.7 is now a dg-functor

$$\underline{h} : T \longrightarrow \widehat{T}.$$

Exercise 4.2.2 *(a) Let R be an associative and unital k-algebra which is conside-red as a dg-category with a unique object. Show that there is an isomorphism in $Ho(dg - cat)$*

$$\mathbb{R}\underline{Hom}(R, Int(C(k))) \simeq L(R).$$

(b) Show that for any two k-algebras R and R', one of them being flat over k we have

$$\mathbb{R}\underline{Hom}(R, L(R')) \simeq L(R \otimes R').$$

Exercise 4.2.3 *Let T be a dg-category. We define the Hochschild cohomology of T by*

$$HH^*(T) := H^*(\mathbb{R}\underline{Hom}(T, T)(id, id)).$$

Let R be an associative k-algebra, flat over k, and considered as a dg-category with a unique object. Show that we have

$$HH^*(R) := Ext^*_{R \otimes R^{op}}(R, R),$$

where the right hand side are the ext-groups computed in the derived category of $R \otimes R^{op}$-modules.

4.3 Existence of Localizations

Let T be a dg-category and let S be subset of morphisms in $H^0(T)$ we would like to invert in $Ho(dg - cat)$. For this, we will say that a morphism $l : T \longrightarrow L_S T$ in $Ho(dg - cat)$ is a *localization of T along S* if for any $T' \in Ho(dg - cat)$ the induced morphism

$$l^* : [L_S T, T'] \longrightarrow [T, T']$$

is injective and its image consists of all morphisms $T \longrightarrow T'$ in $Ho(dg-cat)$ whose induced functor $H^0(T) \longrightarrow H^0(T')$ sends all morphisms in S to isomorphisms in $H^0(T)$. Note that the functor $H^0(T) \longrightarrow H^0(T')$ is only well defined in $Ho(Cat)$ (i.e. up to isomorphism), but this is enough for the definition to makes sense as the condition of sending S to isomorphisms is stable by isomorphism between functors.

Proposition 4.3.1 *For any dg-category T and any set of maps S in $H^0(T)$, a localization $T \longrightarrow L_S T$ exists in $Ho(dg-cat)$.*

Idea of proof (see [28] for details): We start by the most simple example of a localization. We first suppose that $T := \Delta_k^1$ is the dg-category freely generated by two objects, 0 and 1, and a unique morphism $u : 0 \to 1$. More concretely, $T(0,1) = T(0,0) = T(1,1) = k$ and $T(1,0) = 0$, together with the obvious compositions and units. We let **1** be the dg-category with a unique object $*$ and $\mathbf{1}(*,*) = k$ (with the obvious composition). We consider the dg-foncteur $T \longrightarrow \mathbf{1}$ sending the non trivial morphism of T to the identity of $*$ (i.e. $k = T(0,1) \to \mathbf{1}(*,*) = k$ is the identity). We claim that this morphism $T \longrightarrow \mathbf{1}$ is a localization of T along S consisting of the morphism $u : 0 \to 1$ of $T = H^0(T)$. This in fact follows easily from our Proposition 4.1.2. Indeed, for a $C(k)$-model category M the model category M^T is the model category of morphisms in M. It is then easy to check that the functor $Ho(M) \longrightarrow Ho(M^T)$ sending an object of M to the identity morphism in M is fully faithful and that its essential image consists of all equivalences in M.

In the general case, let S be a subset of morphisms in $H^0(T)$ for some dg-category T. We can represent the morphisms S by a dg-functor

$$\bigsqcup_S \Delta_k^1 \longrightarrow T,$$

sending the non trivial morphism of the component s to a representative of s in T. We define $L_S T$ as being the homotopy push-out (see [13] for this notion)

$$L_S T := (\bigsqcup_S \mathbf{1}) \overset{\mathbb{L}}{\underset{\bigsqcup_S \Delta_k^1}{\bigsqcup}} T.$$

The fact that each morphism $\Delta_k^1 \longrightarrow \mathbf{1}$ is a localization and the universal properties of homotopy push-outs imply that the induced morphism $T \longrightarrow L_S T$ defined as above is a localization of T along S. $\qquad\square$

Exercise 4.3.2 *Let Δ_k^1 be the dg-category with two objects and freely generated by a non trivial morphism u between these two objects. We let $S := \{u\}$ be the image of u in $H^0(\Delta_k^1)$. Show that $L_S \Delta_k^1 \simeq \mathbf{1}$.*

Exercise 4.3.3 *Let T and T' be two dg-categories and S and S' be two sets of morphisms in $H^0(T)$ and $H^0(T')$ that contain all the identities. Prove that there is a natural isomorphism in $Ho(dg-cat)$*

$$L_S T \otimes^{\mathbb{L}} L_{S'} T' \simeq L_{S \otimes L_{S'}} T \otimes^{\mathbb{L}} T'.$$

The following proposition describes $Int(M)$ as a dg-localization of M.

Proposition 4.3.4 *Let M be a cofibrantly generated $C(k)$-model category, conside-red also as dg-category $\underline{M}$. There exists a natural isomorphism in $Ho(dg - cat)$*

$$Int(M) \simeq L_W \underline{M}.$$

Idea of proof: We consider the natural inclusion dg-functor $i : Int(M) \longrightarrow \underline{M}$. This inclusion factors as

$$Int(M) \xrightarrow{\ j\ } \underline{M}^f \xrightarrow{\ k\ } \underline{M},$$

where $\underline{M}^f$ is the full sub-dg-category of $\underline{M}$ consisting of fibrant objects. Using that M is cofibrantly generated we can construct dg-functors

$$r : \underline{M} \longrightarrow \underline{M}^f \qquad q : \underline{M}^f \longrightarrow Int(M)$$

together with morphisms

$$jq \to id \qquad qj \to id \qquad id \to ri \qquad id \to ir.$$

Moreover, these morphisms between dg-functors are levelwise in W. This can be seen to imply that the induced morphisms on localizations

$$L_W Int(M) \longrightarrow L_W \underline{M}^f \longrightarrow L_W \underline{M}$$

are isomorphisms in $Ho(dg - cat)$. Finally, as morphisms in W are already invertible in $H^0(Int(M)) \simeq Ho(M)$, we have $L_W Int(M) \simeq Int(M)$. $\qquad\square$

Finally, one possible way to understand localizations of dg-categories is by the following proposition.

Proposition 4.3.5 *Let T be a dg-category and S be a subset of morphisms in $H^0(T)$. Then, the localization morphism $l : T \longrightarrow L_S T$ induces a fully faithful functor*

$$l^* : D(L_S T) \longrightarrow D(T)$$

whose image consists of all T-dg-modules $F : T \longrightarrow C(k)$ sending all morphisms of S to quasi-isomorphisms in $C(k)$.

Idea of proof: This follows from the existence of internal Homs and localizations, as well as the formula

$$D(T) \simeq H^0(\mathbb{R}\underline{Hom}(T, Int(C(k)))) \qquad D(L_S T) \simeq H^0(\mathbb{R}\underline{Hom}(L_S T, Int(C(k)))).$$

Indeed, the universal properties of localizations and internal Homs implies that $\mathbb{R}\underline{Hom}(L_S T, Int(C(k)))$ can be identified full the full sub-dg-category of $\mathbb{R}\underline{Hom}(T, Int(C(k)))$ consisting of dg-functors sending S to quasi-isomorphisms in $C(k)$. $\qquad\square$

Exercise 4.3.6 *Let* $l : T \longrightarrow L_S T$ *be a localization of a dg-category with respect to set of morphisms S in* $H^0(T)$, *and let*

$$\mathbb{L}l_! : D(T^{op}) \longrightarrow D(L_S T^{op})$$

be the induced functor in the corresponding derived categories of modules. Let W_S *be the subset of morphisms u in* $D(T^{op})$ *such that* $\mathbb{L}l_!(u)$ *is an isomorphism in* $D(L_S T^{op})$.

(a) *Show that a morphism* $u : E \longrightarrow F$ *of* $D(T^{op})$ *is in* W_S *if and only if for any* $G \in D(T^{op})$ *such that* $G_x \longrightarrow G_y$ *is a quasi-isomorphism for all* $x \to y$ *in S, the induced map*

$$u^* : Hom_{D(T^{op})}(F, G) \longrightarrow Hom_{D(T^{op})}(E, G)$$

is bijective.

(b) *Show that the induced functor*

$$W_S^{-1} D(T^{op}) \longrightarrow D(L_S T^{op})$$

is an equivalence of categories.

4.4 Triangulated dg-Categories

In this section we will introduce a class of dg-categories called *triangulated*. The notion of being triangulated is the dg-analog of the notion of being Karoubian for linear categories. We will see that any dg-category has a triangulated hull, and this will allow us to introduce a notion of Morita equivalences which is a dg-analog of the usual notion of Morita equivalences between linear categories. The homotopy category of dg-categories up to Morita equivalences will then be introduced and shown to have better properties than the category $Ho(dg - cat)$. We will see in the next lecture that many invariants of dg-categories (K-theory, Hochschild homology ...) factor throught Morita equivalences.

Let T be a dg-category. We recall the existence of the Yoneda embedding (see definition 3.2.7)

$$T \longrightarrow \widehat{T} = Int(T^{op} - Mod),$$

which is quasi-fully faithful. Passing to homotopy categories we get a fully faithful morphism

$$\underline{h} : H^0(T) \longrightarrow D(T^{op}).$$

An object in the essential image of this functor will be called *quasi-representable*.
Recall that an object $x \in D(T^{op})$ is *compact* if the functor

$$[x, -] : D(T^{op}) \longrightarrow k - Mod$$

commutes with arbitrary direct sums. It is easy to see that any quasi-representable object is compact (see exercice 3.2.4). The converse is not true and we set the following definition.

Definition 4.4.1 *A dg-category T is triangulated if and only if every compact object in $D(T^{op})$ is quasi-representable.*

Remark 3. When T is triangulated we have an equivalence of categories $H^0(T) \simeq D(T^{op})_c$, where $D(T^{op})_c$ is the full sub-category of $D(T)$ of compact objects. The category $D(T)$ has a natural triangulated structure which restricts to a triangulated structure on compact objects (see [22] for more details on the notion of triangulated categories). Therefore, when T is triangulated dg-category its homotopy category $H^0(T)$ comes equiped with a natural triangulated structure. This explains the terminology of triangulated dg-category. For more about the relations between the notions of triangulated dg-categories and the notions of triangulated categories we refer to [4]. However, it is not necessary to know the theory of triangulated categories in order to understand triangulated dg-categories, and thus we will not study in details the precise relations between triangulated dg-categories and triangulated categories.

We let $Ho(dg - cat^{tr}) \subset Ho(dg - cat)$ be the full sub-category of triangulated dg-categories. Note that the notation $Ho(dg - cat^{tr})$ suggests that this category is the homotopy category of some model category. We will see that it is, equivalent to, the localization of the category $dg - cat$ along the class of *Morita equivalences*, that will be introduced later on in this section.

Proposition 4.4.2 *The natural inclusion*

$$Ho(dg - cat^{tr}) \longrightarrow Ho(dg - cat)$$

has a left adjoint. In other words, any dg-category has a triangulated hull.

Idea of proof: Let T be a dg-category. We consider the Yoneda embedding (see definition 3.2.7)

$$\underline{h} : T \longrightarrow \widehat{T}.$$

This is a quasi-fully faithful dg-functor. We consider $\widehat{T}_{pe} \subset \widehat{T}$, the full sub-dg-category consisting of all objects which are compact in $D(T^{op})$ (these objects will simply be called *compact*). The dg-category $\widehat{T}_{pe}$ will be called the dg-category of *perfect T^{op}-dg-modules*, or equivalently of *compact T^{op}-dg-modules*. As any quasi-representable object is compact, the Yoneda embedding factors as a full embedding

$$\underline{h} : T \longrightarrow \widehat{T}_{pe}.$$

Let now T' be a triangulated dg-category. By definition, the natural morphism

$$T' \longrightarrow \widehat{T}'_{pe}$$

is an isomorphism in $Ho(dg - cat)$. We can then consider the induced morphism

$$[\widehat{T}_{pe}, \widehat{T}'_{pe}] \longrightarrow [T, \widehat{T}'_{pe}],$$

induced by the resytiction along the morphism $T \longrightarrow \widehat{T}_{pe}$. The hard point is to show that this map in bijective and that $\widehat{T}_{pe}$ is a triangulated dg-category. These two facts can be deduced from the following lemma and the Proposition 4.1.2.

Lemma 4.4.3 *Let T be a dg-category, and $\underline{h} : T \longrightarrow \widehat{T}_{pe}$ be the natural inclusion. Let M be a $C(k)$-model category which satisfies the conditions (1), (2) and (3) of Sect. 3.1. Then the Quillen adjunction*

$$h_! : M^T \longrightarrow M^{\widehat{T}_{pe}} \qquad M^T \longleftarrow M^{\widehat{T}_{pe}} : h^*$$

is a Quillen equivalence.

The proof of the above lemma can be found in [28, Lem. 7.5]. It is based on the fundamental fact that the quasi-representable objects in $D(T^{op})$ generate the subcategory of compact objects by taking a finite number of finite homotopy colimits, shifts and retracts, together with the fact that $\mathbb{L}h_!$ and h^* both preserve these finite homotopy colimits, shifts and retracts.   □

The proof of the proposition shows that the left adjoint to the inclusion is given by

$$\widehat{(-)}_{pe} : Ho(dg - cat) \longrightarrow Ho(dg - cat^{tr}),$$

sending a dg-category T to the full sub-dg-category $\widehat{T}_{pe}$ of $\widehat{T}$ consisting of all compact objects.

For example, if R is a k-algebra, considered as a dg-category with a unique object BR, $\widehat{BR}_{pe}$ is the dg-category of cofibrant and perfect complexes of R-modules. In particular

$$H^0(\widehat{BR}_{pe}) \simeq D_{parf}(R)$$

is the perfect derived category of R. This follows from the fact that compact objects in $D(R)$ are precisely the perfect complexes (this is a well known fact which can also be deduced from the general result [32, Prop. 2.2]). Therefore, we see that the dg-category of perfect complexes over some ring R is the triangulated hull of R.

Definition 4.4.4 *A morphism $T \longrightarrow T'$ in $Ho(dg - cat)$ is called a* Morita equivalence *if the induced morphism in the triangulated hull*

$$\widehat{T}_{pe} \longrightarrow \widehat{T'}_{pe}$$

is an isomorphism in $Ho(dg - cat)$.

It follows formally from the existence of the left adjoint $T \mapsto \widehat{T}_{pe}$ that $Ho(dg - cat^{tr})$ is equivalent to the localized category $W_{mor}^{-1}dg - cat$, where W_{mor} is the subset of Morita equivalences in $dg - cat$ as defined above.

Exercise 4.4.5 *Prove the above assertion: the functor*

$$\widehat{(-)}_{pe} : Ho(dg - cat) \longrightarrow Ho(dg - cat^{tr})$$

induces an equivalence of categories

$$W_{mor}^{-1}Ho(dg - cat) \simeq Ho(dg - cat^{tr}).$$

We can characterize the Morita equivalences in the following equivalent ways.

Proposition 4.4.6 *Let* $f : T \longrightarrow T'$ *be a morphism of dg-categories. The following are equivalent.*

(a) *The morphism* f *is a Morita equivalence.*
(b) *For any triangulated dg-category* T_0, *the induced map*

$$[T', T_0] \longrightarrow [T, T_0]$$

is bijective.
(c) *The induced functor*

$$f^* : D(T') \longrightarrow D(T)$$

is an equivalence of categories.
(d) *The functor*

$$\mathbb{L}f_! : D(T) \longrightarrow D(T')$$

induces an equivalence between the full sub-category of compact objects.

Exercise 4.4.7 *Prove the proposition 4.4.6.*

We finish this section by a description of morphisms in $Ho(dg - cat^{tr})$ in terms of derived categories of bi-dg-modules.

Proposition 4.4.8 *Let* T *and* T' *be two dg-categories. Then, there exists a natural bijection between* $[\widehat{T}_{pe}, \widehat{T}'_{pe}]$ *and the subset of* $Iso(D(T \otimes^{\mathbb{L}} (T')^{op}))$ *consisting of* $T \otimes^{\mathbb{L}} (T')^{op}$-*dg-modules* F *such that for any* $x \in T$, *the* $(T')^{op}$-*dg-module* $F_{x,-}$ *is compact.*

Exercise 4.4.9 *Give a proof of proposition 4.4.8.*

Exercise 4.4.10 (a) *Show that the full sub-category* $Ho(dg - cat^{tr}) \subset Ho(dg - cat)$ *is not stable by finite coproducts (taken inside* $Ho(dg - cat)$).
(b) *Show that the category* $Ho(dg - cat^{tr})$ *has finite sums and finite products.*
(c) *Show that in the category* $Ho(dg - cat^{tr})$, *the natural morphism*

$$T \bigsqcup T' \longrightarrow T \times T',$$

for any T *and* T' *objects in* $Ho(dg - cat^{tr})$. *Note that the symbols* $\bigsqcup$ *and* $\times$ *refer here to the categorical sum and product in the category* $Ho(dg - cat^{tr})$.
(d) *Deduce from this that the set of morphisms* $Ho(dg - cat^{tr})$ *are endowed with natural structure of commutative monoids such that the composition is bilinear. Identify this monoid structure with the direct sum on the level of bi-dg-modules throught the bijection of corollary 1.*

Exercise 4.4.11 *Let* $T \longrightarrow T'$ *be a Morita equivalence and* T_0 *be a dg-category. Show that the induced morphism*

$$T \otimes^{\mathbb{L}} T_0 \longrightarrow T' \otimes^{\mathbb{L}} T_0$$

is again a Morita equivalence (use the lemma 4.4.3).

Exercise 4.4.12 *Let T and T' be two triangulated dg-catgeory, and define*

$$T \widehat{\otimes}^{\mathbb{L}} T' := \widehat{T \otimes^{\mathbb{L}} T'}_{pe}.$$

(a) *Show that $(T, T') \mapsto T \widehat{\otimes}^{\mathbb{L}} T'$ defines a symmetric monoidal structure on $Ho(dg - cat^{tr})$ in such a way that the functor*

$$\widehat{(-)}_{pe} : Ho(dg - cat) \longrightarrow Ho(dg - cat^{tr})$$

is a symmetric monoidal functor.

(b) *Show that the monoidal structure $\widehat{\otimes}^{\mathbb{L}}$ is closed on $Ho(dg - cat^{tr})$.*

Exercise 4.4.13 (a) *Let T and T' be two dg-categories. Prove that the Yoneda embedding $\underline{h} : T \hookrightarrow \widehat{T}_{pe}$ induces an isomorphism in $Ho(dg - cat)$*

$$\mathbb{R}\underline{Hom}(\widehat{T}_{pe}, \widehat{T'}_{pe}) \longrightarrow \mathbb{R}\underline{Hom}(T, \widehat{T'}_{pe}).$$

(b) *Deduce from this that for any dg-category T there is a morphism in $Ho(dg - cat)$*

$$\mathbb{R}\underline{Hom}(T, T) \longrightarrow \mathbb{R}\underline{Hom}(\widehat{T}_{pe}, \widehat{T}_{pe})$$

which is quasi-fully faithful.

(c) *Deduce from this that for any dg-category T there exist isomorphisms*

$$HH^*(T) \simeq HH^*(\widehat{T}_{pe}).$$

5 Lecture 4: Some Applications

In this last lecture I will present some applications of the homotopy theory of dg-categories. We will see in particular how the problems mentioned in *Sect.*1.2 can be solved using dg-categories. The very last section will be some discussions on the notion of saturated dg-categories and their use in the definition of a *secondary K-theory* functor.

5.1 Functorial Cones

One of the problem encountered with derived categories is the non existence of functorial cones. In the context of dg-categories this problem can be solved as follows.

Let T be a triangulated dg-category. We let Δ_k^1 be the dg-category freely generated by two objects 0 and 1 and freely generated by one non trivial morphism $0 \to y$, and **1** be the unit dg-category (with a unique object and k for its endomorphism). There is a morphism

$$\Delta_k^1 \longrightarrow \widehat{\mathbf{1}}_{pe}$$

sending 0 to 0 and 1 to k. We get an induced morphism in $Ho(dg - cat)$

$$\mathbb{R}\underline{Hom}(\widehat{\mathbf{1}}_{pe}, T) \longrightarrow \mathbb{R}\underline{Hom}(\Delta_k^1, T).$$

As T is triangulated we have

$$\mathbb{R}\underline{Hom}(\widehat{\mathbf{1}}_{pe}, T) \simeq \mathbb{R}\underline{Hom}(\mathbf{1}, T) \simeq T.$$

Therefore, we have defined a morphism in $Ho(dg - cat)$

$$f : T \longrightarrow \mathbb{R}\underline{Hom}(\Delta_k^1, T) =: Mor(T).$$

The dg-category $Mor(T)$ is also the full sub-dg-category of $Int(Mor(T^{op} - Mod))$ corresponding to quasi-representable dg-modules, and is called the dg-category of morphisms in T. The morphism f defined above intuitively sends an object $x \in T$ to $0 \to x$ in $Mor(T)$ (note that 0 is an object in T because T is triangulated).

Proposition 5.1.1 *There exists a unique morphism in $Ho(dg - cat)$*

$$c : Mor(T) \longrightarrow T$$

such that the following two $(T \otimes^{\mathbb{L}} Mor(T)^{op})$-dg-modules

$$(z, u) \mapsto Mor(T)(u, f(z)) \qquad (z, u) \mapsto T(c(u), z)$$

are isomorphic in $D(T \otimes^{\mathbb{L}} Mor(T)^{op})$ (In other words, the morphism f admits a left *adjoint).*

Idea of proof: We consider the following explicit models for T, $Mor(T)$ and f. We let T' be the full sub-dg-category of $\widehat{T}$ consisting of quasi-representable objects (or equivalently of compact objects as T is triangulated). We let $Mor(T)'$ be the full sub-dg-category of $Int(Mor(T^{op} - Mod))$ consisting of morphisms between quasi-representable objects (these are also the compact objects in $Ho(Mor(T^{op} - Mod))$ because T is triangulated). We note that $Mor(T)'$ is the dg-category whose objects are cofibrations between cofibrant and quasi-representable T^{op}-dg-modules. To each compact and cofibrant T^{op}-dg-module z we consider $0 \to z$ as an object in T'. This defines a dg-functor $T' \longrightarrow Mor(T)'$ which is a model for f. We define c as being a $C(k)$-enriched left adjoint to c (in the most naive sense), sending an object $c : x \longrightarrow y$ of $Mor(T)'$ to $c(u)$ defined by the push-out in $T^{op} - Mod$

We note that the T^{op}-module $c(u)$ is compact and thus belongs to T'. It is easy to check that c, as a morphism in $Ho(dg - cat)$ satisfies the property of the proposition.

The unicity of c is proved formally, in the same way that one proves the unicity of adjoints in usual category theory. $\square$

The morphism $c : Mor(T) \longrightarrow T$ is a functorial cone construction for the triangulated dg-category T. The important fact here is that there is a natural functor

$$H^0(Mor(T)) \longrightarrow Mor(H^0(T)),$$

which is essentially surjective, full but not faithful in general. The functor

$$H^0(c) : H^0(Mor(T)) \longrightarrow H^0(T)$$

does not factor in general throught $Mor(H^0(T))$.

To finish, proposition 5.1.1 becomes really powerful when combined with the following fact.

Proposition 5.1.2 *Let T be a triangulated dg-category and T' be any dg-category. Then $\mathbb{R}\underline{Hom}(T',T)$ is triangulated.*

Exercise 5.1.3 *Deduce proposition 5.1.2 from exercice 4.4.11.*

One important feature of triangulated dg-categories is that any dg-functor $f : T \longrightarrow T'$ between triangulated dg-categories commutes with cones. In other words, the diagram

$$
\begin{array}{ccc}
Mor(T) & \xrightarrow{\ c\ } & T \\
{\scriptstyle c(f)}\downarrow & & \downarrow{\scriptstyle f} \\
Mor(T') & \xrightarrow[\ c\ ]{} & T'
\end{array}
$$

commutes in $Ho(dg-cat)$. This has to be understood as a generalization of the fact that any linear functor between additive categories commutes with finite direct sums. This property of triangulated dg-categories is very useful in practice, as then any dg-functor $T \longrightarrow T'$ automatically induces a triangulated functor $H^0(T) \longrightarrow H^0(T')$.

Exercise 5.1.4 *Prove the above assertion, that*

$$
\begin{array}{ccc}
Mor(T) & \xrightarrow{\ c\ } & T \\
{\scriptstyle c(f)}\downarrow & & \downarrow{\scriptstyle f} \\
Mor(T') & \xrightarrow[\ c\ ]{} & T'
\end{array}
$$

commutes in $Ho(dg-cat)$ (here T and T' are triangulated dg-categories).

5.2 Some Invariants

Another problem mentioned in *Sect.* 1.2 is the fact that the usual invariants, (K-theory, Hochschild homology and cohomology ...), are not invariants of derived categories. We will see here that these invariants can be defined on the level of $Ho(dg - cat^{tr})$. We will treat the examples of K-theory and Hochschild cohomology.

(a) Let T be a dg-category. We consider $T^{op} - Mod^{cc}$ the full sub-category of compact and cofibrant T^{op}-dg-modules. We can endow $T^{op} - Mod^{cc}$ with a structure of an exact complicial category (see [25]) whose equivalences are quasi-isomorphisms and cofibrations are the cofibrations of the model category structure on $T^{op} - Mod$. This Waldhausen category defines a K-theory space $K(T)$ (see [25]). We note that if T is triangulated we have

$$K_0(T) := \pi_0(K(T)) \simeq K_0(H^0(T)),$$

where the last K-group is the Grothendieck group of the triangulated category $H^0(T)$.

Now, let $f : T \longrightarrow T'$ be a morphism between dg-categories. It induces a functor

$$f_! : T^{op} - Mod \longrightarrow (T')^{op} - Mod.$$

This functor preserves cofibrations, compact cofibrant objects and push-outs. Therefore, it induces a functor between Waldhausen categories

$$f_! : T^{op} - Mod^{cc} \longrightarrow (T')^{op} - Mod^{cc}$$

and a morphism on the corresponding spaces

$$f_! : K(T) \longrightarrow K(T').$$

This defines a functor

$$K : dg - cat \longrightarrow Sp$$

from dg-categories to spectra. It is possible to show that this functor sends Morita equivalences to stable equivalences, and thus defines a functor

$$K : Ho(dg - cat^{tr}) \longrightarrow Ho(Sp).$$

We see it particular that two dg-categories which are Morita equivalent have the same K-theory.

(b) (See also exercice 4.4.13) Let T be a dg-category. We consider $\mathbb{R}\underline{Hom}(T,T)$, the dg-category of (derived) endomorphisms of T. The identity gives an object $id \in \mathbb{R}\underline{Hom}(T,T)$, and we can set

$$HH^{\cdot}(T) := \mathbb{R}\underline{Hom}(T,T)(id,id),$$

the Hochschild complex of T. This is a well defined object in $D(k)$, the derived category of complexes of k-modules, and the construction $T \mapsto HH^{\cdot}(T)$ provides a functor of groupoids

$$Ho(dg - cat)^{iso} \longrightarrow D(k)^{iso}.$$

Using the results of *Sect.* 3.2 we can see that

$$HH^*(T) \simeq Ext^*(T,T),$$

where the Ext-group is computed in the derived category of $T \otimes^{\mathbb{L}} T^{op}$-dg-modules. In particular, when T is given by an associative flat k-algebra R we find

$$HH^*(T) \simeq Ext^*_{R \otimes R}(R,R),$$

which is usual Hochschild cohomology. The Yoneda embedding $T \longrightarrow \widehat{T}_{pe}$, provides an isomorphism in $Ho(dg - cat)$

$$\mathbb{R}\underline{Hom}(\widehat{T}_{pe}, \widehat{T}_{pe}) \simeq \mathbb{R}\underline{Hom}(T, \widehat{T}_{pe}),$$

and a quasi-fully faithful morphism

$$\mathbb{R}\underline{Hom}(T,T) \longrightarrow \mathbb{R}\underline{Hom}(T, \widehat{T}_{pe}).$$

Therefore, we get a quasi-fully faithful morphism in $Ho(dg - cat)$

$$\mathbb{R}\underline{Hom}(T,T) \longrightarrow \mathbb{R}\underline{Hom}(\widehat{T}_{pe}, \widehat{T}_{pe})$$

sending the identity to the indentity. Therefore, we obtain a natural isomorphism

$$HH^{\cdot}(T) \simeq HH^{\cdot}(\widehat{T}_{pe}).$$

We get that way that Hochschild cohomology is a Morita invariant.

(c) There also exists an interpretation of Hochschild homology purely in terms of dg-categories in the following way. We consider two dg-categories T and T', and the Yoneda embedding $\underline{h} : T \hookrightarrow \widehat{T}$. We obtain an induced functor

$$\underline{h}_! : \mathbb{R}\underline{Hom}(T, \widehat{T}') \longrightarrow \mathbb{R}\underline{Hom}(\widehat{T}, \widehat{T}').$$

It is possible to show that this morphism is quasi-fully faithful and that its quasi-essential image consists of all morphisms $\widehat{T} \longrightarrow \widehat{T}'$ which are continuous (i.e. commute with direct sums). We refer to [28, Thm. 7.2] for more details about this statement. This implies that there is an isomorphism in $Ho(dg - cat)$

$$\mathbb{R}\underline{Hom}(T, \widehat{T}') \simeq \mathbb{R}\underline{Hom}_c(\widehat{T}, \widehat{T}'),$$

where $\mathbb{R}\underline{Hom}_c$ denotes the full sub-dg-category of continuous dg-functors. Let now T be a dg-category and consider the $T \otimes^{\mathbb{L}} T^{op}$-dg-module sending (x,y) to $T(y,x)$. This dg-module can be represented by an object in the dg-category

$$L(T \otimes^{\mathbb{L}} T^{op}) \simeq \mathbb{R}\underline{Hom}(T \otimes^{\mathbb{L}} T^{op}, \widehat{\mathbf{1}}) \simeq \mathbb{R}\underline{Hom}_c(\widehat{T \otimes^{\mathbb{L}} T^{op}}, \widehat{\mathbf{1}}),$$

and thus by a continuous in $Ho(dg - cat)$

$$L(T \otimes^{\mathbb{L}} T^{op}) \longrightarrow \widehat{\mathbf{1}}.$$

The image of T, considered as a bi-module sending (x,y) to $T(y,x)$, by this morphism is denoted by $HH.(T) \in D(k) \simeq H^0(\widehat{1})$, and is called the Hochschild homology complex of T. When T is a flat k-algebra R then we have

$$HH.(T) \simeq R \otimes_{R \otimes R^{op}}^{\mathbb{L}} R \in D(k).$$

From its definition, it is not hard to show that $T \mapsto HH.(T)$ is invariant by Morita equivalences.

5.3 Descent

In this section we will see how to solve the non-local nature of derived categories explained in *Sect*. 1.2. For this, let X be a scheme. We have the Grothendieck category $C(\mathscr{O}_X)$ of (unbounded) complexes of sheaves of $\mathscr{O}_X$-modules. This category can be endowed with a model category structure for which the equivalences are the quasi-isomorphisms (of complexes of sheaves) and the cofibrations are the monomorphisms (see e.g. [14]). Moreover, when X is a k-scheme then the natural $C(k)$-enrichment of $C(\mathscr{O}_X)$ makes it into a $C(k)$-model category. We let

$$L(\mathscr{O}_X) := Int(C(\mathscr{O}_X)),$$

and we let $L_{pe}(X)$ be the full sub-dg-category consisting of perfect complexes on X. The K-theory of X can be defined as

$$K(X) := K(L_{pe}(X)),$$

using the definition of K-theory of dg-categories we saw in the last section.

When $f : X \longrightarrow Y$ is a morphism of schemes, it is possible to define two morphisms in $Ho(dg - cat)$

$$\mathbb{L}f^* : L(\mathscr{O}_Y) \longrightarrow L(\mathscr{O}_X) \qquad L(\mathscr{O}_Y) \longleftarrow L(\mathscr{O}_X) : \mathbb{R}f_*,$$

which are adjoints (according to the model we chose $\mathbb{L}f^*$ is a bit tricky to define explicitly). The morphism

$$\mathbb{L}f^* : L(\mathscr{O}_Y) \longrightarrow L(\mathscr{O}_X)$$

always preserves perfect complexes are induces a morphism

$$\mathbb{L}f^* : L_{pe}(Y) \longrightarrow L_{pe}(X).$$

This construction provides a functor

$$L_{pe} : k - Sch^{op} \longrightarrow Ho(dg - cat^{tr}),$$

from the (opposite of) category of k-schemes to the category of triangulated dg-categories. The existence of this functor is the starting point of an extremelly rich source of questions about its behaviour. Following the philosophy of non-commutative geometry according to M. Kontsevich, $Ho(dg - cat^{tr})$ can be considered as the category of *non-commutative schemes*, and the functor L_{pe} above is simply passing from the commutative to the non-commutative setting. Contrary to what this might suggest, at leas as first naive thoughts, the functor L_{pe} is far from being an embedding and its general study leads to very interesting questions.

(a) A first observation is that two schemes X and Y can be such that $L_{pe}(X) \simeq L_{pe}(Y)$ without being isomorphic, as shown by many well known examples of *derived equivalences* (see [23] for more about this). The *fibers* of the functor L_{pe}, that is the set of schemes, up to isomorphisms, having the same dg-categories of perfect complexes, is expected to be finite, at least when we restrict to smooth and projective schemes. It is shown in [1] that these fibers are *discrete* and countable.

(b) The functor L_{pe} sends direct product of k-schemes into tensor product in $Ho(dg - cat^{tr})$, at least under reasonable conditions (see [28], [3]). In other words, L_{pe} is a symmetric monoidal functor, when $k - Sch$ is considered as a symmetric monoidal category for the direct monoidal structure.

(c) The image of the smooth and proper k-schemes inside $Ho(dg - cat^{tr})$ has an explicit description: its objects are smooth and projective k-schemes, and morphisms between two such schemes X and Y are given by quasi-isomorphism classes of perfect complexes on $X \times_k Y$ (see 5.3.2 below). This category is very close to the category of Chow motives, for which morphisms are rather correspondences up to rationnal equivalences. By analogy, $Ho(dg - cat^{tr})$ can be used in order to define a notion of non-commutative motives (see [18]). Constructing *realisations* for these non-commutative motives has lead to the notion of non-commutative Hodge structures (see [16]), and to the construction of the non-commutative Gauss–Manin connexion (see [34]).

We now come back to the descent property. The following proposition will not be proved in these notes. We refer to [12] for more details about the descent for perfect complexes.

Proposition 5.3.1 *Let $X = U \cup V$, where U and V are two Zariski open subschemes. Then the following square*

$$
\begin{array}{ccc}
L_{pe}(X) & \longrightarrow & L_{pe}(U) \\
\downarrow & & \downarrow \\
L_{pe}(V) & \longrightarrow & L_{pe}(U \cap V)
\end{array}
$$

is homotopy cartesian in the model category $dg - cat$.

Let us also mention the following related statement.

Proposition 5.3.2 *Let X and Y be two smooth and proper schemes over $Spec\,k$. Then, there exists a natural isomorphism in $Ho(dg - cat)$*

$$\mathbb{R}\underline{Hom}(L_{pe}(X), L_{pe}(Y)) \simeq L_{pe}(X \times_k Y).$$

For a proof we refer the reader to [28]. It should be emphasised here that the corresponding statement is false on the level of derived categories. More precisely, let $E \in D_{parf}(X \times_k Y)$ and let

$$\phi_E : D_{parf}(X) \longrightarrow D_{parf}(Y)$$
$$F \longmapsto \mathbb{R}(p_Y)_*(E \otimes^{\mathbb{L}} p_X^*(F))$$

be the corresponding functor. The construction $E \mapsto \phi_E$ defines a functor

$$\phi_- : D_{parf}(X \times_k Y) \longrightarrow Fun^{tr}(D_{parf}(X), D_{parf}(Y)),$$

where the right hand side is the category of triangulated functors from $D_{parf}(X)$ to $D_{parf}(Y)$. When X and Y are projective over $Spec\,k$ (and that k is field) then it is known that this functor is essentially surjective (see [23]). In general it is not known if ϕ_- is essentially surjective or not. In any case, even for very simple X and Y the functor ϕ_- is not faithful, and thus is not an equivalence of categories in general. Suppose for instance that $X = Y = E$ and elliptic curve over $k = \mathbb{C}$, and let $\Delta \in D_{parf}(X \times_k X)$ be the structure sheaf of the diagonal. The image by ϕ_- of the objects Δ and $\Delta[2]$ are respectively the identity functor and the shift by 2 functor. Because X is of cohomological dimension 1 we have $Hom(id, id[2]) = 0$, where this hom is computed in $Fun^{tr}(D_{parf}(X), D_{parf}(X))$. However, $Hom(\Delta, \Delta[2]) \simeq HH^2(X) \simeq H^1(E, \mathcal{O}_E) \simeq k$.

5.4 Saturated dg-Categories and Secondary K-Theory

We arrive at the last section of these lectures. We have seen that dg-categories can be used in order to replace derived categories, and that they can be used in order to define K-theory. In this section we will see that dg-categories can also be considered as *coefficients* that can themselves be used in order to define a secondary version of K-theory. For this I will present an analogy between the categories $Ho(dg - cat^{tr})$ and $k - Mod$. Throught this analogy projective k-modules of finite rank correspond to the notion of *saturated dg-categories.* I will then show how to define secondary K-theory spectrum $K^{(2)}(k)$ using saturated dg-categories, and give some ideas of how to define analogs of the rank and chern character maps in order to see that this secondary K-theory $K^{(2)}(k)$ is non-trivial. I will also mention a relation between $K^{(2)}(k)$ and the Brauer group, analog to the well known relation between K-theory and the Picard group.

We start by the analogies between the categories $k - Mod$ of k-modules and $Ho(dg - cat^{tr})$. The true analogy is really between $k - Mod$ and the homotopy theory of triangulated dg-categories, e.g. the simplicial category $Ldg - cat^{tr}$ obtained by simplicial localization (see [29]). The homotopy category $Ho(dg - cat^{tr})$ is sometimes too coarse to see the analogy. We will however restrict ourselves with $Ho(dg - cat^{tr})$, even though some of the facts below about $Ho(dg - cat^{tr})$ are not completely intrinsic and requires to lift things to the model category of dg-categories.

(a) The category $k - Mod$ is a closed symmetric monoidal category for the usual tensor product. In the same way, $Ho(dg - cat^{tr})$ has a closed symmetric monoidal structure induced from the one of $Ho(dg - cat)$ (see *Sect*. 3.2). Explicitly, if T and T' are two triangulated dg-category we form $T \otimes^{\mathbb{L}} T' \in Ho(dg - cat)$. This is not a triangulated dg-category anymore and we set

$$T \widehat{\otimes}^{\mathbb{L}} T' := (\widehat{T \otimes^{\mathbb{L}} T'})_{pe} \in Ho(dg - cat^{tr}).$$

The unit of this monoidal structure is the triangulated hull of $\mathbf{1}$, i.e. the dg-category of cofibrant and perfect complexes of k-modules. The corresponding internal Homs is the one of $Ho(dg - cat)$, as we already saw that $\mathbb{R}\underline{Hom}(T, T')$ is triangulated if T and T' are.

(b) The category $k - Mod$ has a zero object and finite sums are also finite products. This is again true in $Ho(dg - cat^{tr})$. The zero dg-category (with one object and 0 as endomorphism ring of this object) is a zero object in $Ho(dg - cat^{tr})$. Also, for two triangulated dg-categories T and T' their sum $T \bigsqcup T'$ as dg-categories is not triangulated anymore. Their direct sum in $Ho(dg - cat^{tr})$ is the triangulated hull of $T \bigsqcup T'$, that is

$$\widehat{T \bigsqcup T'}_{pe} \simeq \widehat{T}_{pe} \times \widehat{T'}_{pe} \simeq T \times T'.$$

We note that this second remarkable property of $Ho(dg - cat^{tr})$ is not satisfied by $Ho(dg - cat)$ itself. We can say that $Ho(dg - cat^{tr})$ is *semi-additive*, which is justified by the fact that the Homs in $Ho(dg - cat^{tr})$ are abelian monoids (or abelian semi-groups).

(c) The category $k - Mod$ has arbitrary limits and colimits. The corresponding statement is not true for $Ho(dg - cat^{tr})$. However, we have homotopy limits and homotopy colimits in $Ho(dg - cat^{tr})$, whose existence are insured by the model category structure on $dg - cat$.

(d) There is a natural notion of short exact sequences in $k - Mod$. In the same way, there is a natural notion of short exact sequences in $Ho(dg - cat^{tr})$. These are the sequences of the form

$$T_0 \xrightarrow{j} T \xrightarrow{p} \widehat{(T/T_0)}_{pe},$$

where i is a quasi-fully faithful functor between triangulated dg-categories, and $\widehat{(T/T_0)}_{pe}$ is the quotient defined as the triangulated hull of the homotopy pushout of dg-categories

$$\begin{array}{ccc} T_0 & \longrightarrow & T \\ \downarrow & & \downarrow \\ 0 & \longrightarrow & T/T_0. \end{array}$$

These sequences are natural in terms of the homotopy theory of triangulated dg-categories as it can be shown that quasi-fully faithful dg-functors are precisely the *homotopy monomorphisms* in $dg - cat$, i.e. the morphisms $T \longrightarrow T'$ such that the diagonal map

$$T \longrightarrow T \times_{T'}^h T$$

is a quasi-equivalence (the right hand side is a homotopy pull-back). This defines a dual notion of homotopy epimorphisms of triangulated dg-categories as being the morphism $T \longrightarrow T'$ such that for any triangulated dg-categories T'' the induced morphism

$$\mathbb{R}\underline{Hom}(T', T'') \longrightarrow \mathbb{R}\underline{Hom}(T, T'')$$

is a homotopy monomorphisms (i.e. is quasi-fully faithful). In the exact sequences above j is a homotopy monomorphism, p is a homotopy epimorphism, p is the cokernel of j and j is the kernel of p. The situation is therefore really close to the situation in $k - Mod$.

If $k - Mod$ and $Ho(dg - cat^{tr})$ are so analoguous then we should be able to say what is the analog property of being projective of finite rank, and to define a K-group or even a K-theory spectrum of such objects. It turns that this can be done and that the theory can actually be pushed rather far. Also, we will see that this new K-theory might have some geometric and arithmetic significance.

It is well know that the projective modules of finite rank over k are precisely the dualizable (also called rigid) objects in the closed monoidal category $k - Mod$. Recall that any k-module M has a dual $M^\vee := \underline{Hom}(M, k)$, and that there always exists an evaluation map

$$M^\vee \otimes M \longrightarrow \underline{Hom}(M, M).$$

The k-module M is dualizable if this evaluation map is an isomorphism, and this is known to be equivalent to the fact that M is projective of finite rank.

We will take this as a definition of *projective triangulated dg-categories of finite rank*. The striking fact is that these dg-categories have already been studied for other reasons under the name of *saturated dg-categories*, or *smooth and proper dg-categories*.

Definition 5.4.1 *A triangulated dg-category T is* saturated *if it is dualizale in $Ho(dg - cat^{tr})$, i.e. if the evaluation morphism*

$$\mathbb{R}\underline{Hom}(T, \widehat{1}_{pe}) \widehat{\otimes}^{\mathbb{L}} T \longrightarrow \mathbb{R}\underline{Hom}(T, T)$$

is an isomorphism in $Ho(dg - cat^{tr})$.

The saturated triangulated dg-categories can be characterized nicely using the notion of smooth and proper dg-algebras (see [19, 30, 32]). Recall that a dg-algebra B is smooth if B is a compact object in $D(B \otimes^{\mathbb{L}} B^{op})$. Recall also that such a dg-algebra is proper if its underlying complex if perfect (i.e. if B is compact in $D(k)$). The following proposition can be deduced from the results of [32]. We omit the proof in these notes (see however [32] for some statements about saturated dg-categories).

Proposition 5.4.2 *A triangulated dg-category is saturated if and only if it is Morita equivalent to a smooth and proper dg-algebra.*

This proposition is interesting as it allows us to show that there are many examples of saturated dg-categories. The two main examples are the following.

(a) Let X be a smooth and proper k-scheme. Then $L_{pe}(X)$ is a saturated dg-category (see [32]).
(b) For any k-algebra, which is projective of finite rank as a k-module and which is of finite global cohomlogical dimension, the dg-category $\widehat{A}_{pe}$ of perfect complexes of A-modules is saturated.

The symmetric monoidal category $Ho(dg-cat^{sat})$ of saturated dg-categories is rigid. Note that any object T has a dual $T^{\vee} := \mathbb{R}\underline{Hom}(T, \mathbf{1}_{pe})$. Moreover, it can be shown that $T^{\vee} \simeq T^{op}$ is simply the opposite dg-category (this is only true when T is saturated). In particular, for T and T' two saturated dg-categories we have the following important formula

$$T^{op}\widehat{\otimes}^{\mathbb{L}}T' \simeq \mathbb{R}\underline{Hom}(T, T').$$

We can now define the secondary K-group. We start by $\mathbb{Z}[sat]$, the free abelian group on isomorphism classes (in $Ho(dg-cat^{tr})$) of satuared dg-categories. We define $K_0^{(2)}(k)$ to be the quotient of $\mathbb{Z}[sat]$ by the relation

$$[T] = [T_0] + [\widehat{(T/T_0)}_{pe}]$$

for any full saturated sub-dg-category $T_0 \subset T$ with quotient $\widehat{(T/T_0)}_{pe}$.

More generally, we can consider a certain Waldhausen category Sat, whose objects are cofibrant dg-categories T such that $\widehat{T}_{pe}$ is saturated, whose morphisms are morphisms of dg-categories, whose equivalences are Morita equivalences, and whose cofibrations are cofibrations of dg-categories which are also fully faithful. From this we can construct a spectrum, denoted by $K^{(2)}(k)$ by applying Waldhausen's construction, called the *secondary K-theory spectrum of k*. We have

$$\pi_0(K^{(2)}(k)) \simeq K_0^{(2)}(k).$$

We no finish with some arguments that $K^{(2)}(k)$ to show that is non trivial and interesting.

First of all, we have the following two basic properties.

(a) $k \mapsto K^{(2)}(k)$ defines a functor from the category of commutative rings to the homotopy category of spectra. To a map of rings $k \to k'$ we associate the base change $- \otimes_k^{\mathbb{L}} k'$ from saturated dg-categories over k to saturated dg-categories over k', which induces a functor of Waldhausen categories and thus a morphism on the corresponding K-theory spectra.

(b) If $k = colim_i k_i$ is a filtered colimit of commutative rings then we have

$$K^{(2)}(k) \simeq colim_i K^{(2)}(k_i).$$

This follows from the non trivial statement that the homotopy theory of saturated dg-categories over k is the filtered colimit of the homotopy theories of saturated dg-categories over the k_i (see [31]).

(c) The monoidal structure on $Ho(dg - cat^{tr})$ induces a commutative ring structure on $K_0^{(2)}(k)$. I guess that this monoidal structure also induces a E_∞-multiplication on $K^{(2)}(k)$.

Our next task is to prove that $K^{(2)}(k)$ is non zero. For this we construct a rank map

$$rk_0^{(2)} : K_0^{(2)}(k) \longrightarrow K_0(k)$$

which an analog of the usual rank map (also called the trace map)

$$rk_0 : K_0(k) \longrightarrow HH_0(k) = k.$$

Let T be a saturated dg-category. As T is dualizable in $Ho(dg - cat^{tr})$ there exists a trace map

$$\mathbb{R}\underline{Hom}(T,T) \simeq T^{op} \widehat{\otimes}^{\mathbb{L}} T \longrightarrow \widehat{1}_{pe},$$

which is the dual of the identity map

$$id : \widehat{1}_{pe} \longrightarrow T^{op} \widehat{\otimes}^{\mathbb{L}} T.$$

The image of the identity provides a perfect complex of k-modules, and thus an element

$$rk_0^{(2)}(T) \in K_0(k).$$

This defines the map

$$rk_0^{(2)} : K_0^{(2)}(k) \longrightarrow K_0(k).$$

It can be shown that $rk_0^{(2)}(T)$ is in fact $HH.(T)$, the Hochschild homology complex of T.

Lemma 5.4.3 *For any saturated dg-category T we have*

$$rk_0^{(2)}(T) = [HH.(T)] \in K_0(k),$$

where $HH.(T)$ is the (perfect) complex of Hochschild homology of T.

In particular we see that for X a smooth and proper k-scheme we have

$$rk_0^{(2)}(L_{pe}(X)) = [HH.(X)] \in K_0(k).$$

When $k = \mathbb{C}$ then $HH_*(X)$ can be identified with Hodge cohomology $H^*(X, \Omega_X^*)$, and thus $rk_0^{(2)}(L_{pe}(X))$ is then the euler characteristic of X. In other words, we can say that the rank of $L_{pe}(X)$ is $\chi(X)$. The map $rk_0^{(2)}$ shows that $K_0^{(2)}(k)$ is non zero.

The usual rank $rk_0 : K_0(k) \longrightarrow HH_0(k) = k$ is only the zero part of a rank map

$$rk_* : K_*(k) \longrightarrow HH_*(k).$$

In the same way, it is possible to define a secondary rank map

$$rk_*^{(2)} : K_*^{(2)}(k) \longrightarrow K_*(S^1 \otimes^{\mathbb{L}} k),$$

where $S^1 \otimes^{\mathbb{L}} k$ is a simplicial ring that can be defined as

$$S^1 \otimes^{\mathbb{L}} k = k \otimes_{k \otimes_{\mathbb{Z}}^{\mathbb{L}} k}^{\mathbb{L}} k.$$

Note that by definition of Hochschild homology we have

$$HH.(k) \simeq S^1 \otimes^{\mathbb{L}} k,$$

so we can also write

$$rk_*^{(2)} : K_*^{(2)}(k) \longrightarrow K_*(HH.(k)).$$

Using this map I guess it could be possible to check that the higher K-groups $K_i^{(2)}(k)$ are also non zero in general. Actually, I think it is possible to construct an analog of the Chern character

$$Ch : K_*(k) \longrightarrow HC_*(k)$$

as a map

$$Ch^{(2)} : K_*^{(2)}(k) \longrightarrow HC_*^{(2)}(k) := K_*^{S^1}(S^1 \otimes^{\mathbb{L}} k),$$

where the right hand side is the S^1-equivariant K-theory of $S^1 \otimes^{\mathbb{L}} k$ (note that S^1 acts on $S^1 \otimes^{\mathbb{L}} k$), which we take as a definition of secondary cyclic homology (see [33, 34] for more details about this construction).

To finish we show that $K_0^{(2)}(k)$ has a relation with the Brauer group, analog to the relation between $K_0(k)$ and the Picard group. For this, we define $Br_{dg}(k)$ to be the group of isomorphism classes of invertible objects (for the monoidal structure) in $Ho(dg - cat^{tr})$. As being invertible is stronger than being dualizable we have a natural map

$$Br_{dg}(k) \longrightarrow K_0^{(2)}(k)$$

analog to the natural map

$$Pic(k) \longrightarrow K_0(k).$$

Now, by definition $Br_{dg}(k)$ can also be described as the Morita equivalence classes of *Azumaya's dg-algebras*, that is of dg-algebras B satisfying the following two properties

(a)

$$B^{op} \otimes^{\mathbb{L}} B \longrightarrow \mathbb{R}\underline{End}_{C(k)}(B)$$

is a quasi-isomorphism.

(b) The underlying complex of B is a compact generator of $D(k)$.

In particular, a non-dg Azumaya's algebra over k defines an element in $Br_{dg}(k)$, and we thus get a map $Br(k) \longrightarrow Br_{dg}(k)$, from the usual Brauer group of k (see [21]) to the dg-Brauer group of k. Composing with the map $Br_{dg}(k) \longrightarrow K_0^{(2)}(k)$ we get a map

$$Br(k) \longrightarrow K_0^{(2)}(k),$$

from the usual Brauer group to the secondary K-group of k. I do not know if this map is injective in general, but I guess it should be possible to prove that it is non zero in some examples by using the Chern character mentioned above.

Acknowledgements: It is my pleasure to thank the referee for his detailed lecture, and for all its comments, which, I hope, have been very useful in order to make these notes more readable.

References

1. Anel, M., Toën, B.: Dénombrabilité des classes d'équivalences dérivées de variétés algébriques. J. Algebr. Geom. **18**(2), 257–277 (2009)
2. Artin, M., Grothendieck, A., Verdier, J.L.: Théorie des topos et cohomologie étale des schémas- Tome 1. In: Lecture Notes in Mathematics, vol. 269, Springer, Berlin (1972)
3. Francis, J., Ben-Zvi, D., Nadler, D.: Integral transforms and drinfeld centers in derived algebraic geometry. preprint arXiv arXiv:0805.0157
4. Bondal, A., Kapranov, M.: Enhanced triangulated categories, Math. USSR Sbornik **70**, 93–107 (1991)
5. Bourbaki, N.: Eléments de mathématique. Algbre. Chapitre 10. Algèbre homologique. Springer, Berlin (2007)
6. Deligne, P.: Equations différentielles à points singuliers réguliers. Lecture Notes in Mathematics, vol.163. Springer, Berlin (1970)
7. Dugger, D., Shipley, B.: K-theory and derived equivalences. Duke Math J. **124**(3), 587–617 (2004)
8. Elmendorf, A.D., Kriz, I., Mandell, M.A., May, J.P.: Rings, modules, and algebras in stable homotopy theory. In: Mathematical Surveys and Monographs, vol. 47. American Mathematical Society, Providence, RI (1997)
9. Gabriel, P., Zisman, M.: Calculus of fractions and homotopy theory. In: Ergebnisse der Mathematik und ihrer Grenzgebiete, Band 35. Springer, New York (1967)
10. Gabriel, P., Popesco, N.: Caractérisation des catégories abéliennes avec générateurs et limites inductives exactes. C. R. Acad. Sci. Paris **258**, 4188–4190 (1964)
11. Griffiths, P., Harris, J.: Principles of algebraic geometry, Wiley Classics Library. Wiley, New York (1994)
12. Hirschowitz, A., Simpson, C.: Descente pour les n-champs. Preprint math.AG/9807049.
13. Hovey, M.: Model categories. In: Mathematical surveys and monographs, vol. **63**, American Mathematical Society, Providence (1998)
14. Hovey, M.: Model category structures on chain complexes of sheaves. Trans. Am. Math. Soc. **353**(6), 2441–2457 (2001)
15. Joyal, A., Tierney, M.: Strong stacks and classifying spaces. Category theory (Como, 1990). Lecture Notes in Mathematics vol. 1488, pp. 213–236. Springer, Berlin (1991)
16. Katzarkov, L., Kontsevich, M., Pantev, T., Hodge theoretic aspects of mirror symmetry. prépublication arXiv:0806.0107

17. Kelly, G.: Basic concepts of enriched category theory. In: London Mathematical Society Lecture Note Series, vol. **64**. Cambridge Universtity Press, Cambridge (1982)

18. Kontsevich, M.: Note of motives in finite characteristic. preprint arXiv math/0702206.

19. Kontsevich, M., Soibelman, Y.: Notes on A-infinity algebras, A-infinity categories and non-commutative geometry. preprint math.RA/060624

20. May, J.P.: Simplicial objects in algebraic topology. Van Nostrand Mathematical Studies, vol. 11. D. Van Nostrand, Princeton (1967)

21. Milne, J.: Etale cohomology. In: Princeton mathematical series 33. Princeton University Press, Princeton (1980)

22. Neeman, A.: Triangulated categories. In: Annals of Mathematics Studies, vol. 148. Princeton University Press, Princeton, NJ (2001)

23. Rouquier, R.: Catégories dérivées et géométrie birationnelle (d'après Bondal, Orlov, Bridgeland, Kawamata et al.). Séminaire Bourbaki, vol. 2004/2005. Astrisque No. **307**, Exp. No. 946, viii, pp. 283–307 (2006)

24. Schlichting, M.: A note on K-theory and triangulated categories. Invent. Math. **150**, 111–116 (2002)

25. Schlichting, M.: Higher Algebraic K-theory, this volume

26. Simpson, C.: Higgs bundles and local systems. Inst. Hautes tudes Sci. Publ. Math. **75**, 5–95 (1992)

27. Tabuada, G.: Une structure de catégorie de modèles de Quillen sur la catégorie des dg-catégories. In: Comptes Rendus de l'Acadadémie de Sciences de Paris, vol. **340**, pp. 15–19 (2005)

28. Toën, B.: The homotopy of dg-categories and derived Morita theory. Invent. Math. **167**(3), 615–667 (2007)

29. Toën, B.: Higher and derived stacks: A global overview. Algebraic geometry–Seattle 2005. Part 1, 435–487, Proceedings of Symposia in Pure Mathematics, vol. 80, Part 1. American Mathematical Society, Providence, RI (2009)

30. Toën, B.: Finitude homotopique des dg-algèbres propres et lisses. Proc. Lond. Math. Soc. (3) **98**(1), 217–240 (2009)

31. Toën, B.: Anneaux de définitions des dg-algèbres propres et lisses. Bull. Lond. Math. Soc. **40**(4), 642–650 (2008)

32. Toën, B.: Vaquié, M.: Moduli of objects in dg-categories. Ann. Sci. de l'ENS **40**(3), 387–444 (2007)

33. Toën, B.: Vezzosi, G.: Chern character, loop spaces and derived algebraic geometry. In: Algebraic Topology. (Proc. The Abel Symposium 2007). Abel Symposia, vol. 4, pp. 331–354. Springer, Berlin (2009)

34. Toën, B.: Vezzosi, M.: ∞-Catégories monoidales rigides, traces et caractères de Chern preprint arXiv 0903.3292

Lecture Notes in Mathematics

For information about earlier volumes
please contact your bookseller or Springer
LNM Online archive: springerlink.com

I. From Classical Probability to Quantum Stochastic Calculus. Editors: M. Schürmann, U. Franz (2005)

Vol. 1866: O.E. Barndorff-Nielsen, U. Franz, R. Gohm, B. Kümmerer, S. Thorbjønsen, Quantum Independent Increment Processes II. Structure of Quantum Lévy Processes, Classical Probability, and Physics. Editors: M. Schürmann, U. Franz, (2005)

Vol. 1867: J. Sneyd (Ed.), Tutorials in Mathematical Biosciences II. Mathematical Modeling of Calcium Dynamics and Signal Transduction. (2005)

Vol. 1868: J. Jorgenson, S. Lang, $Pos_n(R)$ and Eisenstein Series. (2005)

Vol. 1869: A. Dembo, T. Funaki, Lectures on Probability Theory and Statistics. Ecole d'Eté de Probabilités de Saint-Flour XXXIII-2003. Editor: J. Picard (2005)

Vol. 1870: V.I. Gurariy, W. Lusky, Geometry of Mntz Spaces and Related Questions. (2005)

Vol. 1871: P. Constantin, G. Gallavotti, A.V. Kazhikhov, Y. Meyer, S. Ukai, Mathematical Foundation of Turbulent Viscous Flows, Martina Franca, Italy, 2003. Editors: M. Cannone, T. Miyakawa (2006)

Vol. 1872: A. Friedman (Ed.), Tutorials in Mathematical Biosciences III. Cell Cycle, Proliferation, and Cancer (2006)

Vol. 1873: R. Mansuy, M. Yor, Random Times and Enlargements of Filtrations in a Brownian Setting (2006)

Vol. 1874: M. Yor, M. Émery (Eds.), In Memoriam Paul-Andr Meyer - Sminaire de Probabilits XXXIX (2006)

Vol. 1875: J. Pitman, Combinatorial Stochastic Processes. Ecole d'Et de Probabilits de Saint-Flour XXXII-2002. Editor: J. Picard (2006)

Vol. 1876: H. Herrlich, Axiom of Choice (2006)

Vol. 1877: J. Steuding, Value Distributions of L-Functions (2007)

Vol. 1878: R. Cerf, The Wulff Crystal in Ising and Percolation Models, Ecole d'Et de Probabilités de Saint-Flour XXXIV-2004. Editor: Jean Picard (2006)

Vol. 1879: G. Slade, The Lace Expansion and its Applications, Ecole d'Et de Probabilits de Saint-Flour XXXIV-2004. Editor: Jean Picard (2006)

Vol. 1880: S. Attal, A. Joye, C.-A. Pillet, Open Quantum Systems I, The Hamiltonian Approach (2006)

Vol. 1881: S. Attal, A. Joye, C.-A. Pillet, Open Quantum Systems II, The Markovian Approach (2006)

Vol. 1882: S. Attal, A. Joye, C.-A. Pillet, Open Quantum Systems III, Recent Developments (2006)

Vol. 1883: W. Van Assche, F. Marcellàn (Eds.), Orthogonal Polynomials and Special Functions, Computation and Application (2006)

Vol. 1884: N. Hayashi, E.I. Kaikina, P.I. Naumkin, I.A. Shishmarev, Asymptotics for Dissipative Nonlinear Equations (2006)

Vol. 1885: A. Telcs, The Art of Random Walks (2006)

Vol. 1886: S. Takamura, Splitting Deformations of Degenerations of Complex Curves (2006)

Vol. 1887: K. Habermann, L. Habermann, Introduction to Symplectic Dirac Operators (2006)

Vol. 1888: J. van der Hoeven, Transseries and Real Differential Algebra (2006)

Vol. 1889: G. Osipenko, Dynamical Systems, Graphs, and Algorithms (2006)

Vol. 1890: M. Bunge, J. Funk, Singular Coverings of Toposes (2006)

Vol. 1891: J.B. Friedlander, D.R. Heath-Brown, H. Iwaniec, J. Kaczorowski, Analytic Number Theory, Cetraro, Italy, 2002. Editors: A. Perelli, C. Viola (2006)

Vol. 1892: A. Baddeley, I. Bárány, R. Schneider, W. Weil, Stochastic Geometry, Martina Franca, Italy, 2004. Editor: W. Weil (2007)

Vol. 1893: H. Hanßmann, Local and Semi-Local Bifurcations in Hamiltonian Dynamical Systems, Results and Examples (2007)

Vol. 1894: C.W. Groetsch, Stable Approximate Evaluation of Unbounded Operators (2007)

Vol. 1895: L. Molnár, Selected Preserver Problems on Algebraic Structures of Linear Operators and on Function Spaces (2007)

Vol. 1896: P. Massart, Concentration Inequalities and Model Selection, Ecole d'Été de Probabilités de Saint-Flour XXXIII-2003. Editor: J. Picard (2007)

Vol. 1897: R. Doney, Fluctuation Theory for Lévy Processes, Ecole d'Été de Probabilités de Saint-Flour XXXV-2005. Editor: J. Picard (2007)

Vol. 1898: H.R. Beyer, Beyond Partial Differential Equations, On linear and Quasi-Linear Abstract Hyperbolic Evolution Equations (2007)

Vol. 1899: Séminaire de Probabilités XL. Editors: C. Donati-Martin, M. Émery, A. Rouault, C. Stricker (2007)

Vol. 1900: E. Bolthausen, A. Bovier (Eds.), Spin Glasses (2007)

Vol. 1901: O. Wittenberg, Intersections de deux quadriques et pinceaux de courbes de genre 1, Intersections of Two Quadrics and Pencils of Curves of Genus 1 (2007)

Vol. 1902: A. Isaev, Lectures on the Automorphism Groups of Kobayashi-Hyperbolic Manifolds (2007)

Vol. 1903: G. Kresin, V. Maz'ya, Sharp Real-Part Theorems (2007)

Vol. 1904: P. Giesl, Construction of Global Lyapunov Functions Using Radial Basis Functions (2007)

Vol. 1905: C. Prévôt, M. Röckner, A Concise Course on Stochastic Partial Differential Equations (2007)

Vol. 1906: T. Schuster, The Method of Approximate Inverse: Theory and Applications (2007)

Vol. 1907: M. Rasmussen, Attractivity and Bifurcation for Nonautonomous Dynamical Systems (2007)

Vol. 1908: T.J. Lyons, M. Caruana, T. Lévy, Differential Equations Driven by Rough Paths, Ecole d'Été de Probabilités de Saint-Flour XXXIV-2004 (2007)

Vol. 1909: H. Akiyoshi, M. Sakuma, M. Wada, Y. Yamashita, Punctured Torus Groups and 2-Bridge Knot Groups (I) (2007)

Vol. 1910: V.D. Milman, G. Schechtman (Eds.), Geometric Aspects of Functional Analysis. Israel Seminar 2004-2005 (2007)

Vol. 1911: A. Bressan, D. Serre, M. Williams, K. Zumbrun, Hyperbolic Systems of Balance Laws. Cetraro, Italy 2003. Editor: P. Marcati (2007)

Vol. 1912: V. Berinde, Iterative Approximation of Fixed Points (2007)

Vol. 1913: J.E. Marsden, G. Misiołek, J.-P. Ortega, M. Perlmutter, T.S. Ratiu, Hamiltonian Reduction by Stages (2007)

Vol. 1914: G. Kutyniok, Affine Density in Wavelet Analysis (2007)

Vol. 1915: T. Bıyıkoğlu, J. Leydold, P.F. Stadler, Laplacian Eigenvectors of Graphs. Perron-Frobenius and Faber-Krahn Type Theorems (2007)

Vol. 1916: C. Villani, F. Rezakhanlou, Entropy Methods for the Boltzmann Equation. Editors: F. Golse, S. Olla (2008)

Vol. 1966: D. Deng, Y. Han, Harmonic Analysis on Spaces of Homogeneous Type (2009)

Vol. 1967: B. Fresse, Modules over Operads and Functors (2009)

Vol. 1968: R. Weissauer, Endoscopy for GSP(4) and the Cohomology of Siegel Modular Threefolds (2009)

Vol. 1969: B. Roynette, M. Yor, Penalising Brownian Paths (2009)

Vol. 1970: M. Biskup, A. Bovier, F. den Hollander, D. Ioffe, F. Martinelli, K. Netočný, F. Toninelli, Methods of Contemporary Mathematical Statistical Physics. Editor: R. Kotecký (2009)

Vol. 1971: L. Saint-Raymond, Hydrodynamic Limits of the Boltzmann Equation (2009)

Vol. 1972: T. Mochizuki, Donaldson Type Invariants for Algebraic Surfaces (2009)

Vol. 1973: M.A. Berger, L.H. Kauffmann, B. Khesin, H.K. Moffatt, R.L. Ricca, De W. Sumners, Lectures on Topological Fluid Mechanics. Cetraro, Italy 2001. Editor: R.L. Ricca (2009)

Vol. 1974: F. den Hollander, Random Polymers: École d'Été de Probabilités de Saint-Flour XXXVII – 2007 (2009)

Vol. 1975: J.C. Rohde, Cyclic Coverings, Calabi-Yau Manifolds and Complex Multiplication (2009)

Vol. 1976: N. Ginoux, The Dirac Spectrum (2009)

Vol. 1977: M.J. Gursky, E. Lanconelli, A. Malchiodi, G. Tarantello, X.-J. Wang, P.C. Yang, Geometric Analysis and PDEs. Cetraro, Italy 2001. Editors: A. Ambrosetti, S.-Y.A. Chang, A. Malchiodi (2009)

Vol. 1978: M. Qian, J.-S. Xie, S. Zhu, Smooth Ergodic Theory for Endomorphisms (2009)

Vol. 1979: C. Donati-Martin, M. Émery, A. Rouault, C. Stricker (Eds.), Séminaire de Probablitiés XLII (2009)

Vol. 1980: P. Graczyk, A. Stos (Eds.), Potential Analysis of Stable Processes and its Extensions (2009)

Vol. 1981: M. Chlouveraki, Blocks and Families for Cyclotomic Hecke Algebras (2009)

Vol. 1982: N. Privault, Stochastic Analysis in Discrete and Continuous Settings. With Normal Martingales (2009)

Vol. 1983: H. Ammari (Ed.), Mathematical Modeling in Biomedical Imaging I. Electrical and Ultrasound Tomographies, Anomaly Detection, and Brain Imaging (2009)

Vol. 1984: V. Caselles, P. Monasse, Geometric Description of Images as Topographic Maps (2010)

Vol. 1985: T. Linß, Layer-Adapted Meshes for Reaction-Convection-Diffusion Problems (2010)

Vol. 1986: J.-P. Antoine, C. Trapani, Partial Inner Product Spaces. Theory and Applications (2009)

Vol. 1987: J.-P. Brasselet, J. Seade, T. Suwa, Vector Fields on Singular Varieties (2010)

Vol. 1988: M. Broué, Introduction to Complex Reflection Groups and Their Braid Groups (2010)

Vol. 1989: I.M. Bomze, V. Demyanov, Nonlinear Optimization. Cetraro, Italy 2007. Editors: G. di Pillo, F. Schoen (2010)

Vol. 1990: S. Bouc, Biset Functors for Finite Groups (2010)

Vol. 1991: F. Gazzola, H.-C. Grunau, G. Sweers, Polyharmonic Boundary Value Problems (2010)

Vol. 1992: A. Parmeggiani, Spectral Theory of Non-Commutative Harmonic Oscillators: An Introduction (2010)

Vol. 1993: P. Dodos, Banach Spaces and Descriptive Set Theory: Selected Topics (2010)

Vol. 1994: A. Baricz, Generalized Bessel Functions of the First Kind (2010)

Vol. 1995: A.Y. Khapalov, Controllability of Partial Differential Equations Governed by Multiplicative Controls (2010)

Vol. 1996: T. Lorenz, Mutational Analysis. A Joint Framework for Cauchy Problems In and Beyond Vector Spaces (2010)

Vol. 1997: M. Banagl, Intersection Spaces, Spatial Homology Truncation, and String Theory (2010)

Vol. 1998: M. Abate, E. Bedford, M. Brunella, T.-C. Dinh, D. Schleicher, N. Sibony, Holomorphic Dynamical Systems. Editors: G. Gentili, J. Guenot, G. Patrizio (2010)

Vol. 1999: H. Schoutens, The Use of Ultraproducts in Commutative Algebra (2010)

Vol. 2000: H. Yserentant, Regularity and Approximability of Electronic Wave Functions (2010)

Vol. 2001: T. Duquesne, O.E. Barndorff-Nielson, O. Reichmann, J. Bertoin, K.-I. Sato, J. Jacod, C. Schwab, C. Klüppelberg, Lévy Matters I (2010)

Vol. 2002: C. Pötzsche, Geometric Theory of Discrete Nonautonomous Dynamical Systems (2010)

Vol. 2003: A. Cousin, S. Crépey, O. Guéant, D. Hobson, M. Jeanblanc, J.-M. Lasry, J.-P. Laurent, P.-L. Lions, P. Tankov, Paris-Princeton Lectures on Mathematical Finance 2010. Editors: R.A. Carmona, E. Cinlar, I. Ekeland, E. Jouini, J.A. Scheinkman, N. Touzi (2010)

Vol. 2004: K. Diethelm, The Analysis of Fractional Differential Equations (2010)

Vol. 2005: W. Yuan, W. Sickel, D. Yang, Morrey and Campanato Meet Besov, Lizorkin and Triebel (2011)

Vol. 2006: A. Rouault, A. Lejay, C. Donati-Martin (Eds.), Séminaire de Probabilités XLIII (2011)

Vol. 2007: G. Gromadzki, F.J. Cirre, J.M. Gamboa, E. Bujalance, Symmetries of Compact Riemann Surfaces (2010)

Vol. 2008: P.F. Baum, G. Cortiñas, R. Meyer, R. Sánchez-García, M. Schlichting, B. Toën, Topics in Algebraic and Topological K-Theory (2011)

Recent Reprints and New Editions

Vol. 1702: J. Ma, J. Yong, Forward-Backward Stochastic Differential Equations and their Applications. 1999 – Corr. 3rd printing (2007)

Vol. 830: J.A. Green, Polynomial Representations of GL_n, with an Appendix on Schensted Correspondence and Littelmann Paths by K. Erdmann, J.A. Green and M. Schoker 1980 – 2nd corr. and augmented edition (2007)

Vol. 1693: S. Simons, From Hahn-Banach to Monotonicity (Minimax and Monotonicity 1998) – 2nd exp. edition (2008)

Vol. 470: R.E. Bowen, Equilibrium States and the Ergodic Theory of Anosov Diffeomorphisms. With a preface by D. Ruelle. Edited by J.-R. Chazottes. 1975 – 2nd rev. edition (2008)

Vol. 523: S.A. Albeverio, R.J. Høegh-Krohn, S. Mazzucchi, Mathematical Theory of Feynman Path Integral. 1976 – 2nd corr. and enlarged edition (2008)

Vol. 1764: A. Cannas da Silva, Lectures on Symplectic Geometry 2001 – Corr. 2nd printing (2008)

LECTURE NOTES IN MATHEMATICS ☐ Springer

Edited by J.-M. Morel, F. Takens, B. Teissier, P.K. Maini

Editorial Policy (for Multi-Author Publications: Summer Schools/Intensive Courses)

(a) Lecture Notes aim to report new developments in all areas of mathematics and their applications - quickly, informally and at a high level. Mathematical texts analysing new developments in modelling and numerical simulation are welcome. Manuscripts should be reasonably self-contained and rounded off. Thus they may, and often will, present not only results of the author but also related work by other people. They should provide sufficient motivation, examples and applications. There should also be an introduction making the text comprehensible to a wider audience. This clearly distinguishes Lecture Notes from journal articles or technical reports which normally are very concise. Articles intended for a journal but too long to be accepted by most journals, usually do not have this "lecture notes" character.

(b) In general SUMMER SCHOOLS and other similar INTENSIVE COURSES are held to present mathematical topics that are close to the frontiers of recent research to an audience at the beginning or intermediate graduate level, who may want to continue with this area of work, for a thesis or later. This makes demands on the didactic aspects of the presentation. Because the subjects of such schools are advanced, there often exists no textbook, and so ideally, the publication resulting from such a school could be a first approximation to such a textbook. Usually several authors are involved in the writing, so it is not always simple to obtain a unified approach to the presentation.

For prospective publication in LNM, the resulting manuscript should not be just a collection of course notes, each of which has been developed by an individual author with little or no co-ordination with the others, and with little or no common concept. The subject matter should dictate the structure of the book, and the authorship of each part or chapter should take secondary importance. Of course the choice of authors is crucial to the quality of the material at the school and in the book, and the intention here is not to belittle their impact, but simply to say that the book should be planned to be written by these authors jointly, and not just assembled as a result of what these authors happen to submit.

This represents considerable preparatory work (as it is imperative to ensure that the authors know these criteria before they invest work on a manuscript), and also considerable editing work afterwards, to get the book into final shape. Still it is the form that holds the most promise of a successful book that will be used by its intended audience, rather than yet another volume of proceedings for the library shelf.

(c) Manuscripts should be submitted either to Springer's mathematics editorial in Heidelberg, or to one of the series editors. Volume editors are expected to arrange for the refereeing, to the usual scientific standards, of the individual contributions. If the resulting reports can be forwarded to us (series editors or Springer) this is very helpful. If no reports are forwarded or if other questions remain unclear in respect of homogeneity etc, the series editors may wish to consult external referees for an overall evaluation of the volume. A final decision to publish can be made only on the basis of the complete manuscript; however a preliminary decision can be based on a pre-final or incomplete manuscript. The strict minimum amount of material that will be considered should include a detailed outline describing the planned contents of each chapter.

Volume editors and authors should be aware that incomplete or insufficiently close to final manuscripts almost always result in longer evaluation times. They should also be aware that parallel submission of their manuscript to another publisher while under consideration for LNM will in general lead to immediate rejection.

(d) Manuscripts should in general be submitted in English. Final manuscripts should contain at least 100 pages of mathematical text and should always include

- a general table of contents;
- an informative introduction, with adequate motivation and perhaps some historical remarks: it should be accessible to a reader not intimately familiar with the topic treated;
- a global subject index: as a rule this is genuinely helpful for the reader.

Lecture Notes volumes are, as a rule, printed digitally from the authors' files. We strongly recommend that all contributions in a volume be written in LaTeX2e. To ensure best results, authors are asked to use the LaTeX2e style files available from Springer's web-server at

ftp://ftp.springer.de/pub/tex/latex/svmultt1/ (for summer schools/tutorials).

Additional technical instructions are available on request from: lnm@springer.com.

(e) Careful preparation of the manuscripts will help keep production time short besides ensuring satisfactory appearance of the finished book in print and online. After acceptance of the manuscript authors will be asked to prepare the final LaTeX source files (and also the corresponding dvi-, pdf- or zipped ps-file) together with the final printout made from these files. The LaTeX source files are essential for producing the full-text online version of the book. For the existing online volumes of LNM see: www.springerlink.com/content/110312

The actual production of a Lecture Notes volume takes approximately 12 weeks.

(f) Volume editors receive a total of 50 free copies of their volume to be shared with the authors, but no royalties. They and the authors are entitled to a discount of 33.3% on the price of Springer books purchased for their personal use, if ordering directly from Springer.

(g) Commitment to publish is made by letter of intent rather than by signing a formal contract. Springer-Verlag secures the copyright for each volume. Authors are free to reuse material contained in their LNM volumes in later publications: a brief written (or e-mail) request for formal permission is sufficient.

Addresses:

Professor J.-M. Morel, CMLA,
École Normale Supérieure de Cachan,
61 Avenue du Président Wilson,
94235 Cachan Cedex, France
E-mail: Jean-Michel.Morel@cmla.ens-cachan.fr

Professor F. Takens, Mathematisch Instituut,
Rijksuniversiteit Groningen, Postbus 800,
9700 AV Groningen, The Netherlands
E-mail: F.Takens@math.rug.nl

Professor B. Teissier,
Institut Mathématique de Jussieu,
UMR 7586 du CNRS,
Équipe "Géométrie et Dynamique",
175 rue du Chevaleret
75013 Paris, France
E-mail: teissier@math.jussieu.fr

For the "Mathematical Biosciences Subseries" of LNM:
Professor P.K. Maini, Center for Mathematical Biology
Mathematical Institute, 24-29 St Giles,
Oxford OX1 3LP, UK
E-mail: maini@maths.ox.ac.uk

Springer, Mathematics Editorial I, Tiergartenstr. 17,
69121 Heidelberg, Germany,
Tel.: +49 (6221) 487-8259
Fax: +49 (6221) 4876-8259
E-mail: lnm@springer.com